Texts in Philosophy
Volume 1

The Road Not Taken
On Husserl's Philosophy of
Logic and Mathematics

Volume 9
Language, Knowledge, and Metaphysics. Proceedings of the First SIFA Graduate Conference
Massimiliano Carrara and Vittorio Morato eds.

Volume 10
The Socratic Tradition. Questioning as Philosophy and as Method
Matti Sintonen, ed.

Volume 11
PhiMSAMP. Philosophy of Mathematics: Sociological Aspects and Mathematical Practice
Benedikt Löwe and Thomas Müller, eds.

Volume 12
Philosophical Perspectives on Mathematical Practice
Bart Van Kerkhove, Jonas De Vuyst and Jean Paul Van Bendegem, eds.

Volume 13
Beyond Description: Naturalism and Normativity
Marcin Miłkowski and Konrad Talmont-Kaminski, eds.

Volume 14
Corroborations and Criticisms. Forays with the Philosophy of Karl Popper
Ivor Grattan-Guinness

Volume 15
Knowledge, Value, Evolution.
Tomáš Hříbek and Juraj Hvorecký, eds.

Volume 16
Hao Wang. Logician and Philosopher
Charles Parsons and Montgomery Link, eds.

Volume 17
Mimesis: Metaphysics, Cognition, Pragmatics
Gregory Currie, Petr Koťátko, Martin Pokorný

Volume 18
Contemporary Problems of Epistemology in the Light of Phenomenology. Temporal Consciousness and the Limits of Formal Theories
Stathis Livadas

Volume 19
The Road Not Taken. On Husserl's Philosophy of Logic and Mathematics
Claire Ortiz Hill and Jairo José da Silva

Texts in Philosophy Series Editors
Vincent F. Hendriks   vincent@hum.ku.dk
John Symons   jsymons@utep.edu
Dov Gabbay   dov.gabbay@kcl.ac.uk

The Road Not Taken
On Husserl's Philosophy of Logic and Mathematics

Claire Ortiz Hill
and
Jairo José da Silva

© Individual authors and College Publications 2013.
All rights reserved.

ISBN 978-1-84890-099-8

College Publications
Scientific Director: Dov Gabbay
Managing Director: Jane Spurr

http://www.collegepublications.co.uk

Original cover design by Laraine Welch
Cover artwork by Jacqueline Wegmann

Printed by Lightning Source, Milton Keynes, UK

All rights reserved. No part of this publication may be reproduced, stored in a retrieval system or transmitted in any form, or by any means, electronic, mechanical, photocopying, recording or otherwise without prior permission, in writing, from the publisher.

TABLE OF CONTENTS

Introduction   X

Acknowledgments   XIII

1. On Husserl's Mathematical Apprenticeship   1
   Claire Ortiz Hill

2. Husserl on Geometry and Spatial Representation   31
   Jairo José da Silva

3. Beyond Leibniz, Husserl's Vindication of Symbolic Mathematics   61
   Jairo José da Silva

4. One Dogma of Empiricism   81
   Claire Ortiz Hill

5. Husserl on Axiomatization and Arithmetic   93
   Claire Ortiz Hill

6. Husserl and Hilbert on Completeness and Imaginary Elements   115
   Jairo José da Silva

7. The Many Senses of Completeness   137
   Jairo José da Silva

8. Frege's Letters   151
   Claire Ortiz Hill

9. Reference and Paradox   171
   Claire Ortiz Hill

10. Tackling Three of Frege's Problems: Husserl on Sets and Manifolds *Claire Ortiz Hill*	195
11. On Fundamental Differences between Dependent and Independent Meanings *Claire Ortiz Hill*	219
12. Incomplete Symbols, Dependent Meanings, and Paradox *Claire Ortiz Hill*	241
13. Husserl's Phenomenology and Weyl's Predicativism *Jairo José da Silva*	265
14. Husserl on the Principle of Bivalence *Jairo José da Silva*	285
15. Cantor's Paradise, Metaphysics and Husserlian Logic *Claire Ortiz Hill*	299
16. Gödel and Transcendental Phenomenology *Jairo José da Silva*	325
17. Mathematics and the Crisis of Science *Jairo José da Silva*	345

Appendix I

LETTERS PERTAINING TO HUSSERL FROM CANTOR'S LETTER BOOK III
NIEDERSÄCHSISCHE STAATS- UND UNIVERSITÄTSBIBLIOTHEK, ABTEILUNG
HANDSCHRIFTEN UND SELTENE DRÜCKE, UNIVERSITY OF GÖTTINGEN
Translated by Dr. Ruth Ellen Burke

Letter 238	367
Letter 240	368
Letter 249	370
Letter 250	370
Letter 253	371
Letter 262	372
Letter 264	374
Letter 277	375
Letter 291	376

Appendix II

DOCUMENTS CONCERNING HUSSERL FROM THE NIEDERSÄCHSISCHE
STAATS- UND UNIVERSITÄTSBIBLIOTHEK, ABTEILUNG HANDSCHRIFTEN
UND SELTENE DRÜCKE AT THE UNIVERSITY OF GÖTTINGEN AND THE
GEHEIMES STAATSARCHIV PREUSSISCHER KULTURBESITZ IN BERLIN
Translated by Dr. Ruth Ellen Burke

Letter from H. E. Müller, May 29, 1900	379
Letter from Prof. Baumann, May 29, 1900	381
Letter from Dr. Höpfner, August 30, 1900	381
Letter from W. Fleischmann, May 12, 1905	383
Letter from the Göttingen Philosophy Faculty, May 22, 1905	383
Draft of David Hilbert's *Separatvotum* for Husserl, July 30, 1908	384
Extracts from Hilbert's *Denkschrift* for Leonard Nelson, undated	386
Extracts from Nelson's Letter to Hilbert, December 29, 1916	388
David Hilbert's Private Letter to Becker, July 30, 1918	391
Letter from Hermann Weyl, August 13, 1918	392
"For the Minister," by David Hilbert, unpublished, undated	394
Selected Bibliography	395
Index	413

Jairo José da Silva

INTRODUCTION

We could say, somewhat provocatively, that in some ways Edmund Husserl failed as a philosopher. He cut many paths to phenomenology, by the way of logic (*Experience and Judgment, Formal and Transcendental Logic*) and the empirical sciences (*Crisis of European Sciences and Transcendental Phenomenology*), but none has proven to be the royal road thereto. Despite the different attempts at a theoretical formulation of phenomenology, both before and after his transcendental turn (*Logical Investigations* before, *Ideas I* and *Cartesian Meditations*, afterward), he never developed a systematic presentation of his philosophical insights that he himself judged satisfactory, certainly not in his published works, some of them written in collaboration with students. As he once explained, "there unfortunately dwells within me an intractably critical sense unmindful of my natural inclinations.... By nature bound, intellectually free, so I go... my way... I am... still a poor beginner;... As always, I work, and often with despairing doggedness, as if to rid myself of some of the endless shame of my dullness, unclarity, and ignorance".[1]

A systematic presentation of his philosophical insights is perhaps to be pieced together from the piles of unpublished material that he left, but certainly does not already exist in an organic and satisfiable form. For those who, like myself, want to extract from his writings a systematic philosophy of logic and mathematics, there is no other choice than to face the imposing edifice of the ever growing Husserlian corpus, and the work it demands. Although for decades now, the Husserl Archives in Leuven has been publishing volumes of his *Nachlass*, it is very dismaying to find that few scholars of his philosophy of logic and mathematics have taken the time to familiarize themselves with them. They continue rehashing the same works over and over.

---

[1] Husserl, Edmund, *Briefwechsel, Die Brentanoschule I*, Dordrecht: Kluwer, 1994, 20-21.

Husserl has been misunderstood even by those who appreciated his work, such as Wilhelm Dilthey and the readers of the *Prolegomena*, who thought he had subsequently fallen back into psychologism after having so extensively criticized it. His emphasis on subjectivity never ceased to baffle and confound admirers (such as Kurt Gödel) and detractors (such as Gottlob Frege); so often the idea of intentional constitution, particularly in transcendental phenomenology, has been mistakenly interpreted epistemologically, by realists, and psychologically, by idealists of both traditional and contemporary extraction. (I have mathematical intuitionists in mind, of course).

Although one the most seminal and original thinkers ever, and certainly an intellectual giant of the 20th century, Husserl fell into disgrace with the two main trends of western philosophy of the century, both of which, ironically, he helped to develop. After struggling to create a rigorous scientific philosophy, he had to endure the decaying standards of philosophical rigor and the anti-scientific disposition of the one he considered his natural philosophical heir, Martin Heidegger, who had no qualms about betraying his master in more ways than one. With Heidegger, phenomenological philosophy pursued paths Husserl did not recognize as his. And ironically, and unfortunately, it was to a large extent Heidegerrians who transmitted his legacy to subsequent generations during the last half of the 20th century.

After contributing many original, thought-provoking ideas on issues in the philosophy of logic, mathematics, science and language dear to the analytic sensitivity, in contact with Gottlob Frege, Georg Cantor and David Hilbert, but independently of them, Husserl had the misfortune of being first completely misunderstood and then forgotten by the analytic tradition, and then, when the opportunity for reassessment appeared, to be read through Frege, who for a time was accorded a role in Husserl's philosophical development that he had never played.

A trained mathematician (he held a doctorate in mathematics, and had been the student of Karl Weierstrass, whose assistant he was for a time), Husserl much more than Frege (considering his correspondence with Hilbert on the nature of formal mathematics) or Ludwig Wittgenstein (considering everything he wrote about it), was in the right place at the right time with the right training to understand the new, even revolutionary, developments in mathematics taking place around the turn of the 20th century; but we have to dig into Husserl's writings to extract a philosophy of mathematics in general (not only of arithmetic), which despite his profound insights never attracted the large, enthusiastic audiences that Frege and Wittgenstein did and still do (something which is particularly unfortunate in case of the latter). We cannot even be completely sure that his philosophy of mathematics is finally coming to be appreciated and understood, given some of the interpretations it has been subject to in recent years.

Husserl certainly influenced many thinkers, but not everyone recognized the fact (Rudolf Carnap did not). Kurt Gödel and Hermann Weyl did, but we cannot be completely sure the former really understood him, though the latter certainly did. Husserl's transcendental phenomenology, the epoché, the transcendental ego, conceptual intuition were so mixed with foreign concepts coming from Kant or Leibniz that we cannot easily assess the extent of Husserl's influence on Gödel.

Despite all that, Husserl's philosophy in general, and to no lesser degree his philosophy of logic and mathematics, deserves all the attention it can get, and for many reasons. Husserl took mathematics and logic as he found them: well-established sciences that were being dramatically reshaped in his time and in no need of restraint or "sound" foundations of the kind championed by Brouwer or Frege, but demanding urgent philosophical clarification as to their nature, scope and methods. And precisely because of this, Husserl's philosophy offers alternatives to the traditional foundational schools; logicism, nominalism and formalism are well-beaten tracks that failed to lead to a comprehension of either traditional or the new mathematics, and still less of mathematics as actually practiced. Unlike more traditional philosophies of mathematics, Husserl's philosophy is neither reductionist nor revisionist; it does not try to rewrite mathematics or reduce it to logic or a meaningless game with symbols, but clarifies instead the sense in which it can be both and still be at the service of knowledge, empirical knowledge in particular.

Husserl recognized the obvious kinship of logic and mathematics, and even reserved a chapter of logic for pure mathematics, but in a sense completely different from Frege's. He saw clearly the point of Hilbert's formalism (more clearly than Frege did), according, however, formal mathematics the epistemological relevance other formalists failed to grant it. Although placing intuition as the condition sine qua non of knowledge proper, Husserl extended it to limits that Kant-oriented intuitionists (like Brouwer and the neo-Brouwerians) found, and still find, inaccessible. But even so, Husserl safeguarded a place for purely symbolic, intuition-free mathematics in the overall schema of pure and applied mathematics.

In his *Prolegomena to Pure Logic* (which opens the *Logical Investigations*), Husserl presented the best articulated, most thoroughly argued case for logic as an objective science to be found in the literature. But unlike Frege, he did not see this as the end, but rather the beginning, of a thoroughly philosophical account of logic (or for that matter of any objective, positive science). For him, it was essential to uncover the intentional genesis of objective logic (or of any objective science) in intentional consciousness. The grounding of objectivity in subjectivity constitutes the core of his phenomenology, which brought him so much incomprehension. Although an enemy of psychologism, Husserl was not philosophically naïve to the point of not seeing the role of subjectivity in shaping the sense of the reality facing objective science.

Although critical of some aspects of the new logics of George Boole, Ernst Schröder, and others, for believing that the calculi they created risked not being *logical* calculi at all, but calculi of *classes*, Husserl appreciated and welcomed their work as the immense scientific accomplishments that they were. In logic or mathematics, Husserl was no reactionary, realizing nonetheless, maybe more acutely than others, that these fields required philosophical clarification.

With so much to recommend it, it is difficult to understand why Husserl's philosophy of logic and mathematics, despite the increasing number of philosophers attracted to it, still fights for the place in the sun that it rightfully deserves. This book is our contribution towards putting it there. We believe that Husserl's failure is not his as the immense philosopher he was, but ours, owing largely to prejudices, not the least of which have been those of phenomenologists themselves, which have stood for far longer than they ever should have.

****

Claire and I are grateful to Dr. Helmut Rohlfing Director of the Niedersächsische Staats- und Universitätsbibliothek, Abteilung Handschriften und Seltene Drucke at the University of Göttingen for the permission to publish the translations of Letters 238, 240, 249, 250, 253, 262, 264, 277, 291 from Cantor's *Briefbücher* (Cod.Ms. Cantor 18) and the documents from David Hilbert's *Nachlass* (Cod.Ms. Hilbert 482) found in Appendices I and II. We are also grateful to the Director of the Geheimes Staatsarchiv Preussicher Kulturbesitz in Berlin for permission (Geschäftszeichen 514/05-2.1) to publish the translations of the archival material found in Appendix II. Rev. Fritz Weber and Dr. Marielène Weber helped us by reading and transcribing archival material written in old German script. Dr. Ruth Ellen Burke, German professor at California State University in San Bernardino graciously translated the documents for us. Señora Monica Delgado de Rollinger helped us prepare the index. Much of Claire's research in the archives of the University of Göttingen, the Geheimes Staatsarchiv Preussicher Kulturbesitz in Berlin, and the Husserl Archives Leuven that went into the making of this book was made possible by a fellowship from the National Endowment of the Humanities.

In addition, I wish to thank Marco Ruffino, the editor of *Manuscrito*, for granting me the permission to republish "The Many Senses of Completeness," originally published in volume 23, 2, 2000, pp. 41-60 of that journal. I am also grateful to Tonina Salis for granting me the permission to reprint "Gödel and Transcendental Phenomenology", which was originally published in the *Revue Internationale de Philosophie* 234(4), 2005, pp. 553-74. Likewise, I am grateful to the editor of *Diálogos* for granting me the permission to republish "Mathematics and the Crisis of Science", originally published in 2008 in volume 91, pp. 37-58 of that journal.

Several of the essays anthologized here originally appeared in publications now under the umbrella of Springer Verlag and have all been reprinted here with their kind permission of Berendina Schermers-van Straalen. These include: my "Husserl's Phenomenology and Weyl's Predicativism", originally published in *Synthese* 110, 1997, pp. 277-96; "Husserl's Two Notions of Completeness, Husserl and Hilbert on Completeness and Imaginary Elements in Mathematics", originally published in *Synthese* 125, 2000, pp. 417-438; "Beyond Leibniz, Husserl's Vindication of Symbolic Mathematics", originally published in *Phenomenology and Mathematics*, Mirja Hartimo (ed.) Dordrecht: Springer, 2010, pp. 123-145; "Husserl on Geometry and Spatial Representation", originally published in *Axiomathes* 22, 2012, pp. 5-30 (the versions published here are sometimes less, sometimes more substantially modified versions of the original papers). They also include Claire's: "Frege's Letters", originally published in *From Dedekind to Gödel, Essays on the Development of the Foundations of Mathematics*, Jaakko Hintikka (ed.), Dordrecht: Kluwer, 1995, pp. 97-118; "Tackling Three of Frege's Problems: Edmund Husserl on Sets and Manifolds", originally published in *Axiomathes* 13, 2002, pp. 79-104; "On Husserl's Mathematical Apprenticeship and Philosophy of Mathematics", originally published in *Phenomenology World Wide*, Anna-Teresa Tymieniecka (ed.), Dordrecht: Kluwer, 2002, pp. 76-92; "Incomplete Symbols, Dependent Meanings, and Paradox", originally published in *Husserl's Logical Investigations*, Daniel O. Dahlstrom (ed.), Dordrecht: Kluwer, 2003, pp. 69-93; "Reference and Paradox," originally published in *Synthese*, 138,2, January 2004, pp. 207-32. "On Fundamental Differences between Dependent and Independent Meanings", originally published in *Axiomathes* 20: 2-3, 2010, pp. 313-32; "Husserl on Axiomatization and Arithmetic", originally published in *Phenomenology and Mathematics*, Mirja Hartimo (ed.), Dordrecht: Springer, 2010, pp. 47-71.

Peter Ohlin of Oxford University Press has kindly granted Claire the permission to reprint "Cantor's Paradise, Metaphysics and Husserlian Logic", originally published in *Categories of Being, Essays on Metaphysics and Logic*, Leila Haaparanta and Heikki Koskinen (eds.), Oxford: Oxford University Press, 2012, pp. 217-40. She is also grateful for the permission to reprint "One Dogma of Empiricism", originally published in *Experience and Analysis, Erfahrung und Analyse*, the Proceedings of the International Wittgenstein Conference on held in Kirchberg am Wechsel, August 2004, M. E. Reicher and J. C. Marek (eds), Vienna: ÖBV&HPT Verlag, 2005, pp. 30-38.

Jairo wants to dedicate his contribution to this book to:
Jairo Wolf, Shoshanna, and Uri, for giving him more than he can thank them for, to Michael Beaumont Wrigley, in the tenth year of his premature death, and Therezinha Casari da Silva, his late mother, in praise of love, life and friendship.
Claire wishes to dedicate her contribution to this book to:
Roger Schmidt, her first philosophy teacher and to his wife Ann.

THE ROAD NOT TAKEN

Robert Frost

Two roads diverged in a yellow wood,
And sorry I could not travel both
And be one traveler, long I stood
And looked down one as far as I could
To where it bent in the undergrowth;

Then took the other, as just as fair,
And having perhaps the better claim,
Because it was grassy and wanted wear;
Though as for that the passing there
Had worn them really about the same,

And both that morning equally lay
In leaves no step had trodden black.
Oh, I kept the first for another day!
Yet knowing how way leads on to way,
I doubted if I should ever come back.

I shall be telling this with a sigh
Somewhere ages and ages hence:
Two roads diverged in a wood, and I—
I took the one less traveled by,
And that has made all the difference.

1

Claire Ortiz Hill

ON HUSSERL'S MATHEMATICAL APPRENTICESHIP

Insight into the formative role that Edmund Husserl's early training in mathematics played in the development of his ideas is fundamental to understanding his philosophy as a whole. Besides shedding light on the genesis of phenomenology, which began to take shape in Husserl's reflections on the inability of the logic, psychology, mathematics, and philosophy of his time to respond to certain onerous questions raised by his earliest attempts to secure radical foundations for arithmetic, understanding Husserl's ideas about mathematics sheds needed light on a number of other dimensions of his thought that have puzzled and challenged philosophers in this century. For example, this is precisely where many of the clues are to be found that are needed to answer questions of a controversial nature about seemingly enigmatic aspects of his thought, among them questions regarding the nature and evolution of his views on psychologism, on Platonism, on realism, and the relationship between his formal and his transcendental logic.

Moreover, this is the only way there is to situate and evaluate Husserl's philosophy in relation to the ideas and innovations of the most eminent and influential mathematicians of his time, notably Karl Weierstrass, Georg Cantor, David Hilbert, and Kurt Gödel, or Gottlob Frege and Bertrand Russell, men who often shared Husserl's desire to discover secure, scientific foundations for mathematics and the theory of knowledge, his concern to reform logic, his intent to fight against psychologism, his desire to develop a theory of meaning, his questions regarding the philosophical significance of the latest developments in mathematics, and so on.

Understanding the evolution of Husserl's views on mathematics is therefore essential to establishing his proper place in 20th-century philosophy of logic and mathematics, a field with deep roots in Austro-German ideas about mathematics, logic and philosophy that flowered in English speaking countries in the 20th century, but into which his ideas have never been properly integrated. Given the preeminent role that philosophy of logic and mathematics has played in shaping the way philosophy was done in English-speaking countries in the 20th century, investigations into Husserl's work in this area thus also supply the material essential for the building of any possible bridge between phenomenology and its principal rival, analytic philosophy. And such investigations afford the best possible explanation as to why so many of Husserl's ideas seem so close to those of that antagonistic school, while others remain so plainly diametrically opposed to it.

Under the Influence of Weierstrass

Husserl came to the decision to pursue mathematics as a career during his student years in Berlin, where he enthusiastically threw himself into the study of that most rigorous of disciplines. It was there that from 1877-1881 he attended the courses of the great mathematician Karl Weierstrass. (Schuhmann, 7; M. Husserl; Osborn, 12-14)

Weierstrass' thoroughgoing, systematic treatment, *ab initio*, of the theory of analytic functions had led him to profound investigations into the principles of arithmetic. His scrupulous manner of submitting the foundations of analytic functions to close scrutiny awoke in Husserl an interest in seeking radical foundations for mathematics. Husserl recalled,

> I came to understand the pains he was taking to transform analysis from the mixture of reason and irrational instincts and know-how it was at the time into a pure rational theory. His aim was to expose its original roots, its elementary concepts and axioms on the basis of which the whole system of analysis might be deduced in a completely rigorous, perspicuous way. (Schuhmann, 7; Jourdain, 295-96)

In reaction to the Kantian psychologization of mathematics popular among his contemporaries, Weierstrass was preaching the arithmetization of analysis, the rigorous founding of analysis purely on the basis of the positive whole numbers. Weierstrass was famous for teaching that once one had thus admitted the notion of whole number, arithmetic needed no further postulate, but then could be built up in a purely logical fashion. This would have the effect of depsychologizing and degeometrizing analysis, of liberating it from the insidious appeals to intuitions of space and time that had been imported into it since Kant had proclaimed that mathematical propositions were synthetic a priori. (Coffa; Demopoulos)

Husserl's encounter with Weierstrass had a deep and lasting effect on the future founder of the phenomenological movement. It was from Weierstrass, Husserl would say, that he acquired the ethos of his intellectual endeavours (Schuhmann, 7). Late in his career he would even say that he had sought to do for philosophy what Weierstrass had done for mathematics (Becker, 40-42; Schuhmann, 34). As Andrew Osborn, who actually consulted Husserl about this, explained,

> Through Weierstrass especially, too, the Berlin school placed enormous importance on the rigor of demonstration, a practice that seized hold on Husserl's imagination so that when later he turned to philosophy he sought to find there a strict science similar to that on which Weierstrass insisted, along with the certainty that follows from such strictness and such rigorous proof. (Osborn, 12)

Indeed, closely inspecting the course of Husserl's intellectual career, one continually finds him reworking themes present in Weierstrass' work and striving to apply the very principles that underpinned the mathematician's efforts to rigorize analysis. This is, for example, evident not only in Husserl's early espousal of Weierstrass' conviction that the cardinal number was "the first and most underivative domain, the sole foundation of all remaining domains of numbers" (Husserl 1994, 2), but also in Husserl's struggles with psychologism, his lifelong search for radical foundations for knowledge, his striving to lay bare the original roots, the most primitive concepts and principles of knowledge, to uncover the fundamental building blocks on the basis of which his whole system of philosophy might rest, his ideas about phenomenology as a strict science, his efforts to extend the notion of the analyticity, and so on. The nature of his attraction to Weierstrass' work also explains much about the nature of Husserl's attraction to the work of Franz Brentano, Georg Cantor, Bernard Bolzano, David Hilbert, and even Gottlob Frege.

Husserl was, of course, not alone in being decisively influenced by Weierstrass' thoroughness and systematic approach. What we know of Husserl's reaction to Weierstrass' efforts to rigorize analysis is consonant with the impression that he left on much of the mathematical world of his time. "Mathematicians under the influence of Weierstrass," Bertrand Russell once noted, "have shown in modern times a care for accuracy, and an aversion to slipshod reasoning, such as had not been known among them previously since the time of the Greeks". (Russell 1917, 94)

In Berlin, Husserl was also influenced by Leopold Kronecker who also believed that:

> Sometime we shall succeed in 'arithmetizing', –that is to say, in founding alone on the number-concept in the narrowest sense, and therefore in stripping away again all the modifications and extensions of this concept, which have mostly been caused by the applications to geometry and mechanics, –the whole of arithmetic". (cited Jourdain, 5)

Osborn credited Kronecker with having "sown the first seeds of philosophical understanding" in Husserl "and fostering the interest so aroused". Osborn recounted,

> Husserl found in him a depth of understanding that stirred an echo in his own nature. Kronecker's special field was the philosophy of mathematics and it was through contact with him accordingly that Husserl first came to any appreciation of the philosophic point of view. Reflective by nature, Husserl found a ready interest in the philosophy of mathematics which was for him, as it proved, a very big step in the direction of an interest in pure philosophy. (Osborn, 12)

Osborn speculates that Husserl's interest in Descartes may have first been awakened by Kronecker. (*Ibid.*)

As happy as Husserl was in Berlin, acting upon his father's wishes, he left for Vienna to prepare his doctoral thesis on the calculus of variations. Summoned by Weierstrass to serve as his assistant, Husserl later returned to Berlin. However, he quickly took advantage of an opportunity to return to Vienna to indulge a growing interest in philosophy. (M. Husserl; Osborne, 15)

Husserl Makes Philosophy His Life's Work

Although Husserl had manifested little interest in philosophy during his time in Berlin, it became the minor subject for his doctorate in mathematics in Vienna. During that time, when his interest in philosophy was growing, and he was wondering whether to make mathematics or philosophy his life's work, Husserl began attending the courses of the philosopher Franz Brentano. At first he did so merely out of curiosity, but these courses finally proved to be the decisive factor encouraging him to dedicate himself entirely to philosophy. But for Brentano, Husserl would say, he would not have become a philosopher. (Husserl 1919, 342; M. Husserl; Brück)

The specific reasons for admiring Brentano that Husserl gave actually quite resemble his reasons for admiring Weierstrass. The man in whom the latter had awakened an interest in seeking radical foundations for knowledge was impressed by Brentano's clear, rigorous, insightful, objective, and precise philosophical analyses and ability to transform unclear beginnings into clear thoughts and insights, his "finely dialectical measuring of various possible arguments, his clarifying of equivocations, and retracing of every philosophical concept to its original intuitive sources". "Brentano relatively quickly moved from intuition to theory, to the delimitation of sharp concepts, to theoretical formulation of working problems", Husserl recalled. For him, Brentano was someone entirely devoted to the austere ideal of a strict philosophical science, someone completely certain of his method, who believed that his sharply polished concepts, his strongly constructed and systematically ordered theories, and his all round aporetic refutation of alternative interpretations captured final truths. He "strove constantly to satisfy the highest claims of an almost mathematical strictness". "Sometimes it was the subject matter which overcame me", Husserl recalled, "other times the quite singular clearness and dialectical sharpness of his expositions, the cataleptic power as it were of his way of developing problems and of his theories". It was from Brentano, Husserl acknowledged, that he acquired the conviction that philosophy "was a serious discipline which could and must be dealt with in the spirit of the strictest science". (Husserl 1919, 343-44)

Georg Cantor and Husserl's *Philosophy of Arithmetic*

Having attended Brentano's lectures for two years, Husserl's next career move was to the University of Halle, to prepare his *Habilitationsschrift* under the direction of Carl Stumpf, a member of Brentano's circle (Smith, 21-24) convinced of the great need for cooperation between mathematicians or scientists and philosophers in the area of logic. (Frege 1980a, 171)

Husserl would reside in Halle from 1886 to 1901. These were years during which his ideas were particularly malleable and changed considerably and definitively. In 1887, he completed *On the Concept of Number*. *Philosophy of Arithmetic* was published in 1891. The better part of the subsequent years was spent in the throes of an intellectual struggle in the course of which he abandoned some of the main lessons he had learned from Weierstrass and Brentano and came to write the groundbreaking *Logical Investigations*, in which he began laying the foundations of the phenomenological movement that went on to shape the course of 20th century philosophy in Continental Europe.

Georg Cantor, the creator of set theory, taught at the University of Halle during those years and served on the *Habilitationskommittee* that judged Husserl's *On the Concept of Number* (Gerlach and Sepp eds.). The two became close friends. At the height of his creative powers in the 1880s and 1890s, Cantor had studied in Berlin from 1863 to 1869, where he too had come under the influence of Weierstrass, a fact that explains much of the initial intellectual kinship between Husserl and Cantor, whose ideas overlapped and crisscrossed in a number of respects. (Hill 1997a; Hill 1999)

During Husserl's time in Halle, Cantor was particularly seeking philosophical justification for his theories. He wanted to show how his entire transfinite set theory rested upon sound principles and how the transfinite numbers might be regarded as consistent extensions of the finite reals. He had begun his *Mannigfaltigkeitslehre* explaining to his readers that he had come to a point of realizing that further work on set theory would require extending the concept of real whole numbers beyond previously set bounds and in a direction which as far as he knew no one had searched yet, and he offered this a justification or an excuse for introducing apparently strange ideas. (Cantor 1883)

Cantor was one of the few mathematicians of his time intent upon wedding mathematics and philosophy. Over the years he had grown increasingly interested in philosophy and by the time of Husserl's arrival in Halle was primed to abandon mathematics for philosophy. In 1894 Cantor would write to the French mathematician Charles Hermite that "in the realm of the spirit" mathematics had no longer been "the essential love" of his soul for more than twenty years. Metaphysics and theology, Cantor "openly confessed", had so taken possession of his soul as to leave him relatively little time for, his "first flame", i.e. mathematics. He was now serving God better, he told Hermite, than, owing to his "apparently meager mathematical talents", he might have done through exclusively pursuing mathematics. (Cantor 1991, 350)

Although older, and far less in a position to change course than Husserl was, this did not prevent Cantor from trying to teach philosophy (Cantor 1991, 210, 218) and from seasoning his writings with philosophical reflections and references. In 1883, Cantor had published the *Grundlagen einer allgemeine Mannigfaltigkeitslehre*, a work which, according to its original 1882 foreword, had been "written with two groups of readers in mind—philosophers who have followed the developments in mathematics up to the present time, and mathematicians who are familiar with the most important older and newer publications in philosophy" (Hallett, 6-7). During Husserl's early years in Halle, Cantor published his theories in the *Zeitschrift für Philosophie und philosophische Kritik* because, as he said, he had grown disgusted with mathematical journals. He was in fact trying to integrate philosophy into his mathematical work to such an extent that colleagues warned him that this was liable to harm his reputation. (Dauben, 139, 336 n. 29)

During Husserl's years in Halle, Cantor persisted in clothing his theories about numbers in a metaphysical garb. And he left no doubts as to his philosophical sympathies. In the *Mannigfaltigkeitslehre*, he had emphasized that the idealist foundations of his reflections were essentially in agreement with the basic principles of Platonism according to which only conceptual knowledge in Plato's sense afforded true knowledge (Cantor 1883, 181, 206 n. 6). His own idealism being related to the Aristotelian-Platonic kind, Cantor wrote in an 1888 letter, he was just as much a realist as an idealist (Cantor 1991, 323). "I conceive of numbers", he informed Giuseppe Peano, "as 'forms' or 'species' (general concepts) of sets. In essentials this is the conception of the ancient geometry of Plato, Aristotle, Euclid etc". (Cantor 1991, 365). To Hermite, he wrote that "the whole numbers both separately and in their actual infinite totality exist in that highest kind of reality as eternal ideas in the Divine Intellect" (cited Hallett, 149). Cantor considered his transfinite numbers to be but a special form of Plato's *arithmoi noetoi* or *eidetikoi*, which he thought probably even fully coincided with the whole real numbers (Cantor 1884, 84; Cantor 1887/8, 420). By 'manifold' or a 'set' he explained in the *Mannigfaltigkeitslehre*, he was defining something related to the Platonic *eidos* or *idea*, as also to what Plato called a *mikton* (Cantor 1883, 204 n. 1). For Cantor, the transfinite "presented a rich, ever growing field of ideal research". (Cantor 1887/8, 406)

Cantor considered that his technique for abstracting numbers from reality provided the only possible foundations for his Platonic conception of numbers (Cantor 1991, 363, 365; Cantor 1887/8, 380, 411). Abstraction was to show the way to that new, abstract realm of ideal mathematical objects that could not be directly perceived or intuited. It was a way of producing purely abstract arithmetical definitions, a properly arithmetical process as opposed to a geometrical one with appeals to intuitions of space and time (Cantor 1883, 191-92). He envisioned it as a technique for focussing on pure, abstract arithmetical properties and concepts, which divorced them from any sensory apprehension of the particular characteristics of the objects figuring in the sets and freed mathematics from psychologism, empiricism, Kantianism and insidious appeals to intuitions of space and time to engage in strictly arithmetical forms of concept formation. (ex. Cantor 1883, 191-92; Cantor 1885; Cantor 1887/8, 381 n. 1; Eccarius 1985, 19-20; Couturat, 325-41)

With his theory of abstraction Cantor believed that he was laying bare the roots from which the organism of transfinite numbers developed with logical necessity. In the "Mitteilungen," written during the late 1880s, the embattled mathematician was particularly intent upon proving that his theorems about transfinite numbers were firmly secured "through the logical power of proofs" which, proceeding from his definitions, which were "neither arbitrary nor artificial, but originate naturally out of abstraction, have, with the help of syllogisms, attained their goal" (Cantor 1887/8, 418). Inspired by Weierstrass' famous theory to that effect, he was hard at work demonstrating that the positive whole numbers formed the basis of all other mathematical conceptual formations.

All this was part of a greater strategy aimed at providing his "strange" new transfinite numbers with secure foundations by demonstrating precisely how the transfinite number system might be built from the bottom up (Dauben 1979, Chapter 6). In so doing, he was acting upon a conviction, spelled out in an 1884 letter to Gösta Mittag-Leffler, that the only correct way to proceed was "to go from what is most simple to that which is composite, to go from what already exists and is well-founded to what is more general and new by continually proceeding by way of transparent considerations, step by step without making any leaps". (Cantor 1991, 208)

Husserl's First Forays into Philosophy

Impressed by Karl Weierstrass' work to arithmetize analysis and armed with analytical tools learned from Brentano, Husserl embarked upon a project to help supply radical foundations for mathematics by submitting the concept of number itself to closer scrutiny. *On the Concept of Number* and *Philosophy of Arithmetic* were the result.

Husserl began *On the Concept of Number* writing of the need to examine the logic of the concepts and methods that mathematicians were introducing and using and for a logical clarification, precise analysis, and rigorous deduction of all of mathematics from the least number of self-evident principles. The definitive removal of the real and imaginary difficulties on the borderline between mathematics and philosophy, he deemed, would only come about by first analyzing the concepts and relations that were in themselves simpler and logically prior and then analyzing the more complicated and more derivative ones. (Husserl 1887, 92-95)

The natural and necessary starting point of any philosophy of mathematics, Husserl still believed, was the analysis of the concept of whole number (*Ibid.*, 94-95). He was confident that,

> a rigorous and thoroughgoing development of higher analysis... would have to emanate from elementary arithmetic alone in which analysis is grounded. But this elementary arithmetic has... its sole foundation... in that never-ending series of concepts which mathematicians call 'positive whole numbers'. All of the more complicated and artificial forms which are likewise called numbers, the fractional and irrational, and negative and complex numbers have their origin and basis in the elementary number concepts and their interrelations. (*Ibid.*, 95)

As he undertook his project to provide a more detailed analysis of the concepts of arithmetic and a deeper foundation for its theorems, the still faithful student of Brentano also considered that psychology was the indispensable tool for analyzing the concept of number (Husserl 1975, 33; see Husserl 1891, 16). However, although the psychological analyses of *On the Concept of Number* were almost entirely incorporated into the first four chapters of *Philosophy of Arithmetic*, the enthusiastic espousal of psychologism found in the earlier work is absent from the later one. And Husserl, who had not initially considered Brentano's teachings to be empirical and psychological in any pernicious sense, later confessed that there had been "connections in which such a psychological foundation never came to satisfy" him, that it could bring "no true continuity and unity", that he had grown "more and more disquieted by doubts of principle, as to how to reconcile the objectivity of mathematics, and of all of science in general, with a psychological foundation for logic". (Husserl 1900-01, 42; Husserl 1975, 34)

Husserl also soon abandoned Weierstrass' teaching on the primacy of the cardinal number. In a letter to Stumpf, written in 1890 or 1891, Husserl revealed that the theory that the concept of cardinal number forms the foundation of general arithmetic that he had tried to develop in *On the Concept of Number* had soon proved to be false. By no clever devices, he explained,

> can one derive negative, rational, irrational, and the various sorts of complex numbers from the concept of cardinal number. The same is true of the ordinal concepts, of the concepts of magnitude, and so on. And these concepts themselves are not logical particularizations of the cardinal concept. (Husserl 1994, 13)

Husserl's tergiversation in this regard also becomes apparent through a comparison of the foreword and the introduction to *Philosophy of Arithmetic.* (Husserl 1891, VIII, 5 and note; Hill 1991, 81-85)

The lessons learned from his revered mentors had left him in the lurch. Husserl felt forced to embark upon an independent path. Ten years of hard, lonely work and struggle ensued. He felt that his efforts had brought him "close to the most obscure parts of the theory of knowledge", and that he was standing before "great unsolved puzzles" concerning the very possibility of knowledge in general. He described himself as having been "powerfully... gripped by deep, and by the deepest, problems" (Husserl 1975, 16-17; Husserl 1994, 167, 492-93). His search for answers that he did not believe his early training could provide eventually led him to adopt metaphysical and epistemological views that he had learned to consider odious and despicable. (Hill 1998)

From Bolzano the Mathematician to Bolzano the Philosopher

By his own account, Husserl had always been well positioned to appreciate the work of Bernard Bolzano who, as a mathematician, had already come to his attention as a student of Weierstrass. Husserl had become further acquainted with Bolzano's ideas through Brentano's critical discussions of the paradoxes of infinity in his lectures, and then through Georg Cantor. (Husserl 1975, 37)

Bolzano was a forerunner of the movement to rigorize analysis that would gain momentum later in the 19th century. His pioneering work to rebuild intuitively accepted proofs of theorems in a rigorous way solely on the basis of arithmetical and logical concepts prepared the way for much that Weierstrass would later advocate and undertake. And, as Weierstrass himself acknowledged, Bolzano actually developed much of the theory of real functions in much the same form that, inspired by him, Weierstrass would teach it in his inspiring courses forty years later. (Sebestik, 17, 107 and note; Kline, 948, 950-55; Jourdain, 297; Føllesdal, 7-10; Coffa)

With so many of his mentors impressed by Bolzano's work, Husserl should have been primed to appreciate it. This was not, however, immediately the case. Once acquainted with Bolzano's thought, Husserl recalled, he had "made a point of looking through the long-forgotten *Wissenschaftslehre* of 1837 and of making use of it from time to time with the help of its copious index", but he originally misinterpreted Bolzano's original thoughts about ideas, propositions and truths in themselves as being about mythical entities, suspended somewhere between being and non-being. (Husserl 1975, 37; Husserl 1994, 201-02)

This particular reaction on Husserl's part is understandable, for Brentano inculcated in his students a model of philosophy based on the natural sciences and trained them to despise metaphysical idealism. So, it is easy to see how Husserl, so completely under Brentano's influence in the beginning, might not have quickly warmed to philosophical ideas that Brentano taught his students to disdain (Husserl 1919, 344-45). It was only after Husserl had grown disillusioned with Brentano's empirical psychology that he became receptive to Bolzano's idealism.

Cantor's anti-naturalistic prejudices and deep pro-idealistic convictions must have had a hand in prying Husserl away from empirical psychology and steering him in the direction of Platonic idealism. But it was the study of Hermann Lotze's logic and his reflections on the interpretation of Plato's theory of ideas, Husserl maintained, that provided him with his first major insight and was responsible for his fully conscious, radical turn from psychologism and attendant espousal of Platonic idealism. (Husserl 1975, 36; Husserl 1994, 201)

Lotze's work provided Husserl with the key to understanding the "curious conceptions" of Bolzano that had initially seemed so naive and unintelligible to him. Lotze's talk of truths in themselves gave Husserl the idea to transfer all of mathematics and a major part of traditional logic into the realm of the ideal. It then suddenly occurred to him that the first two volumes of Bolzano's *Wissenschaftslehre* on ideas in themselves and propositions in themselves were to be seen as a first attempt at a unified presentation of the area of pure ideal doctrines and that a complete plan of pure logic was already available there. (Husserl 1994, 201-02; Husserl 1975, 36-38, 46-49)

Though Bolzano's propositions in themselves had originally seemed to Husserl to be metaphysical abstrusities, it then became clear to him that what Bolzano had in mind was basically something quite obvious. By proposition in itself, Husserl now understood what people ordinarily called the sense of a statement, what is explained as one and the same when, for example, different persons are said to have asserted the same thing. Or, again, propositions in themselves were simply what scientists called a theorem, for example the theorem about the sum of the angles in a triangle, which no one would think of considering the product of anyone's subjective experience of judging. This realization demystified Bolzano's teachings for Husserl. (Husserl 1994, 201-02; Husserl 1905, 37)

It then further became clear to Husserl that this identical sense could be nothing other than the universal, the species, which belongs to a certain Moment present in all actual assertions with the same sense and makes that very identification possible, even when the descriptive content of the individual lived experiences of asserting varies considerably otherwise. Interpreted in this way, he found Bolzano's idea that propositions are objects that nonetheless have no existence quite intelligible. They had the ideal being or validity of objects which are universals, the being that is established, for example, in the existence proofs of mathematics. (Husserl 1994, 201-02)

So, although Husserl had come to Halle free of Platonic idealism, he was to leave a committed Platonic idealist, who had come to believe that idealistic systems were of "the highest value", that entirely new and totally radical dimensions of philosophical problems were illuminated in them, that "the ultimate and highest goals of philosophy were opened up only when the philosophical method which these particular systems require is clarified and developed" (Husserl 1919, 345). Every possible effort, Husserl would write, had been made in the *Logical Investigations* "to dispose the reader to the recognition of this ideal sphere of being and knowledge... to side with 'the ideal in this truly Platonistic sense', 'to declare oneself for idealism' with the author" (Husserl 1975, 20). Phenomenology would be an "eidetic" discipline. The "whole approach whereby the overcoming of psychologism is phenomenologically accomplished", Husserl explained, "shows that what... was given as analyses of immanent consciousness must be considered as a pure *a priori* analysis of essence". (*Ibid.*, 42)

This transformation had been prepared, Husserl said, by the study of Leibniz and reflections on his distinction between *vérités de raison* and *vérités de fait* and on Hume's ideas about knowledge about matters of fact and relations of ideas. Husserl had become keenly aware of the contrast between Hume's distinction and Kant's distinction between analytic and synthetic judgments, and this became crucial for the positions that he later adopted. (*Ibid.*, 36)

The early 1890s thus found Husserl striving to develop the true concept of analyticity and to discover the basic philosophical line separating genuine analytical ontology from material, synthetic a priori, ontology, which he believed must be fundamentally distinct from it (*Ibid.*, 42-43). In the *Logical Investigations*, he would condemn Kant's logic as being utterly defective (Husserl 1900-01, *Prolegomena*, §58). Kant, Husserl maintained, had not understood the nature and role of formal mathematics and the way in which he had defined the concept of analyticity was totally inadequate and even utterly wrong (Husserl 1906/07, §23). "Not only", Husserl complained, did Kant "never see how little the laws of logic are all analytic propositions in the sense laid down by his own definition, but he failed to see how little his dragging in of an evident principle for analytic propositions really helped to clear up the achievements of analytic thinking". (Husserl 1900-01, Sixth Investigation, §66)

Persuaded of the inadequacy of Kant's analytic-synthetic distinction, Husserl came to believe that Bolzano's more Leibnizian approach to analyticity and meaning harbored the insights logicians needed to prove their propositions by purely logical means. However, in Husserl's opinion, Bolzano never saw the internal equivalence between the analytic nature of both formal logic and formal mathematics made possible by developments in the field of mathematics that had only taken place after his death. (Husserl 1929, §26; Husserl 1975, 36-38)

By drawing the boundary line existing a priori between mathematics and natural sciences like psychology, Husserl believed that he was drawing the line of demarcation and expanding the domain of the analytical in keeping with the most recent discoveries in mathematics. Analytic logic, Husserl would ultimately explain in *Formal and Transcendental Logic*, is first of all valid as an absolute norm presupposed by any rational knowledge. He said that his

> war against logical psychologism, was in fact meant to serve no other end than the supremely important one of making the specific *province* of analytic logic visible in its purity and ideal particularity, freeing it from the psychologizing confusions and misinterpretations in which it had remained enmeshed with from the beginning. (Husserl 1929, §67)

The value of his criticisms of logical psychologism were, he believed, precisely in his drawing attention to a pure, analytic logic, distinct from any psychology, as being an independent field, like geometry or the natural sciences. Epistemological questions may well arise regarding this pure logic, he considered, but this must not interfere with its independent course, or involve delving into the concrete aspects of the logical life of the consciousness. For that would be psychology. (*Ibid.*)

No psychologistic empiricism, Husserl had come to believe, "can change the fact that pure mathematics is a strictly self-contained system of doctrines which is to be cultivated using methods that are essentially different from those of natural science" (Husserl 1975, 29). He wrote to Brentano in 1905,

> The empirical sciences—natural sciences are sciences of 'matters of fact'.... Pure Mathematics, the whole sphere of the genuine Apriori in general, is free of all matter of fact suppositions.... We stand not within the realm of nature, but within that of Ideas, not within the realm of empirical... generalities, but within that of the ideal, apodictic, general system of laws, not within the realm of causality, but within that of rationality.... Pure logical, mathematical laws are laws of essence.... (Husserl 1905, 37)

Husserl did, though, realize that not all the sciences are theoretical disciplines that, like mathematical physics, pure geometry or pure arithmetic, are characterized by the fact that their systemic principle is a purely analytical one. Sciences like psychology, history, the critique of reason and, notably, phenomenology, he believed, require that one go beyond the analytico-logical model. When they are formalized and one asks what it is that binds the propositional forms into a single system form, one finds oneself facing, Husserl maintained, nothing more than the empty general truth that there is an infinite number of propositions connected in objective ways that are compatible with one another in that they do not contradict each other analytically. (Husserl 1917/18, §54)

In Conflict with Gottlob Frege

Disproportionately more has been written about Husserl and Gottlob Frege than about Husserl and any other mathematician, a fact attributable to the enthusiasm so many feel about Frege's work and also to a genuine desire on the part of a smaller number of philosophers to see the legitimate ties recognized between Husserl's ideas and those that went into the making of analytic philosophy and of 20th century philosophy of logic and mathematics in general. (ex. Haaparanta (ed.); Hill and Rosado Haddock)

Husserl's interest in Frege's work can be traced back to the late 1880s. There is no mention of it in *On the Concept of Number*. However, by 1887 Husserl had obtained a copy of Frege's 1884 *The Foundations of Arithmetic* (Schuhmann, 18), which is thoroughly examined in *Philosophy of Arithmetic*, as a mere glance at the index of names reveals.

It should not be surprising that Husserl picked up and studied the book of that reputedly obscure thinker at that particular point, for members of Brentano's school knew Frege's work (Linke 1947). In particular, in 1882, Carl Stumpf had expressed to Frege his pleasure upon learning that he was "working on logical problems, an area where there is such a great need for cooperation between mathematicians or scientists and philosophers". Stumpf, moreover, suggested to Frege that his work might be more favorably received were he first to explain his ideas in ordinary language, advice that Frege scholars think he might have taken in writing the *Foundations*. Stumpf also, importantly, expressed his conviction that arithmetical, algebraic and geometrical judgments were analytic. (Frege 1980a, 171-72)

In 1885, Cantor had reviewed Frege's *Foundations*. In the review, Cantor decried Frege's recourse to extensions, but praised him for demanding that all psychological factors and intuitions of space and time be banned from arithmetical concepts and principles because that was the only way in which their strict logical purity and validity might be secured (Cantor 1885, 440-41). So two of the three members of Husserl's *Habilitationskommittee* knew Frege's work.

Husserl's reaction to the *Foundations* was mixed. In 1891, he wrote to Frege that of all the many writings he had before him as he worked on *Philosophy of Arithmetic*, he could not name another that he had studied with nearly as much enjoyment as he had his. Husserl acknowledged "the large amount of stimulation and encouragement" he had derived from the book. He had, he explained, derived constant pleasure from the originality of mind, clarity and honesty of Frege's investigations, "which nowhere stretch a point or hold back a doubt, to which all vagueness in thought and word is alien, and which everywhere try to penetrate to the ultimate foundations". Husserl did, however, admit that he had not been able on the whole to agree with Frege's theories and alluded to "fundamental divergences". (Frege 1980a, 64-65)

In *Philosophy of Arithmetic* itself, Husserl called *Foundations* an ingenious, insightful, remarkable attempt to analyze and define the concept of cardinal number that came close to providing the right answer only to move further off the mark. Husserl characterized Frege's ideal as being that of grounding arithmetic on a series of formal definitions out of which all the theorems of that science might be derived in a purely syllogistic manner. Husserl called Frege's goal chimerical and criticized him for wandering off into hyper-subtleties in a sterile way and concluding with no positive result (Husserl 1891, 129-31). Husserl eventually retracted those particular criticisms of Frege's work. (Husserl 1900-01, *Prolegomena* §45 n.)

In comments that Husserl never retracted, he complained that Frege's results were such that one could only be astonished that anyone might have taken them to be true, except for temporarily. All Frege's definitions, Husserl argued, become true propositions when one substitutes the concepts to be defined with their extensions, but then they are absolutely self-evident propositions and worthless. Husserl also made some very perspicacious and timely criticisms of the fundamental role that Frege accorded to Leibniz's principle of substitutivity of identicals. (Husserl 1891, 133-35 and note, 104-05; Hill 1994b)

In case Husserl's judgment may seem unduly harsh, especially in light of the indulgent approach to Frege's ideas that is popular nowadays, it is important to realize that in *Foundations*, Frege himself acknowledged that that his plan to use Leibniz's principle to obtain the concept of number by fixing the sense of a numerical identity (Frege 1884, §§62-65) was liable to produce nonsensical conclusions or be sterile and unproductive (Frege 1884, §§66-67, §105). As a remedy, he cautiously introduced the extensions (Frege 1884, §§68-69 and note, §107) that by the time he published *The Basic Laws of Arithmetic* in 1893 had come to take on such "great fundamental importance" that he said that he could just no longer do without them. (Frege 1893, ix-x; Hill 1997b, 61-68; Hill 1997c, 58-62)

Never one to accept criticism gracefully, Frege lashed back at Husserl in a cranky review of *Philosophy of Arithmetic* (Frege 1894), in which a great deal is going on below the surface (Hill 1994a). For one thing, in the review Frege tellingly chose to attack Husserl's failure to appeal to extensions. Any view according to which a statement of number is not a statement about a concept or about the extension of a concept, Frege said, is naive for "when one first reflects on number, one is led by a certain necessity to such a conception" (Frege 1894, 197). Had Husserl used the term 'extension of a concept' in the way he, Frege, had, they "should hardly differ in opinion about the sense of a number statement". (Frege 1894, 201-02)

By criticizing Frege's use of extensions Husserl had surely struck a sensitive chord in Frege who in *Basic Laws I*, published two years after *Philosophy of Arithmetic*, pinpointed this as the place where any decision about any defects or errors in his logic would ultimately be made (Frege 1893, vii; Hill 1997b, 68-72). Indeed, upon learning of the famous contradiction of the set of all sets that are not members of themselves derivable in *Basic Laws*, Frege immediately replied that it was the law about extensions that was to blame and that its collapse seemed to undermine his foundations for arithmetic (Frege 1980a, 130-32). At least, it was consoling to know, Frege wrote in the appendix to *Basic Laws II*, after Russell's discovery, that everybody who had made use of extensions of concepts, classes, sets in proofs was in the same position that he was. (Frege 1980b, 214)

Frege never believed that his law about extensions recovered from the shock it sustained from Russell's paradox, and he came to rue having used the expression 'extension of concept' which, he finally concluded, easily "can get one into a morass" and leads "into a thicket of contradictions" (Frege 1980a, 55; Frege 1979, 269-70). "Only with difficulty", Frege confessed, "did I resolve to introduce classes (or extents of concepts), because the matter did not appear to me quite secure—and rightly so, as it turned out.... I was constrained to overcome my resistance.... I confess that by acting thus, I fell into the error of letting go too easily my initial doubts" (Frege 1980a, 191). "While I sometimes had slight doubts during the execution of the work, I paid no attention to them. And so it happened that after the completion of the *Basic Laws of Arithmetic* the whole edifice collapsed around me" (Frege 1980a, 55). Frege's review of Husserl actually contains some of the most forceful statements that Frege ever made in favor of extensions, a fact that, in retrospect, but reflects his insecurity about resorting to them.

Telling too is the fact that Frege directly incorporated into his review of Husserl's book criticisms and examples that he had previously used in reviewing Cantor's *Mitteilungen* (Frege 1892, 178-81; Frege 1979, 68-71). For example, Frege had complained that Cantor was "asking for impossible abstractions" (Frege 1892, 179). In his review of Husserl's book, Frege charged, that abstraction would "cleanse things of their peculiarities... in the wash-tub of the mind" where "we can easily change objects by directing our attention towards them or away from them... We attend less to a property and it disappears" (Frege 1894, 197). Suppose, Frege went on,

> that there are a black and a white cat sitting side by side before us. We do not attend to their colour, and they become colourless—but they still sit side by side. We do not attend to their posture, and they cease to sit... but each of them is still in its place. We no longer attend to the place and they cease to occupy one.... By continued application of this procedure, each object is transformed into a more and more bloodless phantom. (Frege 1894, 197-98)

It must surely be assumed, Frege maintained in the review, that the process of abstraction effects some change in the objects and that they become different from the original objects which are either transformed or actually created by the abstraction process. (Frege 1894, 204)

Yet in *Foundations* §34, Frege himself had acknowledged that in the abstraction process the things themselves do not lose any of their characteristics, that one might disregard the properties that distinguish a white cat and a black cat, but the cats would not thereby become colorless. What is more, despite the ridicule he heaped on the abstraction process in his review of Husserl's book, Frege admitted there that Husserl himself did not claim that the mind creates new objects or changes old ones. Frege actually acknowledged that Husserl "disputes this in the most vehement terms" (Frege 1894, 205), which is true. (ex. Husserl 1891, 28-30, 42, 46, 139)

In his review, Frege also accused Husserl of taking "the road of magic rather than of science (Frege 1894, 205). In the posthumous draft of a review of Cantor's *Mitteilungen* (Frege 1979, 68-71), Frege wrote of the "astounding magical properties", "magical effects" and "miraculous powers" in Cantor's work (*Ibid.*, 69). Choosing mice instead of cats, Frege complained there that mathematicians like Cantor find a whole host of things in mice that are unworthy to form a part of number. So they begin abstracting. Ridiculing these mathematicians, Frege writes that for them "everything in the mice is out of place: the beadiness of their eyes no less than the length of their tails and the sharpness of their teeth. So one abstracts from the nature of the mice... one abstracts presumably from all their properties...". (*Ibid.*, 70)

These are but a few reasons why Frege's review should not be considered the straightforward, objective evaluation of Husserl's book that it has popularly been believed to be. Much, in fact, still remains to be rectified regarding Frege's review, the uncritical reading of which has unfairly distorted philosophers' perception of a work that they do not know very well. Husserlians themselves have been only too willing to dismiss *Philosophy of Arithmetic* as the product of an immature, pre-phenomenological phase of Husserl's life and have not risen to its defense.

Husserl and Russell's Paradox

A grasp of the nature of the exchange of ideas that took place among Husserl, Cantor and Frege naturally raises questions as to what exactly Husserl knew about the famous contradiction of the set of all not sets that are not members of themselves and when he knew it.

Bertrand Russell has received the lion's share of the credit for having discovered this "paradox". However, though he was certainly the first to publicize Frege's errors and to bring the point home to Frege himself, the finding was not at all as surprising to the members of the circle of eminent mathematicians that Husserl frequented as has been assumed.

In Halle, Husserl had taken up residence with the antinomies of set theory themselves (Cantor 1991, 387-464; Dauben 240-70) and so was one of the very first to be on hand to witness the paradoxical consequences of Cantor's theories. Remember that it was in studying Cantor's theories that Russell discovered the famous antinomy that put an end to the logical honeymoon he was having when he began writing the *Principles of Mathematics*. (ex. Russell 1903, §§100, 344, 500; Russell 1959, 58-61; Grattan-Guinness 1978, 1980)

Of the mathematical world's reaction to Cantor's theories, Hilbert has written that it was violent and

> took very dramatic forms... purely through the ways in which notions were formed and modes of inference used... contradictions appeared, sporadically at first, then ever more severely and ominously.... In particular, a contradiction discovered by Zermelo and Russell had, when it became known, a downright catastrophic effect in the world of mathematics. Confronted with these paradoxes Dedekind and Frege actually abandoned their standpoint and quit the field. (Hilbert 1925, 375)

Hilbert's mention of Ernst Zermelo in this context brings us to the next point. The paradoxes of set theory were a topic of lively discussion in Göttingen, where Husserl took up residence in 1901 (ex. Peckhaus, 168-95). In a November 7, 1903 letter, Hilbert told Frege that Russell's antinomy was already known to them in Göttingen. Hilbert added that he believed that Zermelo had found it three or four years earlier after having learned of other, even more convincing, contradictions from Hilbert himself as many as four or five years before. Hilbert further commented that the idea that "a concept is already there if one can state of any object whether or not it falls under it" did not seem adequate to him. What would be decisive, he states, "is the recognition that the axioms that define the concept are free from contradiction". (Frege 1980a, 51-52)

Concrete evidence corroborating Zermelo's finding of the paradox is to be found in a note that he sent to Husserl in April 1902, in which he conveyed his proof. Husserl recorded what Zermelo wrote and that record has survived (Husserl 1994, 442). The exchange turned upon certain remarks that Husserl had made in his 1891 review of Ernst Schröder's *Vorlesungen über die Algebra der Logik* (*Ibid.*, 52-91, 421-41). There Schröder had tried to show that bringing all possible objects of thought into a class gives rise to contradictions. In his review, Husserl had written that although Schröder's argument might at first sight appear astonishing, it was actually sophistical. However, importantly, Husserl conceded:

> in the cases where we simultaneously have, besides certain classes, also classes *of* those classes, the calculus may not be blindly applied. In the sense of the calculus of sets as such, any set ceases to have the status of a set as soon as it is considered as an element of another set; and this latter in turn has the status of a set only in relation to its primary and authentic elements, but not in relation to whatever elements *of* those elements there may be. If one does not keep this in mind, then actual errors in inference can arise. (*Ibid.*, 84-85)

In Zermelo's opinion, however, Schröder had been basically right, but his reasoning had been faulty. According to Zermelo's argument as recorded by Husserl: given a set M which contains each of its sub-sets m, m'... as elements, and a set M_0 which is the set of all sub-sets M, which do not contain themselves as elements, it can then be shown that M_0 both does and does not contain itself. (*Ibid.*, 442)

In reflecting on Husserl and "Russell's" paradox, it is also of interest to note that Husserl and Frege exchanged letters between October 1906 and January 1907. Copies of two of the letters that Frege wrote to Husserl have survived. The three letters that Husserl wrote to Frege were, however, been lost during World War II. We do know, however, that those letters dealt with, among other things "the paradoxes". (Frege 1980a, 70; Hill 2000)

So, it is intriguing to note in this regard that preeminent Frege scholar Michael Dummett has contended that "Frege's posthumous writings allow us with high probability to date almost exactly Frege's disillusionment over the attempt, to which he had devoted his life, to derive arithmetic from logic" in mid-1906 (Dummett, 21). Dummett cites the article that Frege had begun writing during that year about the paradoxes of set theory and the inadequacy of certain remedies that were being proposed (Frege 1979, 176-83). "Tantalizingly little of the article survives ...", Dummett concludes, and "very probably it represents the very moment at which Frege came to realize that the attempt was hopeless" (Dummett, 22). Hans Sluga too has concluded that by 1906 Frege "was beginning to think that the theory of sets was undermined by the contradiction. He concluded that there was no use for sets or classes anymore" (Sluga, 169-70). Indeed, nothing that Frege ever wrote after 1906 indicates that he again tried to salvage the specific logical doctrines that he had concluded had led to the paradoxes of set theory. However, as I show in "Frege's Letters", anthologized here, a distressing amount of material on this precise matter has been lost or destroyed.

Husserl left a hundred pages of manuscripts (Husserl, Ms A 1 35.) show that show that he worked directly on the paradoxes of set theory, and on Russell's paradox in particular, in 1912 and the 1920s (Rosado Haddock). As someone who frequented Hilbert's school in Göttingen, and who pointed to the kinship existing between his own manifolds and Hilbert's axiomatic systems (Husserl 1913, §72, Husserl 1929, §31), one may hypothesize, however, that Husserl considered that properly carried out, the axiomatization of set theory might neutralize the contradictions found in Cantorian set theory. Defined and regulated by a complete axiomatic system, sets would thus be apt to play their fruitful role in mathematics, which brings us to our next topic.

Situating Husserl with Regard to
Hilbert and Formalist Theories of Mathematics

Although Husserl retracted the three pages of his criticism of *Foundations* in which he had denied that one could provide sound foundations for arithmetic by deriving theorems from a series of formal definitions in a purely logical fashion, it should be clear from what has been said here that he did not ultimately chose Frege's way. Rather, Husserl's mature philosophy of mathematics would have a formalist flavor.

Once appointed to the University of Göttingen in 1901, Husserl was welcomed into David Hilbert's circle (Reid 1970, 1976; Husserl 1983, XIII). Frege and Hilbert had just been corresponding about truth and logical consistency. In their letters, Hilbert had put his finger on one of the main points dividing Frege and him when he wrote that he was particularly interested in his statement that from the truth of the axioms it follows that they do not contradict one another, because for as long as he had been thinking, writing and lecturing on these things he had been saying exactly the opposite,

> If the arbitrarily given axioms do not contradict one another, then they are true, and the things defined by the axioms exist. This for me is the criterion of truth and existence... only the whole structure of axioms yields a complete definition .... Every axiom contributes to the definition, and hence every new axiom changes the concept. (Frege 1980a, 42, 39-40)

Husserl had access to the Frege-Hilbert correspondence, and partial copies of it, along with notes that Husserl made on it, were found in his *Nachlass*. In his notes, Husserl remarked that Frege had not really understood the meaning of Hilbert's axiomatic foundation for geometry, that it was a matter of a purely formal system of conventions that coincides formally with the Euclidean system. (Frege 1980a, 34-51; Husserl 1970, 447-52)

Now, in the letter to Stumpf from the early 1890s discussed above, Husserl had expressed his frustrations regarding the inability of Brentano's methods to cope with imaginary numbers and his new faith in the *arithmetica universalis* as a part of formal logic understood as a symbolic technique and making up a special, important chapter in logic as technology of knowledge (Husserl 1994, 17). Husserl used the term 'imaginary' in the broadest way to include negative, irrational numbers, fractions, negative square roots (Husserl 1983, 244-49; Husserl 1970, 432) and he called infinite sets 'imaginary concepts' (Husserl 1891, 249). In the case of 'imaginary' numbers like $\sqrt{2}$ and $\sqrt{-1}$, he told Stumpf,

> I first sought to get clear on how operations of thought with contradictory concepts could lead to correct theorems.... Finally I noticed that, through the calculation itself and its rules (as defined for those fictive numbers), the impossible falls away, and a genuine equation remains. Indeed, the procedure of calculation runs its course another time with the same signs, but now referred to valid concepts, and again the result is correct. Thus it is not a matter of the "possibility" or "impossibility" of concepts. Even if I mistakenly imagine that the contradictory exists–even if I hold the most absurd theories concerning the content of the corresponding concepts of number... the calculation remains correct, if it follows the rules. So it must be an accomplishment of the signs and their rules. (Husserl 1994, 15-16)

And, in Husserl's epistolary exchange with Frege in 1891 (Frege 1980a, 61-66), we already find Husserl expressly entertaining a formalist solution to problems about imaginary numbers that were preoccupying him. This was, remember, the stumbling block that led him to abandon Brentano's empirical psychology and Weierstrass' thesis about the primacy of the cardinal number when he was writing the second volume of *Philosophy of Arithmetic*.

Husserl wrote that he had only a rough idea of how Frege would justify the imaginary in arithmetic, since in "On Formal Theories of Arithmetic" (Frege 1885) Frege had rejected the path that Husserl himself had found after much searching. In the passage of the article cited in Husserl's letter (Frege 1980a, 65), Frege had criticized the procedure by which one which but sets down rules by which one passes from the equations given to new ones in the way one moves chess pieces. Unless an equation contains only positive numbers, it no more has a meaning than the position of chess pieces expresses a truth. Now in virtue of these rules, Frege continues his criticism, an equation of positive whole numbers may actually appear. And if the rules are such that true equations can never lead to false conclusions, then only two results are possible: either the final equation is meaningless, or it has a content about which we can pass judgment. The latter will always be the case if it contains only positive whole numbers, and then it must be true, for it cannot be false. If the rules contain no contradictions among themselves, and do not contradict the laws of positive whole numbers, then no matter how often they are applied, no contradiction can ever enter in. Consequently, if the final equation has any meaning at all, it must be non-contradictory, and hence be true. This is a mistake, Frege concludes, for a proposition may very well be non-contradictory without being true.

Husserl's reasons for believing that Frege's logic could not satisfactorily justify the imaginary in arithmetic surely involved Frege's well-known thesis that in a logically perfect language expressions that do not denote objects are unfit for scientific use. The use of signs or combinations of signs without reference was at the heart of Frege's dispute with formalists who, he believed, only manipulated signs without any regard for what those signs might stand for. But, Frege insisted, "logic is not concerned with how thoughts, regardless of truth-value, follow from thoughts... we have to throw aside proper names that do not designate or name an object..." (Frege 1979, 122). In *Foundations* and "On Formal Theories of Arithmetic", he wrote, "I showed that for certain proofs it is far from being a matter of indifference whether a combination of signs–e.g. $\sqrt{-1}$ has a meaning or not, that, on the contrary, the whole cogency of the proof stands or falls with this". (*Ibid.*, 123)

Invited by Hilbert and Felix Klein, Husserl addressed the Göttingen Mathematical Society in 1901 on the subject of imaginary numbers (Husserl 1970, 430-506). Questions regarding imaginary numbers, he explained, had come up in mathematical contexts in which formalization yielded constructions that, arithmetically speaking, were nonsense but which could nonetheless be used in calculations. When formal reasoning was carried out mechanically as if these symbols had meaning, if the ordinary rules were obeyed, and the results did not contain any imaginary components, it seemed that these symbols might be legitimately used. However, this raised significant questions about the consistency of arithmetic and about how one was to account for the achievements of certain purely symbolic procedures of mathematics despite the use of apparently nonsensical combinations of symbols.

By the time Husserl gave his talk, he had concluded that the key to the only possible answer to his questions lie in the theory of complete manifolds that he first expounded in the first volume of the *Logical Investigations* (Husserl 1900-01, *Prolegomena* §70), a conviction that he reaffirmed late in his career in *Formal and Transcendental Logic.* (Husserl 1929, §31)

For Husserl, the general theory of the manifolds, or science of theory forms, was a field of free, creative investigation that was made possible once the form of the mathematical system had been emancipated from its content. Discovering that deductions, series of deductions, continue to be meaningful and to remain valid when one assigns another meaning to the symbols actually freed one to liberate the mathematical system, he considered. Nothing more need be presupposed than the fact that the objects figuring in it were such that, for them, a certain connective supplied new objects and did so in such a way that the form determined is assuredly valid for them.

According to Husserl's theory, manifolds themselves were pure forms of possible theories which, like molds, remain totally undetermined as to their content, but to which thought must necessarily conform in order to be thought and known in a theoretical manner. By using axioms of such and such a form, theories of such and such a form may be developed. The objects are exclusively determined by the form of the interconnections assigned to them, meaning, neither directly inasmuch as individuals, nor indirectly by their kind or species. The interconnections themselves are just as little determined in terms of content as are the objects. Only their form determines them by virtue of the form of the elementary laws admitted as valid for these interconnections, laws that also determine the theory to be constructed, the form of the theories.

One can operate freely within a manifold with imaginary concepts and be sure that what one deduces is correct when the axiomatic system completely and unequivocally determines the body of all the configurations possible in a domain by a purely analytical procedure. A domain is complete, Husserl held, when each grammatically constructed proposition exclusively using the language of this domain is, from the outset, determined to be true or false in virtue of the axioms. In that case, calculating with imaginary concepts can never lead to contradictions. It is the completeness of the axiomatic system that gives one the right to operate freely. It was formal constraints requiring that one not resort to any meaningless expression, no meaningless imaginary concept that were restricting us in our theoretical, deductive work. Husserl considered that the kinship between his conception of the completeness of axiomatic systems and Hilbert's was obvious. (Husserl 1900-01, *Prolegomena*, §70; Husserl 1913, §§71-72; Husserl 1906/07, §§18-19; Husserl 1917/18, Chapter 11)

Kurt Gödel's Secret Admiration for Husserl's Work

Husserl did not, of course, believe that formal logic alone could suffice. For one thing, he believed that once armed with the objective structures of formal logic, philosophical logicians were obliged to go further and come to terms with really hard epistemological questions about subjectivity, logic and mathematics themselves. According to him, philosophical logicians had to see that the logical sense of the formal sciences also includes a sphere of cognitive functioning and a sphere of possible applications. They could only submit to a logic that they had thought through, and thought through with insight, a fact that Husserl believed cried out for thorough epistemological investigations into the subjective and intersubjective processes and the ways in which they inevitably interact with the objective order.

In this respect too, Hilbert and Husserl were in agreement with one another and disagreed with Frege. Upon several occasions, Hilbert expressed his conviction that "counter to the earlier endeavors of Frege and Dedekind... if scientific knowledge is to be possible certain intuitive conceptions and insights are indispensable; logic alone does not suffice" (Hilbert 1925, 392). According to Hilbert, "the efforts of Frege and Dedekind were bound to fail" because

> as a condition for the use of logical inferences and the performance of logical operations something must already be given in our faculty of representation, certain extralogical concrete objects that are intuitively present as immediate experience prior to all thought. If logical inference is to be reliable, it must be possible to survey these objects completely in all their parts, and the fact that they occur, that they differ from one another, and that they follow each other, or are concatenated, is immediately given intuitively, together with the objects, as something that neither can be reduced to anything else nor requires reduction. (Hilbert 1925, 376)

This, Hilbert held, "is the basic philosophical position that I regard as requisite for mathematics and, in general, for all scientific thinking, understanding and communication". (*Ibid.*; see also Hilbert 1922, 202; Hilbert 1927, 464-65)

Now, new support for Husserl's conviction that "logic must not be a mere formal (mathematical) theory... but, as a philosophical logic... requires phenomenological and epistemological elucidations in virtue of which we not merely are completely certain of the validity of its concepts and theories, but also truly understand them" (Husserl 1994, 215) has come from the famous discoverer of the incompleteness of formal systems, Kurt Gödel (Gödel 1995a; Gödel 1995b), who, the philosophical world was surprised to learn from Hao Wang in the 1980s (Wang 1986, 1987, 1996), had been a secret admirer of Husserl's work.

In a posthumous paper, Gödel contended that

> the certainty of mathematics is to be secured not by proving certain properties by a projection onto material systems–namely the manipulation of physical symbols—but rather by cultivating (deepening) knowledge of the abstract concepts themselves which lead to the setting up of these mechanical systems, and further by seeking, according to the same procedures, to gain insights into the solvability, and the actual methods for the solution, of all meaningful mathematical problems. (Gödel 1995b, 383)

Gödel thought that the procedure by which it might be possible to extend knowledge of the abstract concepts in question was most nearly supplied by the systematic method for clarifying meaning prescribed by Husserl's phenomenology where, as Gödel wrote in his *Nachlass*, "clarification of meaning consists in focusing more sharply on the concepts concerned by directing our attention a certain way, namely, onto our own acts in the use of these concepts, onto our powers in carrying out our acts, etc" (*Ibid.*). Gödel viewed phenomenology as "a procedure or technique that should produce in us a new state of consciousness in which we describe in detail the basic concepts we use in our thought, or grasp other basic concepts hitherto unknown to us" (*Ibid.*). According to Gödel, Husserl's theories could "safeguard for mathematics the certainty of its knowledge" and "uphold the belief that for clear questions posed by reason, reason can also find clear answers". (*Ibid.*, 381)

Of additional interest is the fact that Wang has reported that Gödel was particularly interested in Husserl's ideas on axiomatization (ex. Wang 1996, 168, 334), a topic yet to be fully investigated.

Conclusion

In 1929 Husserl published *Formal and Transcendental Logic*, which was the product of decades of reflection upon the relationship between logic and mathematics, between mathematical logic and philosophical logic, between logic and psychology, and between psychologism and his own transcendental phenomenology. One of the stated goals of the book was to redraw the boundary line between logic and mathematics in light of the new investigations into the foundations of mathematics. A second goal was to examine the logical and epistemological issues such developments have raised. (Husserl 1929, 10-17)

In *Formal and Transcendental Logic*, Husserl expressed his conviction that the formalization of large tracts of mathematics in the 19th century had laid bare the deep, significant connections obtaining between formal mathematics and formal logic, and had thus raised profound new questions about the deep underlying connections existing between the two fields. Logic and mathematics, he believed, had originally developed as separate fields because it had taken so long to elevate any particular branch of mathematics to the status of a purely formal discipline free of any reference to particular objects. Until that had been accomplished the important internal connections obtaining between the two fields were destined to remain hidden. However, once large tracts of mathematics had been formalized, the parallels existing between its structures and those of logic became apparent, and the abstract, ideal, objective dimension of logic could then be properly recognized, as it traditionally had been in mathematics. Developments in formalization had thus unmasked the close relationships between the propositions of logic and number statements making it possible for logicians to develop a genuine logical calculus that would enable them to calculate with propositions in the way mathematicians did with numbers, quantities and the like. (*Ibid.*, Chapter 2)

Mathematics, Husserl deemed, has its own purity and legitimacy. Mathematicians are free to create arbitrary structures. They need not be concerned with questions regarding the actual existence of their formal constructs, nor with any application or relationship their constructs might have to possible experience, or to any transcendent reality. They are free to do ingenious things with thoughts or symbols that receive their meaning merely from the way in which they are combined, to pursue the necessary consequences of arbitrary axioms about meaningless things, restricted only by the need to be non-contradictory and coordinated to concepts previously introduced by precise definition. And the same, Husserl contended, was true for formal logic when it was actually developed with the radical purity that is necessary for its philosophical usefulness and gives it the highest philosophical importance. Severed from the physical world, it lacks everything that makes possible a differentiation of truths or, correlatively of evidences. (*Ibid.*, 138, §§23, 40, 51)

However, as theoreticians of science in general, philosophical logicians are obliged to contend with the question of basic truths about a universe of objects existing outside of formal systems. They are called upon to seek solutions to the problems that come up when scientific discourse steps outside the purely formal domain and makes reference to specific objects or domains of objects. They are not free to sever their ties with nature and science, to accept a logic that tears itself entirely away from the idea of any possible application and becomes a mere ingenious playing with thoughts, or symbols that mere rules or conventions have invested with meaning. They must step out of the abstract world of pure analytic logic, with its ideal, abstract entities, and confront those more tangible objects that make up the material world of things. In addition, they are obliged to step back and investigate the theory of formal languages and systems themselves, and their interpretations. (*Ibid.*, §§40, 52)

So, Husserl believed that formal logic required a complement. Once liberated from things and psychologizing subjectivity, pure logic had to find its necessary complement in a transcendental logic that would take into account the connections that philosophical logic inevitably maintains with both knowing subjects and the concrete world. For Husserl, true philosophical logic could only develop in connection with a transcendental phenomenology by which logicians penetrate an objective realm which is entirely different from them. (ex. *Ibid.*, §§40, 42)

However, Husserl always insisted on the primacy of the objective side of logic. He insisted that the subjective order could not be properly examined until the objective order had been and until the objectivity of the structures girding scientific knowledge had been established and demonstrated. He maintained that pure logic with its abstract ideal structures had to be clearly seen and definitely apprehended as dealing with ideal objects before transcendental questions about them could be asked. (*Ibid.*, §§8, 9, 11, 26, 42-44, 92, 98, 100)

It is knowledge of formal logic, he reminded readers in *Formal and Transcendental Logic*, that supplies the standards by which to measure the extent to which any presumed science meets the criteria of being a genuine science, the extent to which the particular findings of that science constitute genuine knowledge, the extent to which the methods it uses are genuine ones (*Ibid.*, §7). The world constituted by transcendental subjectivity is a pre-given world, Husserl explained in *Experience and Judgment*. It is not a pure world of experience, but a world that is determined and determinable in itself with exactitude, a world within which any individual entity is given beforehand in an perfectly obvious way as in principle determinable in accordance with the methods of exact science and as being a world in itself in a sense originally deriving from the achievements of the physico-mathematical sciences of nature. (ex. Husserl 1939, §11; Husserl 1929, 26b)

Husserl was perfectly conscious of the extraordinary difficulties that this dual orientation of logic involved. Since, according to his theories, the ideal, objective, dimension of logic and the actively constituting, subjective dimension interrelate and overlap, or exist side by side, logical phenomena thus seem to be suspended between subjectivity and objectivity in a confused way. In *Formal and Transcendental Logic*, he suggested that almost all that concerns the fundamental meaning of logic, the problems it deals with, its method, is laden with misunderstandings owing to the very fact that objectivity arises out of subjective activity. He even considered that it was owing to these difficulties that, after centuries and centuries, logic had not attained the secure path of rational development. (ex. Husserl 1929, §8)

References

Becker, Oskar. 1930. "The Philosophy of Edmund Husserl", *The Phenomenology of Husserl, Selected Critical Readings*, R. O. Elveton (ed.), Chicago: Quadrangle Books, 1970, 40-72.

Brück, Maria 1933. *Über das Verhältnis Edmund Husserls zu Franz Brentano. Vornehmlich mit Rücksicht auf Brentanos Psychologie*, Würzburg: K. Tritsch,

Cantor, Georg 1883. *Grundlagen einer allgemeinen Mannigfaltigkeitslehre. Ein mathematisch-philosophischer Versuch in der Lehre des Unendlichen*, Leipzig: Teubner. Cited as appears in Cantor 1932, 165-246.

Cantor, Georg 1884. "Principien einer theorie der Ordnungstypen" (dated November 6, 1884), first published as "An unpublished paper by Georg Cantor" by I. Grattan-Guinness in *Acta Mathematica* 124, 1970, 65-107.

Cantor, Georg 1885. "Rezension von Freges *Grundlagen*", *Deutsche Literaturzeitung* 6, 728-29, in Cantor 1932, 440-41.

Cantor, Georg 1887/88. "Mitteilungen zur Lehre vom Transfiniten". *Zeitschrift für Philosophie und philosophische Kritik* 91, 81-125, 92, 240-65. Cited as appears in Cantor 1932, 378-439. Also published as *Gesammelte Abhandlungen zur Lehre vom Transfiniten*. Halle: C.E.M. Pfeffer.

Cantor, Georg 1932. *Gesammelte Abhandlungen*, Ernst Zermelo (ed.), Berlin: Springer.

Cantor, Georg 1991. *Georg Cantor Briefe*, H. Meschkowski and W. Nilson (eds.), New York: Springer.

Coffa, J. Alberto 1982. "Kant, Bolzano and the Emergence of Logicism", *The Journal of Philosophy* 74, 679-89. Reprinted in Demopoulos 1995, 29-40.

Couturat, Louis 1896. *De l'infini mathématique*, Paris: Blanchard, 1973.

Dauben, Joseph 1979. *Georg Cantor, His Mathematics and Philosophy of the Infinite*, Princeton: Princeton University Press.

Demopoulos, William 1994. "Frege and the Rigorization of Analysis", *Journal of Philosophical Logic* 23, 225-46.

Dummett, Michael 1981. *The Interpretation of Frege's Philosophy*, Cambridge MA: Harvard University Press.

Eccarius, W. 1985. "Georg Cantor und Kurd Lasswitz: Briefe zur Philosophie des Unendlichen", *NTM Schriftenr. Gesch. Naturwiss., Technik., Med.* 22, 7-28.

Føllesdal, Dagfinn 1997. "Bolzano's Legacy", *Grazer philosophische Studien* 53 1-10.

Frege, Gottlob 1884. *The Foundations of Arithmetic*, Oxford: Blackwell, 2nd rev. ed., 1986.

Frege, Gottlob 1885. "On Formal Theories of Arithmetic", *Collected Papers on Mathematics, Logic, and Philosophy*, Brian McGuinness (ed.), Oxford: Blackwell, 1984 112-21.
Frege, Gottlob 1892. "Review of Georg Cantor, *Zur Lehre vom Transfiniten: Gesammelte Abhandlungen aus der Zeitschrift für Philosophie und philosophische Kritik*", in Frege 1984, 178-81.
Frege, Gottlob 1893. *The Basic Laws of Arithmetic*, Berkeley CA: University of California Press, 1964.
Frege, Gottlob 1894. "Review E. G. Husserl's *Philosophy of Arithmetic*", In Frege 1984, 195-209.
Frege, Gottlob 1979. *Posthumous Writings,* Oxford: Blackwell.
Frege, Gottlob 1980a. *Philosophical and Mathematical Correspondence*, G. Gabriel et. al. (eds.), Oxford: Blackwell.
Frege, Gottlob 1980b. *Translations from the Philosophical Writings*. 3rd ed., Oxford: Blackwell, 1952.
Frege, Gottlob 1984. *Collected Papers on Mathematics, Logic and Philosophy*. Brian McGuinness (ed.), Oxford: Blackwell.
Gerlach, H. and H. Sepp (eds.) "Es ist keine Seligkeit 13 Jahre lang Privadocent und Tit. 'prof'. zu sein. Husserls hallesche Jahre 1887 bis 1901", *Husserl in Halle*, Bern: Peter Lang, 1994 15-39.
Gödel, Kurt 1995a, "Is Mathematics Syntax of Language?" *Collected Works III*, Solomon Feferman (ed.), New York: Oxford University Press, 334-62.
Gödel, Kurt 1995b. "The Modern Development of the Foundations of Mathematics in the Light of Philosophy", *Collected Works III*, Solomon Feferman (ed.) New York: Oxford University Press, 374-87. The "Introductory Note" by Dagfinn Føllesdal, 364-73.
Grattan-Guinness, Ivor 1978. "How Russell Discovered His Paradox", *Historia Mathematica* 5, 127-37.
Grattan-Guinness, Ivor 1980. "Georg Cantor's Influence on Bertrand Russell", *History and Philosophy of Logic* 1, 61-93.
Haaparanta Leila (ed.) 1994. *Mind, Meaning and Mathematics, Essays on the Philosophical Views of Husserl and Frege*, Dordrecht: Kluwer.
Hallett, Michael 1984. *Cantorian Set Theory and Limitation of Size*, Oxford: Clarendon.
Hilbert, David 1922. "New Grounding of Mathematics, First Report", in *From Brouwer to Hilbert*, Paolo Mancosu (ed.), New York: Oxford University Press, 1998, 198-214.
Hilbert, David 1925. "On the Infinite", in van Heijenoort, 367-92.
Hilbert, David 1927. "The Foundations of Mathematics", in van Heijenoort, 464-79.
Hill, Claire Ortiz 1991. *Word and Object in Husserl, Frege and Russell, the Roots of Twentieth Century Philosophy*, Athens OH: Ohio University Press.
Hill, Claire Ortiz 1994a. "Frege Attacks Husserl and Cantor", *The Monist* 77 (3), 347-57. Anthologized in Hill and Rosado Haddock.

Hill, Claire Ortiz 1994b. "Husserl and Frege on Substitutivity", in Leila Haaparanta (ed.), 113-40. Anthologized in Hill and Rosado Haddock.
Hill, Claire Ortiz 1997a. "Did Georg Cantor Influence Edmund Husserl?" *Synthese* 113 (October), 145-70. Anthologized in Hill and Rosado Haddock.
Hill, Claire Ortiz 1997b. *Rethinking Identity and Metaphysics, On the Foundations of Analytic Philosophy*, New Haven: Yale University Press.
Hill, Claire Ortiz 1997c. "The Varied Sorrows of Logical Abstraction", *Axiomathes*, 1-3: 53-82. Anthologized in Hill and Rosado Haddock.
Hill, Claire Ortiz 1998. "From Empirical Psychology to Phenomenology: Husserl on the Brentano Puzzle", *The Brentano Puzzle*. Roberto Poli (ed.) Aldershot: Ashgate, 151-68.
Hill, Claire Ortiz 1999. "Abstraction and Idealization in Georg Cantor and Edmund Husserl", *Abstraction and Idealization. Historical and Systematic Studies, Poznan Studies in the Philosophy of the Sciences and the Humanities*, F. Coniglione et al (eds.), Amsterdam: Rodopi. Anthologized in Hill and Rosado Haddock.
Hill, Claire Ortiz 2000. "Husserl, Frege and 'the Paradox'", *Manuscrito, Revista Internacional de Filosofia* 23, 2, October 2000, 101-32.
Hill, Claire Ortiz and G. E. Rosado Haddock 2000. *Husserl or Frege? Meaning, Objectivity, and Mathematics*, Chicago: Open Court.
Husserl, Edmund 1887. "On the Concept of Number", *Husserl: Shorter Works*, P. Mc Cormick and F. Elliston (eds.), Notre Dame: University of Notre Dame Press, 1981, 92-120. Published in Husserl 1970.
Husserl, Edmund 1891. *Philosophie der Arithmetik*, Halle: Pfeffer. Also published in Husserl 1970.
Husserl, Edmund 1900-01. *Logical Investigations*, New York: Humanities Press, 1970.
Husserl, Edmund 1905. "Husserl an Brentano, 27. III. 1905", *Briefwechsel, Die Brentanoschule I*, Dordrecht: Kluwer, 1994.
Husserl, Edmund 1906/07, *Einleitung in die Logik und Erkenntnistheorie, Vorlesungen 1906/07*, Ullrich Melle (ed.), Husserliana XXIV, Dordrecht: Martinus Nijhoff, 1984. Translation by Claire Ortiz Hill, *Introduction to Logic and Theory of Knowledge*, Dordrecht: Springer, 2008.
Husserl, Edmund 1913. *Ideas. General Introduction to Pure Phenomenology*, New York: Collier Books, 1962.
Husserl, Edmund 1917/18. *Logik und allgemeine Wissenschaftstheorie*, Husserliana XXX, Dordrecht: Kluwer, 1996.
Husserl, Edmund 1919. "Recollections of Franz Brentano", *Husserl: Shorter Works*, P. McCormick and F. Elliston (eds.) Notre Dame: University of Notre Dame Press, 1981, 342-49. Also translated by Linda McAlister in her *The Philosophy of Brentano*. London: Duckworth, 1976, 47-55.
Husserl, Edmund 1929. *Formal and Transcendental Logic*, The Hague: Martinus Nijhoff, 1969.

Husserl, Edmund 1939. *Experience and Judgment*, London: Routledge and Kegan Paul, 1973.
Husserl, Edmund 1970. *Philosophie der Arithmetik. Mit Ergänzenden Texten (1890-1901)*, Husserliana XII, introduction by Lothar Eley, The Hague: Martinus Nijhoff. In English, *Philosophy of Arithmetic, Psychological and Logical Investigations with Supplementary Texts from 1887-1901*, translated by Dallas Willard, Dordrecht: Kluwer, 2003. Originally published as Husserl 1891.
Husserl, Edmund 1975. *Introduction to the Logical Investigations*, The Hague: Martinus Nijhoff.
Husserl, Edmund 1983. *Studien zur Arithmetik und Geometrie, Texte aus dem Nachlass (1886-1901)*, Husserliana XXI, The Hague: Martinus Nijhoff.
Husserl, Edmund 1994. *Early Writings in the Philosophy of Logic and Mathematics*, Dordrecht: Kluwer.
Husserl, Edmund. Ms A 1 35. Manuscript on Set Theory available in the Husserl Archives in Leuven, Cologne and Paris.
Husserl, Malvine 1988. "Skizze eines Lebensbildes von E. Husserl", *Husserl Studies* 5, 105-25.
Jourdain, P. 1991. "The Development of the Theory of Transfinite Numbers", *Selected Essays on the History of Set Theory and Logic*, Bologna: CLUEB.
Kline, M. 1972. "The Instillation of Rigor in Analysis", *Mathematical Thought from Ancient to Modern Times*. Oxford: Oxford University Press, 947-78.
Linke, Paul 1947. "Gottlob Frege as Philosopher", *The Brentano Puzzle*. Roberto Poli (ed.), Aldershot Ashgate, 1998, 45-72.
Osborn, Andrew 1934. *Edmund Husserl and his Logical Investigations*, 2nd ed. Cambridge MA, 1949.
Peckhaus, Volker 1990. *Hilbertsprogramm und kritische Philosophie, das Göttinger Modell interdiziplinärer Zusammenarbeit zwischen Mathematik und Philosophie*. Göttingen: Vandenhoeck & Ruprecht.
Reid, Constance 1970. *Hilbert*. New York: Springer.
Reid, Constance 1976. *Courant in Göttingen and New York*. New York: Springer.
Rosado Haddock, Guillermo Ernesto 1973. *Edmund Husserls Philosophie der Logik und Mathematik im Lichte der Gegenwärtigen Logik und Grundlagenforschung*, Bonn: Doctoral Thesis.
Russell, Bertrand 1903. *Principles of Mathematics*, London: Norton.
Russell, Bertrand 1917. *Mysticism and Logic*, London: Allen and Unwin, 1986.
Russell, Bertrand 1959. *My Philosophical Development*, London: Unwin.
Schuhmann, Karl 1977. *Husserl-Chronik*, The Hague: Martinus Nijhoff.
Sebestik, Jan 1992. *Logique et mathématique chez Bernard Bolzano*, Paris: Vrin.
Sluga, Hans 1980. *Gottlob Frege*, London: Routledge and Kegan Paul.

Smith, Barry 1994. *Austrian Philosophy: the Legacy of Franz Brentano*, Chicago: Open Court.

van Heijenoort, Jan, ed. 1967, *From Frege to Godel. A Source Book in Mathematical Logic 1879-1931*, Cambridge MA: Harvard University Press.

Wang, Hao 1986. *Beyond Analytic Philosophy, Doing Justice to What We Know*, Cambridge MA: MIT Press.

Wang, Hao 1987. *Reflections on Kurt Gödel*, Cambridge MA: MIT Press.

Wang, Hao 1996. A *Logical Journey, From Gödel to Philosophy*, Cambridge, MA: MIT Press.

2

Jairo José da Silva

HUSSERL ON GEOMETRY AND SPATIAL REPRESENTATION

Introduction

What the subject matter of geometry is has not always been a matter of consensus, not even for the Greeks. Whereas Plato believed that geometry dealt with things that are not of this world, Aristotle thought it had only to do with abstract aspects of ordinary physical space and spatial objects. Neither, however, ever doubted that geometry was in the business of searching for the truth and that it could tell us something about physical space.

Things changed as history unfolded. The intuitive acceptability of the foundations of classical geometry, successfully axiomatized by Euclid, has since antiquity been taken with a grain of salt (for Proclus, the 5th century commentator of Euclid, the non-existence of asymptotic straight lines is not intuitively obvious). For many, the fifth postulate of the Euclidean system seemed, and for good reasons, to be lacking in intuitiveness. It took, however, more than two thousand years to realize that the system of ancient geometry, which in the form of Euclid's axiomatic presentation is known as Euclidean geometry, is not the only possible consistent theory of space, or even the only to have a rightful claim to being the *true* theory of space.

Throughout the 19th century and into the 20th century, the true geometrical structure of physical space remained a much-debated scientific and philosophical question. The issue was further complicated by the fact, not always clearly recognized, that there are different ways of accessing spatial structure, and then different representations of physical space: there is the primitive, unreflective, operative representation, based on sense perception, but also geometrical representation proper, based on geometrical intuition, a refinement of perceptual intuition.

Taking these distinctions into consideration, some questions immediately arise. To mention only two: How do the perceptual and geometrical representations of space relate to each other? What do these representations of space, "naïve" and scientific, have to do with space as it really is (transcendent space, if there is such a thing)?

At the beginning of his philosophical career, Husserl was very much involved with these and other questions concerning mathematical matters. Witnessing to his interests are the many drafts he left, written basically during the last decade of the 19th century, in which a "philosophy of space" is sketched. And since once a subject had captured his interest, Husserl almost never let it rest for good, a phenomenological analysis of the intentional genesis of geometry and geometrical space also appears, and prominently so, in his last published work, *The Crisis of European Science and Transcendental Phenomenology* (Husserl 1936a), henceforth *Crisis*. It is worth noticing that both approaches bear substantial similarity; his ideas on the matter did not change much during the span of a philosophical lifetime. A work that must also be taken into consideration, although its main focus is not on the constitution of space, but on the spatial (rigid) body, is Husserl's *Thing and Space: Lectures of 1907.* (Husserl 1907)

Unfortunately Husserl did not always express his ideas on geometry with the utmost clarity, so much so that scholars often disagree on how to interpret them. Much of the blame should be credited to the sketchiness of Husserl's notes, which were not written to be published, at least not in the form they have reached us. However, if not his answers, at least the questions he raised were always clearly stated. Here are some: Does our intuitive representation of physical space contain *a priori* elements? (If so, in what sense?) Is it at least partially based on conventions or hypotheses about physical reality? Is the geometry of physical space an empirical science? (If so, to what extent?) How do scientific and pre-scientific representations of space relate to one another and to transcendent space? Is physical geometry a conceptual or an intuitive science? What do perceptual and geometrical intuitions have to do with each other? Can the former have any relevance for the latter? Does the mathematical conception of space derive from the intuitive representation of space? (If so, how?) What role do symbols and symbolic reasoning play in geometry, if any?

Although a few of the texts in which Husserl dealt with these questions were published, most, at least as they have come down to us, were not meant for the public eye. *Crisis* and, in particular, "The Origin of Geometry" (Husserl 1936b, which contains an analysis of the intentional "genesis" of geometry and geometrical space) and *Logical Investigations* (where the nature of mathematics in general and pure geometry in particular are discussed) belong to the first group. The second part of *Studien zur Arithmetik und Geometrie*, containing the bulk of Husserl's account of our representation of space from three perspectives, psychological, logical, and metaphysical, to the second. There are also relevant passing remarks here and there in other works, but the truth is that Husserl never produced a study on geometry nearly as complete as *Philosophy of Arithmetic* is for arithmetic. His views on geometry demand interpretation. My goal here is to articulate one.

The Origins

In the hands of its first practitioners, the Egyptians most notably, geometry was basically a technology for measuring lengths, areas and volumes. With the Greeks, it became a science, later axiomatized by Euclid of Alexandria (one of the persons of the sacred trinity of classical geometry with Apollonius and Archimedes). Euclid probably saw his axioms as statements concerning *constructions* that could in principle be carried out in ideal space with ideal straightedges and compasses, given the constructions that we know we can in principle carry out in physical space, with real straightedges and compasses. Euclid was obviously able to appreciate the difference between ideal and real spaces and ideal and real constructions, but he was nevertheless convinced that our intuitions about real space and real constructions are somehow relevant to what we know about ideal space and ideal constructions.

All of Euclid's five postulates tell us that something can be done: 1) we can draw a (straight) line connecting any two given points; 2) we can extend any given line; 3) we can draw a circle with a center at any given point through any other given point; 4) by moving them in space, we can, if necessary, make any two given right angles coincide (i.e. all right angles are equal; 5) if two given lines are intersected by a third so that the interior angles on one side are less than two right angles, we can then extend the given lines on this side till they meet.

The problem with the fifth postulate is obvious. Whereas the remaining four involve finite tasks, the fifth does not. From the first to the fourth, the things to be done are clearly delimited. The line of the first postulate begins and ends in points that are *given*, the extension mentioned in the second is limited by the case at hand, and so on. But we do not know a priori for *how long* we must extend the lines mentioned in the fifth before they meet. What if they do not? We would never know it by simply extending the lines indefinitely, and if not so, how? No one before Carl Friedrich Gauss during the last years of the 18th century seems to have thought that the fifth postulate could actually be *false*. It was only thought that it lacked intuitive support and therefore should be proved.

But how does one prove on the sole basis of finite constructions that a possibly infinite search (for the point of intersection) comes to an end?

The task was of course doomed to failure. Euclid's fifth postulate was eventually shown to be independent of the others. The search for a proof of the problematic postulate, however, led, first Gauss and then Nicholas Lobachevski and Janos Bolyai, to the creation of the first non-Euclidean geometry, what is now-called hyperbolic geometry, in which the *denial* of the fifth Euclidean postulate holds. That the given lines do *not* necessarily intersect implies that there is *more* than one parallel to a given line passing through a given point (the so-called Playfair axiom—there is *only one* straight line parallel to a given line passing through a given point–is logically equivalent to Euclid's fifth postulate). The case in which *no* such line exists is actually forbidden by the remaining postulates (otherwise there should be a line through the exterior point meeting the given line in *two* distinct points, which is impossible). But with convenient alterations in the system, Bernhard Riemann conceived what is now called elliptic geometry, in which no parallel exists to a given line through a given point.

By taking these new geometries seriously (which was not always the case, for some Kantian-minded mathematicians insisted upon considering Euclidean geometry to be the only *true* geometry and the other geometries to be more or less sterile formal exercises), mathematicians substantially altered the traditional comprehension of the nature of mathematics, which was no longer *only* the study of (formal) aspects of given domains (physical space in the case of geometry), but also of purely formal theories, with no determinate domains, and not only for their own sake, but for scientific purposes as well, for *use* in science and mathematics (which, of course, poses the problem as to how formal *inventions* can be *scientifically* useful. But that is an altogether different problem, and this is not the place to deal with it).

Once new geometries were conceivable, it was difficult to avoid asking whether Euclidean geometry was *really* the true geometry of physical space or was only hypothetically or conditionally true. Riemann and Hermann von Helmholtz thought the validity of either Euclidean or non-Euclidean geometry for physical space ultimately rested on empirical facts or hypotheses. We live in a world, they said, in which some bodies, the so-called *rigid* bodies, *seem* able to move freely in space without changing size or shape (without some *physical* reason for so doing, that is), and the *simplest* (but not the only) space in which rigid bodies can so move is Euclidean space (if we presuppose that space is actually *infinite*, not only *unbounded*–a difference our perception cannot appreciate–, both Euclidean and hyperbolic geometries satisfy the principle of free mobility. If we drop infiniteness, all the three non-Euclidean geometries qualify, for free mobility requires only that space be of constant curvature, which Euclidean, hyperbolic and elliptic spaces are–with zero, negative and positive values, respectively). But how can we tell that rigid bodies *really* exist, that there really are bodies that can move freely in space without deforming themselves?

For Helmholtz, we can tell this on a priori grounds (this is a typical transcendental argument). Since, he thought, the rigidity of measuring rods is required by the very *notion* of measurement, the *possibility* of metric geometry requires the constancy of space curvature. Therefore, space has constant curvature out of necessity. Its particular value, positive, negative or zero, however, must, he thought, be empirically decided, by astronomical observations and measurements, for example[1] (or hypothesized, if no conclusive verification is or can be effected). Henri Poincaré held similar views on the a priori character of the constancy of space curvature, but he believed that no experiment would be conclusive for any particular value of this curvature, for the interpretation of experiments depends on physical theories that we have no reason–logical, methodological, epistemological, or metaphysical–to prefer to the detriment of any geometry. For the sake of simplicity, he thought, we can always endow space with zero curvature (and hence, Euclidean structure) and change physics accordingly. This view is usually labeled *conventionalism*. Notice, however, that the representation of space to which Poincaré is alluding relies on means of accessing spatial structure that go beyond the purely intuitive.

Husserl, as we shall see, thought that both constancy of curvature and its particular value, positive, negative or zero, were *empirical* facts, that only experience could, and *would*, decide–not scientifically informed experiences, though, but relevant sensorial perception pure and simple (and this is why a decision is necessarily reached, by default if needed). Husserl seemed to have reasoned thus: Suppose the curvature of space is not, *as a matter of transcendent fact*, constant, but that we remain ignorant of it, for all dimensions change in displacement in such a lawful manner that an *appearance* of constancy remains. Bodies not actually rigid will be, *ex hypothesi*, *experienced* as rigid. Would we not then be *intuitively justified* in believing space had a constant curvature even if superhuman beings observing us from outside *our* world were to tell a different story (because they can see the deformations that we do not)? Suppose that, like the beings inside Poincaré's ball, who mistake a gradient of temperature that changes lengths in some lawful manner for a deformation of space and consequently adopt a non-Euclidean geometry, our perception of space is influenced by physical factors of which we are unaware. In that case, we may be wrong as to the transcendent reality of our perceptual representation of space, but, nonetheless, we are intuitively *justified* in believing space to be how we represent it to be. For Husserl, the *intuitive* representation of physical space cannot depend on *mediate* verifications regarding matters of actual fact. The *scientific* representation

---

[1] Empirical verifications, in this case, involve not only direct perception via the outer senses (vision, touch, etc.), but mathematically informed experiments if necessary.

of space, of course, is an altogether different matter. Husserl believed that we *naturally* and *irrecusably* represent space intuitively in a way best described within reasonable limits of approximation by Euclidean geometry. What remains to be explained is how we do this.

Husserl came to philosophy by the way of mathematics at a time when the question about the geometrical structure of physical space was pressing for a solution and the debate was intense. Husserl was familiar with the problem and technically prepared to contribute.

The Multiple Tasks of a Philosophy of Space

For Husserl, there were three main groups of questions concerning a philosophy of space:

a) Psychological questions: Is our representation of space intuitive (either adequate or non-adequate) or conceptual? If it is conceptual, is it founded or not on intuitions? (These questions, of course, echo that posed by Kant as to the intuitive or conceptual nature of space). Also belonging to this group are questions concerning the origin and development of spatial representation as an adaptive response to the environment (a natural history of spatial representation, so to speak). Husserl distinguished two complementary tasks for these analyses: that of *descriptive* psychology, which is to describe the primary content, including the basic relations, of the spatial representation; and that of *genetic* psychology, which is to determine from which elements, by which psychophysical functions, according to which laws the spatial representation originates. For him, these tasks are not independent; descriptive analyses provide the foundation for genetic analyses.

b) Logical questions: Does our representation of space serve cognitive purposes? If it is intuitive, is it adequate for understanding? If it is conceptual, is it logically founded (i.e. logically and epistemologically *justified*)? In short, is our representation of space, be it intuitive or conceptual, appropriate for knowledge? Is the intuition (or concept) of space an adequate foundation for the science of space? Are geometrical ideas and assertions *a priori* or *a posteriori*? Are geometrical concepts only idealizations of empirical concepts? Does geometry require intuitive procedures? Is the intuitive preferable to the symbolic in geometry? Are geometrical procedures intuitive by ostensive constructions or non-intuitive by pure, conceptual analyses? Is geometry an inductive science (in deductive dress) or deductive by nature?

c) Metaphysical questions: Does our representation of space have metaphysical value, i.e. does it correspond to something transcendently real, or does it only possess "valid foundations" resting on something whose essence escapes us? Is pure geometry only a convenient fiction, despite its appropriateness, when conveniently interpreted, as a theory of physical space as we represent it? Is our spatial representation true to *form*, but not *content* (i.e. formally, but not materially faithful to reality)? For Husserl, logical and metaphysical investigations are intertwined; the former have fundamental relevance for the latter.

As a starting point, it is interesting to see what, for Husserl, are the main features of some of the most important philosophies of space. He reminds us that, for Kant, space is an *a priori* intuition, a condition of possibility of knowledge possessing empirical reality, but transcendental ideality, whose metaphysical correlate is nothingness. The propositions of geometry are synthetic *a priori*; its method is the ostensive construction; and it applies to experience due to the a priori character of the spatial representation. Although Kant's is obviously a theory to be reckoned with, Husserl did not show signs of uncritical alignment with any of those views.

For the perspective that Husserl calls positivism, he tells us, there is a difference between empirical and pure space, and only applied geometry has metaphysical relevance. Positivists recognize, *correctly*, according to him, the *necessity*, founded on the thing itself, of geometrical judgments. For them, Husserl tells us, empirical space is an idealization of intuitive space, and there are no reasons to suppose it has a correlate in reality. From what Husserl says here, one thing stands out—he thinks that geometrical assertions are necessary (which, as we shall see, does *not* mean that they are independent of experience, only that experience *cannot falsify* them). This is clearly in dissonance with views such as Helmholtz's or Poincaré's (if we understand these views as referring to intuitive, pre-scientific space, which was not their original intention). Husserl thought otherwise. For him, *intuitive* spatial representation and the truths based on it were necessary; not because, in line with Kant, they are independent of experience—but because our perception of space, with the properties pertaining to it, is the *only* response to spatial experience that we have managed to develop throughout our biological, psychological and cultural history. For Husserl, we are not free to *intuit* space in any other way; not even exotic spatial experiences can force us to modify our intuitive representation of space. If science decides, as it very well can, in favor of a spatial representation differing from the intuitive one, it will do so by downgrading the epistemological primacy of intuition.

For realism, he said, geometry is an instrument of natural sciences, of which space is a fundamental presupposition. For this perspective, however, Husserl tells us, space is neither representative (i.e. intuitive), nor idealized, but simply a tri-dimensional continuum. It seems that here, for science, Husserl has in mind mathematized or, as he calls it in *Crisis*, "Galilean" science of nature, in which space as experienced is "modeled" or, rather, substituted by abstract, purely formal "spaces".[2]

---

[2] See Chapter 17.

For realists, Husserl continues, geometry itself is a natural science, the *abstract* science of real space.

For Lotze, Husserl said, the fact that space is tri-dimensional and "flat", i.e. has constant zero curvature, is a *logical* necessity (and the corresponding assertions are *analytic*). For Husserl, and this is an important aspect of his philosophy of space, those claims are *false*. The number of dimensions of space and its curvature, Husserl claimed, are *empirical* facts (*a posteriori*). Only experience could teach us about these things.[3] I shall have more to say about this below.

The Many Concepts of Space

There are, for Husserl, different conceptions of space, forming, in his words, a genetic progression: the space of our quotidian life, given to us perceptually; the space of pure geometry, accessible only to geometrical intuition; the space of applied geometry, that of natural science (before, of course, science grew bold enough to consider other representations of physical space, which happened a few years after Husserl wrote his notes); and metaphysical or transcendent space. The space of quotidian life was, for Husserl, neither simply an imprint of external space upon consciousness, mediated by the relevant sensorial systems (visual, tactile, kinesthetic)–we do not simply abstract space from data of sensorial perception–, nor, contra Kant, an a priori mold that we impose on outer experience. Intuitions are the raw material from which we actively *constitute* a representation of space, without the help of any constituted body of knowledge, mathematics or physics, in particular. The constitution of a perceptual representation of space is not based on scientific knowledge. *It takes place at the pre-predicative level of consciousness* (which, it is important to notice, is not, for Husserl, a domain of pure passivity).[4]

---

[3] But can we say from experience that space is tri-dimensional "all over"? As we shall see below, Husserl seemed to admit that, although we cannot access every corner of reality intuitively, we extend the structural properties perceived in our limited chunk of space to space as a whole. Space *as represented* is, for him, a *uniform* extension of space *as directly intuited*. There are no reasons, presumably, for our space-constituting functions to work differently if we were placed in another region of space. The fact that physicists conjecture today that space may have more than three dimensions in subatomic scale does not pose any threat to our *pre-scientific* representation of space; science, Husserl believed, has the right to adopt any *formally correct* explanation of observed phenomena (even incurring the risk of formalist alienation, as denounced in *Crisis*).

[4] There is in Husserl a clear distinction between pure sensorial data (hyletic data) and percepts (perception is an *intentional* experience). In his lectures on thing and space (Husserl 1907), Husserl presented a minute description of the constitution of spatial rigid bodies, and the medium in which they are placed, physical space, in which we are not allowed to forget this distinction. Husserl believes that there are essentially two systems of sensorial data involved in spatial perception, the visual and the tactile (although, in that work, the visual appears with far more relevance), which are molded into spatial percepts by a series of *intentionally motivated* kinesthetic systems working in isolation

The space of pure geometry is an altogether different matter. It is a mathematical domain ruled by concepts that, regardless of their origin, only non-sensorial geometrical intuition—which despite its roots in perception is non-perceptual—is equipped to investigate. The space of applied geometry is the physical space as we represent it taken as a model (or an approximate model) of pure geometry, and, finally, transcendent space is space as it really is, not simply as we represent it to be.

Intuitive space comes first in the genetic progression. Then comes the space of applied geometry, which is intuitive space taken as an approximate model of pure geometry (the space of applied geometry is then a theoretical enlargement of intuitive space). Next is the space of pure geometry, obtained by idealization from intuitive space. And finally, there is metaphysical space, which is an object of pure thought. Each step in the genetic sequence of spaces takes us further from sense perception. Parallel to the sequence of spaces runs that of constituting acts, beginning with raw perception, followed by theoretically enriched perception, then idealization, ending with non-intuitive (empty) conceptualization. There is also the possibility of generating other conceptions of space by freely varying geometric space in one's imagination. In this manner we can obtain all sorts of mathematically interesting spaces, n-dimensional Riemannian spaces in particular (the word "Riemannian" here refers to abstract formal spatial structures, not to elliptic or spherical spaces, which like Euclidean space are particularizations of the generic Riemannian space).

In Husserl's own words, the following are the questions that he wants to answer (it does not hurt to remember): Based on what content can the intuition of space offer a foundation for pure geometry? In particular, how can geometrical intuition have a basis in *perceptual* spatial intuition? To what extent can we give an *objective* meaning to geometry and a transcendent epistemic value to geometrical assertions (i.e. Are geometrical truths true also of transcendent, not only geometric space?)? Does (geometrical) intuition have any metaphysical correlate (i.e. Does *geometrical* intuition correspond to anything metaphysically, that is, transcendently real?)?

Let us take a closer look at these different conceptions of space:

---

and cooperatively; there are basically four of these systems (some terms are Husserl's, some are mine): 1) the oculomotor system, by means of which a non-homogeneous 2-dimensional flat finite space is constituted; 2) the restricted cephalomotor system, by which a non-homogeneous 2-dimensional *curved* space is constituted, limited "above" and "below" by closed lines, like the section of the Earth's surface between the tropics; 3) the full cephalomotor system, by which a 2-dimensional spherical space is constituted (which Husserl calls Riemannian space); and, finally, 4) the (full) somatomotor system, by which Euclidean space is constituted. It is worth noticing that, for Husserl, binocularity does originate depth, but depth is not yet, by and in itself, a third dimension comparable to breadth and height; spatiality requires the subject to be able to move *freely* towards, away from and around the body, and would be constituted independently of binocularity.

a) For Husserl, *intuitive* or *represented* physical space (physical space as we represent it) is more than what meets the eye (or any sensorial system for that matter). It can be better described as the result of sensorial impressions (visual, tactile, kinesthetic) and *innate* and *fixed* psychophysical functions whose task it is to accommodate these impressions in a coherent spatial mold. Our senses offer the raw material that built-in systems (selected in the course of our biological history) put into spatial form. Our retinas, for example, offer two slightly different two-dimensional projections of the multiplicity of outer things (the difference being the binocular parallax); our perceptual system (which for Husserl, remember, involves *more* than what is purely sensorial) eliminates double images by "creating" perceptual "depth". But we do not have to go into these matters here. It is enough to keep in mind that, for Husserl, our intuitive spatial representation is *neither* completely prior to sensorial experience, *nor* completely abstracted from it; the subject has an active role in constituting space from raw spatial perception.

Space is first and foremost represented as a multiplicity, an extended medium in which data of *actual* or *possible* spatial perception–bodies and bodies *in spatial relations*–must be accommodated. We intuit space by intuiting things *in space*; space and the spatial are never represented in themselves, but as *dependent moments* of intuited things (we cannot, for instance, see space, but things in space; and if we see, for example, body B *between* bodies A and C, we *only* see this as an *aspect* of the spatial complex formed by A, B and C). Moreover, any spatial extension is a *part* of space, the encompassing totality.

Although Husserl does not say it explicitly, it follows from this that we represent space as *one*, not many, and, moreover, in one piece (a *connected* manifold, i.e. all positions in space are spatially related to each other). This is an *a priori* character of our representation of space, not because it belongs analytically to the notion of a spatial manifold, but because the manifold of actual and possible spatial *intuitions* (spatial hyletic data), which the spatial manifold must *in-form*, is a manifold connected around a *center*, the perceiving I. Any data of the outer senses (even if they belong to different sensorial systems) must stand in some spatial relation to one another simply because they are all perceptions of *the subject*.[5] Any spatial extension, Husserl says, is in space as a part, a limit (points, lines, surfaces) or a connection or relation among parts (for instance, the space *between* spatial extensions).

---

[5] The contemporary scientific image of space is that of a system of formal relations induced primarily by *physical* relations among physical entities (independently of how *we* happen to perceive or conceive it). It follows that if these entities were separated in clusters, with entities in one cluster having no physical relation to those in other clusters, space could very well be conceived, for scientific purposes, as disconnected into isolated "multiverses", bearing no *spatial* relation one to the other (logical relations such as that of difference, would, of course, still hold).

Space, Husserl also says, is conceived as *infinite*, that is, ultimate parts of space are not *thinkable*. In fact, *finite*, but *closed* manifolds (like the surface of a sphere) do not have ultimate parts either; so, to be rigorous, Husserl should have said that we conceive space only as *unbounded*. However, as we shall see below, since for him space is also represented as torsionless (zero curvature at any point), it must necessarily be infinite. It is worth noticing that space cannot be *intuited* as infinite; we can only intuit *limited* parts of space, but also, simultaneously, *conceive* of an indefinite enlargement of intuited parts, since beyond actual intuitions there is an open *horizon* of possible intuitions.

For Husserl, on the other hand, only experience could have shown us that physical space is *continuous* and *tri-dimensional*[6] (even though, as he admits, discontinuities in transcendent space could exist, undetected by the senses). Nonetheless, he thinks, our space-constituting functions are not biased towards continuity or tri-dimensionality; remaining as we are, we could have come up with a different representation, a space with a different number of dimensions or discontinuities, if experience had led us that way.[7]

What about *flatness* or absence of torsion? We *could* have concluded that space is flat by verifying that the sum of the angles of any arbitrary triangle was equal to two right angles (Thales' angle theorem, which is equivalent to Euclid's fifth postulate)[8]. The problem, of course, is that physical space is not geometrical space; there are no straight lines, perfect triangles or precise measurements in it. However, we might insist, if our measurements–imprecise as they are–oscillated around the "precise" value of two right angles, so that on the average Thales' theorem would be approximately true, this might suffice to convince us of the Euclidean character of space. But, for Husserl, our intuitive representation of physical space is pre-scientific; mathematical or scientific considerations have nothing to do with it. How can it be then that we *do* represent space as a *flat* manifold?

It seems safe to suppose that we know from experience that rigid bodies exist (remember that, for Husserl, we do not have intuitive access to space directly–spatial properties are abstract properties–, but only

---

[6] We could in principle conceive of different organisms with different space-constituting mechanisms, representing space with a different number of dimensions (for example, bodies incapable of motion might represent physical space with only two dimensions–see Husserl 1907 or note 3).

[7] It is also conceivable, I think, that the psychophysical functions responsible for the representation of physical space do in fact make it impossible for us to perceive spatial discontinuities, even if they existed in transcendent reality. The situation might be analogous to our representing movement in fast running discrete sequences of still pictures: closely packed atoms of space might necessarily be perceived as a continuum. But this seems not to be what Husserl thought (although I am not willing to bet on it).

[8] There are other methods, such as trying to draw a square by drawing congruent segments at right angles; only in Euclidean space will this succeed.

via bodies in space)⁹. A rigid body is one that changes neither shape nor size (which are morphological notions given perceptually) by simply moving in space (without some *physical* reasons–excluding, of course, the action of space, which by itself is causally inert–for doing so).

Well, one might object, we *seem* to know plenty of bodies that are for all practical purposes rigid. But how can we be *sure* of this? They may alter shape or size as a consequence of motion without our being able to notice it. If apparently rigid bodies, including our bodies and standard meters, became deformed in motion uniformly, we would not notice any deformation by strictly geometrical means.¹⁰ We, however, do not have to be *sure* that apparently rigid bodies are *really* rigid; our representation of physical space is not based on sound knowledge, being as it is a pre-reflexive response to experience. We do not even raise the question; if bodies *appear* rigid, rigid they are. We simply *experience* rigidity, even if this experience does not correspond to anything real.¹¹ Moreover, since rigid bodies can move freely without changing shape or size (precisely because, they are rigid) they can be compared with respect to shape and size: it is enough to superpose them or just to bring them close to one another and verify this. The notions of congruence (same size and shape) and similarity (same shape, not necessarily same size) are then purely intuitive morphological concepts; bodies in physical space can be *perceptually* related to one another in terms of being (approximately) the same or being different sizes or shapes.

Now, we know that the existence of rigid bodies does not necessarily imply that space is flat (without torsion or curvature at any point); it only tells us that it has *constant* curvature, zero, positive or negative (the Helmholtz-Lie theorem). But if space had a positive or negative curvature that was *noticeable within the range of immediate perceptual experience*, our experience of (bodies in) space would be different from what it is (the difference being a function of the degree of curvature; in an extreme case we would be able to scratch our backs by stretching our arms in front of us!). We then know from experience that space, when its structure is spelled out in geometrical terms (something we do *not* do in ordinary perception, the perceiving subject, simply *as such*, not necessarily being a reflective subject) is either *locally* flat or only slightly non-flat (so slightly

---

⁹ "There is no doubt that the conviction which Euclidean geometry carries for us is essentially due to our familiarity with the handling of that sort of bodies which we call rigid and of which it can be said that they remain the same under varying conditions". (Weyl 1963, 78)

¹⁰ Helmholtz, before Einstein, told us that the metric is not a formal aspect of space, but depends on its material content. So, even if space were not homogeneous and its metric not constant, free mobility would still be valid, since, in Weyl's words, "a body in motion will 'take along' the metric field that is generated or deformed by it". (Weyl 1963, 87)

¹¹ But what sense could be attributed to the hypothesis that the world and everything in it change dimensions in such a way that no change can in principle be noticed? As Weyl tells us (Weyl 1963, 118), a metaphysically real difference that cannot *as a matter of principle* be detected is non-existent.

so that we would not notice it). Either possibility suffices to justify *empirically* our pre-scientific representation of physical space *as a whole* as flat, which is all we need (even, as Husserl admits implicitly, nothing can guarantee we would not observe torsions or deviations from flatness if our senses were more refined). In short, for Husserl, our *perception* of space requires us to represent it as flat.

Helmholtz, among others, admitted that by putting the structure of space to empirical testing it *could* be perceived as non-Euclidean. Husserl was at odds with this view. Although he believed that metaphysical space could, as a matter of *fact*, be non-Euclidean, and science could represent it as so (even if it *were not* in fact so), he did not think we could ever *perceive* space as non-Euclidean. Helmholtz failed to see, Husserl claims, the distinction between *physical* and *psychological* experiences of space. In the latter, the mind is at work making *sense* of (or bestowing sense on) sensorial experiences. For us, the *perception* of physical space only makes sense in the Euclidean mold (Husserl is of course pointing here to the difference between *passive* and *active* experiences, an important distinction in genetic phenomenology that he will make explicit in later works).

To sum up, for Husserl, physical space can be *represented* as a continuous, connected tri-dimensional infinite flat manifold (as well as being homogeneous and isotropic). Our representation of space, however, may not correspond in one respect or another to transcendently real space. Husserl makes a strong claim in this respect; for him, congruence and *all* geometrical properties seem to be in the nature of our *sensations*, not in transcendent reality. Our geometrical-constituting functions, like our tone and color perception systems, may then give us something that is not strictly speaking out there. Transcendent space may be only an analogue[12] of represented space. But, be that as it may, Husserl believes that *we* are *empirically justified* in representing space the way we do; our representation of space is *objectively*, even if perhaps not transcendently, real.

It is important to determine which properties of space belong to it *necessarily*, irrespective of experience or the psychophysical functions involved in the representation of space, in other words, what belongs analytically to our conception of space. From what Husserl said, he believed that any space must necessarily be conceived of as a manifold, which only means that it has parts or components.[13] Once a manifold is given whose parts entertain relevant relations with one another, as is the case of space, it is natural to think that any part of the manifold can be univocally determined in terms of some relations to some of the other

---

[12] It is tempting to read this analogy in terms of the notion of homeomorphism. Represented space may be only a homeomorphic copy of transcendent space.
[13] A manifold is no more, no less than a structured multiplicity of things; we would call it today a structured system of entities.

parts (for example, any sound can be determined by its pitch, intensity and timbre, which are relational determinations), even if this determination is not uniform throughout the manifold, i.e. some notion of dimension must be available *a priori*, even if we think it as varying from region to region. Husserl is not clear on this point; he only insists that the three dimensions of space are empirically given. But I think he would agree that a manifold can be thought as finitely or infinitely dimensional, with constant or variable dimension, but not as dimensionless.

As already observed, Husserl claimed that space cannot be finite, for we cannot *think* of a space with ultimate parts. So, for him, infinity of space (although, as I have pointed out, this argument can only conclude that space is unbounded) is also given a priori; but in a different, *transcendental* sense of *a priori*.

What Husserl had in mind (besides Kant's antinomies) seems to be this: Since space must accommodate the totality of all possible experiences (we only have one space, for we only have one transcendental subject) and we can always conceive of extending the limits of our field of experiences, space cannot be bounded. Infinity is *a priori* not because any spatial manifold must be so conceived, but because our (or the transcendental subject's) field of possible experiences must. For reasons already explained, an analogous sense of *a priori* applies to connectivity. The absence of torsion (constant curvature) and the three dimensions of space, on the other hand, are for Husserl *a posteriori*, in more senses than one: *neither* the notion of space, *nor* our space-constituting functions per se, independently of *complementing* sensorial data, require them. If raw *physical* experience so induced us, Husserl suggested, we would have represented space as four-dimensional or curved. A fortiori, it is not true that any sentient creature in this world that can develop a representation of space will represent it as flat or tri-dimensional (more dramatically, we should not expect extraterrestrial beings to share our geometry: fluid beings in a fluid world would probably have no concept of rigidity, similarity or congruence, and no Euclidean geometry either). Space is represented as tri-dimensional and torsion-free due solely to experience. On the other hand, indefinite divisibility of space seems to follow necessarily from the infinite extensibility of experience, being in this sense also transcendentally *a priori*.[14]

Now, can experience induce us to change our representation of physical space?[15] At first sight, Husserl seems to be forced to accept that it can; after all, experience can take what experience gives. But, from what has just been said, although the psychophysical functions responsible

---

[14] This is the transcendental synthetic a priori in Husserlian dress.
[15] In answering this question Husserl seems to be answering Helmholtz, who argued *against* the non-intuitability of a non-Euclidean space. For Helmholtz, to intuit a non-Euclidean physical space means to *imagine* spatial sense impressions, captured by our sense organs according to the known laws, which would force a non-Euclidean character on space. As I show in the main text, Husserl explicitly denied this possibility.

for the constitution of our spatial representation require, in order to perform their task, an input of sensorial data, they cannot, for Husserl, be restructured by incoming data, being as they are products of our adaptation to the environment. Given that they developed to function in a certain way, they will function in that way no matter what our physical experiences are (which does not mean that in the course of eons our descendents could not develop altogether different space-constituting functions, long after we, their no-longer-spatially-adapted ancestors, have vanished from the Earth). Effects of perspective and like phenomena, for instance, Husserl said, are not seen as distortions of space, but visual *illusions*. If *we* remain what we are, our representation of space will not change, or so Husserl thought.

Again, this can count as an argument against Poincaré's style of conventionalism (if taken to apply to intuitive space) according to which we are to a large extent free to choose this or that spatial mold for our intuitions, for experience by itself is not compelling. Husserl, on the contrary, believed that our representation of physical space, the one we *effectively* possess, is the only we can *perceive*, even if it is not the only one we can *conceive* (in fact, as *thinking*, and not only *perceiving* subjects, we may even decide *against* perception in our scientific representation of space). We may be wrong as to how physical space *really* is, but if we are, we are not free to be right.

What about *radically new* experiences, light rays bending in some region of space without any physical reason for so doing or, more prosaically, otherwise rigid bodies behaving in the strangest ways? Could they force or suggest a change in our spatial representation of the outer world? Again, Husserl answered in the negative. In fact, he explicitly said that if the field of vision were altered, we would say this was no longer the field of vision, but a new experience, no longer space, but something new (drug-induced "trips", with lots of spatial distortions, such as the melting of solids, as a colleague in Berkeley–where else?– once reported, may be interpreted by a ecstatic drug (ab)user as an enlargement of the "doors of perception", giving him access to another space, but more likely, he will when not completely "spaced out" take this for what it is, an hallucination, a *mis*perception of reality). Our concept of space is not altered, but new experiences, new concepts, would be generated (suggesting maybe a different *scientific* representation of physical space). We can also, of course, *always* hypothesize some physical reason to account for the strangeness of observed phenomena; we can even consider this behavior as *evidence* for some unknown physical state or process.

But as Einstein demonstrated, this is not always the best (or the methodologically soundest) solution. Things may be simpler if we just change our conception of space. This is what Poincaré had in mind in saying that applying geometry involved some degree of arbitrariness. The history of science has apparently shown Husserl to have been wrong on this particular point: we *have changed* our representation of physical space, pressed not only by experience, but by *reason* too. But, I must insist, the representation of physical space that Husserl was concerned with, as I said before, is *not* the scientific representation of physical space. Representing space intuitively is prior to any scientific reasoning. Husserl would not disagree that science could for higher reasons exert some violence on our pre-scientific representations. One of the *logical* and *metaphysical* problems posed by a philosophical investigation of space, as mentioned above, is precisely whether our pre-scientific representation of physical space does serve cognitive purposes or has a transcendent correlate. We may realize that it does not and change it accordingly, but we would not ipso facto change our way of *perceiving* space.

We ordinarily say, and Husserl also said, that intuitive space is the realm *par excellence* of the *à peu près*, the approximate, the morphological, the organic (representing physical space perceptually involves *abstraction*, but not *idealization*). But, as Husserl himself noticed in *Crisis*, this is so only by comparison with its ideal counterpart, geometrical space. Intuitions are what they are, and they are not stable vis-à-vis geometrical ideals: they oscillate around ideal points and can only be said to be approximate if we take those points to be the "truly real" reality, something Husserl criticized in his later period as a sort of alienation.[16]

Represented physical space is not geometrical space. It does not contain geometrical points, perfectly flat surfaces or perfectly spherical bodies. But it is not devoid of structure. Physical space, as we represent it, is a proto-geometrical manifold with morphological counterparts to most geometrical structures and relations (proto-metrical and proto-topological relations, in particular). Spatial extensions, Husserl said, can be compared in terms of shape and size. Relations such as equal (or similar) and unequal (or dissimilar) with respect to shape, and larger, smaller and equal, with respect to size, can be established among spatial extensions. But these are *not* relations of *measurement*. Relations of equality or inequality, of shape or size, are based simply on direct *perception* (not measurement). There can be inequality without gradation, Husserl said. We can compare spatial extensions as to size and shape without measurements, to the same extent we can perceptually establish a topological relation of order among them (region A is "approximately" *between* regions B and C).

b) *Geometrical* space is the idealization of intuitively perceived physical space (perceptual space as we represent it to be) taken as a mathematical manifold in its own right. Pure geometry, the science of geometrical space, becomes physical geometry when we take represented space as a(n) (approximate) model of it. In the strictest sense, physical geometry is a mathematical theory of nature, such as dynamics or electromagnetism. Weyl, who often endorsed Husserlian ideas on the

---

[16] See Chapter 17.

nature of mathematics, said (Weyl 1918) that a mathematical theory of some intuitive concept (such as that of continuum) often fails to be completely faithful to intuition, and is at best an approximation whose epistemological value must be put to empirical testing. Husserl seems to have agreed with this. Science can force us to review our representation of physical space. Physical geometry (but not pure mathematical geometry) may be proved wrong, like any theory of nature, mathematized or not.

For Husserl, geometrical space is a structure logically elaborated on the basis of the spatial intuition of pre-scientific consciousness, in which nothing can any longer be said to be perceptually represented or capable of being represented, but is only thinkable. The geometrical intuition that geometers refer to is, according to Husserl, only an ideal; geometrical truths cannot, properly speaking, be perceived. *Perceptual structures* (diagrams, drawings, etc.) can only stand as *symbolic representatives* of geometric structures, for geometrical intuitiveness *can not be perceptually realized*. For Husserl, perceptual representatives represent by perceptually displaying truths that stand in a relation of *analogy* to truths of pure geometry. In other words, even though we *cannot* intuit geometrical truths in the diagrams we draw, we can perceive therein truths standing to geometrical truths as symbols to what is symbolized. When a diagram is drawn and taken to represent (as a symbol for) an abstract structure that is not seen (and cannot be seen), something is perceptually presented that serves as a symbol for something not perceived (but not conceptualized either). Geometrical intuitiveness rests on this relation of analogy.

This last remark is worthy of some comment. If geometrical notions were purely conceptual, geometry would be a conceptual science, where only conceptual intuition would have a place; ostensive constructions in intuitive space would have none or little relevance. But, as Husserl believed, geometrical structures and concepts originate by idealization from mereological analogues (some philosophers, Husserl says, do not accept this because they do not see how idealization can be possible without an ideal). Therefore, ostensive constructions, although not geometrical in the proper sense, have some bearing on geometrical truth. There is a relation of analogy between perceptually displayed truths (concerning represented space) and non-perceptual (and non-conceptual) geometrical truths: what is *shown* in the former (concerning mereological structures) is *thought* in the latter (concerning geometrical structures). This is why we can, in particular, claim an intuitive foundation for geometrical axioms.

On the basis of actual perceptual intuitions (in actual perception or imagination) we can, for Husserl, posit what is not perceived by *idealization*, that is: the unreachable (and not amenable to perception) limit of a sequence of perceptions: geometric points as limits of sequences of vanishing spatial regions; geometric lines as limits of sequences of narrower and narrower perceptual lines; geometric surfaces as limits of sequences of thinner and thinner perceptual surfaces. As explained above, for Husserl, we can reason about objects that are only "emptily" represented by reasoning about their perceptual representatives, provided that we do not allow "irrelevant" properties of the representatives to interfere, that is, properties that do not play any representational role. We must, Husserl said, distinguish between: 1) the actual intuitions given in perception or imagination and; 2) the intuitive procedures depending on intuitive signs that justify the (judicative) content of certain judgments that are *not* intuitive. Here, Husserl claims, intuition and thinking are intimately connected.

Not only the fundamental geometric elements–point, line and plane –are idealized from perceptual correlates. The morphological notions of (approximately the) same size and (approximately the) same shape also generate, by idealization, the geometric notions of similarity and congruence: two geometric structures are congruent (similar respectively) when they have *exactly* the same size and shape (same shape respectively). The notion of congruence captures that of constancy of size and shape independently of place (rigidity), that of similarity, the notion of invariance under change of scale (similarity, not congruence, is the quintessentially Euclidean notion). The morphological notion of order also gives rise, still by idealization, to the corresponding ideal, mathematical (topological) notion of order: point A lies *exactly* in between points B and C. Having points, lines and planes as basic elements, and the relations of belonging, congruence and order (betweenness) as the fundamental notions, Husserl sketches an axiomatization of geometry that is in the main identical to Hilbert's:[17]

Definitions:

D_1) any two points A and B determine a *segment* (denoted by AB); a point C *lies* on AB if, and only if, C is between A and B, C = A or C = B (this definition coincides with Hilbert's). If A = B, AB is a null segment. There is only one segment determined by A and B, which can also be called the *distance* between A and B (this is not yet the *numerical* distance). The distances between A and B and between C and D are *identical* when the segments AB and CD are congruent.

D_2) any two *distinct* points A and B determine a *straight line* (line through A and B). C lies on line through A and B if, and only if, one (and

---

[17] All the definitions (and subsequent assertions) are to a large extent Husserl's own. I only tried to give the ensemble a more coherent (although not logically flawless) presentation, remaining as close as possible to Husserl's original ideas. Although Husserl's approach to the axiomatics of geometry was remarkably similar to that of his colleague Hilbert, in his famous axiomatization of 1899, it was for the most part developed before they became colleagues in Göttingen, in 1901. But this much can be explained: both were buds on the same Paschian branch. Recall that in Hilbert's system the basic elements are also point, line and plane, and the fundamental relations those of incidence (point lies on line or plane, line lies on plane), order (betweenness) and congruence. The notion of continuity (which in Hilbert's system is given by the axiom of completeness and the Archimedean axiom) does not explicitly appear in Husserl's sketchy system.

only one) of the following holds: C is between A and B (C lies on the segment AB), B is between A and C or A is between B and C. If either B is between A and C or A is between B and C, we say that C lies on a *continuation* of segment AB. Obviously, there is only one line through A and B (*the* line determined by A and B). The definition Husserl actually gave is the following: any two points A and B determine a straight line (line through A and B), such that a point C lies on this line if, and only if, AC has either the same or opposite direction with respect to AB. This definition, however, depends on that of direction.

D_3) two straight lines are *parallel* when no point lies on both lines.

D_4) any segment AB can have two possible *directions*, AB or BA, depending on the ordering of A and B. A *direction* is then, for Husserl, a segment in which an ordering was established, i.e. an ordered pair of points. It is not difficult to define equality of direction in terms of our basic notions. If segments AB and CD are not collinear, directions AB and CD are identical when the straight lines through A and B and through C and D (call them *a* and *b* respectively) are parallel and the straight line through B and D does not meet the straight line through A and C in a point in the plane region delimited by the lines *a* and *b* (which is the set of points of the plane lying on a segment with endpoints in *a* and *b* respectively).[18] If AB and CD are collinear, they have the same direction if there is a segment XY not collinear with them that is co-directional with both (see A_4).

The following assertions are, for Husserl, *intuitively* justified:
Assertions:

A_1) there is a unique segment between any two points A and B (see D_1). This is Euclid's first postulate.

A_2) any segment can be indefinitely and uniquely continued (see D_2). This is Euclid's second postulate.

A_3) from any point A there are identical distances (equal to the distance AB) in any direction (i.e. there is a circle with radius equal to AB centered at A). This is Euclid's third postulate.

A_4) given any direction AB and any point C, there is a point D such AB and CD have the same direction (changing "direction AB" into "segment AB" and "have the same direction" into "are congruent", this assertion bears close resemblance to Hilbert's first axiom of congruence). Since sameness of direction, in the case that C does not lie on the line through A and B, implies that there is a line through C parallel to the line through A and B, this asserts in particular that given a point C and a line not through C, there is a parallel to this line passing through this point.

---

[18] I did not find this definition in Husserl, but I do not think he would object to it. It may also be that he thought sameness of direction was a primitive notion.

A₅) distances and directions are mutually independent (points at equal distances can determine segments with different directions; segments of same direction can be determined by points at different distances).

A₆) any distance (or segment) can be divided into equal parts.

It is not difficult to see that in developing Husserl's ideas in a logically satisfactory way (for example, by stating explicitly the axioms of incidence, order, and congruence, among others things) we could come up with a satisfactory axiomatization of Euclidean geometry along Hilbertian lines–one difference between the two is that Husserl seemed to privilege the notion of direction. Husserl himself did not go that far, but from what he says, it is already clear that, for him, it is possible to develop pure Euclidean geometry on intuitively justified axiomatic bases (where, of course, intuition here means *geometrical* intuition).

c) *Mathematical spaces*: these include geometrical space considered from a purely mathematical perspective and all its free variations, the *formal* structures that mathematicians invent, often pressed by necessity or the internal development of mathematics. For the Husserl of the *Logical Investigations*, *pure* mathematics, including *pure geometry*, is part of formal ontology, the investigation of formal structures in which arbitrary objects, no matter which ones, could in principle be accommodated. Mathematical manifolds, i.e. collections of "points" (abstract forms of unspecified objects), can be finite or infinite, discrete or continuous, with or without a topology or a metric.[19] Any manifold has a dimension, uniform or variable, for as I argued before, dimension seems to be a necessary aspect of any mathematical space. Riemannian spaces are *continuous* n-dimensional spaces in which points have coordinates and a local metric is defined by a quadratic differential form so as to render the lengths of any two line segments commensurable to each other. There are also non-Riemannian spaces (Riemannian spaces not being the most general spatial structures), in which the metric is even more general, and spaces with no metric at all.

Pure and Applied Geometries

Represented space is our *Lebensraum*, the space in which we live, pressed by the necessities of life and survival. It is originally the *Ur*-space that radiates from the I (*my* space) and only derivatively a representation of *objective* physical space (*our* space). We experience objective physical space by experiencing bodies "immersed" in it. This experience eventually coalesces into a body of knowledge, a proto-geometry sufficient for the

---

[19] Discrete (finite or infinite) manifolds (nets) have a natural notion of distance: we can define the distance between two points as the smallest number of points one has to go through to reach one from the other. Non-discrete manifolds, continuous ones in particular, on the other hand, have no "natural" notion of distance and can accommodate various ones.

practical demands of life, the embryo of pure geometry. Pure geometry is the scientific investigation of the properties of space, no longer the space of perception, but an idealized version of it. In the process of idealization, mereological concepts, real constructions and perceptual intuitions give rise, and are taken as symbols, to geometrical concepts, ideal constructions and geometrical intuitions respectively. But like any mathematical discipline, pure geometry cannot be confined to the (geometrical) intuitive realm. Projective techniques, for example, require points and lines at infinity that are not amenable to intuitive representation. Even in the case of traditional geometrical methods, concepts may be introduced and geometrical structures considered whose manipulation cannot be accompanied by intuitions. The situation is analogous to arithmetic, where numerical concepts (such as those corresponding to large or "imaginary" numbers) and conceptual operations (summing these numbers, for example) are beyond the reach of proper numerical intuition, being only indirectly accessible via symbolic methods.

A question then claims our interest, a typical question at this moment in Husserl's philosophical development (the mid-1890s): How can we be sure that *pure* geometry, when taken to represent space and *refer* to it, i.e. when transmuted into *physical* geometry (the mathematical theory of represented physical space) is (at least approximately) *true*? For Husserl, this is the most important *logical* question concerning pure geometry. This problem appears as naturally in geometry as in arithmetic, where it had already surfaced and been dealt with in a way that Husserl found satisfactory. (Husserl 1887-1901; Husserl 1901; Husserl 1983; Husserl 1994)

Husserl says explicitly that we must show that all assertions concerning measure and position valid in pure geometry are also valid in physical geometry; valid, that is approximately, in the sense that real numbers of pure geometry will become the approximate numbers of physical measurements. How can we be sure, he asks, that the laws of the ideal science of pure geometry, including those more remotely, or not at all, related to intuition, are *correct* if taken to refer to physical space (as we represent it)? The axiomatic method is the answer.

Axiomatization was then, and by some time already, making a come back after having lost ground to analytic and projective techniques in geometry. This tendency reached its peak with Hilbert, the champion of the axiomatic method (who, incidentally, also explicitly claimed intuitive foundations for his axiomatic system of geometry, "a logical analysis of our perception of space", according to him). Husserl himself, as we have seen, sketched one such axiomatization, based on the intuitively grounded notions of incidence, congruence and betweenness. For him, the axioms of the system were, as we have also seen, *true*; *geometrical* intuition tells us so. Since pure geometry is a *logical* system based on intuitively true axioms, which are approximately true for represented physical space, the theorems of pure geometry must also be true for geometric space and approximately true for represented physical space. Physical geometry is then epistemologically justified (Husserl also considered axiomatization as possibly being the one way of putting general arithmetical systems on sound logical bases).

Axiomatic reasoning may or may not involve intuitions. Euclid's system relied heavily on perceptually grounded geometrical intuitions; proofs were often accompanied by diagrams (although they sometimes are only illustrative) or consisted basically in pointing to supposedly indisputable evidence. Hilbert's or Husserl's system, on the other hand, confined intuitions to the axioms; the theorems of the system can only be guaranteed to be true (of *geometric* space) if derivations in the system are strictly logical: logical consequences of true axioms are necessarily true. Since represented space approximates geometrical space, pure geometry, axiomatized on logically sound bases, is sure to lead to approximately true facts about physical space as we represent it.

As we see, Husserl was definitely not an adversary of symbolic methods in mathematics, not even in geometry; he is not the "intuitionist" thinker he is sometimes depicted to be. He believes symbols have many advantages over intuition, including *heuristic* advantages. Equations, for example, he says, can make abstract relations intuitive. An equation is a relation among symbols that displays abstract relations *ad oculos* (in the sense that what is *shown* in the equation is *thought* abstractly, and may even be hidden in the abstract). As we have seen, for Husserl, drawings, diagrams and constructions in represented *physical* space are privileged geometrical symbols. But, to the extent that not all relations among algebraic symbols are relevant for the abstract content of equations, not every relation among empirical representatives is geometrically relevant. Although we can usually tell which relations are and are not relevant, we can make mistakes. Axiomatization is a safer procedure; it confines geometrical intuition to the axiomatic bases, and from that point on the responsibility for the production of knowledge falls on the shoulders of logic. We cannot go wrong. Although diagrams and other visible symbols can be useful, the symbols of a well-designed symbolic axiomatic system are, for Husserl, better.

Conclusion

If my reading is correct, Husserl's philosophy of geometry forms a coherent whole that can be summarized thus: There is inner and outer experience. The objects of outer experience are localized in space, which is not an *a priori* form of intuition. Space exists transcendently. Spatial structure cannot be directly accessed; it must be mediated by our experience with spatial objects. But it does not offer itself readymade to consciousness, through the senses, as an imprint; we must *constitute* a representation of space. Through our biological, mental and cultural history, we managed to develop elaborate physical and psychological functions for this purpose. Space-constituting functions cannot be modified at will, and the system of spatial relations they provide is in this aspect analogous to color or tonal systems. Experiences that do not fit into this system are not interpreted as spatial experiences.

We must necessarily represent space as a manifold, arguably endowed with some notion of dimension, and there is not much else logically required by the notion of space. Bodies in space stand in relations we call spatial; they can be far or close, move farther or closer, be bigger, smaller or equal to other bodies, be located between other bodies, etc. Space is the condition of possibility of spatial relations; it must be such as to "accommodate" all conceivable bodies and all the spatial relations they can possibly entertain with one another. Everything that bears the mark of exteriority happens in space. So, space must be such as to make this possible. These are what I call the transcendental a priori properties of space (it is clear to me that Husserl recognized them as such). Since the complete field of all possible experiences of the outer senses is infinitely extendable and capable of indefinite refinement, space must be unbounded (bordering on the infinite) and infinitely divisible (bordering on the continuous). Space is also unique; there is only one space, for there is only one field of all possible experiences of the outer senses.

We are not free to represent space however we want or find convenient, to the same extent that we are not free to have a different color or tonal system. Husserl was not to any extent a conventionalist, at least as far as *represented* space is concerned. Contra Helmholtz or Riemann (if somehow unfaithfully interpreted as referring to represented space), he thought that the representation of space does not harbor hidden presuppositions or hypotheses. There are no hypotheses for which our experiences and space-representing functions suffice. Contra Kant, he did not think that intuitive space is an a priori form of sensibility, and contra the empiricists he did not believe that space is extracted in full gear from raw perceptual experience.

The representation of external, physical space that we develop on the basis of psychologically (he would later say intentionally) elaborated experience is that of an infinite connected continuous flat three-dimensional manifold, in which certain relations are immediately perceivable, in particular congruence, similarity and order. Represented space is the material from which higher "intentional" acts, idealization in particular, extract the space of pure and physical geometries (pure geometry is a mathematical science; physical geometry is pure geometry interpreted as the geometry of physical space). By idealization, perceptual intuitions can be refined into geometrical intuition, on the basis of which the axioms of pure geometry are justified. Since geometrical truth flows down the line of axiomatic reasoning, we can be sure that applying pure geometry to physical space leads to truth, at least approximately.

Science may find reasons to choose different representations of physical space, but this, of course, does not mean that we shall ipso facto *perceive* space differently. Only if we, in particular our psychophysical functions, and the way reality filters through our senses, had been different might we have represented space in another way. By choosing space representations that may seem artificial from the point of view of our intuitive spatial representation, however, science may widen the gap between perception and reason that, *if not properly understood and adequately interpreted*, may alienate people from their *Lebenswelt*. The mathematical "refinement" of the intuitively given can induce us to give ontological priority to what is phenomenologically secondary; only phenomenological analyses can establish the correct order of foundation, relocating the subject and its intuitions in the fundamental place that is rightfully theirs. It is not perception that falls irremediably short of "capturing" reality as it is, but the idealization of reality that, by substituting reality as experienced perceptually with a mathematically purified version of it, puts the now idealized reality *in principle* out of the reach of adequate perception. Our ordinary intuitive representation of physical space is, as far as intuition goes, perfectly adequate; it does not need scientific amendment. But, on the other hand, for scientific purposes, science may perhaps find it convenient to consider adopting more refined ways of experiencing reality, different representations of physical space; our perceptual picture of space cannot stand in the way of this.[20] We must nonetheless refrain from interpreting what is only a scientific *methodological* strategy as a *correction* of perception.[21]

I believe that the distinctive aspect of Husserl's treatment of the problems regarding spatial representation and the nature of geometry lies in his carefully distinguishing among different conceptions of space, which were not always clearly segregated by his contemporaries. Helmholtz, Riemann or Poincaré were mainly concerned with whether there is, from a *scientific* perspective, a *correct* geometry of space, and, if not, on what basis this or that spatial property could be attributed to physical space. They did not distinguish between represented space, physical space as our (far from passive) perceptual systems represent it, and the space of physical science, or even transcendent space. Husserl, on the contrary, carefully drew two distinctions: one, Kantian in spirit, between noumenal (or transcendent) and phenomenal (or represented) spaces; another, between pre-scientific and scientific representations of space. But, contrary to Kant, he thought that phenomenal space is dependent upon reality in at least two ways: our space-constituting functions were selected

---

[20] I have in mind Henri Bergson's critique of Albert Einstein's conception of time. (Bergson 1922)
[21] See Chapter 17.

in an adaptive process, but, even so, they cannot completely fulfill their task without some contribution from experience. Space-constituting functions are, precisely, *functions*, whose arguments the outer senses must provide. Our senses and space-constituting functions remaining what they are, Husserl thought, *only one* output, *only one* space representation is possible, that which Euclidean geometry approximately describes.

This does not, however, mean that science, which is not only an affair of perception, but one of reason and convenience, cannot choose any other representation of space deemed more convenient. In fact, by distinguishing between the perceptual and the scientific representations of physical space, Husserl opened the door to spatial representations in science that *are not, and need not be* that of perceptual intuition (later in life he came to believe that we may have to pay a high price for that, the dehumanization of science).

To conclude, it may be illuminating, I think, to compare my reading of Husserl with Weyl's views on space representation, for, as already mentioned, the disciple was intellectually very close to his master.

The first chapter of *Space, Time, Matter* (Weyl 1918) begins with a phenomenological analysis of our intuitive representation of physical space that, albeit succinct, renders Weyl's views very clearly. For him, intuitive space is a *form* of phenomena, which is *perceived* as a continuous extended manifold. Due to the formal character of space the *same* content can, without changing in any way other than its position in space, occupy *any* spatial location, i.e. intuitive space is *homogeneous* (the same everywhere, and supposedly also *isotropic*, the same in any direction): "a space that can serve as a 'form of phenomena' is *necessarily* homogeneous [*my emphasis*]" (Weyl 1963, 86). Spatial regions are said to be *congruent* if they can in principle be occupied by the same content. For Weyl, then, homogeneity of space was a priori (this necessarily follows from space's being a form) and the notion of congruence was intuitively given (in keeping with Husserl). But for Husserl, remember, perceptual intuition was a psychological experience, not a merely physical one; congruence was not simply *given* in *passive* spatial experience. It remains then to be seen whether, for Weyl also, intuition, in particular the intuition of spatial form was a passive experience, or, on the contrary, actively involved the perceiving subject.

In his *Philosophy of Mathematics and Natural Science* (Weyl 1963), particularly §18, entitled "The Problem of Space", borrowing heavily from Helmholtz's *Physiological Optics*, among other writings, Weyl first considers the intuitive representation of space from the point of view of the physiology of vision. It is obvious from his considerations that Weyl can appreciate the difference between mere sense perception and intuition, the physical and psychological experiences Husserl mentions. A quote that he approvingly takes from Johann Gottlieb Fichte makes this clear: "I am originally not only sentient but also intuiting" (*Ibid.*, 127). In this same work he also says, agreeing explicitly with Husserl, that "the data of sensation are animated by 'interpretations', and only in union with them do they perform the 'function of representation'" (*Ibid.*, 119). In short, perceiving is an active act of constitution.

Weyl also agrees with Husserl on the necessity that goes with the intuition of space. He tells us so by quoting from *Ideas I*, §150 (Husserl 1913) on the distinctive aspects of visual experience: "All these facts, allegedly mere contingencies of spatial intuitions that are alien to the 'true', 'objective' space, reveal themselves, except for minor empirical particularities, as essential necessities" (Weyl 1963, 129). For Weyl, as for Husserl, we cannot but represent space *intuitively* the way we do, except for minor, purely empirical details (like the three dimensions of space, which Weyl and Husserl take as a contingent fact–*if* experience had so suggested, our space-representing functions *would have represented* space with a different number of dimensions).

One of the forms of the a priori, Weyl says, consists in non-empirical laws (*Wesensgesetze*) [*laws of essence*] [that] express the manner in which data and strata of consciousness are founded upon each other, but do not claim to involve statements of fact; this line of pursuit culminated in Husserl's phenomenology" (*Ibid.*, 134). For Husserl, as Weyl approvingly explained, the a priori is rooted in essential laws. Since the intuitive representation of physical space involved, for Weyl as well as for Husserl, immutable laws of constitution, the features of intuitive space are largely "except for minor empirical details", a priori (for Weyl these details have to do mainly with the particular value of the curvature of space). These laws, however, although immutable, are not indifferent to spatial experience, having been generated as adaptive *responses* to it. Says Weyl: "the manner in which this intuition as an integrating factor penetrates the sense data and utilizes their material is largely conditioned by experience". (*Ibid.*, 130)

From the intuitively given notion of congruence Weyl defines that of rigidity: a *rigid body* is one that always occupies congruent regions of space. With the notions of congruence and congruent transformation (which conceptually captures the intuitive notion of motions of rigid bodies in space), he defines the concept of *straight line* (the line determined by two distinct points A and B is the set of all points of space that are transformed into themselves by all congruent transformations that take A and B into themselves). Given the notion of straight line, the concept of *betweenness* can easily be defined: a point A of a line divides it into two rays, if B and C are in different rays, A is said to lie between B and C.

Summarizing: for Weyl, geometrical concepts, particularly the fundamental ones of congruence and congruent transformation, are grounded in intuition. Intuitive space is a connected unbounded homogeneous isotropic continuous enveloping manifold. Geometry develops out of our effort to "capture" intuitive space conceptually and, within certain limits, symbolically; the mathematical notion of transformation (group of transformations, respectively) offers itself naturally as that which corresponds conceptually to the intuitive notion of motion (system of motions, respectively). With these concepts, different geometrical systems can be introduced and eventually axiomatized. The three dimensions of intuitive space ('the most fundamental property of space is that its points form a tri-dimensional manifold", *Ibid.*, 84) are accidental. They neither belong analytically to the concept of space, nor are a priori in the sense given above. So, from a strictly mathematical perspective, the notion of space can be generalized to that of an n-dimensional manifold (of which the generic Riemannian manifold is a specific instance).

The structure of intuitive space, Weyl says, *necessarily* satisfies Euclidean geometry (cf. *Ibid.*, 135). For him, "there is no doubt that the conviction which Euclidean geometry carries for us is essentially due to our familiarity with the handling of that sort of bodies which we call rigid and of which it can be said that they remain the same under varying conditions" (*Ibid.*, 78). In sharp contrast with the scientific representation of physical space, we do not have choices on how to represent space *intuitively*. "This view", he continues, "does not contradict physics, in so far as physics adheres to the Euclidean quality of the infinitely small neighborhood of a point O (at which the ego happens to be at the moment)" (*Ibid.*, 135), i.e. physics is free to choose any geometry whatsoever for physical space provided it is locally Euclidean. The gap between physical and intuitive spaces grows as one distances oneself from the "center" where the ego is (carrying its Euclidean intuitive space with it).

Weyl's main interest, however, does not lie in the genesis of our intuitive representation of space, but in the constitution of our *scientific* picture of physical space ("intuitive space and intuitive time are... hardly the adequate medium in which physics is to construct the external world", *Ibid.*, 113). He considers in particular whether our representation of physical space can be put to an empirical test. Answering Helmholtz's claim that any such attempt will always involve physical statements about the behavior of rigid bodies and light rays, Weyl says that a constructive theory can only be put to the test *as a whole*. And if so, we are allowed to suppose, science can adopt any geometrical theory of space making the whole geometry + physics adequate vis-à-vis the available empirical data. This immediately leads to a threefold partition of the concept of space: intuitive space, which has to conform only to immediate sensorial perception; physical space ("the ordering scheme of the real things, which enters as an integral component into the theoretical construction of the world", *Ibid.*, 134), which must represent a compromise between intuitive space and physical theories; and more or less arbitrarily conceived mathematical spaces.

Even a cursory reading of Weyl shows the extent to which he is in agreement with the views that I have attributed to Husserl here: 1) There are many conceptions of space: intuitive space, physical space (the space of physical sciences) and mathematical spaces. There is a genetic sequence of spatial representations, beginning with the intuitive and ending with the purely symbolic representation ("all knowledge, while it starts with intuitive description, tends towards symbolic reconstruction", *Ibid.*, 75). 2) Intuitive space is not simply abstracted from "unprocessed" spatial experience; it is on the contrary actively constituted by the subject from data of spatial perception according to immutable laws; there is a difference between passive perception and active intuition of space. 3) Intuitive space is necessarily Euclidean; we are not free to *perceive* space in any other guise. 4) But, for scientific purposes, we are free to *conceive* space in the way we deem best, provided that the scientific representation of space is locally Euclidean–the geometrical structure of physical space must reach a compromise between intuition and logical, methodological or theoretical requirements of science. 5) For Weyl, in order to have objective validity (for all, not only for me), our representation of space must relinquish content and undergo a purely formal symbolic reconstruction, i.e. only the formal aspects of space, which are the only ones that can be symbolically expressed, are truly objective. We have seen that Husserl also accorded great importance to symbolic languages as means of expressing fundamental geometrical truths and erecting axiomatic systems where derivations must rely on symbolic manipulations only. But, more importantly, to the extent that they can be linguistically expressed at all, geometrical assertions can only express abstract formal properties of spatial relations (such as congruence, order and betweenness). Geometry, whether in axiomatic form or not, despite being given a definite realm–idealized perceptual space–can only convey formal truth. We can intuit geometrical truths indirectly via perceptual intuitions (diagrams and other perceptual structures) because the latter *display*, in a materially filled context, the *same* formal content that the former only *express* linguistically or symbolically. Geometrical intuition requires that formal aspects of perceptual structures be recognized as *identical* to those that are only emptily represented in geometrical assertions (this is how, I think, we can interpret Husserl's claim that perceptual structures can stand as visible symbols for geometrical ones–formal identity providing the basis for the representational relation). In short, geometry, and knowledge in general, involve interplay between subjective intuitions and objective symbolization. 6) Both Weyl and Husserl recognized the ideality of geometry. For the former: "the geometrical statements... are merely ideal determinations, which taken in individual isolation lack any meaning verifiable by what is given. Only here and there does the entire network of ideal determinations touch upon experienced reality, and at these points it must 'check'. That, expressed in the most general terms, may well be called *the geometrical method.*" (*Ibid.*, 132)

References

Bergson, Henri 1922. *Ecrits et Paroles*, vol. 3, Paris: Presses Universitaires de France, 1959, 497–503 (exchange recounted in the *Bulletin de la Société française de philosophie*, 22.3, July 1922, 102–113).

Helmholtz, Hermann von 1866. "On the Factual Foundations of Geometry", in Pesic 2006, 47-52.

Helmholtz, Hermann von 1870. "The Origin and Meaning of Geometrical Axioms", in Pesic 2006, 53-70.

Husserl, Edmund 1887-1901. *Philosophy of Arithmetic, Psychological and Logical Investigations with Supplementary Texts from 1887-1901*, translated by Dallas Willard, Dordrecht: Kluwer, 2003, the translation of *Philosophie der Arithmetik. Mit ergänzenden Texten (1890-1901)*, Husserliana XII, introduction by Lothar Eley, The Hague: Martinus Nijhoff. *Philosophy of Arithmetic* was originally published as *Philosophie der Arithmetik*, Halle: Pfeffer, 1891.

Husserl Edmund 1900-01. *Logical Investigations*, London: Routledge, 2001, translation of *Logische Untersuchungen*, Halle: Niemeyer.

Husserl, Edmund 1901. "Double Lecture on the Transition through the Impossible ("Imaginary") and the Completeness of an Axiom System", included as Essay III in Husserl 1887-1901, 409-52.

Husserl Edmund, 1913. *Ideas: A General Introduction to Pure Phenomenology*, translated by W. R. Boyce Gibson, London: Allen & Unwin, 1931, translation of *Ideen zu einer reinen Phänomenologie und phänomenologischen Philosophie I*, Husserliana II, The Hague: Martinus Nijhoff, 1950.

Husserl, Edmund 1936a. *The Crisis of European Sciences and Transcendental Phenomenology: An Introduction to Phenomenological Philosophy*, Evanston IL: Northwestern University Press, 1970, translation of *Die Krisis der europäischen Wissenschaften und die transzendentale Phänomenologie*, Husserliana VI, The Hague: Martinus Nijhoff, 1954.

Husserl, Edmund 1936b. "The Origin of Geometry", in 1936a, 353-78. English translation of "Der Unsprung der Geometrie als intentional-historisches Problem", Beilage III in Husserliana VI, 365-86.

Husserl Edmund 1983. *Studien zur Arithmetik und Geometrie (1886-1901)*, Husserliana XXI, The Hague: Martinus Nijhoff.

Husserl, Edmund, 1994. *Early Writings in the Philosophy of Logic and Mathematics*, Dordrecht: Kluwer, translation by Dallas Willard of *Aufsätze und Rezensionen 1890-1910*. Husserliana XXII. The Hague: Martinus Nijhoff, 1979.

Husserl, Edmund 1907. *Thing and Space: Lectures of 1907*, Husserliana XVI, translated and edited by Richard Rodcewicz, Dordrecht: Kluwer, 1997.

Pesic, Peter 2006. *Beyond Geometry: Classic Papers from Riemann to Einstein*, New York: Dover Books.
Poincaré, Henri, 1952. *Science and Hypothesis.* New York: Dover Books.
Riemann, Bernhard 1854. "On the Hypotheses Which Lie at the Bases of Geometry", translated by William Kingdon Clifford, *Nature*, vol. VIII, nos. 183, 184, 14-17, in Pesic 2006, 23-40,
Weyl, Hermann 1918. *Space, Time, Matter.* New York: Dover Books, 4th ed., translation of *Raum, Zeit, Materie*, Berlin: Springer Verlag, 7th ed., 1988.
Weyl Hermann 1963. *Philosophy of Mathematics and Natural Science.* New York: Atheneum.

3

Jairo José da Silva

BEYOND LEIBNIZ: HUSSERL'S VINDICATION OF SYMBOLIC KNOWLEDGE

Introduction

From the time that he wrote *Philosophy of Arithmetic* (Husserl 1891e), an expanded version of his *Habilitationsschrift* "On the Concept of Number" (Husserl 1887), at the latest, until the completion in about 1896 of the *Prolegomena to Pure Logic* (Husserl 1900), the first volume of *Logical Investigations* (Husserl 1900-01), or maybe a bit later, when he developed the ideas he presented before the Mathematical Society in Göttingen in 1901 (Husserl 1901a), Husserl struggled with the problem of imaginary elements in mathematics. As he himself tells us, this problem forced him to broaden his philosophical horizons, opening up new perspectives on the role of symbolization in thinking and knowing processes and presenting new questions on the sense and scope of formal logic. In his own words:

> Above all it was its [i.e. arithmetic's] purely symbolic procedural techniques, in which the genuine, original insightful sense seemed to be interrupted and made absurd under the label of the transition through the 'imaginary', that directed my thoughts to the significance and to the purely linguistic aspects of the thinking–and knowing–processes and from that point on forced me to general 'investigations' which concerned universal clarification of the sense, the proper delimitation, and the unique accomplishment of formal logic. (Husserl 1913, 33, *Apud* Moran 2005, 90)

Imaginaries are improper representations, i.e. representations without object that nonetheless pass themselves off as denoting something. Despite this apparent absurdity, imaginaries are in general harmless and often useful in mathematics. How can we explain that we can obtain knowledge by operating "blindly"[1] with symbols according to rules, even when these symbols do not represent anything? This epistemological problem presented itself very early in Husserl's philosophical career and was a dominant factor in the development of his thought. From the first to the last work he published the task of clarifying the sense and delimitating the scope of symbolization and formalization in science was one of Husserl's major concerns. Related problems, such as the interplay of representations without object and intuitions in the dynamics of knowledge, among others, were also prominent in his agenda. In this chapter I want to show how Husserl dealt with the problem of imaginary elements and symbolic knowledge in mathematics and the central role it played in his philosophical development.[2]

Symbolic Knowledge

It is unquestionable that Husserl took the epistemological relevance of symbolic representation for granted. Consider the following quote of 1890:

> [W]ithout the possibility of symbolic representations substituting for more abstract proper representations, difficult to distinguish and handle, *or even representations that are not proper* [*my emphasis*], there would not exist a higher spiritual life, and even less science. (Husserl 1890b, 349)

It is that simple: no symbolic reasoning, no science, in particular, no mathematics. Husserl accorded vital importance to well-designed symbolic systems—of *calculation* as well as *derivation*, that is, *logical* systems—*even if they do not have a proper representational function*, as is paradigmatically the case of imaginary mathematical entities and purely symbolic logical

---

[1] "*Cognitio caeca*" is one of the terms that Leibniz—the man who brought this issue into philosophy—used for symbolic knowledge.

[2] We are talking, basically, of Husserl's philosophical development during, roughly, the last decade of the 19th century. The relevant textual sources are the *Philosophy of Arithmetic* (Husserl 1891e), the *Logical Investigations* (Husserl 1900-01) and minor texts of that period published in Husserl 1886-1901, Husserl 1887-1901, and Husserl 1890-1910. In this chapter, I shall concentrate on those in which Husserl's treatment of the many (basically three) versions of the problem of symbolic knowledge (those concerning, symbolisms with and without interpretation, and the role of imaginary elements in symbolisms with interpretation, respectively) comes out more clearly; namely: *Philosophy of Arithmetic*, *Logical Investigations*, "On the Logic of Signs (Semiotic)" (Husserl 1890b), his review of Ernst Schröder's book *Lectures on the Algebra of Logic* (Husserl 1891b), and the draft for a lecture before the Mathematical Society of Göttingen, "Double Lecture: On the Transition through the Impossible ('Imaginary') and the Completeness of an Axiom System", (Husserl 1901a). Other texts, such as "The Concept of General Arithmetic" (Husserl 1890a), "Arithmetic as an A Priori Science" (Husserl 1891a), a letter to Carl Stumpf (Husserl 1891d), "On Set Theory" (Husserl 1891f), and "Formal and Contentual Arithmetic" (Husserl 1889-90), to mention a few, are either much shorter or not directly concerned with the problems discussed here; they will be taken mostly as subsidiary sources of information, not as *loci classici* of Husserl's treatment of the problem of symbolic knowledge. The translations are my own.

systems (formal theories). However, he did not think that operating with the symbols of a system according to prescribed rules, by and in itself, constituted *knowledge*. A *calculus*, he thought, although a useful *technique*, does not necessarily produce science. According to his review of Ernst Schröder's *Lectures on the Algebra of Logic* (Husserl 1891b), where this distinction is introduced, a *language* is a vehicle for thought, whereas a *calculus* may not be, if it is not *logically* justified. A logical justification must *show* that the calculus in question leads to knowledge not only as a matter of *fact*, but as a matter of *right*. Husserl said that:

> All the artificial operations on signs are in a way at the service of knowledge, but in fact they do not all lead to knowledge in the true and authentic sense of logical comprehension [*Einsicht*]. It is only if the process is itself a logical process, if we have logical comprehension that it must lead to truth, as it is, and because it is so, that its results are not only simply *de facto* true, but the knowledge of truth. (Husserl 1890b, 368-69)

Or still:

> A truly fecund formal logic is constituted first of all as a logic of signs, which, when sufficiently developed, will form one of the most important parts of logic in general (as the art of knowledge). The task of logic is here the same as anywhere: to become master of the natural procedures of the spirit that judges, to examine them, to understand the value they have for knowledge in order to assess with exactitude their limits, extend and range, and establish general rules concerning all this. (Husserl 1890b, 373)

So, this much is clear: As early as 1890 Husserl was already concerned with the task of logically justifying the purely symbolic aspects of mathematics. As a trained mathematician Husserl could not ignore the evidence that most of the *practice* and the *theory* of arithmetic rely on algorithms (for calculations) and formal systems (for theoretical development, particularly in the case of more general concepts of number, i.e. general arithmetic) and that maybe the most interesting new mathematical theories of the 19th century (Bernhard Riemann's theory of manifolds, William Rowan Hamilton's theory of quaternions, Sophus Lie's theory of transformation groups, George Boole's logical calculus,[3]

---

[3] George Boole, whose logical calculus could be interpreted either as a calculus of classes or a calculus of propositions, was aware of the way winds were blowing in mathematics: "They, who are acquainted with the present state of the theory of Symbolical Algebra, are aware, that the validity of the processes of analysis does not depend upon the interpretation of the symbols which are employed, but solely upon the laws of their combination. Every system of interpretation which does not affect the truth of the relations supposed, is equally admissible, and it is thus that the same process may, under one scheme of interpretation, represent the solution of a question on the property of numbers, under another, that of a geometrical problem, and under a third, that of a problem of dynamics of optics" (Boole in Bochenski 1970, §38.17). Although Franz Brentano was an influence to be reckoned with, I believe that Husserl's realization that purely symbolic reasoning has an *essential* role in knowledge was mainly due to his mathematical training and the awareness of the logical relevance of symbolization that came from

etc.) were purely formal. Since he was not willing to discredit those theories as mere playing with symbols or sterile formal exercises (as Kantians considered the non-Euclidean geometries)—quite the opposite, Husserl, as the philosopher he had become, could not avoid taking on the burden of giving mathematical symbolic knowledge, particularly if imaginaries were involved, a proper logical justification.

Meaningful Symbols in *Philosophy of Arithmetic*

Husserl's first extensive treatment of the logical problems posed by symbolic knowledge appeared in *Philosophy of Arithmetic*. There are in fact two variants of this problem there, one concerning the justification of the usual algorithms for carrying arithmetical computations, the other with treating the symbols 0 and 1 as numerical symbols proper alongside the numerals 2, 3, etc. One has to do with "blind" manipulations of meaningful symbols; the other with the use of meaningless symbols as if they had a meaning. Husserl treated these problems differently.

The algorithmic manipulation of numerals in the usual arithmetical operations is certainly not presided over by accompanying intuitions; nonetheless, the numerical symbols involved (excluding 0 and 1) have a denotation and, moreover, the symbolic system constituted by numerals and symbolic operations is an *isomorphic* copy of the system of number concepts and conceptual operations (in Husserl's terminology they are *equiform*).[4] In fact, as Husserl showed in the second part of *Philosophy of Arithmetic*, the existence of such an isomorphism is the reason why purely *symbolic* manipulations can produce *true* statements, thus justifying these operations logically.

A closer inspection of the situation, however, reveals an important fact that Husserl did not emphasize, but that is central for a satisfactory explanation of the usefulness of imaginaries in mathematics: we can obtain information about a domain by handling an isomorphic copy of it only because everything we want to be informed about concerning the domain of interest has to do with what this domain and its isomorphic copy have in common, which is *formal structure*. We can obtain arithmetical *knowledge* by playing with *proper* (i.e. denoting) arithmetical symbols

---

reading Boole and Schröder, among other formal logicians of the time. It is clear from his long review of Schröder, that Husserl knew Boole's calculus well, maybe as "beautifully explained by Venn" (Husserl 1891b, 40). Of course, it was Leibniz (who was also well aware of the fact that symbolic systems admit different interpretations, and that in this resides their main interest) was the first to bring to philosophical attention the fact (and the problem) of symbolic knowledge; Boole, Schröder and Frege were Leibniz' natural heirs. I believe, however (hence the title of this chapter) that Husserl took this problem a step further by considering *purely formal*, non-interpreted systems, not only interpreted ones, as many of his predecessors, including Leibniz, did.

[4] It is worth noting that the notion of isomorphism, as we have it today, had not yet by then come out clearly and distinctively in mathematics, although it was already operational, as the work of Richard Dedekind, for instance, testifies.

algorithmically only because arithmetical truths are *formal*, i.e. they do not concern numbers strictly, but relations among, operations with, and properties of whatever objects *behave like numbers*. Being formal, arithmetical truths can be obtained by investigating any domain whatsoever that is formally identical with the domain of numbers, the system of numerals and purely symbolic operations with numerals in particular.

Nonetheless, for Husserl, the system of arithmetical truths, despite their being *formal*, is articulated internally by a unifying concept–the concept of number as a collection of units upon which we can operate (by inserting or removing units). Although formal, Husserl thought, arithmetical truths referred to *numbers*, not arbitrary number-like entities. According to him, the fact that a system of formal truths–an articulated body of *formal knowledge*, as I call it–must ultimately refer to a *possible* system of objects unified under a concept whose formal properties the system of truths expresses has to do with his persistent concern that symbolic systems must be safeguarded from degenerating into dead and dry formalism alien to knowledge, i.e. mere technicalities alienated from our *living experiences* (*Erlebniße*) and the *Life-World* (*Lebenswelt*), as he would later say.[5] In *Philosophy of Arithmetic* and elsewhere, for him, arithmetic was the science of numbers *as particularizations of the concept of quantity*; a science whose *internal* unity is provided by the concept of number.[6] The unity of a formal domain (i.e. the form of an objectual domain ruled by a theory of a certain form) simply conceived by an act of formal imagination, as we might say, on the other hand, Husserl thought, is only the *external* unity of a system whose elements hang together only in virtue of extrinsic formal relations.

Meaningless Symbols in *Philosophy of Arithmetic*

This leads us to 0 and 1, which, according to Husserl, belong to arithmetic only on the basis of their external or purely formal relations to numbers proper. According to Husserl the symbols 0 and 1 do not denote numbers as he understood them, but are merely intentional number-like entities, to use a later terminology. But the inclusion of these entities in the numerical domain was not adequately justified in *Philosophy of Arithmetic*, and this problem became dramatically serious when he took

---

[5] See Chapter 17 for a detailed analysis of Husserl's critique of technization in *Crisis* (Husserl 1936). According to André de Muralt, Husserl, in *Crisis*, spares symbolic mathematics for, as he says: "[a]lthough mathematics is a symbolic knowledge, it can nevertheless be applied and is therefore not technized on its own account. Its original sense is therefore a logical sense" (Muralt, 183). But as I show in Chapter 17, the applicability of purely symbolic mathematics in science must, for Husserl, satisfy certain conditions so as to ward off "formalist alienation".
[6] Even though, let us keep this in mind, the insights we obtain by inquiring this concept cannot distinguish between numbers proper and anything that just looks like numbers, that is, any domain equiform with (isomorphic to) the domain of numbers.

upon himself the task of providing logical foundations for more general systems of arithmetic (for the never published second volume of *Philosophy of Arithmetic*).

This is how Husserl handled the problem of justifying treating 0 and 1 as numbers. First, he acknowledged the obvious fact: we *need* 0 and 1. In *Philosophy of Arithmetic*, he wrote: "We would quickly be led into embarrassing complications and even serious inconveniences in the theory of numbers if we wanted to keep 0 and 1 apart from numbers proper and abandon giving these two kinds a common denomination". (Husserl 1891e, 145)

Second, Husserl noticed that 0 and 1 are common products of the computation with numbers proper–we meet with them often enough in our numerical calculations. But if we were to dismiss these results as non-sense, the algorithms would become worthless. He says:

> If we consider that a uniform operational activity according to rules is not possible unless all imaginable results of an operation can be treated formally in the same way, it becomes clear why this enlargement of the domain of calculation [*via the introduction of 0 and 1 and all other number concepts, negative, rational, irrational and complex–my addition*] was indeed an important advance towards the establishment of arithmetic. (*Ibid.*, 146-47)

In short, our usual numerical algorithms cannot do without 0 and 1; these imaginaries are, so to speak, engendered by the algorithms, which cannot work without them (a fact already appreciated by 16th century algebraists with respect to negative and complex numbers). The introduction of 0 and 1 in arithmetic on a par with the other numbers "makes it possible an arithmetical algorithm, i.e. a system of formal rules by means of which we can operate in a purely mechanical way in order to solve numerical problems, that is, to find unknown numbers from known numbers and relations among them". (*Ibid.*, 145)

The only reasons Husserl gives for accepting 0 and 1 as numbers (namely: (1) arithmetical operations among numbers proper produce them naturally–it is as if they were *required* as *necessary* completions of the arithmetical domain; and (2) the algorithms for solving numerical problems are worthless without them) are clearly not satisfactory answers from the perspective of a *logical* justification; obviously, Husserl could not accept this lame justification as the final word on the matter–after all, in a strictly analogous way, all number-like entities that Husserl called imaginaries could also be justified. So, what I call *the problem of symbolic knowledge*[7] had already stuck out his head in *Philosophy of Arithmetic*, but it remained unsolved therein, at least as far as imaginary elements were concerned. The problem only got worse when a logical justification for *general* arithmetic was required.

---

[7] How can we *know* anything by simply "playing" with symbols according to pre-fixed rules?

Logical Systems

Husserl's treatment of *interpreted* systems of derivation is similar to his approach to interpreted systems of calculation. The requirements for a logical justification of interpreted axiomatic theories, i.e. theories whose axioms are true by virtue of some sort of intuition into what gives the theory its internal unity are clearly stated in "On the Logic of Signs (Semiotic)". (Husserl 1890b)

According to Husserl, logically sound symbolic reasoning within interpreted systems must fulfill two conditions: 1) "the systematic forms of junction of words [*i.e. the symbolic expressions*] must reflect exactly those of thinking [*i.e. the meaningful judgments*], otherwise the former could never become habitual substitutes for the latter" and 2) "the first part of the system, which contains the premises... must manifestly determine in a purely formal manner, univocally, the part that contains the conclusion... the set of premises determine univocally the conclusion". In short, logical *languages* (not simply *calculi*) must *represent* thinking proper, that is, the formal expressions must stand for meaningful judgments and the formal machinery for drawing conclusions must produce logically sound inferences.[8]

The logical justification of computational algorithms involving *interpreted* symbols depends also on the existence of a representational relation based on strict formal identity, as is the case of arithmetical algorithms. Consider the following quote from another text of 1891, "On the Concept of Operation":

> [A]ny algorithm first establishes a rigorous parallel correspondence between fundamental concepts, fundamental judgments and fundamental chains of reasoning and algorithmic elements. In fact, the objects of the domain, which are represented in an indeterminate manner, are replaced by simple signs, composites of objects, by composites of signs, established by means of signs of operation that correspond to the different concepts of operation; the relations, by signs of relation. Moreover, the fundamental propositions [are replaced] by symbolic conventions telling which are the permitted symbolic modifications (to the extent that they correspond to true judgments) and which are not. Concomitantly, conventional meanings are given to the symbols; hence, algorithmic concepts are one-to-one coordinated with the original concepts. (Husserl 1891c, 418)

But a proper *logical* justification for *non-interpreted* symbolic axiomatic systems and *non-denoting* symbols cannot follow along similar lines.

---

[8] Already clearly discernible in this passage are the tasks Husserl would impose upon formal logic in later works, to elaborate a logical grammar so as to guarantee that the formulas of a logical language are meaningful–and consequently denote states-of-affairs–and a theory of deduction so as to guarantee that formal derivations preserve truth. In fact, according to Dallas Willard, "On the Logic of Signs (Semiotic)" was "apparently Husserl's first systematic effort towards a 'logic', in his special sense, for symbolic calculation". (Willard, XIV)

For in such cases, of course, we cannot speak of a parallelism between representations and represented, since non-denoting symbols do not represent. In other texts of the same and later periods, Husserl made it clear how he thought purely symbolic systems of calculation and derivation could be justified. In few words: as far as *determinate* objectual domains were concerned (for instance, the domain of numbers as answers to the question "how many?"), purely symbolic reasoning (that is, involving non-denoting symbols) was acceptable, if useful, provided it was essentially unnecessary. This, for instance, was the justification of the use of imaginaries that he presented in the 1901 lectures before Göttingen Mathematical Society. The problem with this approach, however (and my criticism of Husserl's handling of the problem of imaginaries resides here) is that it falls short of Husserl's own requirements for a *logical* justification of symbolic mathematics, since it does *not* explain *why* imaginaries are useful (*why* their manipulation leads to truth); it only tells us how they can be rendered harmless.

Imaginary Elements

Earlier Treatment. In his 1891 review of Schröder's *Lectures on the Algebra of Logic,* we can already see a hint of the mature treatment of imaginaries presented in Göttingen ten years later. In that book, in the tradition of Boole, Schröder had presented a calculus interpretable either as a logic of deduction, if the symbols are interpreted as denoting extensions of concepts, or a calculus of classes, if the symbols are interpreted as arbitrary collections of objects.[9] For Husserl, however, Schröder's algebra of logic is nothing but a *calculus*, not *logic* properly speaking; a "technique of consequence", as he says, rather than the *science* of deduction. Interestingly, he compares Schröder's calculus with the *arithmetica universalis*, the theory of the most general concept of number, whose logical justification was by then an important focus of interest for him, and he points out that neither is yet logically justified. As a logical calculus of *extensions*, Husserl thought, Schröder's calculus cannot qualify as a pure theory of deduction, for deductions involve *concepts*, and extensions of concepts cannot determine their concepts, a task only their *contents*, or contents of concepts that are materially equivalent to them, can accomplish. So, said Husserl, "the ideal of a 'logic of extension', i.e. a logic that considers by principle *only* the extension of concepts is without value because it is without object". (Husserl 1891b, 16)

---

[9] Husserl observed correctly that Schröder's calculus has little value when interpreted as a logic of deduction, for the domain of deduction it formalizes is very restricted, whereas, as a calculus of classes, it can have many applications in mathematics. (Husserl 1891b, 42-43)

Husserl thought that in order to be logically justified, as we have already seen, a calculus needs to be adequately correlated with reasoning proper so as to be able to serve as a *substitute* for it. He said:

> the proper task of a calculus is to be, for an entire domain of knowledge, a method of symbolic deduction of consequences; hence, an art for substituting, by means of an appropriate designation of ideas, a calculus for effective deductions, that is, a conversion and a substitution according to rules of signs by signs and then, by virtue of the correspondence between signs and ideas, for obtaining from the final formulas the desired judgments. (*Ibid.*, 21)

The symbols and symbolic transformations of what Schröder presents as a logical calculus, Husserl thought, cannot be adequately correlated with concepts and thinking proper; so, it cannot be a logically justified *logical* calculus (although it could be a logically justified calculus of classes). This is the same strategy of justification that Husserl employed for arithmetical algorithms in *Philosophy of Arithmetic*. But there are some novelties with respect to 0 and 1.

Schröder introduces 0 and 1 in a purely formal way: 0 as the class that can be subsumed under any class, 1 as the class that subsumes any other class. Husserl could not, however, accept those purely formal definitions, for besides avoiding contradiction (*Widerspruch*), he said, a calculus must also avoid *conflict* (*Widerstreit*). That is, it must avoid opening the doors to imaginary entities, i.e. objects that do not exist but are treated, even in deductions, as if they did.[10] This is paradigmatically the case of 0. Husserl just could not conceive of an empty extension. The idea of a class that is contained in any other class, he thought, is absurd, for there are, after all, disjoint classes. 0 does not denote anything, which puts it on an equal footing with $\sqrt{-1}$ in general arithmetic.

There are, Husserl said, only two ways of accepting the symbol 0 as introduced by Schröder: 1) in the logical calculus, i.e. the calculus of classes as extensions of concepts, by giving it a meaning as the extension of the concept of non-existence; 2) in the calculus of identity, i.e. Schröder's calculus of classes in general, not only extensions of concepts, by treating it as a meaningless, perhaps useful, but essentially eliminable symbol (just like $\sqrt{-1}$). He says:

> this 'creative' definition of 0 does not yet give it the right to exist in the system of the calculus...–however, is there anything that can give it such a right? *Of such a thing I cannot find the shadow of a proof. The 0 of the calculus of identity presents the same problem as $\sqrt{-1}$ in the arithmetical calculus* [*my emphasis*]. In one as in the other case, we can only give the correspondent proof by considering the corresponding algorithmic technique. Here, it would be necessary to show that any relation deduced with the help of 0, which involves moreover only symbols that are real [*i.e. meaningful, denoting symbol– my note*], must be a valid

---

[10] "A geometry is still geometry if after having defined square circles it uses them in deductions?" (Husserl 1891b, 31) A *conflict* is an incongruity between a symbolic system and its *intended* objectual domain.

> relation exclusively accorded to the meaning of those symbols and the laws that concern them. *Creative definitions do not contribute anything, even if they preserve the internal consistency of the calculus. The question is not whether the calculus remains consistent, but whether it remains a calculus of classes* [*my emphasis*]. (Husserl 1891b, 33)

Husserl insisted on this. If a calculus is logically justified, it then "stands for" something (even if it can stand for different things), its basic principles and transformation rules are founded on the meaning of what it stands for, and the introduction of meaningless symbols in it can only be justified if these symbols–no matter how useful from a purely algorithmic perspective–are in the end unnecessary as far as the application of the calculus to its *intended* domain is concerned, despite the fact that their incorporation does not generate formal inconsistencies.

It is clear that by the time Husserl had published *Philosophy of Arithmetic* and written the review of Schröder's book, he did not see a calculus as a free creation; calculi only had a surrogate function and could not stand on their own. It is curious that essentially the same arguments appeared in Husserl's 1901 Göttingen talks about imaginary entities in mathematics, when he was already in possession of much more sophisticated views concerning the epistemological value of purely formal symbolic systems *per se*. (According to these views, as we shall see below, such systems provide us with formal knowledge, i.e. knowledge of formal manifolds or structures regardless of their eventual material fillings).

In the review of Schröder, as in those Göttingen talks, however, Husserl was *not* dealing with purely formal symbolic systems *for their own sake*, but with systems that have (in the case of symbolic arithmetic) or are presented as having (Schröder's calculus) *intended interpretations*. In such cases, Husserl thought, the systems cannot conflict with their *intended* domains,[11] admitting by purely formal means the adjunction of entities that just cannot belong to these domains.

Later Treatment. I discuss in detail Husserl's treatment of imaginaries in the Göttingen talks in other places in this book,[12] so I shall be brief here. Basically, he said to his audience in Göttingen in 1901 that the introduction of imaginary elements into a domain is allowed provided that: 1) the (formal) theory extending the (formal) theory of the domain in question (obtained by formal abstraction from the contentual theory of that domain, i.e. the theory of which the domain is the intended model) by means of formal axioms introducing imaginaries in an extended language is consistent, and 2) the formal theory of the domain, written in

---

[11] I shall take this opportunity to say that one of my main criticisms of Husserl's philosophy of mathematics is that it does not take into consideration the fact that a formal system, even when built upon the intuitive apprehension of truths about a *determinate* objective domain, or the concept of which this domain is the extension, *is never only a theory of this domain*, even if it is categorical (for categoricity guarantees only the uniqueness of the formal structure of the domain, not its material content).

[12] See Chapters 6 and 7.

the restricted language without imaginaries, is complete (Husserl's term is 'definite').

If we compare this solution with the then ten-year old solution presented in the review of Schröder, one, and only one difference is noticeable. Whereas in the Schröder review Husserl said what amounts to saying that the extended theory must be *conservative* with respect to the narrower theory (a fact he confessed to be unable to prove), in the Göttingen talks he required the narrower theory to be *complete*–this, of course, implies the conservativeness of the extended theory, provided it is a consistent extension–(a fact that he then thought he knew how to prove, as far as arithmetic is concerned). Indeed, Husserl claimed in the talks that the arithmetic of real numbers was complete and the sketches of what he took to be a proof of the completeness of different systems of arithmetic can be found in his notes for the talk.[13] (Husserl 1901a, 442-43)

In 1901 Husserl had already written the *Prolegomena*, so, we must compare the solution for the problem of imaginaries given in Göttingen with the views on the nature and role of formal mathematics presented in that work.

Formal Ontology

The scientific usefulness of symbolic reasoning is still fully appreciated in the *Prolegomena*. According to Husserl:

> The solution of problems raised within a theoretical discipline, or one of its theories, can at times derive the most effective methodological help from recourse to the categorial type or (what is the same) to the form of the theory, and perhaps also by going over to a more comprehensive form or class of forms and to its laws. (Husserl 1900, §70)

That is, problems raised within a theoretical science–structured as an *interpreted* symbolic theory–can be solved by resorting to its formal *abstractum* (the formal theory obtained by divesting the original theory of any intended reference) or even to formal extensions of it. This obviously includes problems like, say, finding adequate formal procedures for solving arithmetical problems. As Husserl knew very well, the adequate solution of this problem required the extension of the formal manifold determined by the arithmetic of numbers seen as specifications of the notion of quantity to the manifold of complex numbers.[14] Again, formal mathematics can

---

[13] He reasons thus: Numerical equalities and inequalities are decidable on the basis of the axioms; algebraic equalities and inequalities and general assertions are decidable because their numerical instances are decidable (needless to remark that what Husserl understands by "decidable" has nothing to do with our notion of syntactic–or even semantic–decidability).

[14] Another example may be this: the famous problems of ancient Greek geometry, the squaring of the circle, the trisection of the angle and the duplication of the cube could only

provide useful *techniques*. But can symbolic formal theories, *even if not given any interpretations*, lead to *knowledge*?

Clearly, in *Logical Investigations* Husserl thought they could. Formal mathematics, he thought, is a province of formal logic, being pre-occupied with (logical) forms *independently of their eventual material fulfilments*. Formal mathematics studies formal manifolds, which are domains of objects determined only with respect to form, regardless of their particular material content, the properties these objects have and the relations they may entertain with one another considered exclusively as object-forms (in short, mathematical structures in the sense of modern or abstract algebra. To use a metaphor that Berkeley made famous in another context, a formal manifold is the ghost of a departed objectual domain).[15]

Studying the interrelations of formal theories (or, equivalently, the manifolds they determine) is an aspect of the pure theory of manifolds.[16] The key to understanding *why*, say, complex numbers are useful for the theory of real numbers (for instance, in the theory of algebraic equations over the field of real numbers) lies in the *formal* relations between the *forms* of both real and complex number fields. Husserl seemed to be suggesting this in the following quote from §70 of the *Prolegomena*:

> Not less important than... going back to pure form is the closely related ranging of each... form in more comprehensive forms or classes of forms. That we have here a central item in the wonderful, methodological art of mathematics, becomes plain if we look... at the first, simplest case of this sort, the extension of the field of real numbers (i.e. of the corresponding form of theory, the 'formal theory of real numbers') into the formal, two dimensional field of ordinary complex numbers. *In this concept we indeed have the key to the only possible solution of the problem that has not yet been cleared up: how, e.g., in the field of numbers impossible (essenceless) concepts can be methodologically treated like real ones* [my emphasis]. This is not, however, the place to discuss this more closely.

---

be adequately dealt with–if not solved, at least shown to be unsolvable as stated–by going through an elaborate *algebraic*, i.e. formal analysis of geometrical constructions with straight edge and compass and what they can accomplish. This analysis interprets geometrical constructions formally in terms of algebraic field extensions.

[15] Similar ideas were voiced in the sketches for the talks before the Mathematical Society of Göttingen: "[M]athematics in the highest and most general sense is the science of theoretical systems in general, abstracting the objects of theoretical interest of the given theories of different sciences; in no matter which given theory, in no matter which given deductive system, we abstract its subject matter, the particular types of objects it tried to theoretically master, and if we substitute the representations of objects materially determinate by simple formulas, that is, the representation of objects in general that is mastered by such a theory, by a theory of this form, we have then accomplished a generalization that considers the given theories as particular cases of a class of theories, or rather of a form of theory that we consider in a unifying manner and in virtue of which we can say that all these particular scientific domains have, as form is concerned, the same theory" (Husserl 1901a, 430-31). "Mathematics is then, according to its highest ideal, a doctrine of theories, the most general science of deductive systems that are possible in general". (*Ibid.*, 432)

[16] Husserl was very consistent in his characterization of the doctrine of multiplicities (manifolds): it is a science of forms of theories. See, for instance, *Introduction to Logic and Theory of Knowledge* (Husserl 1906/07, §19), a work written ten years after the *Prolegomena*.

Since the Göttingen talks of 1901 were given *after* Husserl had written and published the *Prolegomena*, it is safe to assume that the solution of the problem of imaginaries presented there is the solution alluded to in the quote above. According to the talks, imaginaries can be "methodologically treated" as real numbers because *methodologically* both can be treated as symbols subjected only to formal relations. But a good method is not yet a logically justified method.

In the talks Husserl is clearly making the logical justification of imaginaries rest on a *proof* of their dispensability by means of a *proof* of the completeness of the system to which they are added (since this narrower system has an intended model that must be completely mastered by the intuitive apprehension of its fundamental concept). As long as we consider the arithmetic of real numbers as a theory in the pregnant sense, i.e. as founded on the concept of real number as, say, Cauchy sequences of rational numbers or Dedekind cuts, and its intuitive truths, then complex numbers had to be rendered dispensable by a *proof* of the conservativeness of the formal theory of complex numbers with respect to the formal theory of real numbers (in fact, Husserl said we have to prove the *completeness* of the narrower theory, but this would imply the conservativeness of any consistent extension of it, expressed in a language extending that of the narrower theory). I think this was the sort of task Husserl envisaged for the metatheory of formal systems that he located on the third level of apophantic formal logic.

The solution presented in the Göttingen talks for the problem of imaginaries says that, from a formal perspective, imaginary entities can be treated like real ones, and establishes the *logical conditions* under which treating them so can be allowed, if useful, *considering conceptual knowledge exclusively*. Husserl insists that as long as we are interested in knowing the properties (even only the formal properties) of the *concept* that founds a theory (for instance, the properties of numbers as *numbers strictu sensu*), the use of imaginaries cannot be an essential one.

Although, as Husserl said in *Logical Investigations,* purely symbolic theories are *per se* a form of knowledge, structural or formal knowledge precisely–they provide knowledge of formal manifolds independently of their interpretations, thus belonging to formal ontology–, they must be teleologically oriented towards objectual domains.[17] According to Husserl, formal theories are mere *forms of theories*, not theories in a pregnant sense (the chosen terminology is revealing, a "pregnant" theory is obviously a non-sterile theory); formal theories are like wandering spirits in search of a body to snatch. For him, the creation and study of formal theories for their own sake, independently of intended applications, amounts to toying with what we can call "formalist alienation" (he would later attribute

---

[17] A sign of this orientation is that Husserl sees even formal theories as referring to objects, *formal objects* precisely, indeterminate as to content, but determinate as to form by their theory.

to its excessive technization—of which its mathematization is an aspect—the "crisis" of European science.)[18] As formal theories *of objectual domains*, on the other hand, formal theories can be enlarged to serve *methodological* purposes (whose logical justification depends on the completeness of the restricted theory).

If we admitted that a consistent formal theory always describes the formal aspects of an existing objectual manifold, as is the case of elementary, that is, first-order theories, Husserl's cautions would seem vacuous. But if we assumed that he did *not* take for granted that all *consistent* formal theories have models, his considerations would acquire some relevance, for what would be the point of investigating abstract formal manifolds that could not be given any objectual content, the only support that could remove them from the world of fantasy in which they live?

But it may be that Husserl was saying something more prosaic: that formal theories are only interesting if they can be applied, or, in other words, that formal mathematics must keep its pragmatic sensibility alert so as to avoid indulging in sterile investigations. At first sight, this seems to be an honest and well-intended scruple. But the history of mathematics shows that hardly any formal mathematical device or theory has been created that did not prove its usefulness, the impossible and absurd numbers of the Italian algebraists of the Renaissance, the points at infinity of Johannes Kepler, all the variants of the old Greek geometry, infinitesimals, quaternions, non-commutative algebras, etc. What makes Husserl's scruple undesirable is that we do not and cannot know *a priori* when a mathematical purely "fictional" creation will prove its applicability, in mathematics itself, physics or any other field of knowledge. So, the best strategy is one of tolerance: to grant mathematicians their freedom and wait for the survival of the fittest go to work (the probable reason that we do not know many pure game-like formal theories is that the mathematical community does not take them seriously and they just vanish from sight —the community ultimately sees to it that mathematicians do not indulge in formalism for its own sake, alienated from fundamental cognitive interests).

Being trained as a mathematician at the end of the 19th century made Husserl naturally suspicious of the possible excesses of the formalistic turn this science was undergoing (the proximity and influence of Leopold Kronecker may also have played a role). But he also saw the immense possibilities of purely formal mathematics. So, he endorsed it and lent it epistemological dignity (formal mathematics is a chapter of formal ontology), but with a note of caution (formal theories must be applicable).

Objectual domains of mathematics are in general infinite, so we cannot expect to have an intuitive access to its objects directly, or by inductive means, but only through a concept that unifies that domain as

---

[18] See Chapter 17.

the extension of this concept.[19] In this case, we can intuitively access the infinite domain by intuitively accessing its concept. It may be the case that the formal theory of such a domain is already at our disposal (being previously or independently developed in the realms of formal ontology), in which case the formal theory finds its *raison-d'être* (for instance, the formal theory of complex numbers as the theory of the two-dimensional domain of displacements and their operations).

But consider that we have an objectual domain given from the start as the extension of a concept, such as, for instance, the domain of cardinal numbers as answers to "how many?" Do we have the right to use formal extensions of its formal theory, which may have completely unrelated models, to obtain knowledge, *even of a formal nature*, of such a domain, whose truths must be exclusively derived from the intuition of its ruling concept? Husserl thought that we do not, and that the interference of formal manipulations could only be tolerated under the presupposition that conceptual intuition was in principle capable of providing the contentual theory of the domain with a *complete* set of basic principles and laws.

This was the solution to the problem of imaginary elements in mathematics that Husserl presented in Göttingen. It is worth noticing that if Husserl indeed believed that even consistent formal theories may not have models (or, at least, *interesting* models), the attribution of a *pragmatic* role for formal theories *in general* would vindicate even those that cannot be given any (interesting) objectual domain, thus safeguarding mathematical methodology (of inventing vacuous symbols for solving problems) from falling into sheer nonsense (or formalist alienation). It is obvious that Husserl did not want to give up even the riskier formal procedures of mathematics, but it is also obvious that he was not willing to let the formalistic approach to mathematics be interpreted in a way that would alienate mathematics from a firm compromise with knowledge.

Critical Considerations

I would like to conclude with some final comments about the *correctness* of Husserl's vindication of symbolic knowledge, in particular his treatment of imaginaries, and the role these questions played in the development of his philosophy. With respect to Husserl's cautious treatment of purely symbolic knowledge, I have already expressed my reservations

---

[19] Husserl believed that any *a priori* contentual axiomatic theory is necessarily a *conceptual* theory, that is, the theory of a concept under which the objects of the relevant domain are assembled. In "Arithmetic as an A Priori Science", inquiring on the nature of arithmetic as an *a priori* science, Husserl said that a science of such a nature "does not begin with single facts in order to obtain then possibly true generalities by induction, but immediately with certain generalities that are apodictically certain and immediately evident, which it acquires by simply presenting to itself certain 'fundamental concepts' that give, by means of mediate evidence and certitude, the entire sequence of theorems of this science". (Husserl 1891a, 382)

above. We must let mathematicians do their work. No matter how inapplicable a formal theory may be, if it is consistent, it is the theory of a mathematical formal structure and time will decide whether it is sufficiently interesting to survive.

With respect to imaginaries, I have more serious concerns. I believe that Husserl was so worried about securing mathematics against a possible infection with imaginaries that he put more effort into developing a protective vaccine than into explaining *why* imaginaries are useful when they are. The vaccine, of course, was *completeness*, inoculated via *conceptual intuition*. Husserl believed that *a priori* mathematical contentual theories are in general *conceptual* theories and that imaginaries cannot substitute proper relevant intuition and be *essentially* involved in the business of proving theorems.

But the fact is that, contentual or purely formal, mathematical theories are invariably theories of *forms*, not materially *well-determined* objectual domains, even when these are *categorical* theories.[20] No matter how clearly we intuit the concept of a non-negative integer, for instance, all we get in the process are formal or structural properties of ω-sequences. The intuitively apprehended fundamental facts about numbers, which constitute the axiomatic basis of arithmetic, are also true of no matter which domain of objects that is structured as a ω-sequence, regardless of the nature of these objects. So, there is not much difference between contentual or conceptual, on the one hand, and purely symbolic mathematical theories, on the other; theories of both types are in a sense formal, since their objects are only and invariably *forms* or *structures*.

The fact that even contentual mathematics is a formal science reduces Husserl's distinction between theories and mere forms of theories to one between conceptual or eidetic formal theories (such as arithmetic and physical geometry) and hypothetical formal theories (such as Riemannian n-dimensional geometries), respectively. The reason Husserl insisted on keeping both types apart has to do with epistemological relevance: the former are already theories of something, the latter only describe possible hypothetical forms waiting for objectual domains to appear that can be in-formed by them. But this seems to me to be excessively cautious, since, far from being mere symbolic games, hypothetical formal theories also provide knowledge–formal knowledge, as Husserl himself acknowledged.

And it is precisely because of this that imaginaries can be useful. Despite the fact, for instance, that complex numbers do not measure quantity, introducing them in the domain of numbers and extending the

---

[20] In a letter to Frege, summarized and commented by Husserl, Hilbert says something relevant in this context: "Any theory can be applied to an infinite number of systems of fundamental elements. It suffices to apply a one to one reversible transformation and stipulate that the corresponding axioms for the thing thus transformed are the same (this is the case, for instance, with the principle of duality and my proofs of independence)". (Husserl 1899, 450)

arithmetic of non-negative integers to the arithmetic of complex numbers amounts to imbedding the original structure of arithmetic into a richer structural milieu in which problems originally posed for the restricted numerical domain can be adequately handled–for they may very well be problems about non-negative integers *as complex numbers*, i.e. problems demanding the larger structure in order to be adequately stated and treated.

It is because arithmetic, even though founded on a conceptual intuition, is in the sense explained above a formal theory–i.e. it does not tell us more about numbers proper than the formal theory abstracted from it tells us about number-like entities–that imaginaries can be added to its domain and be useful in solving numerical problems. Imaginaries are after all number-like entities. The process of formal abstraction then amounts to no more than just forgetting that we have an intended domain for a theory, since we cannot fix this domain as the *only* single model of the theory anyway. If we let the theory "talk" about structural aspects of any system of objects that satisfies its axioms, we can approach the phenomenon of imaginaries from the perspective of the interrelations between a given structure and structures extending it.[21]

Let us insist a bit more on this matter. If you look at a painting from the Romanesque or Gothic periods, the elements of the picture do not really fall together in the same space. The invention of perspective by Filippo Brunelleschi solved this problem by introducing an imaginary point on the canvas, which is not really there, but organizes the whole into an articulated unity, a formal substitute of the eye in relation to which all space relations are determined. The point at infinity is not a point of the visual space, but helps to organize it. Imaginary elements work in an analogous manner.

Pictorial perspective influenced some mathematical developments. The geometries of Johannes Kepler and Girard Desargues were great improvements over Greek geometry–whose finitist sensibility could not conceive of points at infinity–because they knew how to take advantage of the increased *formal possibilities* of an enlarged space which was not only potentially infinite, but actually had points *at* infinity. In particular this made possible a treatment of conics much more elegant than Apollonius'.

The answer to the riddle posed by imaginary elements lies on this observation: like the points at infinity of Kepler and Desargues, imaginary elements increase formal possibilities. Or still, they enrich structure. But if this is all they do, their utility can *only* be explained if the domains to which they are added interested us only insofar as their structures were concerned, which would not suffer from the substitution of the elements of the domain by others of a totally different nature. Otherwise the utility of imaginary elements would remain a mystery.

Imaginary elements such as $\sqrt{-1}$, introduced contrary to good sense as bona fide numbers by the Italian algebraists of the 16th century were

---

[21] It is interesting to notice that the interplay of formal domains is, according to Husserl, a topic of study of formal ontology.

useful because they added necessary structure to the original domain of numbers to which they were added. Arbitrary algebraic operations on the system of proper numerical symbols quickly lead outside this domain; imaginary elements added necessary extra room to it so as to allow a bigger range of transformations. Analogously, if we tried to transform a right-hand glove into a left-hand glove using only rotations and translations, we would necessarily fail, but if we are allowed to use inversions, we shall easily succeeded. By so doing, we increase the formal possibilities at our disposal. Since the original problem was essentially formal, it benefited from this enrichment. It is always thus with imaginary elements, they work because they enrich structure, and the problems they help to solve are structural problems.

The problem of finding algorithmic procedures for solving numerical problems is analogous to the problem of transforming a right-hand into a left-hand glove. As long as we allow only operations that are confined to the realm of positive integers (addition, multiplication and exponentiation) it is often difficult and in general impossible. But as long as we admit inversions of operations (subtraction, division and root extraction) we immediately escape that confined space and consequently allow imaginaries to come in. But this solves the problem, which was after all a purely formal one, at first unnecessarily restricted to a domain where it could not be adequately handled.

The Problem of Symbolic Knowledge in the Development of Husserl's Philosophy

As we have seen, the problem of imaginaries was fundamental in making Husserl consider questions such as: the role of representations without object in the general scheme of knowledge; the interplay between conceptual theories (those that are based on intuition) and purely formal theories (whose objects are purely intentional) and, consequently, the interplay between empty intentions and intuitions in the dynamics of human knowledge; the need to enlarge the field of intuition well beyond Kantian limits in order to account for *a priori* mathematical contentual theories; the *logical* relevance of studying formal theories, their properties —such as consistency and completeness, in particular— and their mutual relations, and, consequently, the need for an enlargement of the field of formal logic vis-à-vis the tradition; the need for an adequate study of logical grammar and the theory of deduction so as to guarantee epistemological relevance for manipulation of signs within logical symbolic systems; and many other questions along the same lines. It is clear now why, in trying to come up with a philosophical account of general arithmetic, he wrote the *Logical Investigations*, instead of only the second volume of the *Philosophy of Arithmetic*.

References

Bochenski, Józef 1970. *A History of Formal Logic?* New York: Chelsea Publishing Co.

Husserl, Edmund 1886-1901. *Studien zur Arithmetik und Geometrie, Texte aus den Nachlass*, 1886-1901, Husserliana XXI, Ingeborg Strohmeyer (ed.), The Hague: Martinus Nijhoff, 1983

Husserl, Edmund 1887, *Über den Begriff der Zahl: Psychologische Analysen*, Halle: Heynemansche Buchdrückerei, 1887, in Husserliana XII, 289-339. Published in English as *On the Concept of Number*, in Dallas Willard's translation of Husserl 1887-1901, 305-57.

Husserl, Edmund 1887-1901. *Philosophie der Arithmetik, mit ergänzenden Texten (1890-1901)*, Husserliana XII, introduction by Lothar Eley, The Hague: Martinus Nijhoff, 1970, published in English as *Philosophy of Arithmetic, Psychological and Logical Investigations with Supplementary Texts from 1887-1901*, translated by Dallas Willard, Dordrecht: Kluwer, 2003. The English edition contains the page references to the Husserliana edition.

Husserl, Edmund 1889-90. "Die formal und die wirkliche Arithmetik", published in Husserl 1886-1901, 21-23.

Husserl, Edmund 1890a. "The Concept of General Arithmetic", in Husserl 1994, 1-6, published in German in Husserliana XII, 374-79.

Husserl, Edmund 1890b. "Zur Logik der Zeichen", in Husserl 1887-1901, 340-73, published in English as "On the Logic of Signs (Semiotic)" in Husserl 1994, 20-51. The English edition contains page references to the Husserliana edition.

Husserl, Edmund 1890-1910. *Aufsätze und Rezensionen 1890-1910*, Husserliana XXII, The Hague: Martinus Nijhoff, 1979.

Husserl, Edmund 1891a. "Arithmetic as an A Priori Science", in Husserl 1994, 7-11, in German in Husserl 1887-1901, 380-384.

Husserl, Edmund 1891b. "Besprechung von E. Schröder, *Vorlesungen über die Algebra der Logik I*", in Husserl 1890-1910, 3-43, published in English as "Review of Ernst Schröder's *Vorlesungen über die Algebra der Logik*" in Husserl 1994, 52-91. Originally published in *Göttinger Gelehrte Anzeigen*, 1891, n° 7, 243-78.

Husserl, Edmund 1891c. "On the Concept of Operation", in 1887-1901: 385-408, in German in Husserliana XII, 408-29.

Husserl, Edmund 1891d, "Letter from Edmund Husserl to Carl Stumpf, in Husserl 1994, 12-19, in German in Husserl 1886-1901, 244-51.

Husserl, Edmund 1891e. *Philosophie der Arithmetik*, Halle: Pfeffer. Published in Husserl 1887-1901, 1-283.

Husserl, Edmund 1891f. "Zur Lehre der Inbegriff", Published in Husserl 1887-1901, 385-407, published in English as "Essay I, On the Theory of the Totality", in Dallas Willard's translation of Husserl 1887-1901, 359-83.

Husserl, Edmund 1899. "*Husserl's* Excerpts from an Exchange of Letters between *Hilbert* and *Frege*", published in Husserl 1887-1901, 468-73, in German in Husserliana XII, 447-51.
Husserl, Edmund 1900. *Prolegomena to Pure Logic,* volume I of Husserl 1900-01, translation of *Prolegomena zur reinen Logik.*
Husserl, Edmund 1900-01. *Logical Investigations,* New York: Humanities Press, 1970, translation of *Logische Untersuchungen.* Halle: Niemeyer, also published as Husserliana XVIII, XIX/I-II, The Hague: Martinus Nijhoff, 1975, 1984.
Husserl, Edmund 1901a. "Double Lecture: On the Transition through the Impossible ("Imaginary") and the Completeness of an Axiom System", included as Essay III in Husserl 1887-1901, 409-52. English translation of Husserliana XII, 430-51.
Husserl, Edmund 1906/07. *Introduction to Logic and Theory of Knowledge,* Dordrecht, Springer Verlag 2008, translation by Claire Ortiz Hill of *Einleitung in die Logik und Erkenntnistheorie, Vorlesungen 1906/07,* Ullrich Melle (ed.), *Husserliana* XXIV, Dordrecht: Martinus Nijhoff, 1984. The English edition contains page references to the Husserliana edition.
Husserl, Edmund 1913, *Introduction to the Logical Investigations, A Draft of a Preface to the Logical Investigations,* Eugen Fink (ed.), translated by P. Bossert and C. Peters, The Hague: Martinus Nijhoff, 1975, published in Husserliana XX/1, 272-329.
Husserl, Edmund 1936. *The Crisis of European Sciences and Transcendental Phenomenology: An Introduction to Phenomenological Philosophy,* Evanston IL: Northwestern University Press, 1970, translation of *Die Krisis der europäischen Wissenschaften und die transzendentale Phänomenologie,* Husserliana VI, The Hague: Martinus Nijhoff, 1954.
Husserl, Edmund 1975. *Articles sur la logique (1890-1913),* translation by Jacques English, Paris: Presses Universitaires de France.
Husserl, Edmund 1994. *Early Writings in the Philosophy of Logic and Mathematics,* translated by Dallas Willard, Dordrecht: Kluwer, the translation of Husserl, Edmund 1890-1910 and other short works.
Moran, Dermot 2005. *Edmund Husserl–Founder of Phenomenology,* Cambridge: Polity Press.
Muralt, André de 1974. *The Idea of Phenomenology: Husserlian Exemplarism,* Evanston IL: Northwestern University Press.
Willard, Dallas 1994, "Translator's Introduction" to Husserl 1994, VII-XLVIII.

4

Claire Ortiz Hill

ONE DOGMA OF EMPIRICISM

Introduction

It is well known that Quine argued that modern empiricism was to a large extent conditioned by an ill-founded belief in a fundamental cleavage made by Kant between analytic truths, which are grounded in meaning independently of matters of fact, and synthetic truths, which are grounded in fact. In "Two Dogmas of Empiricism", Quine argued that it was a folly to look for such a boundary and that the idea that there was such a distinction to be drawn at all was "an unempirical dogma of empiricists, a metaphysical article of faith". (Quine 1953)

It is much less well known that Edmund Husserl repudiated the same distinction and that phenomenology was to a large extent conditioned by his determination to overcome the destructive impact of Kant's theory. So, at this intersection of experience and analysis, we find ourselves at yet another crossroads between analytic philosophy and phenomenology. For the original cleavage between the two schools was to a large extent conditioned by reactions to Kant's cleavage.

Husserl's theory of manifolds was an important part of his answer to problems he detected. In what follows, I study its development in terms of the evolution of his ideas about empiricism and analyticity. To provide a context for integrating his theory into mainstream philosophy, I establish connections between it and work on axiom systems, truth in structures, and model theory.

Against Kantian Empiricism and Analyticity

Bolzano, Weierstrass, Brentano, Cantor, Frege, and Husserl all spotted dangers in Kant's distinction and steered their thought in another direction. As Peter Simons has pointed out,

> Bolzano not only gave the definitive criticism of Kant's concept of analyticity, he also proposed a definition of his own which is almost totally acceptable even by today's standard, and which anticipates Quine, among others. According to Bolzano, those propositions (in themselves) are analytic of which at least one constituent concept (idea in itself) can be varied *salva veritate aut falsitate*. Of analytic propositions, logically analytic ones are those of which all non-logical concepts can be varied without change of truth-value. (Simons 1992; Bolzano 1837)

Bolzano's pioneering work to rebuild intuitively accepted proofs of theorems in a rigorous way solely on the basis of arithmetical and logical concepts paved the way for Weierstrass, who taught that arithmetic could be built up in a purely logical fashion from the concept of whole number, which would free analysis from insidious appeals to intuitions of space imported into it since Kant had declared mathematical propositions synthetic a priori. (Sebestik 1992; Kline 1972; Jourdain 1991; Demopoulos 1994). It was Weierstrass who awoke in Husserl a desire to seek radical foundations for knowledge. Husserl saw him as aiming to lay bare the original roots, elementary concepts, and axioms upon which the whole system of analysis might be deduced in a completely rigorous, perspicuous way. At the end of his career, Husserl said he had tried to do for philosophy what Weierstrass had done for mathematics. (Schuhmann 1977)

Brentano saw Kant as having been "misled into a false definition of analytical judgement, according to which an affirmative judgement is supposed to be analytic if its predicate is included in the concept of its subject", an error Brentano saw as being "connected with... the disastrous illusion that mere analytic judgments do not add to our knowledge". Kant himself, Brentano pointed out, inadvertently offered striking evidence against it when he maintained that logic was "supposed to be purely analytic and yet truly a science, and hence an enrichment of our knowledge" (Brentano 1911). Uncertain as to whether to pursue a career in mathematics or in philosophy, the profound impact of Brentano's insightful, clear, rigorous, precise, objective analyses, and his ability to transform unclear beginnings into clear thoughts and insights made Husserl into a philosopher. (Hill 1998)

Inventor of set theory, Georg Cantor was an avowed enemy of psychologism, empiricism, positivism, naturalism, sensualism, skepticism, and Kantianism, which he saw as wrongly locating the sources of knowledge and certainty in the senses or in the "supposedly pure forms of intuition of the world of presentation". Cantor wanted to free mathematicians to engage in strictly arithmetical forms of concept formation and open the way to a new, abstract realm of ideal mathematical objects that could not be directly perceived or intuited. Much of the freedom he believed he enjoyed came from distinguishing between an empirical treatment of numbers and Plato's pure, ideal *arithmoi eidetikoi*. Phenomenology began to take shape during his fifteen years in Cantor's company. (Hill and Rosado Haddock 2000)

Frege held that arithmetic was analytic and that analytic statements could be informative. He denounced what was empirical as subjective, psychological, and fleeting (Frege 1884). He was, as Paul Benacerraf has stressed, no empiricist and establishing the analyticity of arithmetical judgments was not his way of defending empiricism against Kantian attack. (Benacerraf 1981)

Repudiating Kant's Distinction

Husserl left no doubt about the role that his dissatisfaction with Kant's distinction played in the development of phenomenology. He considered Kant's logic to be utterly defective. He had not understood the nature and role of formal mathematics, and the way in which he had defined the concept of analyticity was totally inadequate and even utterly wrong. Not only had he never seen how little the laws of logic are all analytic propositions in the sense of his own definition, but he failed to see how little his dragging in of an evident principle for analytic propositions helped to clear up the achievements of analytic thinking. (Hill 1991)

So, Husserl set out to develop the proper concept of the analytical and to discover the boundary separating genuine analytic ontology from material ontology essentially distinct from it. He said that it was his study of Leibniz' *verities de raison* and *verities de fait* and a new keen awareness of the contrast between Hume's matters of fact and relations of ideas and Kant's analytic and synthetic judgments that undermined his early confidence in empirical psychology, set the stage for his conversion from it, and played an important role in the formulation of the positions he later took. (Husserl 1913)

Abandoning Empirical Psychology

Initially inspired by Brentano, whose ideal was most nearly realized in the exact natural sciences, Husserl first tried to anchor arithmetical concepts in direct experience (Husserl 1891). However, empirical psychology never came to satisfy him. He believed that it proved useful in investigating the origin of mathematical presentations, or in elaborating practical methods that were psychologically determined, but that when it came to going from the psychological connections of thinking to the unity of theory, it could establish no true clarity and continuity. (Husserl 1900-01; Husserl 1913)

During the 1890s, Husserl began veering towards an objective logic, where truth was an analysis of essences or concepts. He concluded that the ultimate meaning and source of all objectivity making it possible for thinking to reach beyond contingent, subjective, human acts, and to lay hold of objective being in itself was found in ideality and in the ideal laws defining it. Everything that was purely logical, was an "in itself", something ideal having nothing to do with acts, subjects, or empirical persons belonging to actual reality. The entire overthrowing of psychologism through phenomenology, he said, showed that his analyses in *On the Concept of Number* and *Philosophy of Arithmetic* would have to be seen as pure, a priori analysis of essence (Husserl 1929; Husserl 1913). In 1905, he wrote to Brentano:

> The empirical sciences–natural sciences, –are sciences of 'matters of fact'.... Pure Mathematics, the whole sphere of the genuine a priori in general, is free of all matter of fact suppositions.... We stand not within the realm of nature, but within that of Ideas, not within the realm of empirical... generalities, but within that of the ideal, apodictic, general system of laws, not within the realm of causality, but within that of rationality.... Pure logical, mathematical laws are laws of essence.... (Husserl 1905)

Husserl came to see Bolzano's theories about presentations and propositions in themselves as an early attempt to provide a unified presentation of the domain of pure ideal doctrines and as providing a complete plan of a pure logic. He had initially viewed them as curious conceptions, unintelligible, mythical entities suspended between being and non-being, but it suddenly became clear to him that Bolzano had not hypostasized them. Rather, they enjoyed the ideal existence or "validity" characteristic of universals. (Husserl 1913; Husserl 1994)

Husserl now saw Bolzano's presentations in themselves as what were ordinarily called the senses of statements, what is said to be one and the same when, for example, different persons are said to have asserted the same thing, or what scientists simply call a theorem (for example, the theorem about the sum of the angles in a triangle), which no one would think of as being someone's experience of judging, and that this identical sense could be none other than the universal present in all actual assertions having the same sense, which makes possible the identification in question, even when the descriptive content of individual experiences of asserting varies considerably in other respects. The ideal entities so unpleasant for empiricistic logic and so consistently disregarded by it, Husserl came to insist, had not been artificially devised either by himself or by Bolzano. They were given beforehand by the meaning of the universal talk of propositions and truths that is indispensable in all the sciences. And Husserl believed that that indubitable fact had to be the starting point of all logic, for science was a web of theories, and so of proofs, propositions, inferences, concepts, meanings, and not of lived experiences. (Husserl 1908-09; Husserl 1994)

Analyticity without Empiricism

Husserl said that his fight against logical psychologism was meant to serve no other end than the supremely important one of making pure, analytic logic visible in its purity and ideal particularity (Husserl

1929). He came to hold that the only concrete, fruitful way of explaining analyticity lie in stressing that in purely logical, formal, analytic propositions or laws, the variables are indefinite, the terms can vary completely freely and arbitrarily. Purely arithmetical theories, the purely analytical theories of mathematics, the traditional theory of syllogism, the pure theory of cardinal numbers, of ordinals, Cantorian sets, and so on were purely logical because their basic concepts expressed reasoning forms free of any cognitive content and they could not be had through sensory abstraction. No epistemological reflection was required. (Husserl 1902/03; Husserl 1994)

Husserl wrote of the highly important divorcing of the formal and the factual, or material, spheres of being. As examples of purely formal concepts based on the empty idea of something in general and connected with it through the axioms of formal ontology, he listed "something", "one", "object", "property", "relation", "plurality", "cardinal number", "order", "ordinal number", "whole", "part", which he contrasted with material concepts like house, tree, color, sound, spatial figure, sensation, feeling, smell, intensity, etc., that express something factual or sensory. (Husserl 1900-01)

For him, analytically necessary propositions were propositions that were true completely independently of any particular facts about their objects, independently of any actual matters of fact, of the validity of positing their existence. Analytic laws were universal propositions containing nothing but concepts as formal concepts and hence were free of any explicit or implicit positing of the existence of individuals. Analytic laws stood in contrast to particular instances of them that resulted when concepts regarding matters of fact and particular thoughts positing individual existence were introduced. He considered the difference between what was merely formal and without factual content easy to see in the difference between laws like that of cause and effect, which is about changes real things undergo, or those about qualities, intensities, extensions, limits, forms of relations, which he contrasted with an analytically necessary proposition like, 'There cannot be a father if there is no child', or purely analytic generalizations like, 'A whole cannot exist without parts'. He stressed that it would be a formal analytical contradiction in terms, to call something a part if there was no whole to which it belonged. (Husserl 1900-01)

Analyticity and Manifolds

By drawing the boundary line existing a priori between mathematics and the natural sciences, Husserl believed he that was delimiting and expanding the domain of the analytical in keeping with the most recent discoveries in mathematics, notably those concerning axiomatization and manifolds. (Husserl 1900-01; Husserl 1929)

When he abandoned the second volume of the *Philosophy of Arithmetic* on the logic of the deductive sciences, he began pushing his thinking beyond the mathematical realm towards a universal theory of formal deductive systems. Volume one had harsh words for Frege's project to found arithmetic on formal definitions out of which all its theorems could be deduced purely syllogistically. Those criticisms of Frege's logic were the only ones that Husserl ever explicitly retracted. (Husserl 1891; Husserl 1900-01; Hill 1991)

Husserl saw in the nascent mathematical theory of manifolds a partial realization of his ideal of a science of possible deductive systems (Husserl 1900-1901; Husserl 1929). In the early 1890s, he compared Cantor's definition of a set or manifold as an aggregate of any elements combined into a whole by a law with Riemann's, or kindred ones, for which manifolds are aggregates of elements that are not just combined into a whole, but are ordered and continuously interdependent. Husserl defined "order" as a concatenation having the special property that each member possesses an unambiguous position in the narrow sense in relation to any arbitrary one and can thus be unequivocally characterized by the mere form of the direct or indirect connection with the last one. Manifolds, he stressed, are not mere aggregates of elements without relations. It is precisely the relations that are essential and distinguish them. (Husserl 1983)

Such reflections led Husserl to detect a certain natural order in formal logic and to broaden its domain to include two levels above the traditional formal logic of subject and predicate propositions and states-of-affairs, which deals with what might be stated about objects in general from a possible perspective. On the second level, it was no longer a question of objects as such about which one might predicate something, but of investigating what was valid for higher order objects dealt with in an indeterminate, general way, not as empirical or material entities, and determined in purely formal terms, removed from acts, subjects, or empirical persons of actual reality. This is an expanded, completely developed analytics where one reasons deductively with concepts and propositions in a purely formal manner since each concept is analytic and each procedure purely logical. (Husserl 1906/07; Husserl 1917/18)

Husserl's third level is that of the science of deductive systems in general, the theory of manifolds, theory forms, logical molds totally undetermined as to their content and not bound to any possible concrete interpretation. Here it is a matter of theorizing about possible fields of knowledge conceived of in a general, undetermined way, simply determined by the fact that the objects stand in certain relations that are themselves subject to certain fundamental laws of such and such determined form, are exclusively determined by the form of the interconnections assigned to them, which are themselves just as little determined in terms of content as are the objects. (Husserl 1906/07; Husserl 1917/18)

Through axiom forms, a manifold of anything whatsoever is defined in an indeterminate, general way. A certain something must by definition stand in a certain relationship to something else in the defining manifold. A set of axioms of such and such a form that are consistent, independent, and purely logical in that they obey the principle of non-contradiction yields the set of propositions belonging to the theory of such and such a form to be developed. The form exists insofar as it is correctly defined, insofar as the axiom forms are ordered in such a way as to contain no formal contradictions, no violation of analytic principles. Whether axioms as truths have existence in any objective, real, or ideal spheres corresponding to the prescribed form is left open. On the basis of the definition of the manifold, we can deduce conclusions, construct proofs, and it is then certain a priori that anything obtained in this way will correspond to something in our theory. (Husserl 1906/07; Husserl 1917/18)

For Husserl, this science of forms of possible theories was a field of free, creative investigation made possible once form was emancipated from content. Once one discovers that deductions, series of deductions, continue be meaningful and remain valid when another meaning is assigned to the symbols, one is freed to reason completely on the level of pure forms. One can vary the systems in different ways. One finds ways of constructing an infinite variety of forms of possible disciplines. (Husserl 1906/07; Husserl 1917/18)

Hilbert, Bourbaki, Model Theory

As more pieces of Husserl's theory of manifolds become available, it is apparent that it fits the definition of that systematic study of classes of structures defined by axioms now known as model theory. As Wilfred Hodges explained in "Truth in a Structure":

> algebraists and geometers have often found themselves studying certain objects which we now call structures, or more loosely models.... a structure is a collection of elements together with certain labelled relations...defined on those elements... When we define a class of structures by giving a set of laws which the structures must obey, these laws are called *axioms* for the class of structures. (Hodges 1985/86; see also Demopolous 1994b; Hintikka 1988)

Hilbert, an enthusiastic supporter of Husserl, studied structures in an abstract way, for their own sake, without specifying the nature of the objects subject to the operations whose rules were described. Famous for having said that it must be possible to replace in all geometrical statements the words 'point', 'line', 'plane', by 'table', 'chair', 'mug', Hilbert freed the axiomatic method from the problem of relating to the existence of objects by demonstrating that justifying a system was not a matter of attaching the axioms to empirical facts, but of demonstrating its consistency. His axiom systems were not systems of statements about a subject matter, but systems of conditions for a relational structure, a form, an abstract object taken as the immediate object of the axiomatic theory (Bernays 1922; Bernays 1967; Weyl 1944; *Revue* 1993; Gray 2000; Cassou-Noguès 2001). Husserl considered the kinship between his manifolds and Hilbert's axiom systems to be evident. (Hill and Rosado Haddock 2000)

For the Bourbakian mathematicians, a mathematical structure is a set of elements whose nature is not specified. In defining a structure, one or several relations involving these elements are given, and it is postulated that the given relation or relations satisfy certain conditions, which are spelled out and are the axioms of the structures envisioned. The axiomatic theory of a given structure is studied by deducing the logical consequences of its axioms, while excluding all other hypotheses about the elements considered, notably, any hypothesis concerning their special nature. (Bourbaki 1971)

Bourbaki wrote:

> From the axiomatic point of view mathematics appears on the whole as a reservoir of abstract *forms*—the mathematical structures; and it sometimes happens, without anyone really knowing why, that certain aspects of experimental reality model themselves after certain of these forms, as if by a sort of preadaptation. It cannot be denied... that the majority of these forms had a well-determined intuitive content at the beginning; but it is precisely by voluntarily emptying them of this content that it has been possible to employ them with all their potential efficacy and to render them capable of new interpretation and a complete fulfillment of their elaborative role. (*Ibid.*)

Bourbaki sees the essential goal of the axiomatic method as concerning the deep-lying intelligibility of mathematics. It teaches one to search for the deeper reason for its discoveries, the common ideas buried under the external apparatus, to single them out and to display them by working to dissociate the principle lines of reasoning figuring in the demonstrations of a theory, taking each of them in isolation and, considering it an abstract principle, unfolding its consequences. It then returns to the theory under study, recombines its components and studies their interactions (*Ibid.*). He likened the inner vitality of mathematics to

> a great city whose suburbs never cease to grow in a somewhat chaotic fashion on the surrounding lands, while its center is periodically reconstructed, each time following a clearer plan and a more majestic arrangement, demolishing the old sections with their labyrinthine alleys in order to launch new avenues toward the periphery, always more direct, wider and more convenient. (*Ibid.*)

He anticipated progress in the invention of new fundamental structures through the revealing of the fecundity of new axioms or new combinations of axioms.

Conclusion

Husserl's theory of analyticity and manifolds was his project for limning the true and ultimate structure of reality through an austere scheme of axiomatization that knows no acts, subjects, or empirical persons or objects belonging to actual reality. Once the pieces of his theory are sewn together and then sewn onto where they belong in philosophy, we can experiment with it as an alternative to Fregeo-Russello-Quineo methods of logics (FRQL), rooted as they are in an unworkable theory of reference to objects and identity. (Hill 1997; Hill 2004)

People spooked by intensions need not apply, however. For Husserl's method for finding clarity with respect to the central traits of reality means fraternizing with creatures of darkness, philosophizing in a metaphysical jungle of essentialism, fending off charges of trafficking in a curiously idealistic ontology that repudiates material objects, stomaching the odium of the a priori. (Quine 1947; Quine 1960; Quine 1976; Russell 1956)

As a disciple of Brentano, Husserl experienced emotions like those of Quine and Russell alluded to above, but finally found empirical psychology unable to provide the unity and continuity indispensable to science and knowledge in general. He surely would have also fled a fragmented world of rabbit parts, river stages and kinship, where the ontologies of physical and mathematical objects are but myths relative to an epistemological view (Quine 1960; Quine 1969). Husserl overcame his feelings of disgust when he came to see intensions as the senses of statements, as what is said to be the same when different persons are said to have asserted the same thing, as what scientists call a theorem, the universal present in all actual assertions having the same sense, which makes identification possible, even when the content of individual experiences of asserting varies considerably (Husserl 1994). He developed a logical point to view in conformity with that. Now, what is so bad about that?

References

Benacerraf, Paul 1981. "Frege: The Last Logicist", in William Demopoulos (ed.), *Frege's Philosophy of Mathematics*, Cambridge MA: Harvard University Press, 1995, 41-67.

Bernays, Paul 1922. "Hilbert's Significance for the Philosophy of Mathematics", in Paolo Mancosu (ed.), *From Brouwer to Hilbert, The Debate on the Foundations of Mathematics in the 1920s,* New York: Oxford University Press, 1998, 189-97.

Bernays, Paul 1967. "Hilbert, David", in Paul Edwards (ed.), *The Encyclopedia of Philosophy* vol. 3, New York: MacMillan, 496-504.

Bolzano, Bernard 1837. *Theory of Science*, Berkeley CA: University of California Press, 1972 (partial translation by R. George).

Bourbaki, Nicolas 1971. "The Architecture of Mathematics", in F. Le Lionnais (ed.), *Great Currents of Mathematical Thought*, vol. 1 of *Mathematics: Concepts and Development,* New York: Dover, 23-43.

Brentano, Franz 1911. "On Attempts at the Mathematicization of Logic", in *Psychology from the Empirical Standpoint*, New York: Humanities Press, 1973, 301-06 (section X of the 1911 appendix).

Cassou-Noguès, Pierre 2001. *Hilbert*, Paris: Les Belles Lettres.

Coffa, J. Alberto 1982. "Kant, Bolzano and the Emergence of Logicism", in William Demopoulos (ed.), *Frege's Philosophy of Mathematics*, Cambridge MA: Harvard University Press, 1995, 29-40.

Demopoulos, William 1994a. "Frege and the Rigorization of Analysis", in William Demopoulos (ed.), *Frege's Philosophy of Mathematics*, Cambridge MA: Harvard University Press, 1995, 68-88.

Demopoulos, William 1994b. "Frege, Hilbert, and the Conceptual Structure of Model Theory", *History and Philosophy of Logic* 15, 211-25.

Dieudonné, Jean 1971. *"David Hilbert (1862-1943)"*, in F. Le Lionnais (ed.), *Great Currents of Mathematical Thought*. vol. 1 *Mathematics: Concepts and Development,* New York: Dover, 304-11.

Frege, Gottlob 1884. *Foundations of Arithmetic*, Oxford: Blackwell, 1986.

Gray, Jeremy 2000. *The Hilbert Challenge*, Oxford: Oxford University Press.

Hill, Claire Ortiz 1991. *Word and Object in Husserl, Frege, and Russell. The Roots of Twentieth Century Philosophy*, Athens OH: Ohio University Press.

Hill, Claire Ortiz 1997. *Rethinking Identity and Metaphysics. On the Foundations of Analytic Philosophy*, New Haven CT: Yale University Press.

Hill, Claire Ortiz 1998. "From Empirical Psychology to Phenomenology: Husserl on the Brentano Puzzle", in Roberto Poli (ed.), *The Brentano Puzzle,* Aldershot: Ashgate, 151-68.

Hill, Claire Ortiz 2004. "Reference and Paradox", *Synthese* 138, 207-32.

Hill, Claire Ortiz and G. E. Rosado Haddock 2000. *Husserl or Frege? Meaning, Objectivity, and Mathematics*, Chicago IL: Open Court.
Hintikka, Jaakko 1988. "On the Development of the Model-Theoretic Viewpoint in Logical Theory", *Synthese* 77, 1-36.
Hodges, Wilfred 1985/86. "Truth in a Structure", *Proceedings of the Aristotelian Society*, n.s. 86, 135-51.
Husserl, Edmund 1891. *Philosophy of Arithmetic. Psychological and Logical Investigations with Supplementary Texts from 1887-1901*, translated by Dallas Willard, Dordrecht: Kluwer, 2003.
Husserl, Edmund 1900-01. *Logical Investigations*, New York: Humanities Press, 1970.
Husserl, Edmund 1902/03. *Allgemeine Erkenntnistheorie 1902/03*, Elisabeth Schuhmann (ed.), Dordrecht: Kluwer, 2001.
Husserl, Edmund 1905. "Husserl an Brentano, 27. III. 1905", in *Briefwechsel. Die Brentanoschule I*, Dordrecht: Kluwer, 1994.
Husserl, Edmund 1906/07. *Einleitung in die Logik und Erkenntnistheorie*, Husserliana XXIV, The Hague: Martinus Nijhoff, 1984.
Husserl, Edmund 1908/09. *Alte und Neue Logik. Vorlesung 1908-09*, Elisabeth Schuhmann (ed.), Dordrecht: Kluwer, 2003.
Husserl, Edmund 1913. *Introduction to the Logical Investigations*, The Hague: Martinus Nijhoff, 1975.
Husserl, Edmund 1917/18. *Logik und allgemeine Wissenschaftstheorie* Ursula Panzer (ed.), Husserliana XXX, Dordrecht: Kluwer, 1996.
Husserl, Edmund 1929. *Formal and Transcendental Logic*, The Hague: Martinus Nijhoff, 1974.
Husserl, Edmund 1983. *Studien zur Arithmetik und Geometrie. Texte aus dem Nachlass (1886-1901)*, Husserliana XXI, The Hague: Martinus Nijhoff.
Husserl, Edmund 1994. *Early Writings in the Philosophy of Logic and Mathematics*, Dordrecht: Kluwer.
Jourdain, Paul 1991. "The Development of the Theory of Transfinite Numbers", in *Selected Essays on the History of Set Theory and Logic*, Bologna: CLUEB.
Kline, Morris 1972. "The Instillation of Rigor in Analysis", in *Mathematical Thought from Ancient to Modern Times*, Oxford: Oxford University Press, 947-78.
Quine, Willard 1947. "The Problem of Interpreting Modal Logic", *Journal of Symbolic Logic* 12, 2, June, 43-48.
Quine, Willard 1953. "Two Dogmas of Empiricism", in *From a Logical Point of View* (2nd ed. rev.), New York: Harper & Row, 1961, 20-46.
Quine, Willard 1960. *Word and Object*, Cambridge MA: M.I.T. Press.
Quine, Willard, 1969. *Ontological Relativity and Other Essays*, New York: Columbia University Press.
Quine, Willard 1976. *Ways of Paradox*, Cambridge MA: Harvard University Press.
Revue Internationale de Philosophie, Hilbert 1993. 47, 186, 4.

Russell, Bertrand 1956. *Logic and Knowledge*, London: Allen & Unwin.
Schuhmann, Karl 1977. *Husserl-Chronik*, The Hague: Martinus Nijhoff.
Sebestik, Jan 1992. *Logique et mathématique chez Bernard Bolzano*, Paris: Vrin.
Simons, Peter 1992. "Wittgenstein, Schlick and the A Priori", in *Philosophy and Logic in Central Europe from Bolzano to Tarski. Selected Essays*", Dordrecht: Kluwer, 361-76.
Weyl, Hermann 1944. "David Hilbert and His Mathematical Work", *Bulletin of the American Mathematical Society* 50, 612-54.

5

Claire Ortiz Hill

HUSSERL ON AXIOMATIZATION AND ARITHMETIC

Introduction

It is well known that efforts to provide what Gottlob Frege once called "a more detailed analysis of the concepts of arithmetic and a deeper foundation for its theorems" (Frege 1879, 8) played a preeminent role in shaping the course of 20th century philosophy. Frege's and Bertrand Russell's efforts to do so generated Analytic philosophy; Edmund Husserl's dissatisfaction with his efforts to do so by applying Franz Brentano's techniques in the *Philosophy of Arithmetic* (Husserl 1891) set him on the path to writing his groundbreaking *Logical Investigations* (Husserl 1901-01) and eventually to phenomenology.

It is also well known that for a time many philosophers were inclined to interpret the evolution of Husserl's ideas about the foundations of mathematics from a Fregean perspective. This view was especially associated with Dagfinn Føllesdal's Master's thesis "Husserl and Frege: A Contribution to Elucidating the Origins of Phenomenological Philosophy" (Føllesdal 1958) and his article "Husserl's Notion of Noema" (Føllesdal 1969). Since the 1970s, it has been combated by me, Guillermo Ernesto Rosado Haddock, J. N. Mohanty and others. (Rosado Haddock 1973, 1982; Mohanty 1974, 1982; Hill 1979; Hill and Rosado Haddock 2000)

Then, once the belief that Husserl was a Fregean subsided, the temptation arose to see links between Husserl's ideas on the foundations of mathematics and Brouwerian Intuitionism. This interpretation is associated the work of Richard Tieszen (Tieszen 1989) and Mark van Atten (van Atten 2007). It is opposed by Jairo da Silva, Rosado Haddock and myself. We consider that even a cursory examination of Husserl's ideas about axiomatization and numbers shows that Husserl's ideas could not in fact be more different from those of Brouwer. However, to my knowledge, the extent to which Husserl rejected the major tenets Brouwerian Intuitionism has never been systematically demonstrated.

Now, the publication of Husserl's logic courses from 1896 and 1902/03 by the Husserl Archives in 2001 has made available the new material necessary to piece together a satisfactory picture of the development of Husserl's theories about axiomatization and the foundations of mathematics, a subject rich in interesting ramifications and implications for the philosophy of logic and mathematics. So, here I propose to take advantage of the raw material now available to piece together a picture of Husserl's ideas about axiomatization and arithmetic that can be used to lay the basic groundwork needed for exploration of those ramifications and implications.

In the course of my exposition of Husserl's theories about axiomatization and arithmetic, I draw attention to specific areas in which they are at odds with Brouwer's main theses. The considerations thus brought to the fore, I argue in the concluding sections, indicate that Husserl's theories were closer to those of Brouwer's opponent, David Hilbert, and belie claims of kinship with Brouwerian Intuitionism understood, in general, as the view that not only the collection of natural numbers, but all of pure mathematics, develops out of the self-unfolding of "the fundamental intellectual phenomenon of the falling apart of a moment of life into qualitatively different things, of which one is experienced as giving way to the other and yet is retained by an act of memory" that Brouwer called the Primordial Intuition of two-ity and considered to be the basis of the whole of Intuitionism (Brouwer 1912, 80; Brouwer 1929, 45-46; Brouwer 1930, 57). I further suggest that now that we are in possession of Husserl's ideas about axiomatization and arithmetic, we need to give them a try in order to determine whether they are just an ingeniously worked out take on many of the issues in the philosophy of mathematics of his time by the father of phenomenology, or a genuine, viable alternative to theories more familiar to philosophers of logic and mathematics.

Husserl's Initial Opposition to the Axiomatization of Arithmetic

Husserl's position in his 1891 *Philosophy of Arithmetic* was resolutely anti-axiomatic. He attacked those who fall into remote, artificial constructions which, with the intent of building the elementary arithmetic concepts out of their ultimate definitional properties, interpret and change their meaning so much that totally strange, practically and scientifically useless conceptual formations finally result. Especially targeted was Gottlob Frege's ideal of the "founding of arithmetic on a sequence of formal definitions, out of which all the theorems of that science could be deduced purely syllogistically". (Husserl 1891, 123-26)

As soon as one comes to the ultimate, elemental concepts, Husserl reasoned, all defining has to come to an end. All one can then do is to point to the concrete phenomena from or through which the concepts are abstracted and show the nature of the abstraction process. A verbal explanation should place us in the proper state of mind for picking out, in inner or outer intuition, the abstract moments intended and for reproducing in ourselves the mental processes required for the formation of the concept. He said that his analyses had shown with incontestable clarity that the concepts of multiplicity and unity rest directly upon ultimate, elemental psychical data, and so belong among the indefinable concepts. Since the concept of number was so closely joined to them, one could scarcely speak of defining it either (*Ibid.*). All these points are made on the only pages of *Philosophy of Arithmetic* that Husserl ever explicitly retracted. (Husserl 1900-01, 179n.)

Four years earlier, in *On the Concept of Number* (Husserl 1887), Husserl had set out to anchor arithmetical concepts in direct experience by analyzing the actual psychological processes to which he thought the concept of number owed its genesis. To obtain the concept of number of a concrete set of objects, say A, A, and A, he explained, one abstracts from the particular characteristics of the individual contents collected, only considering and retaining each one insofar as it is a something or a one. Regarding their collective combination, one thus obtains the general form of the set belonging to the set in question: one and one, etc. and... and one, to which a number name is assigned. (Husserl 1887, 310, 352-56; Husserl 1891, 85-86)

The enthusiastic espousal of psychologism of *On the Concept of Number* is not found in *Philosophy of Arithmetic*. Husserl later confessed that doubts about basic differences between the concept of number and the concept of collecting, which was all that could be obtained from reflection on acts, had troubled and tormented him from the very beginning and had eventually extended to all categorial concepts and to concepts of objectivities of any sort whatsoever, ultimately to include modern analysis and the theory of manifolds, and simultaneously to mathematical logic and the entire field of logic in general. He did not see how one could reconcile the objectivity of mathematics with psychological foundations for logic. (Husserl 1975, 34-35)

Husserl's *Volte-Face*

In sharp contrast to Brouwer who denounced logic as a source of truth (Brouwer 1948, 90-96), from the mid-1890s on, Husserl defended the view, which he attributed to Frege's teacher Hermann Lotze, that pure arithmetic was basically no more than a branch of logic that had undergone independent development. He bid students not to be "scared" by that thought and to grow used to Lotze's initially strange idea that arithmetic was only a particularly highly developed piece of logic. (Husserl 1896, 241, 271; Husserl 1902/03b, 19, 34; Husserl 1906/07, §15)

Many years later, Husserl would explain in *Formal and Transcendental Logic* that his "war against logical psychologism was meant to serve no other end than the supremely important one of making the specific *province* of analytic logic visible in its purity and ideal particularity, freeing it from the psychologizing confusions and misinterpretations in which it had remained enmeshed from the beginning" (Husserl 1929, §67). He had come to see arithmetic truths as being analytic, as grounded in meanings independently of matters of fact. He had come to believe that the entire overthrowing of psychologism through phenomenology showed that his analyses in *On the Concept of Number* and *Philosophy of Arithmetic* had to be considered a pure a priori analysis of essence (Husserl 1975, 42-43). For him, pure arithmetic, pure mathematics, pure logic were a priori disciplines entirely grounded in conceptual essentialities, where truth was nothing other than the analysis of essences or concepts. Pure mathematics, like pure arithmetic, investigated what is grounded in the essence of number. Pure mathematical laws were laws of essence. (Husserl 1905, 37; Husserl 1906/07 §13c)

He told students that it was to be stressed repeatedly and emphatically that the ideal entities so unpleasant for empiricistic logic, and so consistently disregarded by it, had not been artificially devised either by himself, or by Bolzano, but were given beforehand by the meaning of the universal talk of propositions and truths indispensable in all the sciences. This, he said, was an indubitable fact that had to be the starting point of all logic. (Husserl 1908-09, 45)

All purely mathematical propositions, he taught, express something about the essence of what is mathematical, about the meaning of what belongs to it. Their denial is consequently an absurdity. Denying a proposition of the natural sciences, a proposition about real matters of fact, never means an absurdity, a contradiction in terms. In denying the law of gravity, I cast experience to the wind. I violate the evident, extremely valuable probability that experience has established for the laws. But, I do not say anything "unthinkable", absurd, something that nullifies the meaning of the word as I do when I say that 2×2 is not 4, but 5. (Husserl 1906/07, §13c)

Husserl taught that every judgment either is a truth or cannot be a truth, that every presentation either accorded with a possible experience adequately redeeming it, or was in conflict with the experience, and that grounded in the essence of agreement was the fact that it was incompatible with the conflict, and grounded in the essence of conflict that it was incompatible with agreement. For him, that meant that truth ruled out falsehood and falsehood ruled out truth. And, likewise, existence and non-existence, correctness and incorrectness cancelled one another out in every sense. He believed that that became immediately apparent as soon as one had clarified the essence of existence and truth, of correctness and incorrectness, of Evidenz as consciousness of givenness, of being and not-being in fully redeeming intuition.

At the same time, Husserl contended, one grasps the "ultimate meaning" of the basic logical law of contradiction and of the excluded middle. When we state the law of validity that of any two contradictory propositions one holds and the other does not hold, when we say that for every proposition there is a contradictory one, Husserl explained, then we are continually speaking of the proposition in its ideal unity and not at all about mental experiences of individuals, not even in the most general way. With talk of truth it is always a matter of propositions in their ideal unity, of the meaning of statements, a matter of something identical and atemporal. What lies in the identically-ideal meaning of one's words, what one cannot deny without invalidating the fixed meaning of one's words has nothing at all to do with experience and induction. It has only to do with concepts. (Husserl 1902/03b, 33; Husserl 1906/07 §§13a, 50a; Husserl 1908-09, 45; see also da Silva 2005 and Chapter 14 here.)

In sharp contrast to this, Brouwer saw intuitionistic mathematics as deviating from classical mathematics because the latter uses logic to generate theorems and in particular applies the principle of the excluded middle. He believed that Intuitionism had proven that no mathematical reality corresponds to the affirmation of the principle of the excluded middle and to conclusions derived by means of it. He reasoned that "since logic is based on mathematics—and not vice versa—the use of the Principle of the Excluded Middle is not permissible as part of a mathematical proof". (Brouwer 1921, 23; Brouwer 1929, 51-53; Brouwer 1948, 90; Brouwer 1928a; Brouwer, 1928b)

Analysis of the Concept of Number

According to Husserl, only concepts are purely logical that are not limited to a special field of objects, that not only actually figure and can figure in all the sciences, but are common and necessary to all sciences because they belong to what belongs to the ideal essence of science in general. So, all concepts relating to objects in general in the most universal ways, or to thought forms in general in which objects are brought to theoretically objective unity, are purely logical.

In contrast to Brouwer's idea that it is the fundamental phenomenon of mathematical thinking, the intuition of two-oneness that in its self-unfolding creates not only the numbers one and two, but produces the collection of natural numbers and finally all of pure mathematics (Brouwer 1912, 80; Brouwer 1929, 45-46), the concept of number stood as a paradigm of a purely logical concept in Husserl's sense. Each and every thing, he reasoned, can be counted as one. No science is conceivable in which the number concepts cannot find an application. All purely mathematical concepts like unit, multiplicity, cardinal number, order, ordinal number, and manifold are purely logical because they clearly relate in the most universal way to numbers in general and are only made possible out of the most universal concept of object. However, geometry, mathematical mechanics and all mathematico-natural scientific disciplines do not belong in pure logic since their concepts have real content. (Husserl 1902/03b, 31-43, 49)

In his logic courses, Husserl taught that pure number theory is a science that unfolds the meaning of the idea number and arithmetic in a systematic theory of the laws unfolding the meaning of cardinal number, itself the answer to the question: "How many?" He illustrated what he meant by this "unfolding the meaning of the question 'How many?'" Since each and every thing can be counted as one, to conceive (*konzipieren*) the concept of number, or any arbitrarily defined number, we only need the concept of something in general. One is something in general. Anything can be counted as one and out of the units all cardinal numbers built. One and one or two, two plus one, etc. (Husserl 1902/03b, 31-43, 49). One pear and one man, one apple and one pear, one apple and one apple all have the form "one and one...". This form is the concept "one and one" or "two". Anything and anything, remains unchanged. It is different from "one and one and one, etc." (Husserl 1896, 102)

Eminent thinkers like Lotze, Husserl explained on another occasion, correctly recognized cardinal number as a specific differentiation of the concept multiplicity (*Vielheit*) and multiplicity as the most universal logical concept combining objects in general. This most universal concept of multiplicity splits into a series of different special forms and these are the cardinal numbers. Since an apple is not a multiplicity of apples, an A not a multiplicity of As, then an apple or an A cannot be designated by a cardinal number. The first number in the number series is 2 As. If from 2 As, we use definitions to form the new number 2 As and 1 A and designate them as 3 As, likewise 3 As and 1 A as 4 As, etc., then we obtain a series of the so-called natural numbers, infinite in one direction. The totality of numbers is not exhausted in so doing. For, we can also form the concept of the number of numbers of the natural number series, which can easily be shown not to be identical with any number of that series itself. (*Ibid.*, 102, 241-42)

To questions as to how arithmetic came about and how the foundations of arithmetic were provided, Husserl answered that people had analyzed the arithmetic propositions at first given as they were first entertained by people. They found that certain relations were grounded in the concept of number. For instance, any two numbers are either equal or one is larger or smaller than the other. They further noticed that certain combinations were grounded in the concept of number, first of all addition, then multiplication, raising to a higher power and the inverse of these operations, subtraction, division, extracting roots, logarithms. Given with the elementary combinations were certain simple, directly intelligible laws that careful analysis traced back to a certain minimal number of laws no longer reducible to one another. Since these laws lie in the simple meaning of the concepts founding them, they are a priori. They are not propositions about matters of fact drawn from experience, but propositions about relations of ideas obtained by analysis of the universal concepts by merely digging more deeply into their meaning.

The first law of arithmetic, Husserl taught, is a+b = b+a, or 'For any two numbers there is a sum a+b'. Denying its truth would be a contradiction. Anyone who does so uses 'cardinal number' in some other way, does not know what the words mean or is abandoning its concept. It is a matter of a truth that could not possibly be false, of an analytic statement whose denial is self-contradictory. (Husserl 1902/03b, 33, 35; Husserl 1906/07, §13c)

Mathematicians can set down $a+1 = 1+a$ in a single blow as something unconditionally valid and certain because it is part of the meaning of number (of cardinal number in the original sense) for that to be the case, and it would be tantamount to flying in the face of the meaning of the words 'how many' if one wanted to deny it. Likewise, it is part of the meaning of talk of "cardinal numbers" that each number can be increased by one. To say that a cardinal number, a how many, cannot be increased is tantamount to not knowing what one is talking about. It is tantamount to contravening the meaning, the identical meaning, of talk of cardinal numbers. An elementary formula of this kind already contains infinitely many things in it. It gives not one basic law of arithmetic, but a whole series. Infinitely many laws are simply produced from a primitive number proposition like a+b =b+a by the fact that because of their universality a+b are substitutable. (Husserl 1896, 250; Husserl 1902/03b, 33, 35; Husserl 1906/07, §13c)

Each genuine axiom is a proposition that unfolds the idea of cardinal number from some side or unfolds some of the ideas inseparably connected with the idea of cardinal number. These direct arithmetical laws develop directly in the evidence of certainty and this certitude and evidence carries over to all theses in deductive substantiation. And so these basic laws go on to serve as a basis for systematic deductions in which ever new laws are grounded. (Husserl 1896, 39, 243; Husserl 1902/03b, 33, 35, 39; Husserl 1906/07 §13c)

On the basis of its axioms, the theorems of pure arithmetic are derived by pure deduction following systematic, simple procedures. The field branches out into more and more theories and partial disciplines, ever new problems surface and are finally solved by the expending the greatest mathematical acumen and following the most rigorous methods. So it is that all of arithmetic is grounded in the arithmetical axioms. The unending profusion of wonderful theories it develops is already fixed, enfolded in the axioms, and theoretical-systematic deduction effects the unfolding of them. (*Ibid.*)

Such talk of a priori concepts and ideal entities stands in sharp contrast to Brouwer's mockery of what he called the "foolish superstition" to treat words as labels for "fetish-like" concepts which, along with the relations between them, are assumed to exist independently of the causal attitude of human beings. This was how, he thought, people came to believe that certain relations between concepts derived from axioms with the help of logical principles might be treated as ideal truths (Brouwer 1929, 49-50). Brouwer admitted that from certain relations among mathematical entities assumed as axioms, mathematicians deduce other relations in accordance with fixed laws in the conviction that they are deriving truths from truths by logical reasoning, but he maintained that this "non-mathematical conviction of truth or legitimacy has no exactness whatever, and is nothing but a vague sensation of delight arising from the knowledge of the efficacy of the projection into nature of the relations and laws of reasoning". (Brouwer 1912, 78)

Calculating with Concepts and Propositions

Husserl's search for answers raised by his earliest analyses of the concept of number in "On the Concept of Number" and *Philosophy of Arithmetic* led him beyond the confines of the mathematical realm to a universal theory of formal deductive systems in general. He saw that developments in formalization had unmasked close relationships between number statements and the propositions of logic and that this made it possible to develop a genuine logical calculus for calculating with propositions in the way mathematicians do with numbers, quantities, and the like. (Husserl 1900-01, 41-42; Husserl 1975, 16-17, 35; Husserl 1994, 490-91; Husserl 1929, §§23-27)

By 1896, he was teaching that the formal discipline of propositions in general and of concepts in general was a mathematical discipline that was of precisely the same nature, and used the same methods, as familiar mathematical disciplines like arithmetic and that there was nothing at all extraordinary about the idea of calculating with concepts and propositions. Practically speaking, he enthused, arithmetic actually represents the most marvelous tool devised by the human mind for purposes of deduction. It is the science in which the deductive relations are analyzed most carefully. (Husserl 1896, 250, 271-72)

According to Husserl, only the completely unfounded prejudice that the essence of the mathematical lies in number and quantity could explain the rejection of the new mathematical theory of conceptual and propositional inferences. But, what is mathematical in the procedure of arithmetic, he protested, does not hinge upon our having to do with numbers in them. The essence of the mathematical does not lie in being quantitatively determinable, but in establishing a purely apodictic foundation of the truths of a field from apodictic principles. It is a matter of a rigorously scientific, a priori theory that builds from the bottom up and derives the manifold of possible inferences from the axiomatic foundations a priori in a rigorously deductive way. (Husserl 1896, 272-73; Husserl 1902/03b, 231-32, 239-49; Husserl 1906/07, 434)

To the question as to what it is that characterizes calculating in the field of numbers, Husserl answered that the calculating obviously involves operating with the signs, not with the concepts themselves. To solve a problem, to derive a proposition, we must not think at all about the concepts themselves, but by using procedures defined by set rules, we can be content to link signs to signs, replace combinations of signs by other combinations of signs, etc. At first, the result of the calculation is again purely a combination of signs on paper, but, the interpretation of the results of the inference yields precisely the proposition sought. (Husserl 1896, 247)

In a similar fashion, he taught that every purely formal procedure that proceeds strictly deductively can be presented in algebraic forms and when this occurs scientific thinking first wins a free overview of all possibilities of deductive reasoning and that sovereign mastery of all possible problems and ways of solving them that is the prerequisite for the most exact and most universal solution of problems of the field concerned. (Husserl 1896, 272-73; Husserl 1902/03b, 37, 231, 239-49)

In his courses, Husserl gave the details of his theory of inference in terms which, apart from some differences in notation, are familiar and intelligible to us nowadays (Husserl 1896, 250, 254; Husserl 1902/03b, 239-40). Among his laws and principles figured the identity principle–which he considered to be just another way of expressing the principle of contradiction that was preferable for certain goals of inference–and the law of the excluded middle, A or not A, it is not true that not not A and not A = it is not true that A and not A implies A or not A. He thought of his Principle 6:

> If for every M and for every N, it is always true that M and N implies P, then it is always true that P

as especially important because it grounded the mathematical procedure according to which one could manipulate arbitrary number formulas in the calculation as if they were propositions with specifically given numbers. Every inference yields another formula and not just an individual proposition. (Husserl 1896, 265)

Three Levels of Logic

Developments in mathematics also led Husserl to detect a certain natural order in formal logic and to broaden its domain to include two levels above traditional Aristotelian logic. He considered the detection of these three levels of formal logic to be of the greatest importance for the understanding of logic and philosophy.

According to his theory, the lowest level, traditional Aristotelian logic, makes up but a small area of pure logic. A logic of subject and predicate propositions and states of affairs, it deals with what is stated about objects in general from a possible perspective. The purely logical disciplines rising above that logic of subjects and predicates still deal with individual things, but these objects are no longer empirical or material entities. They are removed from acts, subjects, or empirical persons of actual reality. It is no longer a question of objects as such about which one might predicate something, but of investigating what is valid for higher order objects that are determined in purely formal terms and deal with objects in an indeterminate, general way.

The second level is an expanded, completely developed analytics, where one reasons deductively with concepts and propositions in a purely formal manner because each concept is analytic and each procedure purely logical. Husserl located the basic concepts of mathematics, the theory of cardinal numbers, the theory of ordinals, set theory here. Numbers no longer function as independent entities, but are dependent structures. One manipulates signs for which rules having such and such a form are valid, signs which like chess pieces acquire their meaning in the game through the rules of the game. One may proceed mechanically, and the result will prove accurate and justified.

According to Husserl, the third level of formal logic is that of the science of deductive systems in general, the theory of manifolds. Manifolds are pure forms, which, like molds, remain totally undetermined as to their content and not bound to any possible concrete interpretation, but to which thought must necessarily conform in order to be thought and known in a theoretical manner. Axiom forms define a manifold of anything whatsoever in an indeterminate, general way. A set of axioms of such and such a form that are consistent, independent, and purely logical in that they obey the principle of non-contradiction yields the set of propositions belonging to the theory of such and such a form. After formalization, words are completely empty signs only having the purely formal meaning laid down for them by the axiom forms. A certain something must by definition stand in a certain relationship to something else in the defining manifold.

On the basis of the definition of the manifold, we can derive conclusions, construct proofs, and it is then certain a priori that anything obtained in this way will correspond to something in our theory. Only a form is defined. It exists insofar as it is correctly defined, insofar as the axiom forms are ordered in such a way as to contain no formal contradictions, no violation of analytic principles. But whether axioms as truths have existence in any objective real or ideal spheres corresponding to the prescribed form is left open. The theory of manifolds, or science of theory forms, is a field of free, creative investigation made possible once form is emancipated from content. Once it is realized that deductions and sequences of deductions continue to be meaningful and remain valid when another meaning is assigned to the symbols, we are free to reason completely on the level of pure forms where we can vary the systems in different ways. (Husserl 1906-07, §§18-19; Husserl 1917/18, Ch. 11)

Manifolds and Imaginary Numbers

In *Logical Investigations*, Husserl called his theory of complete manifolds the key to the only possible solution to how in the realm of numbers impossible, non-existent, meaningless concepts might be dealt with as real ones (Husserl 1900-01, *Prolegomena* §70). In *Ideas I*, he wrote that his chief purpose in developing his theory of manifolds had been to find a theoretical solution to the problem of imaginary quantities (Husserl 1913, §72 and note).

Husserl saw how questions regarding imaginary numbers come up in mathematical contexts in which formalization yields constructions that, arithmetically speaking, are nonsense, but can be used in calculations. When formal reasoning is carried out mechanically as if these symbols have meaning, if the ordinary rules are observed, and the results do not contain any imaginary components, these symbols might be legitimately used. And this could be empirically verified. (Husserl 1891, 411-13; Husserl 1929, §31; Schuhmann and Schuhmann 2001)

In a letter to Carl Stumpf in the early 1890s, Husserl explained how, in trying to understand how operating with contradictory concepts could lead to correct theorems, he had found that for imaginary numbers like $\sqrt{2}$ and $\sqrt{-1}$, it was not a matter of the possibility or impossibility of concepts. Through the calculation itself and its rules, as defined for those fictive numbers, the impossible fell away, and a genuine equation remained. One could calculate again with the same signs, but referring to valid concepts, and the result was again correct. Even if one mistakenly imagined that what was contradictory existed, or held the most absurd theories about the content of the corresponding concepts of number, the calculation remained correct if it followed the rules. He concluded that this must be a result of the signs and their rules (Husserl 1994, 13, 15-16). The fact that one can generalize, produce variations of formal arithmetic that lead outside the quantitative domain without essentially altering formal arithmetic's theoretical nature and calculational methods brought Husserl to realize that there was more to the mathematical or formal sciences, or the mathematical method of calculation than could be captured in purely quantitative analyses. (Husserl 1900-01, 41-43; Husserl 1975, 35)

Understanding the nature of theory forms, he explained in several texts, shows how reference to impossible objects can be justified. According to his theory of manifolds, one could operate freely within a manifold with imaginary concepts and be sure that what one deduced was correct when the axiomatic system completely and unequivocally determined the body of all the configurations possible in a domain by a purely analytical procedure. It was the completeness of the axiomatic system that gave one the right to operate in that free way. A domain was complete when each grammatically constructed proposition exclusively using the language of the domain was determined from the outset to be true or false in virtue of the axioms, i.e., necessarily followed from the axioms or did not. In that case, calculating with expressions without reference could never lead to contradictions. Complete manifolds have the "distinctive feature that a finite number of concepts and propositions—to be drawn as occasion requires from the essential nature of the domain under consideration—determines completely and unambiguously on the lines of pure logical necessity the totality of all possible formations in the domain, so that in principle, therefore, nothing further remains open within it." In such complete manifolds, he stressed, "the concepts true and formal implication of the axioms are equivalent" (Husserl 1913, §§71-72; Husserl 1900-01, Prolegomena, §70; Husserl 1906/07, §19; Husserl 1917/18, §56; Husserl 1929, §31; Husserl 1891, 439). Completeness is studied in depth in Chapters 6 and 7.

Husserl pointed out that there may be two valid discipline forms that stand in relation to one another in such a way that the axiom system of one may be a formal limitation of that of the other. It is then clear that everything deducible in the narrower axiom system is included in what is deducible in the expanded system, he explained. In the arithmetic of cardinal numbers, Husserl explained, there are no negative numbers, for the meaning of the axioms is so restrictive as to make subtracting 4 from 3 nonsense. Fractions are meaningless there. So are irrational numbers, $\sqrt{-1}$, and so on. Yet in practice, all the calculations of the arithmetic of cardinal numbers can be carried out as if the rules governing the operations are unrestrictedly valid and meaningful. One can disregard the limitations imposed in a narrower domain of deduction and act as if the axiom system were a more extended one (Husserl 1917/18, §56). We cannot arbitrarily expand the concept of cardinal number, Husserl reasoned. But we can abandon it and define a new, pure formal concept of positive whole number with the formal system of definitions and operations valid for cardinal numbers. And, as set out in our definition, this formal concept of positive numbers can be expanded by new definitions while remaining free of contradiction. Fractions do not acquire any genuine meaning through our holding onto the concept of cardinal number and assuming that units are divisible, he theorized, but rather through our abandonment of the concept of cardinal number and our reliance on a new concept, that of divisible quantities. That leads to a system that partially coincides with that of cardinal numbers, but part of which is larger, —meaning that it includes additional basic elements and axioms. And so in this way, with each new quantity, one also changes arithmetics. The different arithmetics do not have parts in common. They have totally different domains, but an analogous structure. They have forms of operation that are in part alike, but different concepts of operation. (Husserl 1891, 435-36)

Husserl concluded that formal constraints banning meaningless expressions, meaningless imaginary concepts, reference to non-existent and impossible objects restrict us in our theoretical, deductive work, but that resorting to the infinity of pure forms and transformations of forms frees us from such conditions and explains why having used imaginaries, what is meaningless, must lead, not to meaningless, but to true results. (Husserl 1917/18, §57) Imaginaries are studied in depth in Chapter 3.

Mathematics and Phenomenology

Husserl wanted to hammer into people's minds a sense of the proper relationship between phenomenology and mathematics. He stressed that all fields of theoretical knowledge are particular instances of manifolds, but not all sciences are theoretical disciplines like mathematical physics, pure geometry, or pure arithmetic whose systemic principles are purely analytical. Theoretical disciplines have a systemic form that belongs to formal logic itself, that must be constructed a priori within formal logic itself and within its supreme discipline the theory of manifolds as part of the overall system of forms of deductive systems that are possible a priori. However, sciences like psychology, history, the critique of reason and, notably, phenomenology require one go beyond the analytico-logical model. When they are formalized and one asks what it is that binds the propositional forms into a single system form, one finds oneself facing nothing more than the empty general truth that there is an infinite number of propositions connected in objective ways that are compatible with one another in that they do not contradict each other analytically. (Husserl 1929, §35a; Husserl 1917/18, §54)

We have the natural sciences of physical and mental nature, the mathematical sciences, logic, including formal logic, the sciences of value, ethics. None of that is phenomenology, Husserl underscored. Transcendental phenomenology has no dealings with a priori ontology, none with formal logic and formal mathematics, none with geometry as a priori theory of space, none with a priori real ontology of any kind (thing, change etc.). Transcendental phenomenology is phenomenology of the constituting consciousness, and consequently not a single objective axiom, meaning one relating to objects that are not consciousness, belongs in it, no a priori proposition as truth for objects, as something belonging in the objective science of these objects, or of objects in general in formal universality. The axioms of geometry do not belong in phenomenology, because phenomenology is not a theory of the essences of shapes, of spatial objects. Essence-propositions about objects do not belong in the phenomenology of knowledge, insofar are they are objective truths and as truths have their place in a truth-system in general. (Husserl 1906/07, 411, 422-23)

The special interest of transcendental phenomenology does not lie in the theoretical concepts and laws to which the sciences are subject. Epistemological interest, transcendental interest, does not aim at objective being and laying down truths for objective being, consequently, not at objective science. What is objective belongs precisely to objective science, and what objective science still lacks for completion is its affair to obtain and its alone. The interest of transcendental phenomenology aims rather at consciousness as consciousness of objects. (Husserl 1906/07, 425)

What Numbers Could Not Be for Husserl

There is no longer any need to prove that Husserl was not a Fregean. Husserl's theory of arithmetic is not grounded in the unworkable theory of identity and reference that forced Frege to introduce the extensions of concepts and axiom of extensionality that he concluded led to Russell's paradox (Hill 1997). Husserl already spurned extensions in *Philosophy of Arithmetic*. In *Formal and Transcendental Logic*, he qualified extensional logic as naive, risky, and doubtful and complained that it had been the source of many a contradiction requiring every kind of artful device to make it safe for use in mathematical reasoning. He condemned the work of extensionalist logicians, as fundamentally misguided and unclear (Husserl 1929, §§23b, 24, 26c; Hill and Rosado Haddock 2000). If, as Quine told us, the notion of essence is the forerunner of the modern notion of intension or meaning, and meaning is what essence becomes when it is divorced from the object of reference and wedded to the word (Quine 1953, 22), then it is clear from the above that Husserl's logic was resolutely intensional.

So Husserl was not a Fregean, but the theory that he was Brouwerian still appeals to some and remains to be countered. In the course of this exposition of Husserl's ideas about axiomatization and arithmetic, I have pointed to some specific areas in which Husserl's and Brouwer's theories on the foundations of mathematics diverge. I wish now to reinforce what I have said by adding the following reflections.

Husserl's theory of the derivation of arithmetic from the unfolding of the concept "How Many?" could not in fact be more different from Brouwer's theory of self-unfolding of mathematics from the mathematical primordial intuition of two-ity. According to Brouwer, mathematics, science and language are the main functions of human activity by which human beings dominate nature and maintain order within it. These three functions originate in three forms of action of the individual human being's will to live: mathematical attention; mathematical abstraction; the use of sounds to impose his or her will on others. Mathematical attention comes into being in two phases. Time awareness, the first, is the fundamental intellectual phenomenon of the separating of a life moment into two qualitatively different parts that unfolds itself to create a time sequence of arbitrary multiplicity by giving birth to temporal two-ity, which can in turn be taken as a element of a new two-ity to create temporal three-ity, and so on. Mathematical attention receives its justification only by the "mathematical act", when "causal attention", the second phase of mathematical attention, enables people to force into being, "indirectly and by cool calculation", a particular event known as the aim that appears later in the sequence of phenomena. For Brouwer, the

> causal coherence of the world is the outward-acting force of human thought, serving a dark function of will, making the world more or less defenseless like the snake that renders its prey powerless through its hypnotic stare of the inkfish through its darkening spray. (Brouwer 1929, 46)

In higher levels of civilization, Brouwer believed, mathematical abstraction enters in to divest two-ity of its material content, whereupon it becomes the empty form that is the common substrate of all two-ities that forms the Primordial Intuition of Mathematics that in its self-unfolding produces,

> not only the numbers one and two, but also all finite ordinal numbers, inasmuch as one of the elements of the two-oneness may be thought of as a new two oneness, which process may be repeated indefinitely; this gives rise still further to the smallest infinite ordinal number ω.... gives rise immediately to the intuition of the linear continuum.... (Brouwer 1912, 80)

...and finally all of pure mathematics. (Brouwer 1929, 45-46; Brouwer 1912, 80; Brouwer 1952, 1200)

In maintaining that temporal two-ity born from time awareness is the basal intuition of all of mathematics Brouwer saw himself as holding resolutely to the Kant's apriority of time. He described intuitionistic mathematics as "an essentially languageless activity of the mind having its origin in the perception of a *move of time*" (Brouwer 1912, 80; Brouwer 1929, 45-46; Brouwer 1952, 1200). In contrast, Husserl taught that numbers could not concern what happens in or to real temporal matters of fact that we call mental experiences of experiencing individuals. He stated unequivocally that Kant had brought pure arithmetic into an entirely inadmissible relationship to time (Husserl 1906/07 §§11, 13b, 23). Theories of number based on intuitions of time were already spurned in *On the Concept of Number* and the *Philosophy of Arithmetic*. (Husserl 1887, 320-29; Husserl 1891, 22-35)

For Husserl, mathematics could not originate in the consciousness or possibly be developed from any intuition whatsoever. He taught that the laws of arithmetic just unfold what is found in the concept of number. They make no pronouncements about acts of counting, causal relationships, experiences of number, but are just about numbers (Husserl 1902/03b, 32). He insisted that the presenting or thinking of a number proposition must be distinguished the number proposition itself. Thinking $2 \times 2 = 4$

is a phenomenon of my consciousness, but it is there when one turns to other objects. If one thinks again that 2 × 2 is 4, then that mental act is new. It is not the same, but what is thought is the same. Countless acts can objectively underlie the same thing, and in this case this identical thing is 2 × 2 = 4. (Husserl 1896, 19)

Brouwer contended that for Intuitionism, mathematical exactness exists in the intellect (Brouwer 1912, 78). In contrast, Husserl insisted that mathematical truth holds whether anyone has reason or not to believe it, or does not believe it, whether anyone sees it or does not see it (Husserl 1906/07 §11). In *Formal and Transcendental Logic*, he said that the problem guiding him originally was in isolating and determining the meaning of a pure analytic logic of non-contradiction was that the evidence of the truths of formal mathematics and formal logic is of an entirely different order than that of other a priori truths in that the former do not involve any intuition of objects or states of affairs whatsoever. (Husserl 1929, Introduction and §§7-8)

Brouwer hoped to make it clear that "intuitionistic mathematics is inner architecture, and that research in foundations of mathematics is inner inquiry" (Brouwer 1948, 96). In contrast, Husserl's formal logic is a blueprint for limning the true and ultimate structure of reality by engaging in pure a priori analyses of essence that know no acts, subjects, or empirical persons, or objects belonging to actual reality. He taught that there was not to be any radical analysis of the psychological origins of the fundamental concepts of mathematics *per se*. He stressed that pure mathematics as pure arithmetic is not concerned with souls (*Seele*). (Husserl 1906/07, §§13c, 18; Husserl 1917/18, ch. 11)

Conclusion

So, if Husserl was neither Fregean nor Brouwerian, or really even a phenomenologist when it came to mathematics, what was he? When we find Husserl teaching that calculating in the field of numbers obviously involves operating with signs and not with the concepts themselves, that to solve a problem, to derive a proposition, one must not think at all about the concepts themselves, but by using procedures defined by set rules, link signs to signs, replace combinations of signs by other combinations of signs, etc. (Husserl 1896, 247), this automatically suggests kinship with the ideas of Husserl's colleague at the University of Göttingen, David Hilbert.

Husserl's teachings about axiomatization, arithmetic, completeness and consistency, the foundations of mathematics also display kinship with Hilbert's ideas, kinship that Husserl himself acknowledged in *Ideas I* §72, *Formal and Transcendental Logic* §§28-36 and *Crisis* §9f and note, where he also made it clear that he considered the fact of this kinship to be significant. In §31 of *Formal and Transcendental Logic*, he even went so far as to say that the close study of his analyses would reveal that the underlying, though inexplicit, reasons which had led Hilbert to attempt to complete a system of axioms by adding a separate axiom of completeness were much the same as those which had played a determinant role in Husserl's own independent formulation of his concept of completeness. In those texts, Husserl explicitly refers back to his theory of complete manifolds in the *Prolegomena* §§69-70 (Husserl 1900-01) and to the then unpublished material from his Göttingen period now available in appendices to his *Philosophy of Arithmetic*.

However, caution also needs to be exercised in uncovering parallels in the ideas of original thinkers. Kinship can be superficial and it is not influence. In this case, it is important to remember that Husserl developed his ideas independently of Hilbert. Husserl's interest in axiomatization, completeness and formalist foundations for mathematics is traceable back to his early years in Halle, before he and Hilbert were together in Göttingen. They originally derived from problems regarding imaginary numbers which first came up while he was trying to complete *Philosophy of Arithmetic*. His 1896 teachings about the axiomatization of arithmetic antedated Hilbert's call to axiomatize arithmetic, which first went out in 1899 in "On the Concept of Number" (Hilbert 1900), Hilbert's first essay on the foundations of arithmetic.

Although Husserl acknowledged kinship as concerns completeness, he said that he developed his concept of completeness independently of Hilbert's axiom of completeness. His earliest ideas on completeness were tied in, not only with his inquiries into the logical foundations of the real number system, but also with a more specifically philosophical quest to clarify the sense of the analytic a priori and develop a pure analytic logic free of any taint of psychologism. Moreover, Husserl criticized Hilbert's appeal to the axiom of completeness in "On the Concept of Number" (*Ibid.*). Husserl said that that kind of completeness can be of no use whatsoever, because it is not legitimate completeness, not something specifically characteristic of axiom systems, because any axiom system can be made quasi complete by appealing to an axiom of that kind. There are, as Jairo da Silva has pointed out, there are many senses of completeness. (da Silva 2000a and Chapter 7 of this book)

When Husserl was appointed to the University of Göttingen in 1900, he was though warmly welcomed into Hilbert's circle (Husserl 1983, XII). As the documents included in Appendix II here show, Husserl's colleagues in the philosophy department at the University of Göttingen did not consider him to "a desirable addition to the faculty". Husserl did, though, have an ardent supporter in the person of Hilbert, who complained that people had not seen, or had not wanted to see, how important it was to support Husserl's efforts. In Hilbert's opinion, Husserl was viewed in professional circles as one of the most prominent and creatively most active scholars in the field of systematic, purely theoretical philosophy.

Hilbert portrayed Husserl as an exception, as someone not tainted by relativism, someone who believed in the possibility of philosophical science and labored to make it a reality. He called the *Logical Investigations* epoch-making. Hilbert considered it no accident that Husserl had come to the mathematical environment cultivated there.

However, nowhere in his published writings on philosophy does Hilbert ever acknowledge having been influenced by Husserl or having exercised influence upon him. Rather, in them, Hilbert affirms the abiding significance of what he called the most general fundamental idea of the Kantian theory of knowledge, namely the philosophical problem of establishing the intuitive *a priori* attitude and, with that, of investigating the prerequisites for the possibility of any conceptual knowledge and at the same time of any experience. Hilbert says that this is essentially what had happened in his own investigations into the principles of mathematics. (Hilbert 1930, 383; Hilbert 1931, 266-67)

Be that as it may, once Husserl's theory of the axiomatization of arithmetic has been pieced together and the relationship of his ideas to Frege's, Brouwer's, and Hilbert's theories on the foundations of mathematics has been clarified, the really important question to be answered is whether his theory is really tenable and viable, whether it works, or whether it is not ultimately just an ingeniously worked out take on many of the issues in the philosophy of mathematics of his time by the father of phenomenology. Now that we have the material we need to piece together Husserl's theory, we need to give it a try. It needs to be tested to see whether it is tenable. That is the next step that needs to be taken.

I am personally of the conviction that such testing will unearth additional arguments to prove that Husserl had a deeper understanding of the issues that went into the investigations into the foundations of mathematics that generated analytic philosophy than analytic philosophers themselves have ever had. As the student and assistant of Karl Weierstrass, the longtime friend and colleague of both Georg Cantor and David Hilbert, Husserl was on the ground floor when it came to the grounding of mathematics. When he was teaching his logic courses, he was already in a position to take into account the shortcomings of both Cantor's and Frege's efforts. He was already perfectly lucid about those of the latter at the time he published *Philosophy of Arithmetic* (Hill 1991; Hill and Rosado Haddock 2000). Husserl's theory is grounded in an analytic derivation of number from the concept of "How Many" and not in the deeply flawed theories about identity and reference with which the mainstream philosophical theories of the foundations of mathematics have struggled for over a hundred years. Husserl's theory makes no appeal to the axiom of extensionality that still blights the axiomatization of set theory to which, following Cantor, Frege, Russell and Whitehead, Zermelo-Fraenkel, Gödel, Quine, philosophers still appeal to ground mathematics.

References

Benacerraf, Paul and H. Putnam (eds.) 1983. *Philosophy of Mathematics, Selected Readings*, Cambridge UK: Cambridge University Press, 2nd ed. rev., 1964.
Brouwer, L. E. J. 1912. "Intuitionism and formalism", in Benacerraf and Putnam (eds.), 77-89.
Brouwer, L. E. J. 1921. "Intuitionist Set Theory", in Mancosu (ed.), 23-27.
Brouwer, L. E. J. 1928a. "On the Significance of the Principle of Excluded Middle in Mathematics, Especially in Function Theory", in van Heijenoort, 335-45.
Brouwer, L. E. J. 1928b. "Intuitionist Reflections on Formalism", in Mancosu (ed.), 40-44; in van Heijenoort, 490-92.
Brouwer, L. E. J. 1929. "Mathematics, Science, and Language", in Mancosu (ed.), 45-53.
Brouwer, L. E. J. 1930. "The Structure of the Continuum", in Mancosu (ed.), 54-63.
Brouwer, L. E. J. 1948. "Consciousness, philosophy, and mathematics", in Benacerraf and Putnam (eds.), 90-96.
Brouwer, L. E. J. 1952. "Historical background, principles, and methods of intuition", in Ewald (ed.), 1197-1207.
da Silva, Jairo 2000a. "Husserl's Two Notions of Completeness, Husserl and Hilbert on Completeness and Imaginary Elements in Mathe-matics", *Synthese* 125, 417-38.
da Silva, Jairo 2000b. "The Many Senses of Completeness", *Manuscrito*, 23, 2, 41-60, anthologized here, Chapter 7.
da Silva, Jairo 2005. "Husserl on the Principle of the Excluded Middle", in *Husserl and the Logic of Experience*, G. Banham (ed.), Palgrave Macmillan, 51-81.
Ewald, William (ed.) 1996. *From Kant to Hilbert. A Sourcebook in the Foundations of Mathematics*, Vol. II, Oxford: Clarendon Press.
Føllesdal, Dagfinn 1958. "Husserl and Frege: A Contribution to Elucidating the Origins of Phenomenological Philosophy", in Haaparanta (ed.), 3-47, translation by Claire Ortiz Hill of his Norwegian Master's Thesis: *Husserl und Frege: Ein Beitrag zur Beleuchtung der Enstehung des phänomenologische Philosophie*. Oslo: Ascheloug.
Føllesdal, Dagfinn 1969. "Husserl's Notion of Noema". *Journal of Philosophy* 66, 680-87.
Føllesdal, Dagfinn 1988. "Husserl: fifty years later... the Noema twenty years later". *Proceedings of the Eighteenth World Congress of Philosophy* held in Brighton, England in August 1988.
Frege, Gottlob 1879. "*Begriffsschrift*, a formula language, modeled upon that of arithmetic, for pure thought", in *From Frege to Gödel, A Source Book in Mathematical Logic, 1879-1931*, Jean van Heijenoort (ed.), Cambridge MA: Harvard University Press, 1-82.

Hilbert, David 1900. "Über den Zahlbegriff", *Jahresbericht der Deutschen Mathematiker Vereinigung* 8, 180-84, translated as "On the Concept of Number", in Ewald (ed.), 1089-95.
Hilbert, David 1930. "Logic and the Knowledge of Nature", in Ewald (ed.), 1157-65.
Hilbert, David 1931. "The Grounding of Elementary Number Theory", in Mancosu (ed.), 266-67. Also in Ewald (ed.), 1148-56.
Hill, Claire Ortiz 1979. *La logique des expressions intentionnelles*, Mémoire de Maîtrise, Université de Paris-Sorbonne, now published online.
Hill, Claire Ortiz 1991. *Word and Object in Husserl, Frege and Russell, the Roots of Twentieth Century Philosophy*, Athens OH: Ohio University Press.
Hill, Claire Ortiz 1995. "Husserl and Hilbert on Completeness", in *From Dedekind to Gödel, Essays on the Development of the Foundations of Mathematics*, Jaakko Hintikka (ed.), Dordrecht: Kluwer, 143-63, Chapter 10 of Hill and Rosado Haddock 2000.
Hill, Claire Ortiz 1997. *Rethinking Identity and Metaphysics, On the Foundations of Analytic Philosophy*, New Haven CT: Yale University Press.
Hill, Claire Ortiz 2002a. "On Husserl's Mathematical Apprenticeship and Philosophy of Mathematics", in *Phenomenology World Wide*, Anna-Teresa Tymieniecka (ed.), Dordrecht: Kluwer, 76-92, anthologized here, Chapter 1.
Hill, Claire Ortiz 2002b. "Tackling Three of Frege's Problems: Edmund Husserl on Sets and Manifolds", *Axiomathes* 13, 79-104, anthologized here, Chapter 10.
Hill, Claire Ortiz and G. E. Rosado Haddock 2000. *Husserl or Frege? Meaning, Objectivity, and Mathematics*, La Salle: Open Court.
Hintikka, Jaakko 1997. "Hilbert Vindicated", *Synthese* 110, no. 1: 15-36.
Husserl, Edmund 1887. *On the Concept of Number*, published in Husserl 1891 and in the English translation.
Husserl, Edmund 1891. *Philosophie der Arithmetik, mit ergänzenden Texten,* Husserliana XII, The Hague: Martinus Nijhoff, 1970, in English, *Philosophy of Arithmetic, Psychological and Logical Investigations with Supplementary Texts from 1887-1901*, translated by Dallas Willard, Dordrecht: Kluwer, 2003.
Husserl, Edmund 1896. *Logik, Vorlesung 1896*, Dordrecht: Kluwer, 2001.
Husserl, Edmund 1900-01. *Logical Investigations*, translated by J. N. Findlay, London: Routledge & Kegan Paul, 1970.
Husserl, Edmund 1901a. "Double Lecture: On the Transition through the Impossible (Imaginary') and the Completeness of an Axiom System", in Husserl 1891, 409-57.
Husserl, Edmund 1901b. "Essay IV, The Domain of an Axiom System/Axiom System-Operation System", in Husserl 1891, 475-92.
Husserl, Edmund 1902/03a. *Allgemeine Erkennthistheorie, Vorlesung 1902/03*, Elisabeth Schuhmann (ed.), Dordrecht: Kluwer, 2001.

Husserl, Edmund 1902/03b. *Logik, Vorlesung 1902/03*, Elisabeth Schuhmann (ed.), Dordrecht: Kluwer, 2001.
Husserl, Edmund 1905. "Husserl an Brentano, 27. III. 1905", *Briefwechsel, Die Brentanoschule I*, Dordrecht: Kluwer, 1994.
Husserl, Edmund 1906/07. *Einleitung in die Logik und Erkenntnis-theorie, Vorlesungen 1906/07*, Ullrich Melle (ed.) Husserliana XXIV, Dordrecht: Martinus Nijhoff, 1984, translation by Claire Ortiz Hill, *Introduction to Logic and Theory of Knowledge*, Dordrecht: Springer, 2008.
Husserl, Edmund 1908/09. *Alte und neue Logik, Vorlesung 1908/09*, Elisabeth Schuhmann (ed.), Dordrecht: Kluwer, 2003.
Husserl, Edmund 1913. *Ideas, General Introduction to Pure Phenomenology*, translated by Boyce Gibson, New York: Collier Books, 1962. Translation of his *Ideen zu einer reinen Phänomenologie und einer phänomenologischen Philosophie I*, Husserliana III, The Hague: Martinus Nijhoff, rev. ed., 1976 (1913).
Husserl, Edmund 1917/18. *Logik und allgemeine Wissenschaftstheorie, Vorlesungen 1917/18, mit ergänzenden Texten aus der ersten Fassung 1910/11*, Husserliana XXX, Ursula Panzer (ed.), Dordrecht: Kluwer, 1996.
Husserl, Edmund 1929. *Formal and Transcendental Logic*, The Hague: Martinus Nijhoff, 1978.
Husserl, Edmund 1936. *The Crisis of European Sciences and Trans-cendental Phenomenology*, Evanston IL: Northwestern University Press, 1970.
Husserl, Edmund 1975. *Introduction to the Logical Investigations, A Draft of a Preface to the Logical Investigations*, Eugen Fink (ed.), P. Bossert and C. Peters (eds.), The Hague: Martinus Nijhoff.
Husserl, Edmund 1983. *Studien zur Arithmetik und Geometrie. Texte aus dem Nachlass 1886-1901*, Husserliana XXI, Ingeborg Strohmeyer (ed.), The Hague: Martinus Nijhoff.
Husserl, Edmund 1994. *Early Writings in the Philosophy of Logic and Mathematics*, translated by Dallas Willard, Dordrecht: Kluwer.
Husserl, Edmund, Ms A 1 35. Manuscript on set theory available in the Husserl Archives in Leuven, Cologne and Paris.
Mancosu, Paolo (ed.) 1998. *From Brouwer to Hilbert, the Debate on the Foundations of Mathematics in the 1920s*, New York: Oxford University Press.
Mohanty, J. N. 1974. "Husserl and Frege: A New Look at Their Relationship", *Research in Phenomenology* 4, 1974, 51-62.
Mohanty, J. N. 1982, *Husserl and Frege*. Bloomington IN: Indiana University Press.
Quine, Willard 1953; "Two Dogmas of Empiricism", *From a Logical Point of View* (2nd rev.), New York: Harper & Row, 1961, 20-46.
Rosado Haddock, Guillermo Ernesto 1973. *Edmund Husserls Philosophie der Logik und Mathematik im Lichte der gegenwärtigen Logik und*

Grundlagenforschung. Bonn: Dissertation, Rheinische Friedrich-Wilhelms-Universität.

Rosado Haddock, Guillermo 1982. "Remarks on Sense and Reference in Frege and Husserl", *Kant-Studien* Vol. 73, 4, 425-39 and chapter 2 of Hill and Rosado Haddock 2000.

Schuhmann Elisabeth and Karl Schuhmann, 2001, "Husserl Manuskripte zu seinem Göttinger Doppelvortrag von 1901", *Husserl Studies* 17: 87-123.

Tieszen, Richard 1989. *Mathematical Intuition, Phenomenology and Mathematical Knowledge*, Dordrecht: Kluwer.

van Atten, Mark 2007. *Brouwer Meets Husserl. On the Phenomenology of Choice Sequences*, Dordrecht: Springer.

6

Jairo José da Silva

HUSSERL AND HILBERT ON COMPLETENESS AND IMAGINARY ELEMENTS IN MATHEMATICS

Introduction

In 1901, Edmund Husserl addressed the Mathematical Society in Göttingen in order to present his views on a problem that had occupied him, in one form or the other, since the beginning of 1890:[1] the problem of imaginary entities in mathematics. In those lectures, this problem is twofold: 1) When is an object "imaginary" from the perspective of a formal axiomatic system (the *ontological* problem)? 2) How is one to justify the use of "imaginary" elements in mathematics (the *epistemological* problem)? In order to deal with these questions, Husserl appealed to the notion of *formal* domain of objects determined by a formal axiomatic system, and *two* notions of completeness of a formal system of axioms, which he called *relative* and *absolute definiteness*.

These concepts offered him the key to the solution of both the ontological and the epistemological problems.[2] The notion of absolute definiteness is identical with David Hilbert's notion of *deductive* or *syntactic completeness*, whereas the notion of relative definiteness is a particular case of it, being nothing more than completeness relative to a

---

[1] This is the date Husserl himself gave in a note to *Ideas I* §72. Husserl was by then involved with the problem of imaginary entities in mathematics, which was to be dealt with in the sequel to *Philosophy of Arithmetic* (Husserl 1891b). In his review of Schröder's 1890 *Vorlesungen über die Algebra der Logik* (Husserl 1891a), Husserl presented a sketch for a solution of the problem of imaginary entities–in this case the null class in Schröder's calculus of classes–, which is essentially identical to the solution presented to the problem of imaginaries in arithmetic in the Göttingen talks of 1901 (Husserl 1901a). Also, in his letter to Frege dated July 18, 1891 Husserl hints at a formalistic solution of the problem of imaginary numbers in arithmetic. (Frege 1980, 65)

[2] Besides also providing an equivalent of the notion of truth in formal contexts: in absolutely definite systems the notion of derivability is a formal equivalent of the notion of truth.

particular set of expressions.[3] The notion of relative definiteness, which depends on the notion of the domain of formal objects determined by a formal system of axioms, is nonetheless the central notion for the solution of both the ontological and the epistemological problems.

In this paper, I intend to show why Husserl developed these notions and to explain their relevance for the problem of imaginary elements in mathematics. I shall also show how Husserl's notions of definiteness are related to Hilbert's *axiom* of completeness–proposed in *On the Concept of Number* (Hilbert 1900)–via the notion of a formal domain of objects, thus offering a possible explanation for the "close relation" that Husserl considered evident between them and the notion of completeness involved in Hilbert's axiom of completeness.[4] Let us go into the details now.

Consider the set N of natural numbers, that is, the smallest set containing 0 and closed with respect to the successor operation (S). We can define the operation of addition ($+$) of natural numbers recursively as follows: for any number n: $n+0=n$; $n+(Sm)=S(n+m)$.

Suppose now that we "enlarge" this domain by introducing "imaginary" elements in the following way: for each number n, let $-n$ be the only "imaginary" element such that $n+(-n)=0$ (*). These elements are "imaginary" for they are non-existent from the perspective of N; the equation (*) is, after all, without solution in N. The problem we face is the following: if we merely add the negative numbers to N, the equation (*) is without meaning, for, as defined, addition only makes sense in N; if we restrict, as we must, addition to its domain, then (*) is impossible. We could, of course, formally extend the operation of addition, but in this case it no longer means what it meant before. The meaning of the concept of number would also change, for negative "numbers" are not numbers in the same sense natural numbers are, i.e. possible answers to the question "How many?"

After the publication of *Philosophy of Arithmetic* (Husserl 1891b), in which numbers, exclusively in the guise of *natural numbers*, are introduced as instances of the concept of quantity, Husserl had to deal with more general concepts of number, such as negative, rational, irrational and complex numbers, which are "imaginary" from the perspective of the "natural" numbers, so as to give them solid philosophical foundations, as Husserl understood it, in the never published second volume of *Philosophy of Arithmetic*. The problems raised above were among those Husserl had to deal with.

In the course of his ensuing investigations, Husserl discovered, independently of Hilbert as is clear from Husserl's writings, two notions of definiteness for a system of axioms that are both very close to Hilbert's

---

[3] I shall use the terms 'definite' and 'definiteness' when referring to Husserl's notions and 'complete' and 'completeness' when referring to Hilbert's notions.
[4] Husserl expressed his belief in the self-evidence of this correlation in a footnote to §72 of *Ideas I*. (Husserl 1913)

concept of completeness. I shall show here that one of Husserl's notions coincides with Hilbert's (as Husserl himself realized), and that both of Husserl's notions involve an idea of maximality, of the *formal multiplicity* (or *manifold*),[5] the domain of formal objects determined by a formal axiomatic system), in the case of one of the notions, or the system itself, in the case of the other. That explains the connection that Husserl saw between his notions of definiteness and the concept of completeness involved in Hilbert's *axiom* of completeness.

Husserl's investigations on this topic influenced the whole of his later philosophy. In the particular case of his philosophy of mathematics, Husserl's characterization, Bourbakian *avant la lettre*, of mathematics as the science of formal systems[6] was probably born in the course of those investigations. In general, it is arguable that they played a more important role in his turning away from *psychologism* than Frege's criticisms (Frege 1894) of Husserl's first book, *Philosophy of Arithmetic*. As we shall see, in order to justify the epistemological relevance of "imaginary" numbers, Husserl chose to embrace a variant of formalism in which these numbers are seen as mere practical tools. Independently of the problem of imaginaries, which appear when different conceptions of number are mingled together, the more general concepts of number, as Husserl realized, could not be given the "genetic" treatment he favored in *Philosophy of Arithmetic*. The fact that Husserl had already given up such an approach to the foundations of mathematics well before Frege's review of *Philosophy of Arithmetic* indicates, I believe, that Frege did not have a major role in Husserl's conversion to antipsychologism.[7]

Note that in these investigations some of the central themes of Husserl's epistemology, which are also present in *Philosophy of Arithmetic*, and which would be even more important in his later philosophy, are already clearly detectable, in particular the interplay between intuited and purely intentional objects in the dynamics of knowledge. Natural numbers are clearly at least partially intuitable whereas imaginary numbers are, from their perspective, purely intentional.

However, in this paper I intend to concentrate on the genesis of Husserl's notions of definiteness and his solution to the problem of the imaginary in mathematics, as presented in the drafts for the Göttingen talks and related texts of the same period (Husserl 1887-1901), contrasting both the solution and the notions with Hilbert's views on these

---

[5] 'Multiplicity' and 'manifold' are used here as synonymous. They both translate Husserl's term *Mannigfaltigkeit*.
[6] "Mathematics is, in the highest and broadest sense, the science of theoretical systems in general, abstracting from what is being theorized about in given theories of different sciences" (Husserl 1901a, 430). "Mathematics is, in its highest ideal, a theory of theories, the most general science of possible systems in general". (*Ibid.*, 431)
[7] In connection with this question see Hill 1991.

matters.⁸ In order to solve the problem of imaginary elements Husserl had to tackle some very technical questions of mathematical logic that were far from clear at the time. For example: what do formal theories refer to, what is their (formal) "domain", which universe do they describe?⁹ When can a formal domain be extended, and how? Under what conditions can knowledge of a larger domain be transferable to a more restricted one? I shall show here how Husserl answered these questions. When dealing with them Husserl was as much a mathematician as a philosopher, and it is precisely this aspect of Husserl's intellectual activity that I want to highlight here.

The Problem of Imaginary Entities

By an "imaginary" entity Husserl understood an entity of a domain of entities that, from the point of view of the meaning attached to this domain, cannot exist. This is how he understood the problem they give rise to:¹⁰

> Let a domain of objects be given, in which certain forms of operation and relation, determined according to the particular nature of the objects, are expressed in a certain system of axioms A. Based on this system, hence on the particular nature of the objects, certain forms of operation do not have any real meaning, that is, these forms of operation are absurd. How can what is absurd be used in computations, how can it be employed in deductive reasoning as if it were consistent? How can it be explained that one can operate with absurdities according to rules, and that, if no absurdity appears inside propositions, the propositions obtained are true? (Husserl 1901a, 433)

Consider the example of the natural numbers to which we arbitrarily add the negative numbers. We can prove, using a suitable system of axioms for N (our A, which is an *interpreted* theory), that for any natural numbers n, m, k: $n+k=m+k \Rightarrow n=m$. But we can also prove the same thing using negative numbers and their defining property (*) together with some general properties of addition: $n+k=m+k \Rightarrow (n+k)+(-k) = (m+k)+(-k) \Rightarrow n+(k+(-k))=m+(k+(-k)) \Rightarrow n+0=m+0 \Rightarrow n=m$. Except

---

⁸ As mentioned above (note 1), Husserl was already considering a formalist solution to the problem of imaginary elements in mathematics in 1890, years before moving to Göttingen and entering into Hilbert's circle in 1901. So, my comparison of Husserl's and Hilbert's notions of completeness for a formal axiomatic system does not intend to establish the originality of Husserl's ideas, a fact that I believe does not need to be established, but simply to highlight his widely ignored contribution to a theory of formal systems by comparing them with Hilbert's much better known ideas on the same subject.
⁹ In §§69-71 of the *Prolegomena to Pure Logic* (Husserl 1900), first volume of the *Logical Investigations* (Husserl 1900-01), a work written during the period in which he was struggling with the problem of imaginary entities, Husserl presents a better version of the notion of formal manifold, a version that he considered definitive. Also, Husserl suggested (Husserl 1900, §70) that the notion of formal manifold contains "the key" to the problem of imaginary entities. However, Husserl did not discuss this problem there.
¹⁰ Translations of quotations from Husserl 1887-1901 in the text are my own.

for the first and two last equations in the above derivation, the others are meaningless in N. Nonetheless, the fact proven is true. Husserl's problem was to explain under what circumstances this is acceptable.

In fact, Husserl had a still more basic problem to solve before explaining how negative numbers can be used in proving true facts about natural (positive) numbers, namely, in what sense negative numbers are *numbers*? Husserl noticed that "if we understand by numbers the answers to the question 'how many?' the sequence of [*natural*] numbers is then the closed multiplicity of instances that are possible in the sphere of the 'how many'" (*Ibid.*, 434). Therefore, negative numbers are not genuine numbers at all. The only way, Husserl believes, in which this problem can be tackled and derivations involving absurdities, like the one we carried out above, can be justified is by abandoning the concept of number as we understand it and moving into a sphere of purely formal reasoning. The axioms that are derived from our understanding of the concept of number are now to be seen as an implicit definition of a purely formal concept, that of positive integer. With appropriate modifications, this system of axioms can also be used to define, without contradiction, the notion of integer, since the meanings of these concepts are given exclusively by their formal definitions. This is how Husserl sketches a first approach to the solution of the problem of imaginary numbers:

> We can well abandon the concept of number, and, through the formal system of definitions and operations that are valid for the numbers, define a new, purely formal, concept, that of positive whole number. This formal concept of positive number, to the extent that it is itself delimited by the definition, can be enlarged by new definitions without contradiction. (*Ibid.*, 435)

But this cannot be the whole solution. Suppose that A is a system of axioms characterizing a certain formal context (the formal domain of A), and let B be a system extending A. Suppose that B is consistent, then nothing can be derived from B that contradicts anything derived from A. Is it acceptable to use B to derive properties of the formal domain of A? Is consistency the only criterion of acceptability, as Hilbert had claimed (claims which were well known to Husserl)? These are the questions that Husserl had to face in order to solve the problem of imaginary entities. Hilbert's solution, which conditions the acceptability of imaginaries exclusively to the consistency of the system in which they are defined, was considered by Husserl, but abandoned as incomplete. How do we know, he asks, that what is not inconsistent with A is also "true in the domain of A" (a concept we shall investigate shortly)?[11] Certainly, Husserl thought,

---

[11] An interpreted axiomatic system, that is, a system that is intended to describe a given mathematical structure, such as the intuitively given structure of natural numbers, has a natural domain, precisely the structure that is designed to describe. But for a non-interpreted or purely formal axiomatic system there is no such given domain. In order to tackle the problem of imaginaries Husserl had to provide a "natural" notion of domain for non-interpreted formal systems.

a proposition that makes sense for A and is derived from B does not necessarily have to be true in the domain of A unless already provable in A. A had to be exclusively responsible for what is true in its domain, and not depend on arbitrary extensions of itself in order to settle this question.

Before presenting what I believe to be the solution that Husserl presented to the problem of imaginary entities in mathematics and how they can be vindicated, let me present and discuss a reading of Husserl's solution (which was actually presented by some authors),[12] which is at first sight consistent with his texts, but that can be shown, on a closer analysis, to be incorrect. Suppose that A is a system of axioms that is extended by another system B, that is, B has at least all the axioms of A and possibly some more. Suppose also that A is *consistent* and *complete*, that is, given any assertion in the language in which the axioms of A are written–$L(A)$, the language of A–A does not prove this assertion *and* its negation, but *either* the assertion *or* its negation is provable in A. Moreover suppose that B is also *consistent*, although not necessarily complete and that in B we can define "imaginary" elements. Can an assertion written in $L(A)$–and so supposedly "making sense" from the perspective of A–that is proved in B–therefore "depending" on imaginary entities–be accepted? The solution some authors interpret as being Husserl's goes like this: Yes, because the use of imaginary elements in the proof of the assertion in question is in principle unnecessary. Since A is complete, either this assertion or its negation is provable in A, if the *negation* can be proved then B is inconsistent, for B is an extension of A and therefore proves everything that A proves, contrary to our hypothesis. Therefore the assertion in question *can* be proved in A, showing that the use of imaginary elements is indeed in principle unnecessary.

Here is why this solution does not work. Since A is complete and the existence of imaginary elements cannot be proved in it, then the *nonexistence* of these elements must be provable. Therefore, no *consistent* extension of A can exist in which the existence of these imaginary elements can be proved. For instance, suppose for the sake of argument that A is a complete system for the arithmetic of natural numbers. In A we can prove "*if $n \neq 0$, then there is no m such that $n + m = 0$*", but if we define the negative numbers in an extension B of A, we can prove "*for any n, there is an m such that $n + m = 0$*" in B. Therefore B is inconsistent.

What is going on here is the following. The system B, having extra axioms, *changes the meaning* of the symbols of $L(A)$ and consequently of any assertion written in $L(A)$. In other words, it is not the case that any assertion written in $L(A)$ is interpreted by A and B in the same way. The puzzle that led Husserl to consider the notion of a formal domain (also multiplicity or manifold) determined by an axiomatic system can

---

[12] For instance, Majer 1997, a variant of the same solution can be found in Hill 1995, 144.

be expressed as follows: how to guarantee that the *meaning* that A attributes to the symbols of $L(A)$ is preserved by B? His answer can be interpreted in the following way: Assertions of $L(A)$ preserve their meaning in systems that extend A provided these assertions refer exclusively to the elements of the domain of A, that is, *exclusively* to the (formal) objects A "has in mind".

Husserl's solution to the problem of imaginary elements has, I believe, the following form: Given systems A and B such that A and B are consistent and B extends A, let D be the formal manifold determined by A (see below) and suppose that A is complete *relative* to the assertions of $L_D(A)$, i.e. the assertions of $L(A)$ with all variables restricted to D. Now, if any of these assertions (i.e. assertions of $L_D(A)$) is proved in B, it can also be accepted from the perspective of A, for we are sure that its meaning, as determined by A, was not altered by B and moreover A can also in principle prove it. So, in order to solve the problem of imaginary entities, Husserl only needs a restricted notion of completeness, as is clear from the relevant texts, not the full-blown notion of completeness due to Hilbert. However, his solution depends on the notion of a formal manifold determined by a system of axioms, to which we turn our attention now.

Domains of Formal Systems

With respect to the definition of the (formal) domain of a formal system, Husserl's notes from 1900-01 are anything but clear. Two different notions can be found there with the same name, one concerns what Husserl calls "formal objects" (I shall refer to this as the *formal ontological* concept of domain), the other concerns judgments (which I shall refer to as the *apophantic* concept of domain). The ontological concept is by far the most difficult to define, and will be the focus of my attention for most of this section. In what follows, if nothing is said to the contrary, by "domain", I mean exclusively the formal ontological concept of domain.

Generally speaking, by the domain of a formal system Husserl understood "the sphere of existence" defined by the system. But for Husserl, the elements of a domain are, this much at least is clear, not specified objects but, rather, unspecified "formal objects", or also, as he sometimes calls them, "forms of objects". By "filling" these forms with the appropriate "stuff", we have the different objects the system in question refers to by means of the formal objects of its domain. The idea is that these "formal objects" are structures of some sort that are transformed into genuine objects when "filled" by the appropriate matter. Formal objects are "unsaturated" objects (appropriating Frege's terminology) that a formal theory singularizes, but only with respect to their formal properties. Their unsaturated character is what makes them formal objects, i.e. forms that can be "filled" in different ways.

Consider a system of (non-interpreted) axioms A. According to Husserl, all the formal objects that A can specify in a *non-ambiguous* way, that is, all the formal objects whose "existence" A presupposes or implies belong to the domain of A (Husserl 1901b, 470-71). Let us suppose that A is expressed in a first-order language and define the *formal ontological domain* of A (also called the *formal multiplicity* or *manifold* determined by A) as the set of all formal objects whose "existence" the theory can prove. Of course, this definition will not take us very far if we do not have a proper definition of the notion of formal object. This is how I propose, based on Husserl's suggestions, to interpret this concept. Formal objects are determined by two types of linguistic entities: i) terms without variables expressed in $L(A)$, the language of A; and ii) formulas of $L(A)$ in one free variable. The formal object denoted by the term t is thought of as constructed from the formal objects denoted by the constants occurring in it by means of the operations involved in the term t (it is in this sense an object that is "constructed" from previously given objects). Objects denoted by terms are all in the domain of A, for A proves $\exists x(x=t)$, for any term t. Any formula $\varphi(x)$ with one free variable expressible in the language of A produces a formal object provided A proves $\exists x \varphi(x)$. In this case there is an object in the formal domain of A satisfying φ.[13]

In other words, if the language of A has symbols for constants then these symbols denote objects in the domain, for they are non-ambiguous names of (formal) objects required by A. If the language of A has symbols for *single-valued* functions (or operations), then all the objects denoted by terms (with their usual logical meaning) generated from these symbols and terms denoting objects already in the domain are also in the domain, for these terms also denote objects whose existence is required by A (the operations are defined in the domain determined by A). The above restriction to single-valued operations, explicitly imposed by Husserl, is easy to understand. A term generated by a multi-valued operation does not single out a unique object, so it does not have the uniqueness we usually associate with objects. Finally, if $\varphi(x)$ is a formula of the language of A in one free variable (we can think of the formula as a description of a purported object), and if A proves $\exists x \varphi(x)$ there must be a formal object in the domain of A satisfying $\varphi(x)$. We can add a new constant c to the language of A to name it and a new axiom to A, namely $\varphi(c)$.

But this is not all. Husserl explicitly says that the domain of a formal system is the collection of all formal objects determined by the system *on the basis of a collection of formal objects arbitrarily given*. So,

---

[13] It is tempting to see Husserl's construction of the formal domain of a theory along the lines of Leon Henkin's construction of a model for a consistent first-order theory from constants.

the domain of a formal system of axioms is relative to the set of formal objects we begin with, that is, a *basis*. So, given a set of new constants, which we add to the language of A (and which we think of as denoting the elements of the basis), in order to obtain the domain of A relative to this basis we proceed as above considering these new constants as *bona fide* constants of $L(A)$.

We must keep in mind that formal objects are not linguistic entities. Formal objects are *denoted* by *terms* and *singularized* by *descriptions*, but they cannot be *reduced* to them. On the other hand, they are not objects in the usual sense. Formal objects are *forms*. Similarly, the formal domain determined by a system of formal axioms is not simply a collection of names; it transcends language, being a domain of object-forms the system *describes*.

And since A *describes* its domain, the axioms of A are *true* in the domain of A. However, we cannot simply identify the notion of a formal domain with the modern notion of a model for the system. A model contains ("materially filled") objects; a formal domain contains ("materially empty") *formal* objects. Moreover, a system may have many different models, but it has only one formal domain (Husserl always refers to *the* domain of a formal axiomatic system).

Husserl at least once considered abandoning the notion of a formal domain for the more natural concept of a collection of proper objects satisfying, under appropriate interpretation, the axioms of a system, as the following quotation indicates:

> Domain of a system of axioms. We restrict ourselves to systems of axioms that have a domain. (Why not simply: set of objects that satisfy the system of axioms.)... Everything remains correct if we simply take the word domain in its natural sense: objects that satisfy the axioms. (Husserl n.d., 457)

Nonetheless, the original notion of formal domain is meant as the collection of formal objects *required to exist* by an axiomatic system.

Husserl, of course, does not explain in Tarskian terms the notion of A being true in its formal domain; he takes it as an intuitively clear notion (we, however, would not find it too hard to show, by following Henkin's procedure, that A is indeed true in its domain). For my purpose here, which is to clarify Husserl's notions of definiteness, it is enough to establish the following:

> 1. Any axiomatic system determines a (unique) domain of purely formal objects, or forms of objects.
> 2. No formal object belongs to this domain that is not required to exist by the axioms of the system.
> 3. All concepts and operations involved in the axioms of the system, and whose meaning is given by the system, are interpreted in the domain of the system in such a way that it is possible to say that all axioms of the system are true in this domain.

Although this is the most general notion of formal domain, it is not the only one Husserl presents in the texts under analysis. He also had a more restricted one, which he calls a *mathematical* or *constructive manifold*. A system of axioms A determines a constructive manifold, he says, if its domain contains only "constructible" objects, i.e. if A proves $\exists x \varphi(x)$, then there must be a term t involving only constants of the basis such that A proves $\varphi(t)$.

In Husserl's words:

> A system of axioms defines a mathematical or 'constructive' multiplicity if the sphere of formal objects that it defines has the property that any object existing in this sphere on the basis of the axioms is uniquely determined by its operational relation to objects 'given in a determined way', that is, uniquely established and no longer specifically distinguishable, hence if all objects in general that exist in the domain and that are thought of according to any concept defined in virtue of the axioms are part of the objects that are obtained from those objects that are given in a determined way by definite concept formations, by definite junctions and relations. Any concept of object that is defined by the axioms is either a concept of object of the domain of determined objects, or does not have an object (in the domain), i.e. among the objects whose existence the system of axioms can establish (prove) there is no object corresponding to the concept. (*Ibid.*, 452-53)

In different, hopefully clearer, words. A system of axioms defines a *mathematical* or *constructive multiplicity* when any formal object whose existence can be inferred from the axioms is generated from the elements of the basis. Again in the words of Husserl, a system of axioms determines a constructive multiplicity when:

> all assertions of existence that are valid on the basis of the axioms, even if they do not determine their objects uniquely or even if they only determine an infinite class of objects, find their equivalents in the sphere of existence due to uniquely determining constructions, hence when all objects whose existence follows from the axioms are among those that can be constructively, hence uniquely, produced from no matter which previously given objects. (Husserl 1901b, 471).

A still more restrictive notion requires that the basis in question be *finite*. In this case we say that A defines a *finite* multiplicity. (*Ibid.*, 471-72)

Having available the notion of formal domain (or manifold, or multiplicity) determined by an axiomatic system, Husserl was then in a position to tackle the following problem: Suppose that A and B are two systems of axioms such that $A \subset B$, i.e. B contains all axioms of A together with some extra axioms that are consistent with but not derivable from A (A and B may have the same language). Can the domain of A be changed by B? Can B, for instance, enlarge the domain of A by proving certain existential statements in the language of A that A itself was incapable of proving? Before we go on with this discussion, one observation is necessary here. If we understand by the domain of a formal system the *totality* of formal objects that this system determines, then the notion of an "extension" of the domain of a *specific* formal system appears to be devoid of sense. But the senselessness is, in this case, only apparent. When Husserl talks about an *extension* of the domain of the system A he means, in fact, the domain of *another* formal system B that extends A by the introduction of *new* axioms. Let us keep in mind that these new axioms may be written in the same language of A, but need not be.

Before I proceed to the next section, let me introduce the apophantic notion of domain. As mentioned above, besides the formal ontological notion, Husserl had yet another notion of the domain of a formal axiomatic system, which I have called the apophantic notion. This is how I propose to understand this notion: given a system A, by the *apophantic domain of A*, I mean the collection of all statements that A can either prove or disprove, i.e. the apophantic domain of a system is the collection of statements this system can decide. If an assertion belongs to the apophantic domain of a system, then it is either true on the basis of the axioms of the system, if they can prove it, or it is false on the basis of these axioms, if they can prove its negation. Therefore, the principle of the excluded middle is valid for the apophantic domain of any axiomatic system.

Extensions of Domains, Imaginary Elements and the Notion of Relative Definiteness

The concept of the formal ontological domain of a system of axioms A allows us to define what is, from the perspective of A, an imaginary element, that is, to solve what I have called the ontological problem. An *imaginary* element is simply any element that is not in the ontological domain of A, no matter what the basis. In other words, it is an element that, from the point of view of A, does not exist. To introduce an imaginary element in the domain of A is simply to extend A by a consistent system of axioms B whose ontological domain includes the domain of A together with this new imaginary element. This can be done, for instance, by adding new constants to $L(A)$ and new axioms to A stating the properties the new formal objects they denote must satisfy, or also, by adding to A assertions of $L(A)$ that do not belong to the apophantic domain of A, that is assertions that A cannot either prove or disprove. In this way, expressions of $L(A)$ of the form *"there is an x such that..."* that could not be proved by A may now be proved in the extended system. Note that in this case, even if the defining expression that introduces a new element into the domain of B is a sentence of $L(A)$, this new element is imaginary from the perspective of A, for A cannot prove that it exists. This again shows us that the meanings of the assertions of $L(A)$ are not determined once and for all, but depend on the system in which they can be proved. Symbols have their meanings constantly open to new interpretations, provided by new systems of axioms. A fixed system cannot guarantee that the meaning of its symbols is not reinterpreted by an arbitrary extension of the system, but it can guarantee that, at least with respect to the elements of its ontological domain, this meaning is given once and for all, or so Husserl seems to think. This leads us to the notion of relative definiteness, which is the key notion for solving what I have called the epistemological problem. Husserl writes:

> A system of axioms is "definite" in a relative manner if any proposition that has a sense according to it is decided within the limits of the domain of this system. (Husserl 1901a, 440)

There are two questions related to the above "definition" that must be answered before it makes any sense: 1) When does a proposition "have a sense" according to a system of axioms? 2) When is a proposition "decided in the limits of the domain of a system"?

The obvious answer to 1) is that a proposition has a sense according to a system of axioms if it belongs to the language of this system. As we have already explained above, by extending the axioms of a system, we can reinterpret the meaning of the symbols of its language. Although it is obvious that any assertion of $L(A)$ has a sense according to A, A cannot rule out different senses for the same assertion. This leads Husserl, I believe, to impose the crucial restriction contained in 2). Instead of considering assertions of the language of A in general, he considers only those that refer exclusively to the elements of the domain of A. In this way, we can be sure that the meaning of the assertion in question is fixed once and for all. For any proposition P of $L(A)$, the restriction of P to the domain $D-P_D$—which is also, supposedly, a proposition of $L(A)$, denotes a state of affairs *in the domain of A*, therefore it is not affected by arbitrary extensions of A. It is far from obvious, however, that the restriction of the variables of a proposition of $L(A)$ to the domain of A can be accomplished by the expressive powers of this language. In fact, in most cases, this will not be possible. To investigate this problem here would take us far afield. So we shall simply suppose, for the sake of exposition, that the restriction of all variables in any assertion of the language of a system A to the domain of A can be accomplished by an assertion of this language.

Now we can answer the question 2) above. A proposition is decided within the limits of the domain of a system when its restriction to the domain of this system is decided by the system, that is, it is either a consequence of the system or is in contradiction with it, in the sense that the negation of this proposition is a consequence of the axioms of the system. Hence, this is how I propose we should understand Husserl's notion of relative definiteness: a system of axioms A is *definite relative to its domain D* if for any proposition *P* in *L(A)*, either P_D or its negation is a consequence of the axioms of *A*. We can now solve the epistemological problem: when is it legitimate to operate with imaginary elements?

Suppose that *A* and *B* are two consistent systems of axioms, and that *B* extends *A*. Suppose also that the ontological domain of *B* includes properly that of *A*, i.e. *B* has imaginary elements (from the perspective of *A*). If *B* proves a proposition *P* of *L(A)*, then the imaginary elements of *B* contributed not only to the proof of *P*, but possibly also to changing the meaning of the concepts *A* implicitly defines—in which case *A* does not prove *P*. But if we restrict all the variables of *P* to the domain *D* of *A*, then the proposition P_D thus obtained refers exclusively to the formal manifold *A* determines, i.e. there is no implicit reference to imaginary elements. Hence the concepts involved in P_D have the sense *A* gives them; therefore *A* must decide this proposition. These were the considerations, I believe, that led Husserl to reject Hilbert's solution of the problem of imaginary entities in mathematics, a solution that, as we know, requires only that imaginaries do not introduce inconsistencies, in favor of more stringent requirements, namely that *A* must be definite relative to its domain, i.e. *A* must either prove P_D or $\neg P_D$ for any proposition *P* of *L(A)*. In other words, *B* cannot prove any assertion of *L(A)* that, when restricted to the domain of *A*, *A* itself cannot *decide* (prove or disprove). In particular, if *B* proves P_D then *A* must prove it too. That is, imaginary elements are not necessary in order to prove assertions *exclusively involving objects of the domain of A*. *A* must, so to speak, be master of its domain. Notice that this version of Husserl's solution is free from the problems, pointed out above, that arise from the requirement that *A* be complete. If *A* proves $\neg P_D$ this will not conflict with the fact that *B* proves *P*, because P_D and *P* are two different propositions, in the sense that they "refer" to different domains.

Let me now offer some pieces of textual evidence for my reading. Defining the notion of relative definiteness for a system of axioms Husserl said:

> [the system of axioms] defines a manifold in such a way that for each proposition involving the concepts defined and purely logical concepts, it is objectively determined whether it holds or does not hold *for this manifold*, instead of remaining indeterminate (*Ibid.*, 447) [*my emphasis*]

It is also clear from the context in which this quotation occurs that Husserl considers the requirement of relative definiteness as a necessary and sufficient condition for the legitimate use of imaginary entities (defined in extensions of the system in question).

Also with respect to the notion of relative definiteness:

> I define a system of axioms that formally defines a domain of objects in such a way that every question that has a meaning for this domain of objects is thus answered by this system of axioms, or that every proposition that has a

meaning according to the axioms, *if we restrict ourselves exclusively to the objects that the axioms establish as existing*, either is a consequence of the axioms or contradicts them. If P is a proposition that says: P holds for the manifold whose existence is proved by A, then this proposition is, *for this manifold*, either true or false on the basis of A.

The manifold contains in this way all the objects whose existence is proved by A, which does not rule out that the same axioms hold for a larger manifold, but do so in such a way that the extra elements are not defined as existing or proved to exist by the axioms (without the inclusion of new axioms) (Husserl n.d. 454) [*my emphasis*]

Another remark is needed here, if we take seriously the suggestion that the symbols of a language are not really the same if interpreted by different axiomatic systems, as I believe Husserl did,[14] it is not quite right to say that a system *B extends* another system *A*, even if *B* contains what may look like the same axioms of *A* plus some extra axioms. Although some axioms of *B* may have the same form as the axioms of *A*, they are not really the same, for the axioms in *B* that are not in *A* change the meaning of those axioms of *B* that both *A* and *B* apparently share. A more adequate form of expression is, I believe, to say that *B* extends *A* when *B* incorporates the axioms of *A* by reinterpreting them in a larger context.

It is clear from Husserl's notes, and I tried to do him justice in my reading, that his earliest attempt to solve the problem of imaginary entities in mathematics involves a complex, and far from clear, interplay between syntactic and semantic notions, and therefore cannot be formulated in a purely formalist way, as could Hilbert's answer to the same problem. Husserl's refusal to accept Hilbert's solution (a purely formal solution that requires of imaginary elements only that they do not introduce inconsistencies in the system to which they are added) seems to indicate that Husserl could still not admit an axiomatic system as something completely independent of a domain of objects, if only the ghostly formal domain it determines. It seems that, despite his criticism of Frege's inability to accept Hilbert's new conception of a formal system of axioms (Husserl 1899, 448-49),[15] Husserl himself was not yet completely free from the same prejudices in favor of the traditional concept of axiomatic system that he criticizes in Frege. However, by admitting, contrary to Frege, that the system takes precedence over the domain it *defines*, and not vice-versa, the domain preceding the system that merely *describes* it, Husserl is, so to speak, halfway between Hilbert's new ideas and Frege's old prejudices. Another interpretation, more sympathetic to Husserl, would be to consider his ideas concerning formal domains and the concept of relative definiteness as belonging to a formal semantics

---

[14] "Every axiom brings a contribution to the definition". (Husserl 1901a, 449)

[15] Husserl was granted access to a correspondence between Hilbert and Frege on the subject of formal axiomatic systems, which he summarizes and comments upon (Husserl 1899). Husserl criticizes the descriptive role Frege assigns to axioms, siding with Hilbert, who views axioms as nothing but implicit definitions. As we have seen, this move is fundamental in Husserl's approach to the problem of imaginary elements.

which would be a precursor of modern model theory. But these questions should not concern us here.

In fact, Husserl's concept of relative definiteness is equivalent to the notion of a system of axioms A that simultaneously *completely* characterizes and *completely* describes a manifold M of objects, which are nonetheless only formally determined by this system. By a complete description we mean that any possible question that we can raise *about this manifold* can be answered by A. If we allow extensions of A, then the "aboutness" mentioned above, I claim, can only be obtained by somehow restricting all the variables of the question we are asking to the manifold M. Otherwise, the question loses its "grip" on M, for it can be, and often is, reinterpreted differently in extensions of A.

In Husserl's words:

> [A] system of axioms is "definite" if it circumscribes a domain of objects as existing, and in such a way that *for this domain* no new axiom is possible. (Husserl n.d., 457) [*my emphasis*]

To say that no *new* axiom is possible *for the domain* of a system is, obviously, equivalent to saying that this system decides any proposition referring to its domain.

Absolute Definiteness and Hilbert's Axioms of Completeness

As I mentioned above, in order to extend the domain of an axiomatic system A we must adjoin new axioms to it. But this is only possible if there is some proposition in $L(A)$ that A does not decide, assuming that we require the new axioms to be expressible in the same language of the system A. Otherwise, if A decides any proposition that can be expressed in $L(A)$, then any possible candidate for a new axiom of A is already a consequence of A, hence either it cannot be a new independent axiom, or it is contradictory with A, in case A proves its negation. When a system of axioms has the property that any proposition that can be expressed in its language is either a consequence of the axioms of the system or is in contradiction with it, Husserl calls this system *absolutely definite*:

> a system of axioms is "definite" in an absolute manner if any proposition that has a sense according to it is decided *in general* (Husserl 1901a, 440) [*my emphasis*].

Obviously, if a system A is absolutely definite, it is also definite relative to its domain, for the propositions expressed in $L(A)$ that refer exclusively to the domain of the system are (supposedly) among the propositions that can be expressed in $L(A)$, which can all be decided in A. It is clear that Husserl is aware of the identity between his notions of absolute definiteness and Hilbert's notion of completeness:

So, "definite" in an absolute manner [equals] complete in the sense of Hilbert. (*Ibid.*, 440)

If a system is definite only relative to its ontological domain, no new proposition referring to this domain can be added to the system, but this does not rule out the possibility that propositions that do not refer exclusively to its domain could be consistently adjoined to the system. However, if a system is absolutely definite, no new axiom can be consistently added to it, that is:

If not only no axiom (that acquires a sense from the axioms already given) can be added for the objects of the domain, but if in general no new axiom can be added. (*Ibid.*, 440)

Of course, the enlargement to which we are referring considers only assertions in the language of the system. Otherwise, if we admit sentences whose symbols do not belong to the language of the system, we can always enlarge the original system by adding to it sentences it can neither prove nor disprove.

The fact that a relatively definite system *completely determines* and *completely describes* a domain of formal objects, so that this domain does not admit being extended and does not admit new axioms, together with the fact that the apophantic domain of an absolutely definite system is the largest possible, i.e. it coincides with the totality of all assertions that can be expressed in the language of this system, show that both notions of definiteness are closely associated with the idea of a system whose domains (formal ontological or apophantic) do not admit being extended, unless the system itself is modified. The connection between the notion of absolute definiteness, which as we saw implies relative definiteness, and the impossibility of extending the domain of an axiomatic system without altering it is clearly indicated by Husserl in the passage below:

But this implies that the multiplicity (the domain) cannot be enlarged in such a way that for the enlarged domain the same system of axioms valid for the old [domain] is [still] valid. (*Ibid.*, 440)

This helps to explain, I believe, Husserl's famous remark in a footnote to §72 of *Ideas I*:

The close relation of the concept of definiteness to the "Axiom of Completeness" introduced by D. Hilbert for the Foundations of Arithmetic will be apparent without further remark on my part to every mathematician. (Husserl 1913, §72n.)

In fact, relative definiteness is already sufficient for the inextensibility of the domain of a system. In order to extend the domain of a system A it is necessary to add new axioms to A. If A is definite relative to its domain, these axioms cannot refer exclusively to the domain of A. Indeed, their restriction to this domain can be *disproved* in A. Therefore the introduction of new elements in the domain of A creates new "situations" in this domain that A cannot describe. Hence, if A is definite relative to its domain, we cannot extend the domain of A and still have it described by A.

Hilbert's *axiom* of completeness is nothing but a way of selecting among all possible models for a system of axioms the one that cannot be enlarged by the introduction of new elements. Let us call a model for a system of axioms *complete* if no object can be adjoined to it such that the enlarged structure is still a model of the system. An axiomatic system satisfies Hilbert's axiom of completeness if any of its models is complete in the sense just defined. Now, if a system of axioms is definite in an absolute, or even relative way, its domain is complete in this sense, as explained above. Although obviously neither Husserl's notion of absolute nor his notion of relative definiteness are the same as or equivalent to Hilbert's notion of semantic completeness involved in the axiom of completeness, all these notions capture the same idea: the inextensibility of objective domains, formal domains in the case of Husserl, 'material' domains in the case of Hilbert. Despite the common opinion among scholars[16] that Husserl confuses the notions of semantic and syntactic completeness, I think that my presentation and discussion of Husserl's two concepts of definiteness makes it evident that there is indeed a relation between these concepts and Hilbert's axiom of completeness.

As far as I know, the only example Husserl gives of an absolutely definite axiomatic system is not a very good one. According to Husserl, any system of arithmetic is absolutely definite, and the reason for this is that formulas of the language of arithmetic can always be reduced to systems of equations and inequalities, which are decidable:

> Any arithmetic, however narrow it may be, that of positive or real integers, rational positive or rational numbers in general, etc., any arithmetic is defined by a system of axioms such that we can prove from them that: any proposition in general constructed exclusively from the concepts that (taken axiomatically) are established by the axioms, any such proposition is in the domain, that is, it is either a consequence of the axioms or it is in contradiction with them. The proof of this assertion consists in noticing that any definite formation of operation [*any variable-free term, my note*] is a natural number and that any natural number is in a relation of magnitude determinable on the basis of the axioms with respect to any other number. If a proposition has in general a meaning in virtue of the axioms, it is either valid as a consequence of the axioms or it is not valid for it contradicts the axioms.... Any numerical equation is true if it can be transformed [*via the axioms, my note*] into an identity, otherwise it is false. Any algebraic formula is then also decided, for it is decided for any numerical instance. (Husserl 1901a, 443)

---

[16] Suzanne Bachelard in particular (Bachelard 1968, 58-63), see also Jan Sebestik's postface to Cavaillès 1947, 135.

This quotation is a bit puzzling for, despite Husserl's claim that *any* arithmetical system is *absolutely* definite, his justification for this relies on the decidability of *quantifier-free* formulas only. So, we seem to have a problem of interpretation. But notice that the final sentence of the passage quoted above says that "[a]ny algebraic formula is... also decided, for it is decided for any numerical instance". It is clear, then, that Husserl calls a general statement 'decidable' when its *instances* are decidable. This way of understanding the notion of decidability, certainly not the same one that is behind Kurt Gödel's theorem, bears some resemblance to Henri Poincaré's notion of verifiability for mathematical propositions (cf. his *Science and Hypothesis*, (Poincaré 1902)). So, since Husserl does not understand the concept of decidability in the same way as Gödel, we should not blame Husserl for making a claim that apparently so blatantly contradicts the famous, although at that time not yet proved, Gödel incompleteness theorem.

Final Considerations

In *Philosophy of Arithmetic*, Husserl had already faced the problem of justifying computations with big numbers, numbers that could not be obtained by a process of abstraction from manageable collections, that is, reasonably small collections that actually could be counted by a human being. Imaginary numbers, however, pose a still more difficult problem. If "huge" natural numbers are non-intuitable due only to our minds' limited powers, imaginary numbers are not accessible to intuition in a much stronger way. No matter how much we free our mental processes of their natural limitations, imaginary numbers cannot be generated by abstraction from given collections of objects. Imaginary numbers are *only* symbols. Symbolic reasoning is justified in *Philosophy of Arithmetic* by appealing to a form of parallelism between intuitive and symbolic manipulations that guarantees that for any symbolic computation there is an intuitive construction that *could*, if we abstract from our *inessential* mental limitations, give it intuitive support. In the case of imaginary numbers nothing of the sort is the case and Husserl had to find another way of justifying their use. He chose to see those "numbers" as pure symbols to which nothing could correspond in our intuitions, and computations with them as mere formal manipulations. The question now is whether or not they can contribute to our *knowledge* about *genuine numbers*, that is, numbers that correspond to a notion of quantity.

The requirement that imaginary numbers must be definable in consistent extensions of relatively definite systems amounts to saying that they cannot introduce new elements into the formal ontological domain of the system in whose extensions they are definable or contribute to the description of this domain. In other words, imaginary numbers cannot disturb the systems for which they are precisely what they are, imaginary, either *ontologically* or *epistemologically*. Still another way of saying the same thing is that no object can exist and, in general, no assertion referring exclusively to a certain domain can be proved solely in virtue of the introduction of imaginary entities.

The notion of *absolute* definiteness, Husserl says, escaped him at first because of his focusing exclusively on the problem of imaginary numbers, for which the restricted notion sufficed (Husserl 1901a, 443). But once the broader notion was available, Husserl immediately recognized its utility for a much more important problem posed by formalism: What can correspond to the notion of truth in *formal* axiomatic systems? The answer to this question was then clear to him: a proposition of the system is true if it is derivable in the system and false if it is in contradiction with it (i.e. its negation is derivable in the system). In order to preserve the validity of logical laws within the context of formal systems, *tertium non datur* in particular, these systems must, of course, be absolutely definite, or complete in Hilbert's sense. Their consistency suffices to guarantee the validity of the law of non-contradiction.

In *Ideas I*, a notion of definiteness is presented again, but with an important difference. This time a mathematical *structure* is under consideration, and it is called a *definite manifold* or a *mathematical manifold* if its complete theory (that is, the set of all sentences that are true in this structure) is *finitely* axiomatizable. That is, the manifold is described by a complete and finite theory or, as Husserl puts it, it is "mathematically, exhaustively definable".

> It [*a definite manifold-my note*] has the following distinctive feature, that a *finite number of concepts and propositions*-to be drawn as occasion requires from the essential nature of the domain under consideration-*determine completely and unambiguously on lines of pure logical necessity the totality of all possible formations in the domain*, so that *in principle*, therefore, *nothing further* remains *open* within it. (Husserl 1913, §72)

A theory is called definite if its *domain* is definite in the above sense: "I also refer to a system of axioms which on purely analytic lines 'exhaustively defines and describes' a manifold in the sense described as a definite system of axioms" (*Ibid;*). Husserl seems to be, again, presenting the notion of a system of axioms that is definite relative to its domain. In the earlier texts of the Göttingen talks a *theory* is definite relative to its domain if it *defines* and *completely describes* a domain of formal objects, whereas in the later text of *Ideas I* a *domain of objects* (possibly only formal objects) is definite if there is a finite theory that *exhaustively defines* and *completely describes* it. Despite some differences in presentation, we still have the same notion of relative definiteness. It is interesting to notice the similarity between the notion of definiteness of a theory presented in *Ideas I* as the condition that guarantees that a domain is exhaustively defined and described by a finite theory, with the notion of a finite mathematical or constructive manifold presented above, as that domain constructively and completely determined from a finite basis.[17]

As already observed, both Husserl's concept of relative definiteness and Hilbert's concept of completeness associated with the axiom of completeness involve the idea of a domain of objects that is exhaustively determined and completely described by a theory, therefore a domain that cannot be enlarged unless this theory is modified. As we have already observed, any adjunction of new elements to the domain of a relatively definite system (by means of new axioms) creates situations in that domain that are inconsistent with the former theory.

The notion of (absolute) definiteness appears again in *Formal and Transcendental Logic* (Husserl 1929) as an ideal presiding over nomological (Euclidean) theories. It seems to me that Husserl introduces this ideal for two basic reasons. On the one hand, definiteness is obviously the ideal to which any theory aspires: to provide complete description of the domain, or domains, of (either purely formal or materially filled) objects that this theory refers to. On the other, definiteness, together with consistency, are required in order to ensure the validity of classical logical laws, the law of the excluded middle in particular, in formal contexts.

References

Bachelard, Suzanne 1968. *A Study of Husserl's Formal and Transcendental Logic*, Evanston IL: Northwestern University Press.

Cavaillès, Jean 1947. *Sur la logique et la théorie de la science*, Paris: Vrin, 1997.

Ewald, William 1996. *From Kant to Hilbert. Readings on the Foundations of Mathematics*, Oxford: Oxford University Press.

Frege, Gottlob 1980. *Philosophical and Mathematical Correspondence*, Gottfried Gabriel et. al. (eds.), Oxford: Blackwell.

Frege, Gottlob 1894. "Review E. G. Husserl's *Philosophy of Arithmetic*", in *Collected Papers on Mathematics, Logic and Philosophy*, Brian McGuinness (ed.), Oxford: Blackwell.

Hilbert, David 1900. "On the Concept of Number", in Ewald (ed.), 1089-1095, translation of "Über den Zahlbegriff", *Jahresbericht der Deutschen Mathematiker Vereinigung* 8, 180-84.

Hill, Claire Ortiz 1995. "Husserl and Hilbert on Completeness", in *From Dedekind to Gödel*, Jaakko Hintikka (ed.), Dordrecht: Kluwer, also anthologized in Hill and Rosado Haddock.

Hill, Claire Ortiz 1991. *Word and Object in Husserl, Frege and Russell: The Roots of Twentieth-Century Philosophy*, Athens OH: Ohio University Press.

---

[17] See Chapter 7.

Hill, Claire Ortiz and Guillermo Rosado Haddock 2000, *Husserl or Frege? Meaning, Objectivity and Mathematics,* Chicago: Open Court.

Husserl, Edmund 1887-1901. *Philosophy of Arithmetic, Psychological and Logical Investigations with Supplementary Texts from 1887-1901,* translated by Dallas Willard, Dordrecht: Kluwer, 2003, the translation of *Philosophie der Arithmetik, mit ergänzenden Texten (1890-1901),* Husserliana XII, introduction by Lothar Eley, The Hague: Martinus Nijhoff. *Philosophy of Arithmetic* was originally published as Husserl 1891b.

Husserl, Edmund 1891a. "Besprechung von E; Schröder, *Vorlesungen über die Algebra der Logik I*", in Husserl 1890-1910, 3-43, published in English as "Review of Ernst Schröder's *Vorlesungen über die Algebra der Logik*", in Husserl 1994, 421-41. Originally published in *Göttinger Gelehrte Anzeigen,* 1891, n° 7, 243-78.

Husserl, Edmund 1891b. *Philosophie der Arithmetik,* Halle: Pfeffer, published in Husserl 1887-1901, 1-283.

Husserl, Edmund 1899. "*Husserl's* Excerpts from an Exchange of Letters between *Hilbert* and *Frege*", published in Husserl 1887-1901, 468-73, in German in Husserl 1887-1901, 447-51.

Husserl, Edmund 1900. *Prolegomena to Pure Logic,* volume I of Husserl 1900-01, New York: Humanities Press, 1970, translation by J. N. Findlay of *Prolegomena zur reinen Logik. Logische Untersuchungen,* Erster Teil, Halle: Niemeyer.

Husserl, Edmund 1900-01. *Logical Investigations,* New York: Humanities Press, 1970, translation of *Logische Untersuchungen.* Halle: Niemeyer, also published as Husserliana XVIII, XIX/I-II, The Hague: Martinus Nijhoff, 1975, 1984.

Husserl, Edmund 1901a. "Double Lecture on the Transition through the Impossible ("Imaginary") and the Completeness of an Axiom System", included as Essay III in Husserl 1887-1901, 409-52. Cited in the text are my translations of Husserliana XII, 430-47.

Husserl, Edmund 1901b. "The Domain of an Axiom System/Axiom System–Operation System", included as Essay IV in Husserl 1887-1901, 475-92. Cited in the text are my translations of Husserliana XII, 470-88.

Husserl, Edmund, 1913. *Ideas: A General Introduction to Pure Phenomenology,* translated by W. R. Boyce Gibson, New York: Collier, 1962 (1931), translation of *Ideen zu einer reinen Phänomenologie und phänomenologischen Philosophie I,* Husserliana II, The Hague: Martinus Nijhoff, 1950.

Husserl, Edmund 1929. *Formal and Transcendental Logic,* The Hague: Martinus Nijhoff, 1969, translation of *Formale und transzendentale Logik,* Husserliana XVII, The Hague: Martinus Nijhoff.

Husserl, Edmund 1994. *Early Writings in the Philosophy of Logic and Mathematics,* translated by Dallas Willard, Dordrecht: Kluwer.

Husserl, Edmund n.d., "Drei Studien zur Definitheit und Erweiterrung eines Axiomensystems", cited in the text are my translations of Husserliana XII, 452-69.

Majer, Ulrich 1997. "Husserl and Hilbert on Completeness: A Neglected Chapter in Early Twenty Century Foundation of Mathematics", *Synthese* 110, 37-56.

Poincaré, Henri 1902. *Science and Hypothesis*, New York: Dover Books, 1952, translation of *La Science et l'Hypothèse*, Paris: Flammarion, 1968.

Scanlon, John 1991. "Tertium non Datur: Husserl's Conception of a Definite Multiplicity", in *Phenomenology and the Formal Sciences*, T. M. Seebohm, D. Føllesdal and J.N. Mohanty (eds.), Dordrecht: Kluwer.

Sebestik, Jan, "Postface", in Cavaillès 1947, 91-142.

7

Jairo José da Silva

The Many Senses of Completeness

Widely neglected and often misinterpreted, Husserl's investigations into the theory of axiomatic systems were born out of his struggle to provide sound logical-epistemological foundations and justification for purely symbolic reasoning. Since at least 1890, that is, even before the publication of his *Philosophy of Arithmetic* (Husserl 1887-1901), Husserl had been hard at work trying to solve the problems concerning the nature and role of imaginary entities in mathematics. Important as these questions were, they were but the tip of a much broader and philosophically relevant problem: How can a calculus be justified on both logical and epistemological grounds? What sort of knowledge, if any, does a purely formal theory (formal mathematics in particular) provide?

A solution to this problem, upon which Husserl worked for most of the last decade of the 19th century, was presented in the *Logical Investigations* (Husserl 1900-01). But first Husserl had to abandon his earlier "genetic" approach to mathematics (which at that stage had a distinctly psychologist flavor, completely lost later), reevaluate and reshape the analytic-synthetic dichotomy, distinguish between formal and material *a priori*, and totally remodel the domains of formal logic in order to accommodate symbolic reasoning and account for its role in science. In *Logical Investigations*, purely symbolic reasoning, embodied in formal axiomatic systems, is justified as a provider of (formal) knowledge about possible (but not necessarily actual) realms of objects.

I shall be mainly concerned here with the different notions of completeness of an axiomatic system that Husserl put forward in his struggle to provide what he considered adequate logical and epistemological foundations for symbolic reasoning in science. He thought that some sort of completeness was a necessary condition for formal domains (domains of formal systems) to be extendable by imaginary numbers, or for formal theories to stand as possible formal abstracta of theories proper, i.e. theories of well-determined objectual domains.

I shall analyze Husserl's concepts of completeness as introduced in *Ideas I* (Husserl 1913) and compare them with the notions of definiteness presented in the Göttingen talks of 1901 (Husserl 1901).[1] In *Ideas I*, he presented two different although equivalent versions of the notion of completeness for *interpreted* axiomatic systems. One of these is semantic in character; the other is syntactic (what shows that instead of confusing semantics and syntax, as some commentators believe, he was in fact clearly distinguishing them). In the Göttingen lectures, he also presented *two* notions of completeness, for purely formal systems this time, that are the exact equivalents of the notions of *Ideas I* adapted for formal domains and formal systems. However, given the definition of the concept of formal manifold in the Göttingen talks, the two notions of definiteness introduced there are not equivalent. Husserl used his concepts of completeness, in particular, to find a solution for the problem of imaginary entities, and, in general, an answer to the broader question concerning the foundations of symbolic reasoning in science.

For the sake of completeness (*dès que nous y sommes*), I shall also discuss the relation between Husserl's notions of completeness for axiomatic systems and David Hilbert's notion of completeness related to his axiom of completeness for geometry and the theory of the real numbers[2]. Is there any connection between the two? In a footnote to §72 of *Ideas I*, Husserl says that there is. However, many scholars think that Husserl was wrong in this belief, that Hilbert's axiom of completeness simply selects a *maximal model* for the (consistent) system of axioms to which it is added, whereas the notion of syntactic completeness (which is the notion of completeness that Husserl found germane to Hilbert's axiom) only implies that a given syntactically complete theory *itself* cannot be consistently extended by the adjunction of new axioms. Nothing is stated with respect to its *models*.

Of course, those scholars are right on this point. But Husserl, I claim, did not have in mind models of formal theories as we understand them today, but the *formal domain* determined by a formal theory, and there is indeed a close connection between a formal theory and its formal domain that has to do with Hilbert's axiom: if the former is syntactically complete the latter is maximal. And it is precisely an idea of maximality (of a domain described by a theory) that is behind Hilbert's axiom of completeness. Therefore, Husserl is quite right when he sees a "close relation" between his notion of definiteness and the notion of completeness related to Hilbert's axiom.

---

[1] See Chapter 6. Here, the notions of definiteness of the Göttingen talks will be contrasted with those of *Ideas I*.
[2] See Chapter 6. This question is, of course, related to the still open problem of the mutual influences between Husserl and Hilbert after Husserl moved to Göttingen in 1901.

A fundamental notion in our discussions is that of a formal domain of (formal) objects determined by a formal theory. Husserl does not seem willing to consider, as Hilbert did, theories from a purely syntactic perspective. For him, formal theories, albeit not theories of *given* objectual domains (domains of objects), are nonetheless theories of formal domains (domains of object-forms) *determined* by them. It is as if formal theories preserve the "intentional drive" (so to speak) of either the theories proper from which they are formally abstracted or the consciousness that simply invents them.

The Husserlian concept of completeness (or definiteness-*Definitheit* in the original German), for either purely formal or interpreted systems, involves the objective domains, purely formal, dependent on the system, or materially determined, independent of the system, these systems describe. Obviously, an interpreted theory has a domain of objects that the theory is *about*. So, in order to extend the notion of completeness for non-interpreted systems, Husserl had to answer the question: What is the domain of a purely formal theory? Before concerning ourselves with his answer to this question, let me first introduce Husserl's notion of definiteness as presented in *Ideas I*, §72. (Husserl 1913)

Definiteness in *Ideas I*

In *Ideas I* Husserl presented the following notion of completeness for a given domain of objects:

> [a definite manifold] has the following distinctive feature, that a *finite number of concepts and propositions*–to be drawn as occasion requires from the essential nature of the domain under consideration–*determines completely and unambiguously* on lines of pure logical necessity the totality of all possible formations in the domain, so that *in principle*, therefore, *nothing further* remains *open* in it. (Husserl 1913, §72)

What Husserl seems to have had in mind is the following: A domain is a *definite* or *mathematical manifold* when the set of all assertions that are true in it (the theory of the domain, written in some appropriate language) is finitely axiomatizable. A definite manifold is, for Husserl, one that can be completely mastered by a finite theory written in a language "extracted" from the domain itself. He said:

> A manifold of this type has the distinctive property of being "*mathematically, exhaustively definable*". The "definition" lies in the system of axiomatic concepts and axioms, and the "mathematically-exhaustive" herein that the defining assertions in relation to the manifold imply the greatest conceivable prejudgment– nothing further is left undetermined. (*Ibid.*)

Husserl also presented what he took for an equivalent formulation of this definition:

> Every proposition constructed out of the designed axiomatic concepts, and in accordance with any logical form whatsoever, is either a pure formal implication of the axioms, or formally derivable from these as the opposite of what they imply, that is, formally contradicting the axioms; the contradictory opposite would then be a formal implication of the axioms. (*Ibid.*)

That is, a mathematical domain D is *definite* when there is a language L strong enough to allow us to say whatever we want to say about D, in which we can write a finite theory T consisting of true assertions about D, such that any assertion A of L is either a logical consequence of T or is in contradiction with it (i.e. the negation of A is a consequence of T.) In other words, D is *definite* when there is a finite, *syntactically complete* theory T consisting of assertions true in D.

The first version of the definition clearly implies the second, for if A is any assertion of L, A is either true or false in D. In the first case, A is a consequence of the theory T that finitely axiomatizes the set of all assertions of L that are true in D, in the second case, the negation of A is true in D. So it is a logical consequence of T.

Let us check the converse. Let T be a syntactically complete finite theory with respect to the set of all assertions of L (all assertions that "make sense" in D.) Now, in order to show that T is a finite axiomatization of the set S of all assertions of L true in D, we must show that any element of S is a consequence of T, and conversely. The axioms of T are, of course, elements of S (T is chosen, after all, as an appropriate set of true assertions in D.) In order for the consequences of T to be also in S, we must presuppose that all the logical (i.e. syntactic) consequences of assertions true in D are also true in D. Husserl appears to have taken this for granted. In order to show the converse, let A be true in D. According to the second version of the notion of a definite manifold, either A or its negation is a consequence of T. In the first case we are done. In the second case, according to the above presupposition, the negation of A must be true in D, which is a contradiction.

Hence, the equivalence of Husserl's two formulations of the concept of definiteness for manifolds depends on the fact that "formal implications" of true assertions are themselves true assertions, i.e. the correctness of the notion of formal implication, which Husserl implicitly took for granted.

For Husserl, the notion of definite manifold embodies a mathematical ideal: the "mastering" of a potentially *infinite* realm of truth by a *finite* theory on lines of pure logical necessity. A definite manifold is one in which "true" and "formal implication of the (finitely many) axioms", and "false" and "formally contradicting the (finitely many) axioms" are equivalent, that is, it is a context in which the notions of truth and falsehood not only have purely formal equivalents, but can be grounded on a *finite* basis of fundamental truths.

Having defined the concept of definiteness for mathematical manifolds, Husserl could, still in the same paragraph, extend this notion to interpreted *theories*. For him an (interpreted) finite theory[3] is definite just when it is the theory of a manifold that is definite in the above sense. He wrote:

> I also refer to a system of axioms which on pure analytic lines "exhaustively defines" a manifold in the way described as a *definite system of axioms*. (*Ibid.*)

Since Husserl had just introduced the notion of definiteness for manifolds in two different, but equivalent ways, it is clear that we can imagine two possible ways of defining definiteness for interpreted theories. The first introduces definiteness *semantically*: a finite theory T is definite when it is the complete theory of a certain mathematical domain, its intended model (i.e. if there is a language L in which a domain D can be described such that T is a finite axiomatization of the set of all assertions of L that are true in D). Definiteness, in this version, means *semantic* completeness in the following sense. An interpreted theory T is semantically complete, with respect to a certain *particular* model D, when any true assertion *about* D is a theorem of T.

Another version is purely syntactic, an interpreted theory is definite when it is *syntactically* complete, in the sense that it decides any assertion written in its language (again, any assertion that "makes sense" in the domain it describes.) These two versions are equivalent, for the same reasons that the two versions of definiteness for manifolds are equivalent.

But now the problems begin. After introducing these definitions, in the same paragraph, Husserl made two claims: (1) that the notion of definiteness for *interpreted* theories just given can be extended to *non-interpreted*, purely formal axiomatic systems:

> The definitions remain as a system even when we leave the material specification of the manifold fully undetermined, thus making a generalization of the formalizing type. The system of axioms is thereby transformed into a system of axiomatic forms, the manifold into a form of manifold, and the discipline relating to the manifold into a form of discipline. (*Ibid.*)

And (2) that the notion of definiteness for non-interpreted theories has a "close relation" to the axiom of completeness that Hilbert introduced in the axiomatization of geometry and the theory of real numbers:

> The close relation of the concept of definiteness to the "Axiom of Completeness" introduced by D. Hilbert for the Foundations of Arithmetic will be apparent without further remark on my part to every mathematician. (*Ibid.*, n. 6)

---

[3] An interpreted theory is, for Husserl, one that describes a given specific domain, or, as we would say today, a theory with an *intended* model.

For Husserl, a "generalization of the formalizing type", or formal abstraction, of a theory amounts to completely eliminating any determinate objective reference the theory may have. That is, transforming the theory into a materially empty theory-form. What then, in this case, plays the role of the domain described by the theory?

More puzzling still is the connection that Husserl sees between his and Hilbert's notions of completeness. Hilbert's axiom of completeness is only a way of selecting among all the possible realizations of a theory (its models) that which is maximal with respect to inclusion (i.e. complete in a certain sense.) The fact that T is a definite system of axioms does not imply that it is complete in Hilbert's sense (it is enough to take any finite axiomatization of an infinite domain by a complete elementary theory; this theory is definite but does not satisfy Hilbert's axiom of completeness as a consequence of the Löwenheim-Skolem theorem). Also, a theory may be Hilbert complete but not definite (take Hilbert's own axiomatization of the theory of real numbers; this axiomatization includes Hilbert's axiom of completeness, but it is not definite). So, it might seem quite clear that Husserl is mistaken on this point, as has indeed been claimed by many scholars. (For instance, Bachelard, 59-63; Cavaillès, 70-72; Sebestik, 135)

But was he really? Firstly let us note that Husserl made the connection between his notion of completeness and Hilbert's only *after* having indicated (admittedly in an extremely vague way) that the notion of definiteness could be extended to non-interpreted, purely formal axiomatic systems. So, it seems that the connection that he has in mind is between the definiteness of *non-interpreted* theories and the inextensibility of their formal domains, *not* any of their material instantiations (models.)

So, in order to accomplish the extension of the notion of completeness to non-interpreted theories and see the connection Husserl referred to in this famous footnote, we must give some attention to the notion of formal domain, to which we now turn.

Formal Domains

In order to accomplish the extension of the concept of definiteness to non-interpreted theories, we must turn to a couple of talks that Husserl gave in Göttingen in 1901[4] (Husserl 1901) upon Hilbert's invitation. Husserl probably had those talks in mind when he indicated in *Ideas I* the possibility of extending the notion of definiteness to purely formal theories.

The problem is that the notion of definiteness for interpreted theories explicitly mentions the manifolds these theories describe. But, we may think, non-interpreted theories do not have intended interpretations. Husserl,

---

[4] The drafts for those talks have been published. See Husserl 1901. See also Chapter 6 for a more detailed treatment of the notion of formal domain.

however, would not concur. For him, purely formal theories also have intended domains, which they describe. But these are not domains of given determinate objects. Rather, they are domains of indeterminate objects only formally determined by their theories. Husserl called these domains *formal* manifolds, and the forms of objects they contain, *formal* objects. So, a non-interpreted theory describes a formal manifold, or equivalently, a formal manifold is the objective correlate (the denotation, we may say) of a purely formal theory.

If we transfer the notions of definiteness for interpreted theories to non-interpreted theories by simply substituting formal manifolds for intended interpretations, we get the following. A finite non-interpreted theory is definite when: a) it is the complete theory of its formal manifold or, equivalently, it decides any assertion referring to its formal manifold; b) it is syntactically complete. However, now, depending on how the notion of formal manifold is defined, these two variants may not be equivalent. To complicate matters, Husserl has two non-equivalent characterizations of the notion of formal manifold: one is presented in the *Logical Investigations*, and is the one he uses, to the best of my knowledge, everywhere else except in the Göttingen talks, the other is precisely the notion presented in the talks.

Let us take a look at the version presented in the Göttingen talks first. As Husserl told his audience in Göttingen, a formal manifold is the collection of all formal objects that a formal theory somehow characterizes and proves to exist. If a formal theory proves that an object exists satisfying a certain property expressed in its language, then a formal object exists in the formal manifold determined by this theory satisfying this property, and conversely. Two fundamental questions concerning formal domains ask: 1) what formal objects (the elements of formal domains) are and 2) what the conditions are for formal objects to belong to their respective formal domains. In Göttingen, Husserl presented clear-cut answers to these questions: a formal object is an indeterminate "something", characterized by a certain property $F(x)$, which belongs to the formal domain determined by a formal theory T if the formula $F(x)$ belongs to the language of T and, moreover, T proves "there is an x such that $F(x)$". A formal object belongs to the formal domain of T, if, and only if, it can be formally characterized in the language of T and proven to exist in T.

A formal object is a materially indeterminate "something" that, according to a theory, exists satisfying a certain property expressible in the language of the theory in question that formally determines it. Or, in other words, a "something" that, although completely indeterminate with respect to its nature, is determined by the theory with respect to its *form*: "It" is a "something" formally characterized by a certain formula that the theory requires to be satisfied. The formal domain of a formal theory is simply the totality of such formal objects.

Unfortunately Husserl was not very clear as to when, in general, an assertion of a formal language *refers* to a formal manifold. It would seem that any assertion of the language of a formal theory refers to the formal manifold determined by this theory. But Husserl made it clear that he thought that the assertions referring to a formal manifold constituted a subset of the set of all assertions that are expressible in the language of the theory.[5]

So, and this is the only thing that interest us here, for a formal theory to be definite, according to the first sense above (sense (a)), given the notion of formal manifold of the Göttingen talks, it suffices that it is syntactically complete with respect to a subset of the assertions expressible in the language of the theory, precisely those assertions that refer to its formal manifold. Hence, a formal theory can be definite with respect to its formal domain, but not be syntactically complete, although any syntactically complete theory will, *a fortiori*, be definite with respect to its formal domain.

This means that the two versions of definiteness for non-interpreted systems mentioned earlier are non-equivalent if we take the notion of formal manifold given in the Göttingen talks. Indeed, the text of those talks makes it clear that Husserl recognized that there are two non-equivalent notions of definiteness for formal theories. One he calls *relative definiteness*, for theories that are definite only with respect to their formal domains. The other he calls *absolute definiteness*, for theories that are simply syntactically complete. So, the two equivalent versions of definiteness for interpreted theories of *Ideas I* give us, with the concept of formal manifold of the Göttingen talks, the non-equivalent notions of relative and absolute definiteness that Husserl presented to the Göttingen listeners.

As we have seen, in the Göttingen talks, Husserl clearly states a necessary and sufficient condition for a formal object to belong to a formal domain, but he is not as clear with respect to the conditions for an assertion of the language of a formal theory to refer to its formal domain. In *Logical Investigations*, however, the situation is reversed. Husserl seems to accept that *any* assertion of the language of a theory refers to its formal domain, but he is silent as to what a formal object is and what the conditions are for a formal object to belong to a formal domain. In *Logical Investigations* a formal manifold is introduced simply as the objective correlate of a formal theory. Its objects remain completely indeterminate with respect to matter and are determined exclusively with respect to form by the formal theory. The operations to which these objects are submitted and the relations under which they fall are indeterminate with respect to their nature, but how they "act" on formal objects, which give these objects their formal properties, is clearly expressed by the theory.

---

[5] See Chapter 6 for details.

Given a formal theory T, written in a certain formal language L, considering now the presentation of the notion of a formal domain of *Logical Investigations*, we still have the following two questions: 1) When does a sentence of L refer to the formal domain of T? 2) When does a formal object belong to the formal domain of T? Since in *Logical Investigations*, Husserl does not make any distinction between a relative and an absolute notion of completeness, I believe we are safe in supposing that he also does not distinguish between assertions referring to the formal domain of T and assertions of L in general. So, I claim, in *Logical Investigations* Husserl implicitly accepted that *any* sentence of L refers to the formal domain of T.

Now, since the Göttingen talks were delivered *after* the ideas of *Logical Investigations* were already established, I also believe we can safely suppose that Husserl did not consider the conditions explicitly stated in those talks for a formal object to belong to a formal domain (or something close to them) to be inconsistent with the characterization of formal domains given in *Logical Investigations*. Therefore, I propose to give to question (2) above the same answer that Husserl gave to this question in the talks: A formal object belongs to the formal domain of T if, and only if, it can be formally characterized and proven to exist in T.[6]

Considering now that *any* assertion expressible in the language of a formal theory is supposed to refer to its formal domain, to say that a theory is definite with respect to its formal domain is simply to say that it is syntactically complete. Therefore, taking into consideration the notion of formal manifold of *Logical Investigations*, the two notions of definiteness for non-interpreted systems presented above, more than just being equivalent, actually coincide. Hence, a purely formal system is definite just when it is syntactically complete.

Let us turn now to the question of the relation of Husserl's notion of definiteness to Hilbert's axiom of completeness.

Husserl's Notion of Definiteness and Hilbert's Axiom of Completeness

In order to see the connection between the notion of definiteness and the idea of inextensibility that presides over Hilbert's axiom of completeness we must answer the following question: How can a formal manifold be extended? No matter which presentation of the notion of

---

[6] Another reading would be to let the object-constants of the formal theory refer to specific formal objects and object-variables vary over formal objects without individualizing them by defining formulas provably satisfiable in the theory. In this case, formal objects would simply be referents of object-denoting symbols of the language, variables, constants, and terms. The notions of semantic completeness with respect to the (arguably vaguely characterized) formal domain of a theory would, then, be indistinguishable from that of syntactic completeness of the theory itself.

formal manifold we choose, that of the Göttingen talks or *Logical Investigations*, a *new* formal object can only be adjoined to a formal manifold if the theory of this manifold is enlarged by new axioms. This is due to the fact that new formal objects can only be added to a formal manifold if new formal expressions, which require these new objects to exist, can be proved. And this requires an enlargement of the system by the addition of new axioms.

Now the connection between Husserl's notion of definiteness and Hilbert's axiom of completeness is evident, as Husserl said it was. Since a formal manifold can be extended if, and only if, its axiom system can also be extended, the formal domain of a definite formal theory is inextensible. Unless, of course, we enlarge the language of this theory in order to define and refer to new formal objects, and redefine operations in order to take care of these new entities, as we shall see in what follows.

Definiteness, Imaginary Elements and the Vindication of Symbolic Reasoning

As we have said before, it was the problem of imaginary entities that pushed Husserl into an investigation of the theory of formal systems. So, in the end, what was his answer to this problem? In the Göttingen talks he says that an imaginary entity is, from the perspective of a formal axiomatic system, an entity that does not belong to the formal domain of this system. Moreover, imaginary entities can be adjoined to a formal domain as convenient, although dispensable auxiliary devices, provided that the theory of this manifold is definite with respect to it.[7]

If we consider the notion of definiteness for theories that Husserl seems to consider definitive (i.e. syntactical completeness), imaginary objects can be adjoined to a formal domain provided the theory of this domain is syntactically complete. As mentioned earlier, an extension of a formal domain can only be accomplished by extending its theory. So, the introduction of imaginary objects into a formal domain requires the enlargement of the language used to describe it with new symbols (symbols for new operations, new variables for the imaginary entities, etc.) New axioms will then be introduced in order to define the imaginary entities and extend to them the operations defined previously only for the elements of the narrower domain. The narrower axiomatic system, and *a fortiori* its enlargement, remain definite only with respect to the assertions of the narrower language, and so they are in a sense only relatively definite (as Husserl stated in the Göttingen talks, the solution of the problem of imaginary entities requires only a restricted notion of definiteness).

---

[7] See Chapters 3 and 6.

Regardless of its technical merits, Husserl seems to have attached only restricted philosophical relevance to this solution of the problem of imaginary elements,[8] for the Göttingen talks are the only place, as far as I know, where he discusses it in some detail. The reason seems clear. Husserl had already found in *Logical Investigations* a solution to the broader question concerning the logical-epistemological justification of symbolic reasoning in general: symbolic thought is scientifically justified to the extent that it provides formal *a priori* knowledge. But definiteness still has a role to play here. Since formal theories are seen as theories of possible objective domains, definiteness is required of formal theories as an ideal of completion. Non-definite theories do not give us everything that can be known *a priori* of possible domains of objects exclusively with respect to their form.

There is in Husserl a clear connection between a syntactic and a semantic notion of completeness: to a definite formal theory corresponds a formal manifold that cannot be extended by the expressive powers of the language of this theory alone. Or, in other words, formal domains of definite formal theories are maximal with respect to inclusion, provided we do not extend the language of the theory in question by adjoining new symbols to it.

Of course, it takes some good will to accept that a *formal* domain is a *semantic* counterpart of a formal theory. The fact is that, for Husserl, it is not possible to conceive of a theory, even a purely formal theory, independently of a domain, even if only a formal domain, this theory refers to. Let us see why.

Formal Ontology[9]

For Husserl, the reference of a true sentence is a state-of-affairs, a *fact*. Names, symbols for relations, operations and similar syntactic components of sentences are each supposed to refer to a component of a state-of-affairs. In the process of formalization, terms, relation and operation symbols are devoid of any pre-determinate reference whatsoever. So, what does the formalized sentence refer to?

Any state-of-affairs can be seen as composed of matter (the *what*) and form (the *how*). Its matter is the specific objects, relations and operations it is made of; its (logical) form is the *how* these components are combined in this particular state-of-affairs. For example, the form of the state-of-affairs given by $2 < 5$ is expressed by aRb, where R is a relation-constant and a and b object-constants. This form expresses that two indeterminate objects stand in an indeterminate binary relation.

---

[8] Which, nonetheless, maintains its mathematical relevance, as Husserl stressed in many occasions.
[9] See my "Husserl's Conception of Logic". (da Silva 1999)

We can say, and this is the answer to the question above, that, for Husserl, formal sentences denote *forms* of states-of-affairs. Hence, a formal theory, viewed simply as a collection of sentence-forms, denotes the *form* of an objective domain, or, in other words, a domain characterized exclusively with respect to form. This form is characterized as follows: we select a language having symbols for indeterminate objects (object-variables and object-constants), operations (operation-variables and operation-constants), relations (relation-variables and relation-constants), properties (property-variables and property-constants), and the like, and express in this language whatever formal properties are considered valid for whatever entities "fill in" the symbols of this language; for instance, the commutativity of one particular operation, the reflexivity of another, etc. The collection of these formal expressions constitutes a formal theory, whose objective correlate, its reference, is a formal domain (or, equivalently, a form of domain). Precisely the kind of formal domains we discussed above. My point here is only that, for Husserl, the introduction of formal domains as correlates of formal theories is a straightforward consequence of his ideas concerning the denotation of true sentences. True sentences denote states-of-affairs, theories proper—collections of true sentences—denote a "world", formal theories denote formal domains, "world-forms".

For Husserl, this perspective opens regions in the domains of logic that are not usually considered to belong to it. In general logic is viewed as concerned only with statements (or propositions, or any other truth-bearer we may prefer), how they are properly formed and can be composed in order to form complex statements (logical grammar), how some statements can be derived from others (theory of deduction), etc. In short, logic is supposed to be concerned, in one way or another, as Frege wanted it, with truth. This is why it is usually thought to be focused exclusively on truth-bearers. But for Husserl logic is *also* a theory of objects, provided that these objects are considered exclusively as the matter on which logical forms are imprinted. Or better, for Husserl, logic is also a theory of logical forms themselves. This is what he called *formal ontology*.

Formal ontology is concerned, among other things, with the study of particular formal domains, their properties, how they are related to other formal domains, and the like. Of course, this is not a new science, but just what formal mathematics has been doing at least since Grassmann and Riemann, the mathematicians who introduced the idea of the mathematical study of formal manifolds. So, for Husserl, formal ontology includes the whole of formal mathematics.

Of course, there must be a close relation between formal theories and their domains, since whatever can be said about a formal domain must be derivable in the theory to which this domain is a correlate. In general, Husserl sees a close connection between the logic of statements and formal ontology. We do not need to enter into the details of this connection here; all I want to stress is that to any meta-logical assertion of the logic of statements corresponds a similar assertion of formal ontology. Logical rules, laws and principles are always twofold. For Husserl, they are supposed to refer, by a shift of perspective, either to statements or to states-of-affairs, or else formal theories and formal manifolds. So, it is to be expected that a property of theories such as syntactic completeness necessarily translates into a property of formal domains. In this case the property of being maximal with respect to the expressive powers of the language in question. And precisely *this* is the property that Husserl considered germane to the idea of completeness behind Hilbert's axiom of completeness.

Hence, by adopting the right perspective on the necessary correlation between formal theories and their domains, we can see why Husserl claimed that there is a "close relation" between his notion of definiteness and Hilbert's notion of completeness related to the axiom of completeness.

Concluding

I believe that by showing that Husserl's notions of completeness and the different versions in which he presents them correspond to careful distinctions he made between semantics and syntax, theories and manifolds, interpreted theories and formal theories, the mathematical problem concerning imaginary entities and the philosophical problem concerning symbolic reasoning, I have shown that Husserl's notions and his remarks concerning their relation to Hilbert's axiom of completeness are free of the sort of confusion that some scholars have attributed to them.

Although the presentation of the notion of definiteness in §72 of *Ideas I* (Husserl 1913) can be read either semantically or syntactically, for Husserl introduces it first for *interpreted* theories, he also considers its extension to purely *formal*, non-interpreted theories. In this case the notion of definiteness corresponds to *syntactic* completeness. Since to any formal theory there corresponds a formal domain that inherits from it certain properties, the property, for a formal theory, of being maximal with respect to the inclusion of new (formal) axioms (i.e. syntactic completeness) translates, for its formal domain, into the property of inextensibility that Husserl thought as essentially the same as Hilbert tried to capture with his axiom of completeness.

I also believe my analysis helps to bring out at least some of the reasons Husserl had for abandoning his project for a second volume of *Philosophy of Arithmetic* and presenting the logical-epistemological theories he put forward in *Logical Investigations*.

References

Bachelard, Suzanne 1957. *A Study of Husserl's Formal and Transcendental Logic*, Evanston IL: Northwestern University Press, 1968.

Cavaillès, Jean 1947. *Sur la logique et la théorie de la science*, Paris: Vrin, 1997.

da Silva, Jairo José 1999. "Husserl's Conception of Logic", *Manuscrito* 22, no. 2, 367-97.

Husserl, Edmund 1887-1901. *Philosophy of Arithmetic, Psychological and Logical Investigations with Supplementary Texts from 1887-1901*, translated by Dallas Willard, Dordrecht: Kluwer, 2003, the translation of *Philosophie der Arithmetik, mit ergänzenden Texten (1890-1901)*, Husserliana XII, introduction by Lothar Eley, The Hague: Martinus Nijhoff. *Philosophy of Arithmetic* was originally published as *Philosophie der Arithmetik*, Halle: Pfeffer, 1891.

Husserl, Edmund 1900. *Prolegomena to Pure Logic,* volume I of Husserl 1900-01, translation of *Prolegomena zur reinen Logik*.

Husserl, Edmund 1900-01. *Logical Investigations,* London: Routledge, 2001, translation of *Logische Untersuchungen*, Halle: Niemeyer.

Husserl, Edmund 1901. "Double Lecture on the Transition through the Impossible ("Imaginary") and the Completeness of an Axiom System", included as Essay III in Husserl 1887-1901, 409-52. Cited in the text are my translations of Husserliana XII, 430-47.

Husserl Edmund, 1913. *Ideas: A General Introduction to Pure Phenomenology*, translated by W. R. Boyce Gibson, London: Allen & Unwin, 1931, translation of *Ideen zu einer reinen Phänomenologie und phänomenologischen Philosophie I,* Husserliana II, The Hague: Martinus Nijhoff, 1950.

Sebestik, Jan 1997, "Postface", in Cavaillès, 91-142.

8

Claire Ortiz Hill

FREGE'S LETTERS

Gottlob Frege's isolation is almost legendary. Michael Dummett portrayed him as having been a man who was perhaps too original to have ever been capable of working with others, "of sailing on any sea on which other ships were in sight...", and someone who "never seems to have learned from anybody, not even by reaction..." (Dummett 1981a, 661). Dummett's Frege felt isolated, misunderstood and unlistened to in the philosophical and mathematical world of his time, led a "life of disillusionment and frustration" (*Ibid.*, xxxi). In addition to this, we have Bertrand Russell's well-known claims that Frege's work had gone virtually unnoticed until Russell discovered it in 1900. (Russell 1946, 858, for example)

There is certainly a good deal of truth in such accounts of Frege's life. Born in 1848, Frege entered the University of Jena at the age of twenty-one. He left Jena for the couple of years he needed to be obtain his doctorate at Göttingen and then returned to Jena when he was twenty-six, remaining there until he retired forty years later (Sluga 41-42; Dummett 1981a, xxxi). Outside of his own work, little, if anything, of lasting interest transpired there in his field during his tenure. And in any case, Frege apparently eschewed all personal contact with the academic community, both in Jena and elsewhere. Although Jena is just a few hours by train from the Universities of Göttingen, Berlin, Leipzig and Halle where Karl Weierstrass, David Hilbert, Georg Cantor, Ernst Zermelo and others were making breakthroughs in Frege's field, he was little inclined to travel. Hilbert chided him about this in a letter saying that it was a shame that Frege had missed two recently held professional meetings. "Since rail travel is so comfortable today", Hilbert wrote, "personal communication is surely preferable to the written kind" (Frege 1980b, 52). From Frege's correspondence, we also know that he turned down opportunities to speak at the International Congress of Philosophy held in Paris in 1900 (*Ibid.*, 6-7) and to take part in a Mathematical

Congress in Cambridge. In the latter case, he cited "weighty reasons" for his going to Cambridge, but mysteriously cited "something like an insuperable obstacle" that was keeping him from it (*Ibid.*, 170). Lothar Kreiser has even suggested that Frege seemed almost to have made it a principle not to speak at all to others about mathematics or logic (Kreiser 1973, 523), behavior that is perhaps not so surprising on the part of one who was so worried about "the many illogical features... at work in language" (Frege 1979, 266), and who so often and so vociferously voiced his complaints about the imperspicuous and imprecise character of investigations conducted in words. (Frege 1980b 33, 57-58).

Perhaps it was this aloofness on Frege's part that has made it seem possible that he was unknown and unappreciated until Bertrand Russell discovered him and that he was, as Michael Dummett claims, totally oblivious to the work of others (Sluga 1980, 69-76). Frege's correspondence, however, tells another story. Frege actually left behind a valuable record in the form of letters of exchanges he had engaged in with the leading people in his field. Bertrand Russell was perhaps just too young in 1900 to have heard of the older Frege who by 1900 had acquired enough of an international reputation to have, since the 1880s, engaged in significant written, if not personal, exchanges with, among others, the Italian logician Giuseppe Peano (Frege 1980b, 108-29), the future phenomenologist Edmund Husserl (*Ibid.*, 61-65), the French philosopher Louis Couturat (*Ibid.*, 6-8), the extensionalist logician Ernst Schröder (Dummett 1981a, 661), mathematicians David Hilbert (Frege 1980b, 31-43) and Georg Cantor (Cantor 1885, 728-729; Frege 1885, 1030), and several noted members of Franz Brentano's Austrian school. (Frege 1980b, 61, 99, 171)

Most of the exchanges cited above were epistolary and many of the letters exchanged, as well as some of the information we have about them, are found in the English edition of Frege's correspondence, *which is abridged*. Unfortunately, many of the letters to and from Frege have been lost. Brian McGuinness, the abridger of the English edition, however, consoles readers who might mourn the loss by advising them that it "would be unphilosophical, and perhaps improbable, to suppose that what we have lost must have been more interesting than what remains to us" (*Ibid.*, xvii). He further informs readers that the abridged English text only includes the actual surviving texts considered by him to be of scientific interest and that he has eliminated what he dismisses as "laundry lists". In fact, he has left out what information we still have about the missing letters, omitting in particular the dates we have for them and what in one case he calls "tantalizingly jejune notes" made by Heinrich Scholz, who had actually read much of the now missing material. In addition, McGuinness admits to having "heavily abridged" the notes found in the German edition along with much pertinent information about the letters (*Ibid.*, xvii). For example, he has eliminated twelve of the twelve and a half pages of introduction to Russell's correspondence with

Frege (Frege 1976, 200-11), and five pages of information about each the Frege-Wittgenstein correspondence (*Ibid.*, 264-68) and the Frege-Löwenheim correspondence (*Ibid.*, 157) so that anyone wishing anything more than a superficial introduction to questions raised or resolved by studying Frege's correspondence is obliged to turn to German editions.

In fact, very close study of the content of the letters we do have and of what we actually know about the letters we do not have makes McGuinness' remarks seem presumptuous and his abridging unscholarly. Frege wrote "detailed, substantial" letters (his correspondents commented upon this: Frege 1980b, 23, 64, 106) in which he tried to explain some of his most basic and important ideas. In them, he discoursed at length about such matters as sense and meaning (*Ibid.*, 61-64, 79-80, 152-54, 163-65), the paradoxes of set theory (*Ibid.*, 130-69), identity (*Ibid.*, 95-98, 113-16, 126-27), and the foundations of geometry (*Ibid.*, 32-51, 90-93). The portion of the correspondence presently available in published form has already played an important role in discussions concerning the influence Frege may have had on Edmund Husserl's thought and the originality of Frege's distinction between *Sinn* and *Bedeutung* (Mohanty 1982). Frege's correspondence records his reactions to Russell's discovery of the paradox and to the solutions Russell proposed for it (Frege 1980b. 130-69). It has shed light on the origins of the analytic tradition in philosophy. It has shown the extent to which Russell and Frege disagreed (*Ibid.*, 78-84, 135, 155-56, 159, 163, 165, 168-69) and Frege's keenness in pointing out problems in Russell's reasoning (*Ibid.*, 166). It clearly shows that Frege's works were known, read and appreciated in the 19th century (*Ibid.*, 6, 64-65; Frege 1976, 145). An 1882 letter indicates that Frege was already aware then of the very problems associated with determining the extension of a concept, with language's fatal tendency to transform concept words into proper names, and with confusions about concepts and objects (Frege 1980b, 100-01) that he, once he had struggled with Russell's paradox, would ultimately blame for the failure of his project to logicize arithmetic (*Ibid.*, 55, 191). The same letter also clearly shows (*Ibid.*, 99, 171) that Russell was wrong to think with regard to the *Begriffsschrift* that he was "the first person who ever read it more than twenty years after its publication". (Russell 1919, n. 25)

Moreover, what the letters divulge about what Frege thought on these matters is often consistent in quality with the discussions found in his published writings, to which the letters provide an illuminating complement. Frege had, in fact, wanted to publish his correspondence with Hilbert concerning the foundations of geometry (Frege 1980b, 48) "because", he said, "of the importance of the questions discussed in it" (*Ibid.*, 92), and "because reading this correspondence would have been the most convenient way of introducing someone into the state of the question, and it would have saved me the trouble of reformulating it" (*Ibid.*, 32). Frege and Löwenheim had at one time intended to publish their twenty-letter exchange of ideas on formalism (Frege 1976, 158, 161). Letters Frege and Peano exchanged actually were published in the *Rivista di matematica* in 1896 (Frege 1980b, 112-20). In addition, Frege left among his papers an impressive number of drafts of the letters he sent,[1] a fact consistent with his striving for rigor and exactness which indicates that he took his letter writing seriously. Fortunately, some of these drafts (or often fragments of them) have survived, often comprising all we now know about letters of which no other trace remains.

So I am arguing that it is rather probable to suppose that what we have lost was quite interesting and that it too could shed light on questions Frege's work raises, which may have remained unresolved until now because we have not been in full possession of the facts. So, in the next pages I propose to engage in the "unphilosophical" task of systematically trying to piece together what we actually can know about the philosophical content of the letters that have been lost. I must begin by going over the story of the fate met by Frege's literary estate.

The Story of the Letters

In 1919 a chemist, Ludwig Darmstaedter, asked Frege whether he might contribute letters addressed to him to an autograph collection. Frege actually selected several letters to be set aside for this collection and upon his death in 1925, Alfred Frege, his adopted son, handed "a not very extensive collection of letters to Frege" over to the Prussian National Library that housed Darmstaedter's collection (*Ibid.*, xi-xii). Although the English edition of Frege's has been cleansed of most of the information regarding the fate of the individual letters, by piecing together the notes in the unabridged German edition we find that a little more than half of the one hundred and twenty-one letters published there were part of this autograph collection, which seems to have remained largely intact.

As one might expect of a person choosing letters for an autograph collection, Frege principally chose letters that were signed by well-known people, but which were relatively void of philosophical or personal content. Often these were the "requests for an offprint, refusals to print an article, apologies for not writing" that McGuinness has qualified as "laundry lists" (*Ibid.*, xvii). With only one or two exceptions, the collection consists entirely of letters written to Frege, not by him. By far the most important letters Frege contributed were those that Bertrand Russell had written to him, including the 1902 letter in which Russell informed Frege of the famous paradox. Frege's reasons for consigning Russell's letters to him to an autograph collection are not known.

---

[1] Reading the notes in the English and German editions of the correspondence one finds information about Frege's drafts of letters to Russell, Wittgenstein, Jourdain, Hilbert, Peano, Pasch, Huntington, Zsigmondy, Linke, Jones, and Speiser, among others.

In the 1930s the German logician Heinrich Scholz began collecting Frege's unpublished writings and correspondence with the intention of publishing them in a three volume work. He succeeded in finding the letters written to Frege that had been preserved in the autograph collection. At the International Congress of Scientific Philosophy held in Paris in 1935, he appealed for help in locating letters written by Frege to others (*Ibid.*, xi-xv). Again he was fortunate enough to acquire valuable material. In exchange for photographic copies, Bertrand Russell provided him with nine letters (Frege 1976, 200). Leopold Löwenheim made available another ten letters (*Ibid.*, 159). Additional letters came from Edmund Husserl (*Ibid.*, 93), David Hilbert (*Ibid.*, 57) and other distinguished correspondents.

From Frege's adopted son, Alfred Frege, Scholz managed to acquire most, perhaps all, of the philosophically interesting papers and letters the older Frege had left in his son's care. Besides Frege's unpublished writing, the papers included letters that Löwenheim, Wittgenstein, Husserl, Hilbert and others had written that Frege had not judged suitable for an autograph collection, and many drafts Frege had made of the letters he wrote. Scholz also examined and sent back to Alfred Frege some letters of a purely personal nature. Wittgenstein declined to make Frege's letters to him available. Others had not kept their correspondence. (Frege 1980b, xii-xiii)

Not including the autograph collection, Scholz was all in all able to obtain approximately one hundred letters. In collaboration with others, he then went to work preparing the material for publication (Frege 1979, x). However, World War II intervened with disastrous consequences. For safekeeping, Scholz fatefully placed the materials he had collected in the University of Münster library. The building was bombed on March 25, 1945, and Scholz' collection was lost.[2] Fortunately, he had typed copies of some of the material. He had also catalogued and made notes on the content of other materials of which there is now no longer any trace. His lists and the copies he made of about fifteen of the letters survived the war (Frege 1980b, xiii wrongly seems to imply that they all survived), as did the autograph collection. However, more than three-fourths of the letters Scholz had gathered, along with any copies he may have made of them, have vanished. For example, Scholz had worked particularly hard on Frege's correspondence with Löwenheim. Unfortunately, the twenty Frege-Löwenheim letters that Scholz had managed to acquire, a complete transcript of all twenty letters, plus two forty-seven page thick typed copies of the correspondence up to the beginning of the fourteenth letter, some drafts of letters and comments on a 1908 letter from Frege to Löwenheim all disappeared from Scholz' Münster archive (Frege 1976,

---

[2] Interestingly, this loss is not mentioned in Scholz' wartime correspondence with his colleague Helmut Hasse found in Niedersächsische Staats- und Universitätsbibliothek, Abteilung Handschriften und Seltene Drücke, University of Göttingen.

159). None of this would be apparent to readers of the English edition, which only mentions Löwenheim once in passing. (Frege 1980b, xvii)

Scholz resumed his work after the war and, following his death in 1956, others eventually set about to retrieve, as much as possible, what had been lost and complete what he had begun (Frege 1979, xii-xiii). Efforts made to determine whether anything of value might have remained in Alfred Frege's possession proved to be in vain. He was killed in 1944, while serving as a soldier in France. The house he had inherited from his adopted father was requisitioned for use by the Soviet army, repeatedly plundered and then used to house refugees. Alfred Frege's own house was completely destroyed by bombing in 1943. (Frege 1980b, xiv-xv)

Using what evidence remained at hand, extensive efforts were also made to locate additional letters. These endeavors were sometimes crowned with success and over the years approximately fifty more letters were found that had been preserved in other archives, published, or kept by Frege's correspondents or their families.[3]

Frege's correspondence was finally published in 1978, more than twenty years after Scholz died and more than forty years after he had begun his work on it. All in all, scholars have been able to relocate more than half of the approximately two hundred and fifty that have been identified as having been written to or by Frege. Unfortunately, over a hundred letters and their copies seem to be irremediably lost, and even where copies or drafts survive, they often contain frustrating gaps.[4]

Three Periods

Losing the letters of a philosopher, even those of an important philosopher, is not in and of itself a reason for great dismay. Much, perhaps most, of what learned people write in letters is of little abiding interest. I have, however, tried to provide some reasons why Frege's correspondence may be an exception to this. Now I want to argue that the close study of the letters we have and what we can know about the letters we do not have actually discloses some very interesting, and not immediately apparent, facts about the latter. To make my point I need first to divide Frege's thought into the following three periods (Dummett 1981b, 7-27; Dummett 1981a, 629-64):

---

[3] The relevant notes in the English and German editions show that letters from Frege to Marty, Husserl, Jourdain and Hilbert were found in various archives; letters to Liebmann and Peano had been published; Dingler's wife had copies of her husband's correspondence; Russell had photographic copies of Frege's letters to him.

[4] For a more complete account of the fate of the letters see: Veraart 1976; Scholz 1936; Kreiser 1974 and Kreiser 1973. I am obliged to give approximate figures since so many letters fall into more than one category. For example, many letters that were lost were recovered, or partially recovered through photocopies, drafts, typescripts, some of the letters Scholz inventoried were only known through other letters, letters Frege wrote to Wittgenstein were discovered at a later time, etc.

1. The first period extends from the publication of Frege's *Begriffsschrift* in 1879 until June 1902, when Russell first wrote to Frege about the contradiction now commonly known as Russell's paradox. Almost all Frege's works were written and published during this period (Dummett 1981a, 685-86; Frege 1980b, 130). Even the 1903 *Basic Laws II* was already in the press when Frege received Russell's letter. Approximately one fourth of the nearly two hundred and fifty letters Frege is known to have exchanged date from this period. With a few quite interesting exceptions,[5] the surviving letters were among those relegated by Frege to the autograph collection, and so have survived and been published.

2. The second period encompasses the brief time from June 1902 through 1906, during which, as Dummett has argued (Dummett 1981b, 21-23), Frege continued his logical work still hoping to find solutions to the questions Russell's finding raises. *Basic Laws* II appeared in 1903, and while working on *Basic Laws III* (Dummett 1981a, 658), Frege published little else (*Ibid.*, 686). We know of approximately fifty letters exchanged during this period. Two thirds of these belonged to the autograph collection that housed the ten important letters Russell sent to Frege between 1902 and 1904. These have survived along with a letter and a postcard found by Ivor Grattan-Guinness at the Mittag-Leffler Institute in Sweden, copies that Scholz made of two letters that Frege wrote to Husserl, and Russell's photographic copies of Frege's letters. The remaining letters were collected by Scholz and are lost.

3. The third period ranges from 1907 until Frege's death in 1925. During the year 1906, Frege apparently abandoned his work on the third volume of the *Basic Laws*, utterly persuaded that his logical work was irremediably flawed (Dummett 1981a, 657-58; Dummett: 1981b, 9-10, 21-27). Well over half of the letters to and from Frege that have been inventoried over the years were written during these years in which Frege wrote very little else. Frege in fact wrote substantially more pages of letters during these years than he did pages of published works. Though the letters we know something of only represent a portion of Frege's epistolary output during this time, they alone already indicate that (Dummett 1981a, 664, 686; Dummett 1981b, 22-23). Moreover, the fact that the dates of twenty-one letters and cards from Frege to Wittgenstein found in Austria in the late 1980s do not correspond with the dates Scholz recorded for the fifteen letters from Wittgenstein to Frege once housed in his Münster Archive indicates that many Frege-Wittgenstein letters are still missing.[6]

---

[5] Scholz' successors were able to locate letters Frege had written to Edmund Husserl, Anton Marty, David Hilbert, and Heinrich Liebmann.
[6] Compare the list in Frege 1989, 'Briefe an Ludwig Wittgenstein', *Grazer philosophische Studien,* 33/34, 8 with the list in Frege 1976, 265-68.

Two-thirds of the approximately one hundred and fifty letters that we presently know Frege exchanged during the last nineteen years of his life were acquired by Scholz and they, along with all but a handful of the copies Scholz was able to make (not all of them complete) were lost in the Münster bombing. Unfortunately, there is reason to believe that it is precisely the most interesting letters that are missing from these last two, crucial periods, for which we have so little other indication of what Frege was thinking. It is, after all, only reasonable to conjecture that Frege would have selected less important letters for an autograph collection, keeping the letters he deemed to be particularly significant among papers which he thought would one day be valued more highly than they were at the time of his death (Frege 1979, IX). In what follows, however, I wish to put such conjectures aside and set out some philosophical reasons for considering the lost letters to be significant.

Subjects Treated

The principal reason, I contend, for believing that the missing letters dealt with matters of particular interest to 20th century philosophy is that, studying the letters written after 1902 that have survived and reflecting on the notes Scholz left regarding the letters he once had in his possession, one cannot help but notice that the now missing materials repeatedly dealt with certain specific, intimately related subjects of prime importance. These subjects are: the paradoxes of set theory and possible solutions to them; extensionality and classes; the differences between concepts and objects; identity (or equality since Frege considered them to be the same); and Frege's opinion of the work of his contemporaries.

Interestingly, many other materials on these same subjects have disappeared in distressing ways, a fact that I shall partially document as I go along.

The way these subjects are interconnected and their significance for Frege at that crucial time in his life during which he abandoned his logical work becomes apparent when one considers that when Russell informed him of the famous paradox of set theory, Frege immediately traced the origin of the contradiction to his axiom of extensionality, Basic Law V of the first volume of *Basic Laws* (Frege 1980b, 130-32). In an appendix to the second volume of *Basic Laws,* then in the press, Frege proposed a solution which involved the modification of Basic Law V (Frege, 1980a, 214-24). For Frege, Russell's contradiction indicated:

> that the transformation of an identity into an identity of ranges of values (sect. 9 of my *Basic Laws)* is not always permissible, that my law V (sect. 20, p. 36) is false... the collapse of my law V seems to undermine not only the foundations of my arithmetic but the only possible foundation of arithmetic as such. And yet, I should think it must be possible to set up conditions for the transformation of the generality of an identity into an identity of ranges of values so as to retain the essential of my proofs. (Frege 1980b, 132)

Frege himself had never been completely satisfied with Basic Law V and he more than once admitted that he had formulated it because he saw no other way of logically grounding arithmetic other than by appealing to the extensions of concepts he reticently began using in §68 of his 1884 *Foundations* (Burge 1984). In the brief overview of that book that figures on the closing pages of the work, Frege acknowledged that appealing to extensions would not "meet with universal approval" and claimed to "attach no decisive importance even to bringing in extensions of concepts at all" (Frege 1884, §105). When he published the first volume of *Basic Laws* in 1893, he was more explicit about his reservations: "If anyone should find anything defective", he wrote there, "he must be able to state precisely where, according to him, the error lies.... A dispute can arise, so far as I can see, only with regard to my Basic Law concerning courses of values (V)...". This is where, he believed, the decision would be ultimately made. (Frege 1893, 3-4)

By transforming "a sentence in which mutual subordination is asserted of concepts into a sentence expressing an equality", Basic Law V would permit logicians to pass from a concept to its extension, a transformation that, Frege held, could "only occur by concepts being correlated with objects in such a way that concepts which are mutually subordinate are correlated with the same object". However, Frege never believed that any proof could be supplied that would sanction such a transformation "in which concepts corresponded to extensions of concepts, mutual subordination to equality". So he devised Basic Law V to mandate that it was true. (Frege 1980a, 214; Frege 1979, 181-82)

In 1912, having already given up trying to save his logic, Frege wrote for an article by Philip Jourdain:

> And now we know that when classes are introduced, a difficulty, (Russell contradiction) arises... . Only with difficulty did I resolve to introduce classes (or extents of concepts) because the matter did not appear to me to be quite secure –and rightly so as it turned out. The laws of numbers are to be developed in a purely logical manner. But numbers are objects.... Our first aim was to obtain objects out of concepts, namely extents of concepts or classes. By this I was constrained to overcome my resistance and to admit the passage from concepts to their extents.... I confess... I fell into the error of letting go too easily my initial doubts. (Frege 1980b, 191)

Six years earlier, Frege had written of this transformation that he considered to be so vital to this theories that if it had been "possible for there to be doubts previously, these doubts have been reinforced by the shock the law has sustained from Russell's paradox" (Frege 1979, 182). Frege was convinced that "everybody who in his proofs has made use of extensions of concepts, classes, sets", was in the same position he was. (Frege 1980a, 214; Sluga, 162-75)

Smart Bombs

In order to carry my argument concerning the importance of the lost letters further, I need to take a close look at the fate of specific letters dating from the last two periods and at how the important subjects discussed above figured in Frege's epistolary exchanges once he was confronted with Russell's paradox.

In June 1902, Frege received Russell's letter regarding the paradox. Frege replied, pinpointing the exact place he believed the error to be. Over the next two years, Russell and Frege exchanged several letters in which each proposed and critically discussed possible solutions to the problems Russell's finding unearthed. Scholz acquired all the letters known to have been part of this exchange and his collaborators, Friedrich Bachmann and Marga Titz, set to work writing about them. The letters Russell had furnished were lost in the Münster bombing and only sixteen pages of Bachmann and Titz's commentary on them have survived. In 1963, however, Scholz' successors managed to retrieve the photographic copies that Scholz had made for Russell, so these Frege letters now number among the few letters from Scholz' archive that are available today. (Frege 1976, 200, 210-11)

Immediately upon receiving Russell's letter regarding the contradiction, Frege replied writing that, though he was surprised beyond words and left thunderstruck by the discovery of the contradiction, "it may perhaps lead to a great advance in logic, as undesirable as it may seem at first sight" (Frege 1980b, 132). In September 1902, Frege replied to a letter from Philip Jourdain, who was then writing a book on the history of set theory. In the letter, Frege mentions that Russell had called his attention to the fact that Basic Law V was in need of restriction, but Frege still displays the same confidence that a satisfactory solution would be found (*Ibid.*, 73). This letter (a draft of which once existed in the Münster archive) and a March 1904 postcard Frege sent to Jourdain were eventually found in the Mittag-Leffler Institute in Sweden (Frege 1976, 109-12). In the postcard, Frege alludes to a letter from Jourdain. An entry in Jourdain's notebooks records a letter written to Frege at that time about "getting over Russell's and Burali-Forti's contradiction by limitation on conception of class–so that mathematical conceptions can apply to it". This letter is now missing. (Frege 1980b, 74n.)

During those years, Jourdain also corresponded with Georg Cantor (Grattan-Guinness, 1971), Russell, Ernst Zermelo, Giuseppe Peano, John Venn, Arthur Schönflies about their work, and with G. H. Hardy regarding Cantorian set theory. Unfortunately, although many of the letters survived, Jourdain's manuscripts have vanished and anything Jourdain's widow may have had in her possession was destroyed in 1940 and 1946. (Grattan-Guinness 1977, 6-7)

In 1902, Frege also replied to a letter sent by the American mathematician, Edward Huntington (Frege 1980b, 57). A draft of this letter was housed in Scholz' archive and a typed copy of it survived the war (Frege 1976, 88). The copy, however, contains a frustrating gap precisely at the point at which Frege discusses his project of basing arithmetic solely on logic and the justification of inference by appealing to purely logical laws. Frege's letter picks up again with the words: "what I call endless, namely the number of all finite numbers, and show that endless is not a finite number" (Frege 1980b, 57). These words tantalizingly recall Russell's words in a second 1902 letter to Frege which tells of having been led to the contradiction while studying Cantor's proof that there is no greatest cardinal number. (Russell 1903, §100; Frege 1980b, 133-34)

In January 1903, Moritz Pasch, a mathematician interested in the logical and methodological problems of mathematics (Frege 1980b, 103), wrote to Frege to inquire as to how he proposed to explain infinitely distant points. According to a later letter from Pasch, Frege responded in a letter containing an "in depth discussion of infinitely distant elements". Though none of Frege's letters to Pasch have ever been found, a fragmentary draft that Frege composed in reply to Pasch's question was once in Scholz' hands. According to Scholz' notes regarding that portion of the letter, Frege held that one could define infinitely distant points by considering extensions of concepts as classes as set out in §68 of *The Foundations of Arithmetic* (Frege 1980b, 106; Frege 1976, 172-73). As mentioned above, this was the very section of *Foundations* in which Frege introduced the extensions of concepts and classes that became codified in the problematic Basic Law V of the first volume of *Basic Laws of Arithmetic.* It is also in a note to this section that Frege brings up two possible objections to his identification of concepts with their extensions: The first objection, he maintains, would be that in so identifying them he would seem to be contradicting his previous statement that the individual numbers are objects, not concepts. The second objection would concern the fact that concepts can have identical extensions without coinciding.

Two 1903 letters and a postcard from Alwin Korselt to Frege were preserved in the autograph collection. In the first letter, Korselt proposes a solution to Russell's paradox concerning which he concludes: "All we need to do is hold on to the fact that a class is not a concept but the object of a concept and that a concept is not a class, a fact which you demonstrated in your essay 'On Concept and Object' (Frege 1980b, 86). Frege immediately replied to Korselt's letter, but the reply was unfortunately among the letters once entrusted to Scholz' care and is now lost. According to Scholz' notes, Frege's letter concerned the unacceptableness of Korselt's proposed solution (Frege 1976, 142; Frege 1980b, 85). Any attempts to find letters from Frege or relevant information regarding their exchange among

Korselt's papers have proved fruitless as it seems that Korselt's papers were burned by their owner in 1962. (Kreiser 1973, 522)

A 1903 postcard from David Hilbert to Frege also made its way into the autograph collection. In it, Hilbert acknowledged receipt of *Basic Laws II* and informed Frege that Russell's paradox was already known to his circle in Göttingen as Ernst Zermelo had discovered it three or four years earlier. Hilbert further claimed that he himself had "found other convincing even more convincing contradictions as long as four or five years ago". (Frege 1980b, 51)

Four years earlier, Hilbert had actually been corresponding with Frege on the significance of contradictions in axiomatic systems (Frege 1980b, 34-51), and Scholz managed to obtain from Hilbert three that letters Frege had written to him on that subject in 1899 and 1900. Frege had hoped these letters would be published (*Ibid.*, 48, 92). Scholz copied them and returned the originals to Hilbert. The copies survived the war, but the originals are nowhere to be found (Frege 1976, 57). Scholz also apparently acquired a late 1899 letter Hilbert had written to Frege that was part of this same exchange. This letter has disappeared, but copies Hilbert and Frege made of parts of it have survived and have been published (Frege 1980b, 38-43). Edmund Husserl also had access to the letters and notes he made on three of the letters have been published. (Husserl 1899)

In 1900, Frege sent Heinrich Liebmann copies of the correspondence with Hilbert and wrote to him criticizing Hilbert's views on axiomatization, once again bringing up the subject of contradictions and charging Hilbert, among other things, with blurring the distinction between first and second level concepts. Frege then went on to discourse on the radical difference between concepts and objects and the essentially predicative nature of the former. "An object can never be predicated of anything. When I say 'the evening star is Venus', I do not predicate Venus but *coinciding with Venus*" (Frege 1980b, 50, 91-92). These letters were published in 1940 and so have survived. Apparently, Liebmann never replied.

During 1906, Frege exchanged letters with another philosopher and mathematician who was participating in Hilbert's and Zermelo's discussions of the foundations of mathematics and the paradoxes of set theory, Edmund Husserl (Reid, 21, 33-34, 41, 47, 90; Schuhmann, 158). Husserl had thoroughly studied Frege's *The Foundations of Arithmetic* in his 1891 *Philosophy of Arithmetic,* a book written in the company of Georg Cantor and Carl Stumpf (Schuhmann, 19, 22; Husserl 1913, 37; Schmit, 40-48, 58-62), a man Frege had once appealed to for help in making the *Begriffs-schrift* known (Frege 1980b, 171-72). The most significant of Husserl's 1891 criticisms of Frege's work had been directed toward §§62-68 of *Foundations*, where Frege had spelled out the very views on identity and extensionality that led to the formulation of Basic Law V and that Frege apparently decided were irremediably flawed in 1906. In 1900, Husserl had moved to Göttingen where, as a 1902 note from Zermelo shows, he was privy to Zermelo's discovery of the paradox that came to bear Russell's name (Rang and Thomas, 15-22). In 1912 and 1920, Husserl, in fact, worked intensively on finding a solution to Russell's paradox. (Rosado Haddock 1973, 145-50; Schmit, 114-17)[7]

From Alfred Frege and Husserl, Scholz managed to acquire five Husserl-Frege letters dating from late October 1906 to mid-January 1907. Scholz made typescripts of the two letters Frege wrote to Husserl. These typescripts have survived and are remarkably similar in content to the brief survey, now published with posthumous writings, that Frege apparently made in 1906 of his own logical doctrines (Frege 1979, 197-202). However, the three letters that Husserl wrote to Frege have disappeared. According to Scholz' notes these letters dealt with, among other things "the paradoxes". (Frege 1980b, 70; Hill 2000)

In *The Interpretation of Frege's Philosophy,* Michael Dummett argued that Frege's posthumously published writings strongly suggest that in 1906 he became persuaded that attempts to answer the questions that Russell's paradox raises would meet with failure. During that year, Frege had begun to write an article in which he discussed Schönflies' and Korselt's work on the paradoxes of set theory and the inadequacy of the remedies they were proposing for them. Dummett considers that the presence in Frege's outline for the article 'Concepts which coincide in their extension although this extension falls under the one but not under the other' indicates that Frege was then still pursuing the solution to Russell's paradox proposed in the appendix to the second volume of the *Basic Laws*. The short twenty line plan Dummett refers to in fact indicates that Frege's article would have dealt exclusively with the themes I have been discussing. The outline plainly states that Russell's contradiction cannot be eliminated in Schönflies' way, that the remedy from extension of second-level concepts is impossible, and that set theory is "in ruins". The unfinished article itself specifically deals with Russell's paradox, and problems with extensions and Basic Law V. At one point, Frege alludes to the shock Basic Law V had sustained from Russell's paradox, but suggests that his readers put these doubts temporarily aside and carry out the operation the problematic law would mandate (Frege 1979, 176-83). It seems, however, that at this point, Frege was himself no longer able to put his doubts aside. "Tantalizingly little of the article survives...", Dummett wrote, "very probably it represents the very moment at which Frege came to realize that the attempt was hopeless". (Dummett: 1981a, 168-76)

Nothing Frege wrote after 1906 indicates that he ever again tried to salvage the specific logical doctrines he concluded had led to the paradoxes of set theory (Sluga 1980, 170-71; Scholz: 1936, 28-29). Indeed, very little on that subject survives at all. Less than half of the material Scholz had in his possession before the war is now available (Dummett: 1981b, 7-8) and many of the materials missing specifically dealt with the reasons

---

[7] See Chapter 10.

Frege's logic leads to Russell's paradox and Frege's despair concerning proposed solutions to the problems. With the exception of the incomplete draft of the Schönflies article discussed above and other incomplete draft of a 1925 article, which is discussed below, all the unpublished writings dated after 1906 in which Frege refers to Schönflies, Korselt, extensions, identity, and Russell's paradox are missing (Veraart 98-101). For example, missing since the bombing are a piece entitled 'Basic Law V Replaced By Basic Law V'" and a 1906 piece entitled 'Two Noteworthy Concepts'. According to Scholz' report to the 1935 International Congress of Scientific Philosophy these represented two attempts on Frege's part to reconstruct his system once confronted with Russell's paradox and were connected with the solution proposed in the appendix on the paradox that Frege wrote for the second volume of *Basic Laws*. (Veraart, 97, 98; Scholz, 28-29)

Moreover, Frege's earlier unpublished writings on the same subjects are missing. As I noted above, Frege had already expressed reservations about extensions in his 1884 *Foundations* and in the 1893 volume of *Basic Laws*. This means he must have been struggling with the matter as he wrote the three celebrated articles, 'Function and Concept', 'On Sense and Meaning', and 'On Concept and Object', which appeared in 1891 and 1892. All indications as to the nature of this struggle, however, disappeared in the Münster bombing. Manuscripts Scholz once had of those articles, along with excerpts from three of Frege's letters to Hilbert that Frege had filed with them are now lost. Two bundles of papers dating from that time which contained Frege's thoughts on Schröder's work on extensional logic have disappeared. According to the contents listed by Scholz, the papers discussed precisely the issues that have been the focus of the present paper. About the manuscript, Scholz had noted that there were sections on, among other things: "Identity and the corresponding second level relation. What two concept-words mean is then and only then the same when the extensions of their respective concepts coincide" (Veraart: 1976, 91, 94; Burge, 1984 discusses this thoroughly). The only portions of these papers that have survived have been published posthumously as 'Comments on Sense and Meaning' (Frege 1979, 118-25). In the now published material, Frege discourses at length about extensionality, arguing that, as Basic Law V would mandate, "concepts differ only in so far as their extensions are different.... If an object falls under a concept, it falls under all concepts with the same extension... just as proper names can replace one another *salva veritate*, so too can concept-words, if their extension is the same". (*Ibid.*, 118)

Another missing sheaf of papers from the 1880s contained critical questions regarding §§63-69 of *Foundations*, the very sections on identity and extensions that Husserl had criticized in 1891. These papers also apparently contained reflections concerning conceiving numerical equality as strict identity, and notes on Frege's early attempts to define extensions of concepts (Veraart, 95). Tyler Burge has conjectured "that in the

Foundations manuscript Frege was reconsidering the whole question of whether numbers were objects..., contemplating a contextual definition of such singular terms (roughly in the spirit of Russell's 'no-class' theory)", similar to an alternative Frege would consider in his 1902 appendix on Russell's paradox. (Burge, 13-14; Sluga, 169)

The absence of so many documents concerning Frege struggles with some of the most fundamental issues his logical work raises only aggravates the loss sustained by the destruction of approximately hundred post-1906 letters, many of which plainly dealt with these same matters. From a copy of a draft of a letter that Frege wrote to philosopher Paul Linke in 1919, we know that in a now lost 1916 letter, Linke had written to Frege asking him for his views as to "whether the mathematical equals sign means equality or identity" (Frege 1976, 152-53). A copy of a partial draft of a letter that Frege sent to the mathematician Karl Zsigmondy survived the destruction of Scholz' archive (*Ibid.*, 269). In the draft, Frege says that his efforts to "get clear about what we mean by the word 'number'... have been a complete failure", and discusses his conviction that mathematics "regards numbers as objects, not as properties. It uses number words substantivally, not predicatively" (*Ibid.*, 176, 178). In addition, Scholz inventoried, but lost, a 1921 card from Carnap concerning Frege's article 'Concept and Object' (*Ibid.*, 16). A now missing 1920 card that Korselt wrote to Frege concerned axioms as definitions (*Ibid.*, 144), recalling Frege's exchange with Hilbert.

Particularly regrettable is the loss of all twenty letters that Leopold Löwenheim and Frege wrote to each other from 1908 to 1910. In the late thirties, Löwenheim wrote to Paul Bernays that the exchange had concerned Frege's and Thomae's views on formalism and, in particular, the comparison with chess as found in §90 of the second volume of *Basic Laws*. According to Scholz' report before the International Congress of Scientific Philosophy, Löwenheim and Frege had engaged in an extensive epistolary exchange which they had intended to publish and in which Löwenheim succeeded in convincing Frege of the possibility of establishing secure foundations for formal arithmetic. As mentioned above, the entire correspondence, along with remarks and drafts Frege had made, and all Scholz' copies were lost. Interestingly, the last letter Löwenheim wrote to Frege was found unopened among Frege's papers. (Frege 1976, 158, 161)

Other materials that could shed light on the nature of this exchange are also missing. For example, Scholz' inventory of Frege's papers shows that the Münster archive once housed several sheaves of papers written between 1907 and 1910 that contained Frege's view's on Thomae's and Korselt's work on formalism. These now missing papers contained a detailed 1907 article written by Frege on Thomae's views and a letter refusing to publish it (Scholz, 29; Veraart, 99). One, apparently inconsequential letter from Thomae to Frege has also been lost (Frege 1976, 258). Any hopes of finding additional clues as to the nature of the exchange among Thomae's papers were dashed when they, like Korselt's papers, were destroyed. In the last days of World War II, Thomae's daughter burned all his papers (Kreiser 1973, 522). Attempts to locate relevant material from among Löwenheim's papers have likewise proved utterly futile; for the last seventeen years of his life, it was presumed that he had died in a Nazi concentration camp, and anything of interest that may have remained in his possession would most probably have been lost when his Berlin apartment was destroyed in a bombing raid in 1943. (Frege 1976, 158-59)

Ten letters Frege and Jourdain exchanged between 1909 and 1914 have survived. Almost all of them belonged to the autograph collection and are inconsequential. Much of the correspondence concerned Jourdain's various efforts to make Frege's work known in the English-speaking world. In particular, Jourdain was preparing an article on Frege's logical and mathematical theories. Frege sent Jourdain his comments, most of which Jourdain published (Frege 1980b, 179-206). In a 1913 letter, Jourdain referred to an unknown letter in which Frege "spoke about working at a theory of irrational numbers". Jourdain notes that he and Wittgenstein were rather disturbed by this "because the theory of irrational numbers −unless you have got a quite new theory of them−would seem to require that the contradiction has been previously avoided" (*Ibid.*, 76-77). There is no indication as to whether Frege replied to this. Two drafts of the last letter that Frege is known to have written to Jourdain were found in the autograph collection (Frege 1976, 126, 129). In them Frege had discoursed at length on problems with Russell's *Principia* (Frege 1980b, 78-84). Scholz at one time had a copy of Jourdain's last letter to Frege. According to Scholz' notes, the 1914 letter from Jourdain to Frege concerned Wittgenstein and Russell's *Principia.* (Frege 1976, 133)

So few letters from the last eighteen years of Frege's life have survived that their number almost doubled with late discovery of twenty-one cards and letters written by Frege to Wittgenstein between 1914 and 1920. In 1936, having acquired twenty-four letters that Wittgenstein (or his sister on his behalf) had written to Frege, Scholz wrote to Wittgenstein inquiring into any letters from Frege that Wittgenstein might still have. Wittgenstein replied saying that, though he did have some cards and letters from Frege, they would not be of any value to a collection of Frege's writings as they were purely personal in nature and devoid of philosophical content (Frege 1976, 265). In the 1980s, fifteen of these cards and six letters from Frege to Wittgenstein were found and published. The last four letters are of interest to philosophers in that they contain Frege's quite harsh pronouncements regarding Wittgenstein's *Tractatus,* which Frege did not hesitate to condemn as unclear and incomprehensible. He, in fact, had nothing good to say about it at all. (Frege 1989, 19-26)

Wittgenstein had wanted Frege to recommend the work to Bruno Bauch and Arthur Hoffmann for publication. The five letters from Hoffmann to Frege and the eight letters that Bauch wrote to Frege that Scholz once had are lost, but it is apparent from Scholz' notes that some of them concerned the *Tractatus* (Frege 1976, 8-9, 81-82). C. K. Ogden finally translated the *Tractatus* into English and it was published in 1921. Scholz inventoried a late 1921 letter from Ogden to Frege, though there is no indication that the letter concerned Wittgenstein (Frege 1976, 168). Also missing from Scholz' archive are four pages of notes by Frege on some of Wittgenstein's views and a package of drafts of replies to Wittgenstein that included drafts of two letters Frege wrote in response to Wittgenstein's requests for his opinion of the *Tractatus*. (*Ibid.*, 265)

In 1925, Bruno Bauch had further occasion to correspond with Frege. This time their correspondence concerned an article entitled "The Sources of Knowledge in Mathematics and the Mathematical Sciences", which Richard Hönigswald had asked Frege to write. Bauch had agreed to act as an intermediary between Frege and Hönigswald. All six letters that Hönigswald and Frege are known to have exchanged were in Scholz' possession and are missing. Scholz, however, made a copy of a letter that Hönigswald wrote to Frege on April 24th and of Frege's reply. These copies have survived, along with part of the article Frege was preparing.

On April 24th, Hönigswald wrote Frege to thank him for the manuscript that Bauch had forwarded and to ask Frege whether he might not expand his discussion of certain particularly significant topics. Hönigswald specifically requested more information on the paradoxes of set theory and Frege's reasons for believing set theory to be untenable, on functions, axiomatization, and the concept of number. Finally, he asked for Frege's view regarding Russell's, Zermelo's and Hilbert's work. Frege responded immediately by letter to Hönigswald's questions. On May 7th, having received Frege's reply, Hönigswald wrote to him once again concerning the expanded version of the article. This last letter is missing (Frege 1976, 87), as are the additional pages Frege had prepared in compliance with Hönigswald's request (Veraart, 92). Frege died two months later. The unfinished article was published for the first time in his posthumous writings. (Frege 1979, 267-74)

Conclusion

There is, of course, nothing very conclusive that one can say about the contents of documents which have had a history like the one described above. Moreover, pure speculation as to the significance of such documents is really pointless. In this case, however, more can be known than first meets the eye, and that is what I have tried to show here.

Frege was a man so impressed by the inherently defective character of natural language that he made it his life's work to free thought from its fetters. His apparent preference for written exchanges is wholly consistent with his convictions in this regard. He wrote long, detailed letters on important subjects. He made drafts of his letters. He wrote many of his letters with the intention of publishing them and letters of his that have been retrieved and published have already proven to be worthy complements to his published writings.

Dummett wrote of Frege's total obliviousness to the work of others and maintained that "there is not a trace in Frege's published or unpublished writing of any notice on his part of the work that was going on in the field he had opened up". In particular, Dummett cites the "many profound contributions from Russell and Whitehead, Hilbert, Zermelo, and Löwenheim" (Dummett 1981a, 661). Though Dummett does well to note the dearth of materials available to scholars, it should be clear from what I have said above that Frege was not oblivious to the work of his contemporaries. However, it is precisely those documents that could shed light on Frege's opinion of their work that are missing.

A close look at Frege's correspondence in fact gives lie to many other false ideas about Frege's isolation. Russell was neither the first to read Frege's *Begriffsschrift* nor to discover his other works. Frege's work was known to his contemporaries and he did exchange ideas with them. The paradoxes of set theory did not, as Russell has stated, follow "from premisses which were not previously known to require limitations" (Russell and Whitehead 1927, 59). From the very beginning Frege himself had reservations about the ideas that went into the making of Basic Law V, and Husserl (as Alonzo Church was for a long time alone in pointing out, Church, 63-65) was studying problems in Frege's basic premises ten years before Russell even read Frege.[8]

Of prime importance, as well, must be the fact that practically all the missing letters, and many of the other materials, destroyed in the bombing of the Münster library, were written after Russell's discovery of the paradox and broadly concerned Frege's views on what the problem was and why he found the solutions that his contemporaries were proposing unacceptable. These letters presumably could have provided information about a subject concerning which distressingly little can now be known because so little remains to indicate what Frege was thinking about matters of the most vital importance to him at this most crucial time in his intellectual career. What indeed could Frege himself have considered more important than those very issues which brought his life's work to an end and led him to conclude that his efforts were a complete failure?

References

Burge, Tyler 1984. "Frege on Extensions of Concepts from 1884 to 1903", *The Philosophical Review* XCIII (I) (January), 3-34.
Cantor, Georg 1885. 'Review of Frege's *Foundations*', *Deutsche Literaturzeitung* VI (20), 728-29.

---

[8] A subject studied in detail in Hill 1991.

Church, Alonzo 1944. "Review of M. Farber, *The Foundations of Phenomenology*", *Journal of Symbolic Logic* 9, 63-65.
Dummett, Michael 1981a. *Frege: Philosophy of Language*, London: Duckworth.
Dummett, Michael 1981b. *The Interpretation of Frege's Philosophy*, Cambridge MA: Harvard University Press.
Frege, Gottlob 1893. *Basic Laws of Arithmetic*, Berkeley CA: University of California Press, 1964.
Frege, Gottlob 1885. "Reply to Cantor's Review of *Grundlagen der Arithmetik*", in his *Collected Papers*, 122, translation of "Erwiderung auf Cantors Rezension der *Grundlagen der Arithmetik*", *Deutsche Literaturzeitung* 6, 28, Sp. 1030.
Frege, Gottlob, 1976. *Nachgelassene Schriften und Wissenschaftlicher Briefwechsel*, vol. 2, Hamburg: Meiner.
Frege, Gottlob, 1979. *Posthumous Writings*, Oxford Blackwell.
Frege, Gottlob 1980a. "Appendix to *Basic Laws of Arithmetic II*", in Peter Geach and Max Black (eds.), *Translations from the Philosophical Writings*, Oxford: Blackwell, 214-24.
Frege, Gottlob 1980b. *Philosophical and Mathematical Correspondence*, Oxford: Blackwell.
Frege, Gottlob 1884. *The Foundations of Arithmetic*, Oxford: Blackwell, 1986.
Frege, Gottlob 1989. "Briefe an Ludwig Wittgenstein", *Grazer philosophische Studien*, 33/34.
Grattan-Guinness, Ivor 1971. "The Correspondence Between Georg Cantor and Philip Jourdain", *Jahresbericht der Deutschen Mathematiker-Vereinigung* 73, 111-30.
Grattan-Guinness, Ivor 1977. *Dear Russell-Dear Jourdain: A Commentary on Russell's Logic Based on his Correspondence with P. Jourdain*, London: Duckworth.
Hill, Claire Ortiz 1991. *Word and Object in Husserl, Frege and Russell: The Roots of Twentieth-Century Philosophy*, Athens OH: Ohio University Press.
Hill, Claire Ortiz 2000. "Husserl, Frege and 'the Paradox'", *Manuscrito, Revista Internacional de Filosofia*, 23, 2, 101-32.
Husserl, Edmund 1891. *Philosophy of Arithmetic, Psychological and Logical Investigations with Supplementary Texts from 1887-1901*, translated by Dallas Willard, Dordrecht: Kluwer, 2003. Translation of *Philosophie der Arithmetik, mit ergänzenden Texten*, Husserliana XII, The Hague: Martinus Nijhoff, 1970.
Husserl, Edmund 1899. "*Husserl's* Excerpts from an Exchange of Letters between *Hilbert* and *Frege*", published in Husserl 1891, 468-73, in German in Husserliana XII, 447-51.
Husserl, Edmund 1913. *Introduction to the Logical Investigations: A Draft of a Preface (1913)*, The Hague: Martinus Nijhoff, 1975.

Kreiser, Lothar 1973. "'Review of *Nachgelassene Schriften*", *Deutsche Zeitschrift für Philosophie,* 21, 523.
Kreiser, Lothar 1974. 'Zur Geschichte des wissenschaftlichen Nachlasses Gottlob Freges', *Ruch Filozoficznej* 33(1), 42-47.
Mohanty, J. N. 1982. *Husserl and Frege,* Bloomington IN: Indiana University Press.
Rang, Bernard, and W. Thomas 1981. "'Zermelo's Discovery of the Russell Paradox'", *Historia Mathematica* 8 (February), 15-22.
Reid, Constance 1976. *Courant in Göttingen and New York,* New York: Springer.
Rosado Haddock, Guillermo Ernesto 1973. *Edmund Husserls Philosophie der Logik und Mathematik im Lichte der Gegenwärtige Logik eine Grundlagenforschung,* Bonn: Dissertation Rheinischen Friedrich-Wilhelms-Universität.
Russell, Bertrand 1903. *Principles of Mathematics,* New York: Norton.
Russell, Bertrand 1919. *Introduction to Mathematical Philosophy,* London: Allen and Unwin.
Russell, Bertrand 1946. *A History of Western Philosophy,* London: Allen and Unwin.
Russell, Bertrand and Alfred N. Whitehead 1927. *Principia Mathematica to *56*; Cambridge UK: Cambridge University Press, 2nd ed., 1964.
Schmit, Roger 1981. *Husserls Philosophie der Mathematik,* Bonn: Bouvier.
Scholz, Heinrich 1936. "Der wissenschaftliche Nachlass von Gottlob Frege", in *Actes du congrès international de philosophie scientifique.* vol. VIII: *Histoire de la logique et de la philosophie scientifique,* Paris: Hermann, 24-30.
Schumann, Karl 1977. *Husserl-Chronik,* The Hague: Martinus Nijhoff.
Sluga, Hans 1980. *Gottlob Frege,* London: Routledge and Kegan Paul.
Veraart, Albert 1976. "Geschichte des wissenschaftlichen Nachlasses Gottlob Freges und seiner Edition. Mit einem Katalog des ursprünglichen Bestands der nachgelassenen Schriften Freges", in M. Schirn (ed.), *Studien zu Frege* vol. 1, Stuttgart-Bad Canstatt: Frommann-Holzboog, 49-106.

9

Claire Ortiz Hill

REFERENCE AND PARADOX[1]

Introduction

It is one of the great merits of Frege's work that he created a system so clear that, to borrow a phrase from Matthias Schirn "all expressions wear their logical form on their sleeves" (Schirn 1996, 122). It was the first purpose of his formal language, he explained in his Preface to the *Begriffsschrift*, "to provide us with the most reliable test of the validity of a chain of inferences and to point out every presupposition that tries to sneak in unnoticed, so that its origin might be investigated" (Frege 1879, 6). In the preface to his *Basic Laws of Arithmetic I*, he explained that his ideal had been to set out his reasoning in such a way that one could see upon what the whole construction rested. He said that he had "drawn together everything that can facilitate a judgment as to whether the chains of inference are cohesive and the buttresses firm" and added that if "any one perchance finds anything faulty, he must be able to indicate exactly where, to his thinking, the mistake lies". He believed that he had essentially attained his goal. (Frege 1893, vi-vii)

That being so, we should not be obliged, as George Boolos suggested in "Whence the Contradiction?", to "guess at Frege's trains of thought", only to conclude that "we cannot explain how the serpent entered Eden", that Frege had been "hoodwinked" (Boolos 1996, 249-50). We should be able to start where Frege started, test with him the validity of his chain of inferences and discern the source of the problems whose origin we wish to investigate.

---

[1] An first version of this paper, entitled "Les paradoxes ensemblistes et les malheurs de la logique modale" was given on January 17, 2000 at the Institut d'histoire et philosophie des sciences, Université de Paris Sorbonne-Panthéon.

I propose to do that here. I investigate the origins of the contradiction derived by Russell within the system of Frege's *Basic Laws I* in order to show that the source of antinomies, paradoxes, contradictions associated with modal and intensional logics is found there too. I first investigate the theory of reference of Frege's *Foundations of Arithmetic* (Frege 1884), then turn to study the philosophical arguments that compelled Russell to adopt a description theory of names and an eliminative theory of descriptions. I finally contend that such investigations shed needed light on issues surrounding the hows and whys of the New Theory of Reference that grew up in connection with modal and intensional logics. This is part of a broader quest to draw the implications of Kurt Gödel's observation that close examination shows that the set-theoretical paradoxes "are a very serious problem, not for mathematics, however, but rather for logic and epistemology". (Gödel 1964, 258)

Frege on Picking Out Self-subsistent Objects and Recognizing Them as the Same Again

"The prime problem of arithmetic", Frege affirmed at the end of the appendix to *Basic Laws II* on Russell's paradox, "may be taken to be the problem: How do we apprehend logical objects, in particular numbers?" (Frege 1903, 224)

Now this problem is indeed acute in Frege's case because for him the reference of words was "at every point... the essential thing for science" (Frege 1979, 123). Frege had to have objects because that was the way he had designed his logic. "Only in the case of objects can there be any question of identity (equality)", he held (*Ibid.*, 182, 120). And he had put identity statements at the very heart of his project to provide foundations for the theorems of arithmetic (Frege 1884, x, §§55-67, 106). He insisted that "in science and wherever we are concerned about truth, we are not prepared to rest content with sense, we also attach a reference to proper names and concept-words; and if through some oversight, say, we fail to do this, then we are making a mistake that can easily vitiate our thinking" (Frege 1979, 118, 122). It was Frege's need for objects that had compelled him to introduce the extensions and Basic Law V, his law about extensions, which he believed led to Russell's contradiction.

Frege's problem arises in *Foundations of Arithmetic* in this form. Having determined to his satisfaction that "the content of a statement of number is an assertion about a concept", Frege started to define 0 as the number that "belongs to a concept, if the proposition that a does not fall under that concept is true universally, whatever a may be". He started to define 1 as the number that "belongs to a concept F, if the proposition that a does not fall under F is not true universally, whatever a may be, and if from the propositions 'a falls under F' and 'b falls under F' it follows universally that a and b are the same". In that case, he concludes, the number ($n + 1$) would have the number that "belongs to a concept F, if there is an object a falling under F and such that the number n belongs to the concept 'falling under F, but not a'". (Frege 1884, §55)

However, Frege promptly confessed: "It was only an illusion that we have defined 0 and 1; in reality we have only fixed the sense of the phrases 'the number 0 belongs to' 'the number 1 belongs to'; but we have no authority to pick out the 0 and 1 here as self-subsistent objects that can be recognized as the same again" (*Ibid.*, §56). He explains that:

> strictly speaking we do not know the sense of the expression 'the number n belongs to the concept G' any more than we do that of the expression 'the number ($n + 1$) belongs to the concept F'... we can never–to take a crude example–decide by means of our definitions whether any concept has the number JULIUS CAESAR belonging to it, or whether that same familiar conqueror of Gaul is a number or is not. Moreover we cannot by the aid of our suggested definitions prove that, if the number a belongs to the concept F and the number b belongs to the same concept, then necessarily $a = b$. Thus we should be unable to justify the expression '*the* number which belongs to the concept F', and therefore should find it impossible in general to prove a numerical identity, since we should be quite unable to achieve a determinate number. (*Ibid.*)

That attempt had failed, he explained in one of the concluding sections of the book, "because we had defined only the predicate which we said was asserted of the concept, but had not given separate definitions of 0 or 1, which are only elements in such predicates. This resulted in our being unable to prove the identity of numbers". (*Ibid.*, §106)

Now, it does not take very much imagination or logical acuity to see what great problems with inference would inevitably immediately arise were Frege not to rise to the occasion and find a definition of number not saddled with such a blatant basic problem of denotation and identity. However, Frege did not dream of leaving things there and so the challenge represented by the Julius Caesar problem itself could never have lived on to haunt philosophers' consciences were not shades of it to be found also haunting Frege's subsequent attempts to define numbers. They are. So Richard Heck was not exaggerating when he wrote that we shall "not fully understand Frege's philosophy until we understand the enormous significance the question how we apprehend logical objects, and the Caesar problem, had for him". (Heck 1993, 287)

Numbers as Self-subsistent Objects for Frege

"Precisely because it forms only an element in what is asserted", Frege argued as he developed his philosophy of arithmetic in *Foundations*, "the individual number shows itself for what it is, a self-subsistent object". For him, this was indicated by the presence of the definite article 'the' in expressions like 'the number 1', which served "to class it as an object" and in arithmetic "this self-subsistence comes out at every turn, as for example in the identity $1 + 1 = 2$". (*Ibid.*, §57; §106)

Any appearance that numbers were not to be construed substantivally, Frege averred, could "always be got around". To illustrate this point, he showed how the statement 'Jupiter has four moons' could be rewritten to read 'the number of Jupiter's moons is four', where the word 'is' is to mean 'is identical with' or 'is the same as'. This, he reasoned, yields an identity statement affirming that both of those expressions referred to the same object that was being called 'Four' or 'the Number of Jupiter's moons'. The second example he proposed was "it is the same man that we call Columbus and the discoverer of America". (*Ibid.*, §§57-58)

However, this self-subsistence of numbers, Frege insisted, "is not to be taken to mean that a number word signifies something when removed from the context of a proposition...". (*Ibid.*, §60). So the problem became one of defining "the sense of a proposition in which a number word occurs". And, for Frege, the answer lie in the very fact that number words stand for self-subsistent objects, which he considered, "is enough to give us a class of propositions which must have a sense, namely those which express our recognition of a number as the same again. If we are to use the symbol a to signify an object, we must have a criterion for deciding in all cases whether b is the same as a...". (*Ibid.*, §62; §106)

In the case at hand, Frege continues, "we have to define the sense of the proposition 'the number which belongs to the concept F is the same as that which belongs to the concept G; that is to say, we must reproduce the content of this proposition in other terms, avoiding the use of the expression 'the Number which belongs to the concept F'. This, he concludes, will provide the needed general criterion for the identity of numbers; when "we have thus acquired a means of arriving at a determinate number and of recognizing it again as the same, we can assign it a number word as its proper name". (*Ibid.*, §62)

Frege's Criterion for Identity

By defining numerical identity or equality in terms of one-to-one correlation, David Hume had provided such a means of arriving at a determinate number and of recognizing it again. However, Frege found that Hume's principle too raised "certain logical doubts and difficulties, which ought not to be passed over without examination". It "is not only among numbers that the relationship of identity is found", Frege reasoned. So it seemed to him "to follow that we ought not to define it specially for the case of numbers", but we "should expect the concept of identity to have been fixed first" and the concept of number "to be determined in the light of our definition of numerical identity". The aim was "to construct the content of a judgement which can be taken as an identity such that each side of it is a number". (*Ibid.*, §63)

He began doing this by showing how one could rewrite 'line *a* is parallel to line *b*' as an identity statement reading 'the direction of line *a* is identical with the direction of line *b*' "through removing what is specific in the content of the former and dividing it between *a* and *b*'. The "only trouble with this is", he admits, "that this is to reverse the true order of things.... Our convenient proof is only made possible by surreptitiously assuming, in our use of the word 'direction', what was to be proved". (*Ibid.*, §64)

Frege realized that his definition of the proposition 'line *a* is parallel to line *b*' is to mean the same as 'the direction of line *a* is identical with the direction of line *b*', "departs to some extent from normal practice, in that it serves ostensibly to adapt the relation of identity, taken as already known, to a special case, whereas in reality it is designed to introduce 'the direction of line *a*', which only comes into it incidentally". So, he tested his procedure against the well-known laws of identity. For this, he chose Leibniz' principle that "things are the same as each other, of which one can be substituted for the other without loss of truth", which he then adopted as his own definition of identity. However, while writing that well-known principle of substitutivity of identicals right into the very foundations of his logic, Frege decided to adjust Leibniz' principle to meet his own ends and to translate sentences of natural languages in such a way as to do away with the differences between being identical (complete agreement in all respects) and equal (only agreement in this respect or that). Although Leibniz' law defines identity, complete coincidence, Frege insisted that for him there was no between equality and identity:

> Whether we use 'the same', as Leibniz does, or 'equal', is not of any importance. 'The same', may indeed be thought to refer to complete agreement in all respects, 'equal', only to agreement in this respect or that; but we can adopt a form of expression such that this distinction vanishes. For example, instead of 'the segments are equal in length', we can say 'the length of the segments is equal', or 'the same', and instead of 'the surfaces are equal in color', 'the color of the surfaces is equal'... in universal substitutability all the laws of identity are contained. (*Ibid.*, §65)

Here, Frege has changed statements of sameness of concrete properties predicated of concrete objects into statements that affirm the equality-identity of abstract objects, in this case surfaces and lengths. He believes he has thus transformed statements about objects that are equal under a certain description into statements expressing a complete identity. By erasing the difference between identity and equality, he is in fact asserting that being the same in any one way is equivalent to being the same in all ways. All the other properties serving to distinguish those objects from one another, or from other objects equal to them in the same respect, are "abstracted" out of the picture; deleted on paper, those properties marking any difference between the mere equality and the full identity of the objects are presumably expected to simplify matters by vanishing entirely from the reasoning.

Reference, Identity and Substitution in Dickens' *A Tale of Two Cities*

At this point, I wish to introduce an example from *A Tale of Two Cities* by Charles Dickens that can be used to judge the cohesiveness of the chains of inferences and the firmness of the buttresses involved as we proceed from Frege's very old theory of reference to the new theory of reference. The story goes somewhat like this. The Marquis de St. Evrémonde voluntarily relinquishes a title and a rank bound to a system that is frightful to him in order to live otherwise and elsewhere earning his living by giving instruction in the French language and literature in England under the name of Charles Darnay.

The Reign of Terror finds Darnay him France. He is arrested and imprisoned in the prison of La Force and tried by the people's Tribunal who want to know whether he is the Marquis de St. Evrémonde, at heart and by descent an aristocrat. The Tribunal finds him guilty of having been born into an obnoxious family of aristocrats, a notorious oppressor of the People, enemy of the Republic (and of equality). Along with 51 other prisoners, he is condemned to die within 24 hours and imprisoned at the Conciergerie. No one has ever escaped from the Conciergerie and it is necessary that 52 heads fall corresponding to those of 52 prisoners condemned to death. Darnay is prisoner no. 22.

For personal reasons, Sydney Carton, who refers to himself as a self-flung away, wasted, drunken, poor creature of misuse, who should never be better than he was, should but sink lower, and be worse, and whom Dickens describes as "the idlest and most unpromising of men", "the jackal", "the fellow of no decency", of careless and slovenly, if not debauched appearance, decides to replace de St. Evrémonde and be guillotined in his place.

The day on which the 52 heads of the 52 condemned prisoners were to fall, Carton enters the vermin haunted cell of the condemned man and exchanges cravats, boots, and coats with him. Carton takes the ribbon from the prisoner's hair and tells him to shake out his hair like his own, drugs the prisoner, and calls a spy to have the man he has replaced taken to a coach. Carton then dresses himself in the clothes the prisoner had laid aside, combs back his hair, and ties it with the ribbon the prisoner had worn. A gaoler eventually comes to find Citizen Evrémonde. It is a dark winter day, and what with the shadows within, and what with the shadows without, one could but dimly discern the prisoners. As prisoner 22, Carton takes Evrémonde's place in the tumbril and dies in his place.

We have our identity:
Charles Darnay = Charles de St. Evrémonde
Our descriptions:
The Marquis de St. Evrémonde; the occupant of cell 22 of the Conciergerie; the guillotined person no. 22, etc.

Our statements:
The Marquis de St. Evremond has hair on his head.
Charles Darnay is the author of certain crimes against the people
The man presumed to be Charles de St. Evremond = Sydney Carton
52 is the number of prisoners found guilty that day by the Tribunal.
52 is the number of prisoners that day in the tumbrils.
52 is the number de these prisoners actually guillotined that day.
Charles Darnay = prisoner 22
Citizen Evrémonde = prisoner 22
Sydney Carton = prisoner 22,
Several one to one correspondences, among them:
The number of condemned people from the prison of La Force (F) and the number of people guillotined (G).
And antinomies resulting from substitution, for example:
Citizen Evrémonde is prisoner no. 22
Prisoner 22 was guillotined.
Citizen Evrémonde was guillotined.

Applying Frege's Reasoning to Our Life and Death Situation

So, taking Frege's reasoning into the cells of the Conciergerie, we have: (1) The Marquis de St. Evrémonde has the same destiny as prisoner 22, which translated in Frege's way yields: (2) The destiny of the Marquis de St. Evrémonde is identical to the destiny of prisoner 22.

Whether we use 'the same', complete agreement in all respects, or 'equal', only agreement in this respect or that, Frege stressed, is not of any importance because "we can adopt a form of expression such that this distinction vanishes" (Frege 1884, §65). Holding that "it is actually the case that in universal substituability all the laws of identity are contained", Frege maintained that to justify our proposed definition of the destiny of a man, we should have to show that it is possible, if

(1) The Marquis de St. Evrémonde has the same destiny as prisoner 22
    to substitute
'the destiny of the Marquis de St. Evrémonde'
everywhere for
'the destiny of prisoner 22'.

The task is simplified, he admits, by the fact that we are being taken initially to know nothing that can be asserted about the destiny of a man except one thing, that it coincides with the destiny of prisoner 22. "We should thus have to show only that substitution was possible in an identity of this one type, or in judgement-contents containing such identities of this kind as constituent elements". (*Ibid.*)

Recognizing Objects When Given in a Different Way

Frege was perfectly conscious that left unmodified the procedure just described was liable to lead to false or nonsensical conclusions, or be sterile and unproductive (Frege 1884, §§66-67). To illustrate this, he carried the reasoning involved in his example of the identity of two lines one step further. Using our life or death example, according to Frege's reasoning, in proposition (2) 'The destiny of Marquis de St. Evrémonde is identical to the destiny of prisoner 22', the destiny of the Marquis de St. Evrémonde plays the part of an object, and our definition affords us a means of recognizing this object in case it could happen to crop up in another guise (*in einer andern Verkleidung*), say as the destiny of prisoner 22, but this means does not provide for all the cases. It does not, still imitating Frege, for example, decide for us whether England is the same as the destiny of a man condemned to death. The definition does not say whether the statement:

(3) 'The destiny of the Marquis de St. Evrémonde is identical with q'

is to be affirmed or denied except in the one case where q is given in the form of 'the destiny of prisoner 22'.

Left as it was his definition was unproductive, Frege further judged, because in adopting this way out, we would be presupposing that an object could only be given in a single way.

"All identities would then amount simply to this", Frege then wrote, "that whatever is given to us in the same way is to be reckoned as the same. This is, however, a principle so obvious and sterile as not to be worth stating. We could not, in fact, draw from it any conclusion which was not the same as one of our premises". Surely though, he concluded, identities play such an important role in so many fields "because we are able to recognize something as the same again even although it is given in a different way" (*Ibid.*, §67; §107).

Apprehending Logical Objects as Extensions

Frege introduced extensions in order to overcome the undesirable consequences that he saw gathering about his theory of number. He hoped that they would guarantee that an identity holding between two concepts could be transformed into an identity of extensions, and conversely (Frege, 1884, §§66-67; also §107). In that case, parroting Frege's reasoning once again, if, in accordance with (1), The Marquis de St. Evrémonde has the same destiny as prisoner 22, then the extension of the concept "man with the same destiny as that of the Marquis de St. Evrémonde" is identical to the "the extension of the concept person having the destiny of prisoner 22". And conversely, if the extensions of the two concepts just named are identical, then the Marquis de St. Evrémonde has the same destiny as prisoner 22. To apply this means of definition to the case at hand, we must substitute concepts for men, and for sharing the same destiny, the possibility of correlating one to one the objects that fall under the one concept with those that fall under the other. (*Ibid.*, §68)

In his next book, *Basic Laws I*, Frege argued that the generality of an identity could always be transformed into an identity of courses of values and conversely, an identity of courses of values may always be transformed into the generality of an identity. By this he meant that if it is true that (x) F(x) = G(x), then those two functions have the same extension and that functions having the same extension are identical. (Frege 1893, §§9, 21; Frege 1903)

Frege never believed that any proof could be devised to sanction such a transformation. So he created Basic Law V to mandate the view of identity, equality, and substitutivity that he required. By transforming "a sentence in which mutual subordination is asserted of concepts into a sentence expressing an equality", his law would permit logicians to pass from a concept to its extension, a transformation which, Frege considered, could "only occur by concepts being correlated with objects in such a way that concepts which are mutually subordinate are correlated with the same object" (Frege 1979, 182; Frege 1979, 118). His law, however, authorized various illicit logical moves which yielded contradictory results by letting logicians to put their symbols to wrong uses and allowing type ambiguities to creep into reasoning unnoticed. The most famous of these was, of course, the contradiction derived by Russell: both (w∈w) ∧ ~ (w∈w).

Russell's Paradox, Extensions, and Grasping Logical Objects

Upon learning of Russell's paradox, Frege tested the validity of the chain of inferences leading to the contradiction and determined that it was his Basic Law V that was at the origin of the contradiction. The contradiction indicated to him that the transformation of the generality of an identity into an identity of ranges of values was not always permissible, that the law was false, that his explanations did not suffice to secure a reference for his combinations of signs in all cases (Frege 1980, 131-32). He confessed that he had been long reluctant to use classes, but that that was the only answer that he had found to the question as to how we apprehend logical objects. (*Ibid.*, 140-41)

Frege studied the problem in an appendix to *Basic Laws II*. There was nothing, he found, to stop one from transforming an equality holding between two concepts into an equality of extensions in conformity with the first part of his law, but from the fact that concepts are equal in extension one cannot infer that whatever falls under one falls under the other. The extension may fall under only one of the two concepts whose extension it is. This can in no way be avoided... (Frege 1903, 214n. f, 218-23). "If in general, for any first-level concept, we may speak of its extension, then the case arises of concepts having the same extension, although not all objects that fall under one fall under the other as well" (*Ibid.*, 221). This is, of course, the drama of substitution, reference and identity of the prison of the Conciergerie, for:

If F is the set of 52 prisoners of La Force condemned to die within 24 hours

G, the set of these 52 prisoners actually guillotined

F = G and (x) F(x) implies that G(x)

There was a one to one correspondence between the extension of concept F and that of concept G, but that does not mean that (x) if F(x) then G(x). No matter how attractive it may seem at first to obliterate distinctions between identity and equality, the differences between *x* and *y* when they are joined together by the equals sign to make an informative statement do not just go away because we make a law that stipulates that equality is to function as identity. Extensionality alone can not keep someone from slipping into the company of the prisoners of La Force and being guillotined in the place of one of them. Frege, as he himself realized, still did not have a firm hold on the reference.

The Definite Article and Pseudo-objects

When specifically asked about the causes of the paradoxes of set theory, Frege explained that the "essence of the procedure which leads us into a thicket of contradictions" consisted in regarding the objects falling under F as a whole, as an object designated by the name 'set of Fs', 'extension of 'F', or 'class of Fs' etc. He wrote that the paradoxes of set theory

> arise because a concept e.g. fixed star, is connected with something that is called the set of fixed stars, which appears to be determined by the concept—and determined as an object. I thus think of the objects falling under the concept fixed star combined into a whole, which I construe as an object and designate by a proper name, 'the set of fixed stars'. This transformation of a concept into an object is inadmissible, for the set of fixed stars only seems to be an object, in truth there is no such object at all. (*Ibid.*, 54; 55)

"The definite article", he explained, "creates the impression that this phrase is meant to designate an object, or, what amounts to the same thing, that 'the concept star' is a proper name, whereas 'concept star' is surely a designation of a concept and thus could not be more different from a proper name". (Frege 1979, 270)

Russell also came to see the problems raised by the logical behavior of expressions having the form 'the So-and-so', which seem to denote objects and often figure in identity statements as if they do. He also saw a link between problems with descriptions and the paradoxes he was trying to evade. Clear access to reference was needed and the paradoxes showed both him and Frege that they were still in need of an effective technique for guaranteeing it.

In his famous article "On Denoting" Russell reasoned that: "If **a** is identical to **b**, whatever is true of the one is true of the other, and either may be substituted for the other in any proposition without altering the truth or falsehood of that proposition". Substituting our Reign of Terror example into Russell's well known train of thought, we have: Now the Tribunal wished to know whether Darnay was the Marquis de St. Evrémonde; and in fact Darnay *was* the Marquis de St. Evrémonde. Hence we may substitute *Darnay* for the *Marquis de St. Evrémonde*, and thereby prove that the Tribunal wished to know whether Darnay was Darnay. However, an interest in the law of identity can hardly be attributed to that revolutionary Tribunal (Russell 1905, 47-48; Russell 1959, 63-64). This and other puzzles about denoting led Russell to develop his theory of definite descriptions, which in the context of this chapter may be called "the old theory of reference".

Drawing Objects Out of Descriptions and Laying Hold of the Extension of a Class

In his preface to the 1980 edition of *Naming and Necessity*, Saul Kripke asks how Russell came to propose a theory plainly incompatible with our direct intuitions of rigidity of reference. In response to his own question, Kripke first suggests that one reason is that Russell did not himself consider modal questions and the question of the rigidity of names in natural language was rarely explicitly considered after him. Kripke secondly answers that it seemed to Russell that various philosophical arguments made a description theory of names and an eliminative theory of descriptions necessary, seemed to him to compel adoption of his theory (Kripke 1980, 14). I examine Kripke's second answer next and shall take up his first one later.

So what philosophical arguments might have compelled Russell to adopt his famous theory? To understand this, let us begin with the obvious fact that Russell inherited Frege's problems. In the preface to *Principia Mathematica*, Russell explains that a "very large part of the labour involved in writing the present work has been expended on the contradictions and paradoxes which have infected logic and theory of aggregates" (Russell 1927, vii) and in the introduction that his "system is specially framed to solve the paradoxes which, in recent years, have troubled students of symbolic logic and the theory of aggregates". (*Ibid.*, 1)

Russell said that once he had sat down to remove the source of that infection, he devoted almost all his time to problems with denotation that he believed tied to them. "The whole theory of definition, of identity, of classes, of symbolism, and of the variable is wrapped up in the theory of denoting", he had written in *Principles of Mathematics* (Russell 1903, §56). And Russell always said that his theory of definite descriptions represented his first breakthrough in his efforts to find a solution to the paradoxes. (Russell 1944, 13-14; Russell 1959, 60-61; Grattan-Guinness 1972, 106-07; Grattan-Guinness 1975, 475-88; Grattan-Guinness 1977, 70, 79-80, 94 and note; Kilmister 1984, 102, 108, 123, 138)

As we saw, to "lay hold upon the extension of a concept", Frege had proposed transforming "a sentence in which mutual subordination is asserted of concepts into a sentence expressing an identity". Since only objects could figure in identity statements, he realized that he would have to find a way of correlating objects and concepts which correlated mutually subordinate concepts with the same object. He suggested that this might be achieved by translating language asserting mutual subordination into statements of the form 'the extension of the concept X is the same as the extension of the concept Y in which the descriptions would then be regarded as proper names as indicated by the presence of the definite article. But by permitting such a transformation, Frege realized, one is conceding that such proper names have references (Frege 1979, 181-82). However, it was precisely that sort of recipe that Russell's paradox had cast doubt upon. Russell would try to circumvent problems that that procedure caused through a new theory of classes based on his theory of definite descriptions. In so doing, he hoped to realize Frege's goal of correlating classes with extensions in such a way that concepts that are mutually subordinate would be correlated with the same objects. (re. Frege 1903, 214)

The Need for a Single Object to Represent an Extension

Russell recognized: "the necessity of discovering some entity determinate for a given propositional function, and the same for any equivalent propositional function. Thus 'x is a man' is equivalent (we will suppose) to 'x is a featherless biped', and we wish to discover some one entity which is determined in the same way by both these propositional functions" (Russell 1903, §486). He was convinced that

> without a single object to represent an extension, Mathematics crumbles. Two propositional functions which are equivalent for all values of the variable may not be identical, but it is necessary there should be some object determined by both. Any object that may be proposed, however, presupposes the notion of *class*... an object uniquely determined by a propositional function, and determined equally by any equivalent propositional function. (*Ibid.*, §489)

On the other hand, the contradiction of the classes that do not belong to themselves had persuaded Russell that one could not in general suppose that objects having a certain property constitute a class that is an entity distinct from the objects making it up. So, he set out in search of a way of dealing with classes as symbolic fictions by which one could avoid having to assume that there are classes without being compelled to assume that there are no classes. (Russell 1919, 184)

Parallels that Russell spotted existing between the problems arising when classes are treated as objects and those arising when descriptions are treated as names suggested to him how classes might be analyzed away much as descriptions had been by his 1905 theory of definite descriptions. Just as descriptions are bound to the particular way of characterizing the object, so classes are formed by specifying the definite property giving the class. And just as two different descriptions could be true of the same object, so a single class of objects could be defined in diverse ways, each one corresponding to another sense of the class name.

Classes were false abstractions, Russell decided, in the same sense as 'the present King of England' or 'the present King of France'. So he sought to extend his ideas about analyzing away descriptions to include class symbols. He reasoned that since:

> we cannot accept "class" as a primitive idea. We must seek a definition on the same lines as the definition of descriptions, i.e. a definition which will assign a meaning to propositions in whose verbal or symbolic expression words or symbols apparently representing classes occur, but which will assign a meaning that altogether eliminates all mention of classes from a right analysis of such propositions. We shall then be able to say that the symbols for classes are mere conveniences, not representing objects called "classes", and that classes are in fact, like descriptions, logical fictions, or (as we say) "incomplete symbols". (*Ibid.*, 181-82)

According to Russell's theory of definite descriptions: "There is a term c such that ϕx is always equivalent to 'x is c'" (*Ibid.*, 178). That being so the putative identity statement 'Darnay is the Marquis de St. Evrémonde' might be rewritten 'Darnay inherited the title the Marquis de St. Evrémonde, and is always true of c that if c inherited the title Marquis de St. Evrémonde, then c est identical to Darnay (Russell 1905, 55). In this rendering of the sentence, what was not identical has been made to be equivalent. A symbol deemed equatable with 'Darnay' has gone proxy for the description and, note well, this symbol will generally be "obedient to the same formal rules of identity as symbols which directly represent objects" (Russell 1927, 83). This means of drawing objects out of descriptions provided Russell with a practical model of how to make non-entities function as entities without incurring contradictory results. The theory of definite descriptions was his way of making an object fit to go proxy for what was said about it. (*Ibid.*, 187)

So a procedure explicitly analogous to the theory of definite descriptions was duly integrated into *Principia Mathematica* and *Introduction to Mathematical Philosophy*. Russell said that he was satisfied that although the real meaning of the notation that he had ultimately adopted was very complicated, it had "an apparently simple meaning which, except at certain crucial points, can without danger be substituted in thought for the real meaning" (*Ibid.*, 1). He had avoided the contradictions arising from the supposition that classes are entities and acquired a means of laying hold of the extension of a class. Henceforth, incomplete symbols, like descriptions and classes, would obey the same formal rules of identity as symbols directly representing objects.

Further Guaranteeing Substitutivity

Russell was aware, however, that classes or descriptions could only obey the same formal rules of identity to the extent that we consider only the equivalence of the values of variables (or of constants) resulting from it and not their identity (Russell 1927, 83) and that brought him right back to the problems with reference, identity and substitutivity that had compelled Frege to introduce extensions. In addition, the contradictions most closely connected to identity and the formal equivalence of functions proved particularly difficult to uproot.

So Russell took aim at Frege's theory of identity. To cope with contradictions arising from necessary talk of 'all properties', or 'all functions', Russell introduced the axiom of reducibility, which was "equivalent to the assumption that 'any combination or disjunction of predicates is equivalent to a single predicate'" (Russell 1973, 250; 1927, 58-59) and would yield most of the results otherwise requiring recourse to all functions or all properties, legitimize much reasoning apparently dependent on such notions. (Russell 1927, 56)

Russell considered that the axiom embodied all that was really essential in his theory of classes (Russell 1919, 191; Russell 1927, 58). "By the help of the axiom of reducibility", Russell affirmed, "we find that the usual properties of classes result. For example, two formally equivalent functions determine the same class, and conversely, two functions which determine the same class are formally equivalent" (Russell 1973, 248-49). He came to believe classes themselves to be mainly useful as a technical means of achieving what the axiom of reducibility would effect (Russell 1919, 191). It seemed to him "that the sole purpose which classes serve, and one main reason which makes them linguistically convenient, is that they provide a method of reducing the order of a propositional function" (Russell 1927, 166). Classes were producing contradictions. They should be replaced with this axiom which seemed to him "to be the essence of the usual assumption of classes" and to retain "as much of classes as we have any use for, and little enough to avoid the contradictions". (Russell 1927, 166-67; Russell 1956, 82; Russell 1919, 191)

Examination shows that Russell leaned on the axiom of reducibility at every crucial point in his definition of classes in *Principia Mathematica* (Russell 1927, 75-81). He believed that many of his proofs "become fallacious when the axiom of reducibility is not assumed" (*Ibid.*, xliii). He also believed that without the axiom, or its equivalent, one would be compelled to regard identity as indefinable and to admit that two objects might agree in all their predicates without being identical (*Ibid.*, 58).

"But in virtue of the axiom of reducibility it follows that, if $x = y$ and x satisfies ψx, where ψ is any function... then y also satisfies ψy". And this effectively made his definition of identity as powerful as if he had been able to appeal to all functions of x (*Ibid.*, 168; also 57; Russell 1973, 243). In particular, by resorting to the axiom of reducibility one might avoid a difficulty with the definition of identity which Russell explained as follows:

> We might attempt to define "x is identical with y" as meaning "whatever is true of x is true of y", i.e., ϕx always implies ϕy". But here, since we are concerned to assert all values of "ϕx implies ϕy" regarded as a function of ϕ, we shall be compelled to impose upon ϕ some limitation which will prevent us from including among values of ϕ values in which "all possible values of ϕ" are referred to. Thus for example "x' is identical with "a" is a function of x; hence, if it is a legitimate value of ϕ in "ϕx always implies ϕy", we shall be able to infer, by means of the above definition, that if x is identical with a, and x is identical with y, then y is identical with a. Although the conclusion is sound, the reasoning embodies a vicious-circle fallacy, since we have taken "$(\phi)(\phi x$ implies $\phi a)$" as a possible value of ϕx, which it cannot be. If, however, we impose any limitation upon ϕ it may happen, so far as appears at present, that with other values of ϕ we might have ϕx true and ϕy false, so that our proposed definition of identity would plainly be wrong. (Russell 1927, 49)

So, if the axiom of reducibility could be true, we could have the properties of identity and equality that we need. Russell thought of it as a generalized form of Leibniz' principle of the identity of indiscernibles (Russell 1919, 192; 1927, 57; Russell 1973, 242). He was finally back to square one, back to the reasons why Frege's theory of identity had made him appeal to extensions in the first place.

But if the axiom of reducibility could do the trick, then Leibniz' law or Basic Law V could have. The problems were not uprooted. They began surfacing again, in another guise, when Willard van Orman Quine engaged in his campaign against intensional and modal logics.

Reference, Modality, and Paradox

In 1946 and 1947, three articles by Ruth Barcan (Marcus) on quantified modal logic appeared in the *Journal of Symbolic Logic* (Barcan 1946a, 1946b, 1947). In them, she stipulated that: x = y, iff necessarily x = y, a form of rigid reference. She also distinguished between identity in the strong sense and in the weak sense. Those papers were formal. We can put some flesh on the bones, however, by looking at her reasoning as mirrored in the reactions of others.

Quine reacted immediately (Quine 1947 a, b, c). As is well known, he had a wealth of objections to modal and intensional logics. To begin with, I want to look at one whose error he eventually acknowledged (Quine 1962). In his review of "The Identity of Individuals in a Strict Functional Calculus of Second Order", he observed that Barcan had defined two relations of "identity" between individuals, a weak one holding between

x and y wherever (F) (Fx ⊃ Fy), and a strong one holding only where (F) (Fx strictly implies Fy). And he remarked that:

> As to be expected, only the strong kind of identity is subject to a law of substitutivity valid for all the modal contexts.
> It should be noted that only the strong identity is therefore interpretable as identity in the ordinary sense of the word. The system is accordingly best understood by reconstruing the so-called individuals as "individual concepts". For example, the physical planet which is both the evening star and the morning star should not be reckoned as a value of the individual variables, lest it turn out not to be identical (in the full sense) with itself. On the other hand, two distinct *concepts* of Evening Star and Morning Star are available as values of the variables without fear of paradox.... It should be noted further that the primitive idea of abstraction need not have been assumed in this system, for it could have been introduced by contextual definition.... (Quine 1947c, 95-96)

In "The Problem of Interpreting Modal Logic", Quine had contended that Barcan's version of quantified modal logic was committed to a curiously idealistic ontology that repudiates material objects (Quine 1947a, 43, 47). He had sought to show that it had

> queer ontological consequences. It leads us to hold that there are no concrete objects (men, planets, etc.), but rather that there are only, corresponding to each supposed concrete object, a multitude of distinguishable entities.... It leads us to hold, e.g. that there is no such ball of matter as the so-called planet Venus, but rather at least three distinct entities: Venus, Evening Star, and Morning Star. (*Ibid.*, 47)

Frederic Fitch and Arthur Smullyan rose quickly to the defense of Barcan's position. In "The Problem of the Morning Star and the Evening Star", Fitch argued that there was really no reason to resort to individual concepts in order to make quantified modal logic consistent with the facts of astronomy, that modal logicians are free to deal directly with concrete individuals and to use the identity relation in relation to them. They might choose to regard the Morning Star and the Evening Star as proper names of individuals, or as descriptive phrases in the sense of *Principia Mathematica* *14. "Both of these ways are orthodox, natural, and available", he pointed out (Fitch 1949, 138). Smullyan showed that Quine's argument fails and that no paradox arises if the Morning Star and the Evening Star are considered as being proper names of the same individual. (Smullyan 1947, 1948; Fitch 1949, 138-39)

For Fitch, the source of the apparent paradox seemed to lie in the ambiguity of the scope of the descriptions. As he explained, the "usage of ordinary language with respect to descriptive phrases is so vague that... it is usually not possible to take at random various sentences of ordinary language and interpret them all correctly by some uniform use of Russell's theory of descriptions..." (Fitch 1949, 140). He also concluded that in Barcan's system "the axiom of extensionality need not be used at all, or only in such a way that $(f = g)$ follows from $\Box \, (fx \equiv_x gx)$". (*Ibid.*, 141)

The message conveyed did not fall on fertile ground. It was not fashionable or welcome at all. As is well-known, Quine, the fashionable philosopher of the 1950s and 1960s, fought hard to contain logical reasoning within the confines of strong extensional calculi at all costs and made quashing modal and intensional logics one of the principal planks of his philosophical program. His attacks, of which Marcus was quite often the target, were very influential and appeared in his most influential works, for example "Reference and Modality" (Quine 1953a), "Three Grades of Modal Development" (Quine 1953b), the chapter of *Word and Object* (Quine 1962) entitled "Flight from Intension", *Ontological Relativity* (Quine 1969). Figuring on his list of the "varied sorrows of modal logic" were referential opacity, failure of substitutivity, of existential quantification, of inference, riddles about descriptions, a bizarre ontology, confusions of use and mention, intensions, essentialism.

Meanwhile, defying Quine's opprobrium, Marcus continued to work to draw attention to ambiguities regarding equality and identity that had slipped into logical reasoning. She endeavored to point out the extent to which extensional systems depend on direct or indirect restrictions forbidding intensional functions and the extent to which they reduce identity to a weaker form of equivalence. She called for lucidity regarding the differences between identity and weaker forms of equivalence that explicit or implicit principles of extensionality would ignore. She called a principle extensional if it either "(a) *directly, or indirectly imposes restrictions on the possible values of the functional variables such that some intensional functions are prohibited or* (b) *it has the consequence of equating identity with a weaker form of equivalence*" (Marcus 1960, 46). She urged "that the distinctions between stronger and weaker equivalences be made explicit before, for one avowed reason or another, they are obliterated" (*Ibid.*, 55), a request which would seem to be perfectly in keeping with the requirements of any logic priding itself on its clarity, and devised to keep ambiguity from slipping unawares into reasoning.

A meeting of the Boston Colloquium in 1962 marked a new step in the history of these ideas (Wartofsky 1963, Quine 1993, Marcus 1993). In her paper on "Modalities and Intensional Languages", expounding on the themes discussed above, Marcus defended quantified modal logic (Marcus 1993, 3-35). In his reply to her, Quine explained his conviction that modern modal logic was conceived in sin (Quine 1962, 177-79). He condemned her conclusion that identity, substitutivity, and extensionality are things that come in grades as unacceptable (*Ibid.*, 180). He saw trouble in the contrast that she developed between proper names and descriptions (*Ibid.*, 181). He once again challenged the "champion" of quantified modal logic to explain away his puzzlement about the logical behavior of statements like '9 = the number of the planets' in modal contexts (*Ibid.*, 183-84). He complained about the pernicious way in which Marcus distinguished between necessary properties and contingent properties, about "the invidious distinction between some traits of an object as essential to it... and other traits of it as accidental" that opens the door to essentialism. (*Ibid.*, 184)

Though Quine's perspective remained dominant and intensional and modal logics were despised and derided for decades, some philosophers began to take a more daring attitude toward limning the true and ultimate structure of reality and began venturing into proscribed territory beyond the narrow boundaries that strong extensional calculi would impose on philosophical reasoning.

Among those making pioneering contributions to the field were Jaakko Hintikka (Hintikka 1962, Hintikka 1969, Hintikka 1975; Hintikka and Davidson 1969), Stig Kanger, Saul Kripke and Dagfinn Føllesdal (Humphreys and Fetzer 1998). They worked to increase the depth and utility of the standard languages and to develop intensional languages capable of analyzing the many non-extensional statements deemed unfit for study because they complicated matters by not conforming to the rigid standards for admission into what even Frege had feared could be a sterile logical world.

The development of possible worlds logic eventually proved to be a particularly effective device for exposing logical form and viewing the inner workings of analytic philosophy's brave new logic. Explorers of possible worlds made discoveries that helped to confirm the results of Marcus' earlier ventures into the discredited world of modality and intensionality. So, more and more reasons for not shoving reasoning into an extensional mold began to mount as those courageous enough not to flee at the sight of intensional and modal phenomena pulled deep issues underlying the puzzles, contradictions and paradoxes to the surface and increasingly demonstrated the insufficiencies of strongly extensional systems.

For one thing, possible worlds logic put a spotlight on problems associated with extensionality, identity, failures of substitutivity and existential generalization and so helped prepared the ground for a wider acceptance of ideas that went into the making of a "new" theory of reference as opposed to Russell's "old" theory of definite descriptions. According to this "new" theory, there is a direct reference relation between names, proper names in particular, and their references; names refer to objects directly and not through descriptions.

In the 1980 preface to *Naming and Neccesity*, Saul Kripke, one of the philosophers whose name has been associated with this new theory of reference,[2] told of the evolution of his own ideas on the matter. Most

---

[2] Quentin Smith undertook to correct what he called the fundamental, widespread idea that Saul Kripke was the author of this theory. According to Smith, the New Theory of Reference was essentially developed by Marcus. Kripke was influenced by the ideas of Marcus heard during the Boston Colloquium in 1962. Scott Soames of Princeton defended Kripke. Soames called Smith's allegation scandalous, shameful, grotesque. According to Soames, Smith was claiming that the entire profession had let itself be duped by Kripke;

of his views, he recalled, had grown out of earlier formal work on model theory of modal logic and were formulated around 1963 to1964 (Kripke 1980, 3).[3] He described the realization that inaugurated his work of 1963-64 in the following way:

> Eventually I came to realize... that the received presuppositions against the necessity of identities between ordinary names were incorrect and that the natural intuition that the names of ordinary language are rigid designators can in fact be upheld. Part of the effort to make this clear involved the distinction between using a description to give a meaning and using it to fix a reference. (*Ibid.*, 5)

He explained that his model theoretic work on modal logic had confirmed his conviction that the principle of the indiscernibility of identicals was self-evident and made it completely clear that the alleged counter-examples involving modal properties always turned out to turn on confusions. "It was clear that from $(x) \square (x = x)$ and Leibnitz's law that identity is an 'internal' relation: $(x) (y) (x = y)$ implies $\square (x = y)$.... If 'a' and 'b' are rigid designators, it follows that '$a = b$', if true, is a necessary truth", he wrote. (*Ibid.*, 3)

Conclusion

In this chapter, I have drawn together evidence to connect sources of inconsistency that Frege himself discerned in his foundations for arithmetic to the origins of the paradox derived by Russell in Frege's *Basic Laws I* and to paradoxes, puzzles, riddles antinomies, contradictions that became major issues in later stages of analytic philosophy. Specifically, this has involved linking Frege's unsatisfied need to specify referents in the way his system requires, the philosophical arguments that compelled Russell to adopt a description theory of names and a eliminative theory of descriptions, and the resurfacing of issues surrounding identity, substitutivity, paradox, descriptions, and rigid reference in the debates concerning modal and intensional logics.

Marcus, among others, anticipated important aspects of the theories in question, but Kripke played the "seminal" role. Wishing to see the philosophical issues examined on objective ground, Jaakko Hintikka published Smith's and Soames's papers in *Synthese* and published a volume of Synthese Library on the issue (Humphreys and Fetzer, 1998), into which are integrated the ideas of Føllesdal, Stig Kanger, Sten Lindström, John Burgess. The book as a whole presents a wealth of facts and arguments in defense of the various positions taken. There is, however, always something elusive about questions of influence and the origins of ideas that is not captured in attention to details. And this is all the more true when, as is the case here, major issues and important figures are under discussion. However, in spite of the acrimonious tone introduced by Soames and sustained by him and Burgess, the book as a whole provides thought-provoking material.

[3] Kripke had attended the 1962 Boston Colloquium. He had already done promising formal work in quantified modal logic and actively participated in the discussions. He queried Marcus about the necessity of identity and the differences between proper nouns and descriptions, about quantification, about essentialism. (Marcus 1993, 25-35)

In his introduction to *Frege's Philosophy of Mathematics*, William Demopoulos reflects that once the Julius Caesar problem is taken to show that Frege's contextual definition fails to specify the referents of numerical expressions, it is natural to want to know whether the problem does not appear in an analogous way in later stages of Frege's work. In particular, Demopoulos suggests, the same problem appears to iterate to extensions, so that it becomes reasonable to ask whether Frege's introduction of extensions really overcomes the difficulty that the Julius Caesar problem posed for numbers while not itself succumbing to a similar objection. Demopoulos further notes that this matter certainly needs to be addressed in connection with *Basic Laws of Arithmetic* where Frege introduces extensions in a way that is formally exactly analogous to his contextual definition of number. (Demopoulos 1995, 8-12)

I have maintained that Frege's appeal to extensions did not overcome the difficulty that the Julius Caesar problem posed for him and that the problem reappeared in different guises both in later stages of Frege's philosophy and in later stages of analytic philosophy. I further suggest that, studied from the angle of Frege's unsatisfied need to specify referents in the way his system requires and the connections that this has with the philosophical issues chronicled in this chapter, Ruth Barcan Marcus' work on identity, necessity, and reference, right from the beginning in the 1940s, proves most lucid, far-reaching, and replete with indications regarding the implications of the set-theoretical paradoxes for logic and epistemology.

Looking at the charges that Quine leveled against Marcus within the broader context that I have outlined here helps shed light on the issues raised and makes it improbable that, as Scott Soames once categorically stated, there "is no way that the formal system of Marcus' early papers could have significant consequences about ordinary names and descriptions in natural language" (Humphreys and Fetzer 1998, 71). Soames himself acknowledged that: "Although Quine's arguments were mistaken, they were enormously influential and they baffled large numbers of the profession for decades" (*Ibid.*, 16, 14).[4] As I see it the very nature of those very public and influential complaints established the significance of the formal system of Marcus' early papers to later theories about ordinary names and descriptions in natural language.

In her doctoral dissertation, Ruth Barcan Marcus already had her finger on the pulse of the problem of rigid reference. Postulating the necessity of identity is another response to the demands that Frege and Russell had tried to meet with the law of substitutivity of identicals, the introduction of extensions, Basic Law V, the theory of definite descriptions, the axiom of reducibility. Fitch's observation, cited above, that in Barcan's system there is lack of need for an axiom of extensionality merits study in this regard.

---

[4] Did they let themselves be duped?

Likewise, Marcus' insistence that the apparent opacity of intensional contexts lies in the way that logicians like Quine use the terms 'identity', 'true identity', 'equality' in their attempts to blind themselves and others to differences between identity and weaker forms of equivalence has far-reaching implications. For the "paradoxes" associated with modal logic arise, not because someone has fiddled with the reference process in some improper way, thereby thwarting access to the objects we want to put in our theories and quantify over, but because, like the identity statements that Frege put identity at the heart of his logic, modal contexts force a bifurcation in the reference of the statements that they govern and so display intensions eclipsed in other forms of discourse. Differences between identity and weaker forms of equivalence always obtain in any non-trivial case in which the difference between the signs in an identity statement corresponds to some intensional difference, some difference in the mode of presentation of the reference.

Frege and Russell both had to struggle with the fact that descriptions are opaque in perfectly extensional contexts. Like Frege's senses, descriptions serve to illuminate a single aspect of the thing that they would refer to. It is this partial illumination that is behind the impression of referential opacity that so vexed Quine. While a definite description of an object can alone suffice to fix the reference univocally, no description, however definite it may be, will suffice to define the reference to the point of blocking all the other descriptions identifiable with the same object which could come up to perturb the inference and change our truths into falsehoods or give truth to a lie.

In the 1970 lectures that went into the making of *Naming and Necessity*, Kripke explained some of these same ideas in particularly accessible, very ordinary language. However, in my opinion the interest of the new theory of reference ultimately lies in its connection to the broader questions discussed above, which are more threatening than those dealt with in *Naming and Necessity*, because a full understanding of the nature and deepest implications of Quine's attacks on modal and intensional logics leads right into the heart of problems in the foundations that Frege laid for his logic. And as Edmund Husserl once warned the Göttingen Mathematical Society, the development of the sciences, has constantly shown that lack of clarity in the foundations ultimately wreaks its vengeance, that if certain levels of progress are reached, further progress is fettered by errors due to obscure methodological ideas. (Husserl 1901, 431-32)

References

Barcan, Ruth 1946a. "The Deduction Theorem in a Functional Calculus of First Order Based on Strict Implication", *The Journal of Symbolic Logic* 11, 1, 115-18.

Barcan, Ruth 1946b. "A Functional Calculus of First Order Based on Strict Implication", *The Journal of Symbolic Logic* 11, 1, 1-16.

Barcan, Ruth 1947. "The Identity of Individuals in a Strict Functional Calculus of Second Order", *The Journal of Symbolic Logic* 12, 1, 12-16.

Boolos, George 1996. "Whence the Contradiction?", *Proceedings of the Aristotelian Society*, Supplementary volume LXVII (1993), 213-33. Also in *Frege: Importance and Legacy,* Matthias Schirn (ed.), Berlin: de Gruyter, 1996, 234-52.

Demopoulos, William (ed.) 1995. *Frege's Philosophy of Mathematics*, Cambridge MA: Harvard University Press.

Dummett, Michael 1996. "Reply to Boolos", in *Frege: Importance and Legacy,* Matthias Schirn (ed.), Berlin: de Gruyter, 253-60.

Fitch, Frederic 1949. "The Problem of the Morning Star and the Evening Star", *Philosophy of Science* 16, 131-41.

Frege, Gottlob 1879. "Begriffsschrift, a formula language, modeled upon that of arithmetic for pure thought", in *From Frege to Gödel: A Source Book in Mathematical Logic,* Jean van Heijenoort (ed.), Cambridge, MA: Harvard University Press, 1967, 1-82.

Frege, Gottlob 1884. *Foundations of Arithmetic.* Oxford: Blackwell, 1986.

Frege, Gottlob 1892. "On Sense and Reference", in *Translations from the Philosophical Writings*, Peter Geach and Max Black (eds.), Oxford: Blackwell, 1980, 56-78.

Frege, Gottlob 1893. *Basic Laws of Arithmetic*, Berkeley CA: University of California Press, 1964.

Frege, Gottlob 1903. "Frege on Russell's Paradox", in *Translations from the Philosophical Writings*, Peter Geach and Max Black (eds.), Oxford: Blackwell, 1980, 214-24.

Frege, Gottlob 1979. *Posthumous Writings*, Oxford: Blackwell.

Frege, Gottlob 1980. *Philosophical and Mathematical Correspondence*, Oxford: Blackwell.

Gödel, Kurt 1964. "What is Cantor's Continuum Problem", *Kurt Gödel Collected Works*, volume II, Solomon Fefermann et al. (eds.), New York: Oxford University Press, 254-70.

Grattan-Guinness, Ivor 1972. "Bertrand Russell on His Paradox and the Multiplicative Axiom: An Unpublished Letter to Philip Jourdain", *Journal of Philosophical Logic* 1, 103-10.

Grattan-Guinness, Ivor 1975. "Preliminary Notes on the Historical Significance of Quantification and the Axiom of Choice in Mathematical Analysis", *Historia Mathematica* 2, 475-88.

Grattan-Guinness, Ivor 1977. *Dear Russell-Dear Jourdain, A Commentary on Russell's Logic Based on his Correspondence with Philip Jourdain*, London: Duckworth.
Heck, Richard 1993. "The Development of Arithmetic in Frege's *Grundgesetze der Arithmetik*", *The Journal of Symbolic Logic* 58, 579-601.
Hintikka, Jaakko 1962. *Knowledge and Belief*, Ithaca NY: Cornell University Press.
Hintikka, Jaakko 1969. *Models for Modalities*, Dordrecht: Reidel.
Hintikka, Jaakko 1975. *The Intentions of Intentionality and Other New Models for Modalities*, Dordrecht: Reidel.
Hintikka, Jaakko and Donald Davidson (eds.) 1969. *Words and Objections, Essays on the Work of W. V. Quine*, Boston: Reidel.
Humphreys, Paul and James Fetzer, (eds) 1998. *The New Theory of Reference, Kripke, Marcus, and Its Origins*, Dordrecht: Kluwer.
Kaplan, David and Richard Montague (eds.) 1960. "A Paradox Regained" *Notre Dame Journal of Formal Logic*, 1 no. 3 (July), 79-90.
Kilmister, C. W. 1984. *Russell*. London: The Harvester Press.
Kripke, Saul 1972. *Naming and Necessity*, Oxford: Blackwell, 1980.
Kripke, Saul 1975. "Outline of a Theory of Truth", *Journal of Philosophy* 72, 690-716, in *Recent Essays on Truth and the Liar Paradox*, Robert Martin (ed.), Oxford: Clarendon, 1984, 53-81.
Marcus, Ruth Barcan 1960. "Extensionality", *Mind* 69, 55-62, also in *Reference and Modality*, Leonard Linsky (ed.), Oxford: Oxford University Press, 1971, 44-51.
Marcus, Ruth Barcan 1993. *Modalities*, New York: Oxford University Press.
Montague, Richard 1962. "Syntactical Treatments of Modality with Corollaries on Reflexion Principles and Finite Axiomatizability", *Proceedings of a Colloquium on Modal and* Many-Valued Logics, Helsinki, 23-26 August, 1962, Acta Philosophica Fennica XVI (1963), 153-67.
Quine, Willard van Orman 1947a. "The Problem of Interpreting Modal Logic", *Journal of Symbolic Logic* 12, 2 (June), 43-48.
Quine, Willard van Orman 1947b. "Review of Ruth C. Barcan, 'The Deduction Theorem in a Functional Calculus of First Order Based on Strict Implication'", *Journal of Symbolic Logic* 12, 95.
Quine, Willard van Orman 1947c. "Review of Ruth C. Barcan, 'The Identity of Individuals in a Strict Functional Calculus of Second Order'", *Journal of Symbolic Logic* 12, 95-96.
Quine, Willard van Orman 1953a. "Reference and Modality", in *From a Logical Point of View*, 2nd (ed.) rev., New York: Harper & Row, 1961.
Quine, Willard van Orman 1953b. "Three Grades of Modal Development", *Ways of Paradox*, Cambridge MA: Harvard University Press, 1976, 158-76.
Quine, Willard van Orman 1960. *Word and Object*, Cambridge MA: M.I.T. Press.

Quine, Willard van Orman 1962. "Reply to Professor Marcus", in *Ways of Paradox*, Cambridge MA: Harvard University Press, 1976, 177-84.

Quine, Willard van Orman 1969. *Ontological Relativity and Other Essays*, New York: Columbia University Press.

Quine, Willard van Orman 1976. *Ways of Paradox*, Cambridge MA: Harvard University Press.

Quine, Willard van Orman 1994. "Promoting Extensionality", *Synthese* 98, 143-51.

Russell, Bertrand 1903. *Principles of Mathematics*, London: Norton.

Russell, Bertrand 1905. "On Denoting", *Logic and Knowledge*, London: Allen & Unwin, 1956, 41-56.

Russell, Bertrand 1919. *Introduction to Mathematical Philosophy*, London: Allen & Unwin.

Russell, Bertrand 1927. *Principia Mathematica to *56*. 2nd (ed.), Cambridge UK: Cambridge University Press, 1964.

Russell, Bertrand 1944. "My Mental Development", *The Philosophy of Bertrand Russell*, Paul Schilpp (ed.), Evanston IL: Northwestern University Press, 3-20.

Russell, Bertrand 1956. *Logic and Knowledge*, London: Allen & Unwin,

Russell, Bertrand 1959. *My Philosophical Development*, London: Allen & Unwin, 1975.

Russell, Bertrand 1973. *Essays on Analysis*, D. Lackey (ed.), London: Allen & Unwin.

Schirn, Matthias 1996. "On Frege's Introduction of Cardinal Numbers as Logical Objects", in *Frege: Importance and Legacy*, Matthias Schirn (ed.), Berlin: de Gruyter, 114-73.

Schirn, Matthias (ed.), 1996. *Frege: Importance and Legacy*, Berlin: de Gruyter.

Smullyan, Arthur 1947. "Review of Quine's 'The problem of interpreting modal logic'", *Journal of Symbolic Logic* 12, 139-41.

Smullyan, Arthur 1948. "Modality and Description", *Journal of Symbolic Logic* 13, 31-37.

Wartofsky, Marx (ed.) 1963. *Proceedings of the Boston Colloquium for the Philosophy of Science 1961/1962*, Dordrecht: Reidel.

10

Claire Ortiz Hill

TACKLING THREE OF FREGE'S PROBLEMS: EDMUND HUSSERL ON SETS AND MANIFOLDS

Introduction

Gottlob Frege was the first to admit that the way he introduced sets in *The Foundations of Arithmetic* (Frege 1884) and endeavored to legitimate them in *Basic Laws of Arithmetic* (Frege 1893) was problematical. On several occasions, he admitted that it was only with difficulty that he had done so because the matter had not seemed to him to be quite secure. He even alluded to having been in a certain way forced to introduce them in the way he did almost against his will because he saw no other alternative.[1]

In this chapter, I study three problems associated with Frege's use of sets: 1) that of reference to non-existent, impossible, imaginary objects; 2) the introduction of extensions; and 3) Frege's true *bête noire*, the paradox publicized by Bertrand Russell. I address these three problems within a discussion of the struggle to overcome the shortcomings of set theory engaged in by Edmund Husserl, a philosopher not generally perceived—even by Husserl scholars who should know better—as having been conversant with set theory at all. In so doing, I hope to stimulate investigation into important theories of Husserl that stand to have interesting applications to philosophy nowadays.

Indeed, although the fact that investigations into set theory and the philosophy of arithmetic by mathematicians turned philosophers played a pre-eminent role in the shaping of 20th century western philosophy is perfectly obvious to anyone familiar with the origins of analytic philosophy, the same cannot be said of philosophers' perceptions of the origins of the

---

[1] Frege expresses this in numerous places, for example in Frege 1884, §§67-73, 107 and Frege 1893, vii, ix-x; Frege 1980a, 55, 130-32, 191; Frege 1980b, 214. About the only place that he expresses enthusiasm about his appeal to sets is in his review of Husserl's *Philosophy of Arithmetic.* (Frege 1894, 201-02)

phenomenological school. However, Husserl was one of the very first to experience the direct impact of challenging problems with set theory and his phenomenology first began to take shape while he was struggling to solve such problems (Hill and Rosado Haddock 2000). First, however, something needs to be said about some terminological obstacles.

Terminological Considerations

Terminologically, it is absolutely necessary to remember that Husserl's theories about sets and manifolds developed during the fifteen years that he spent in the company of Georg Cantor, the creator of set theory and the author of the *Mannigfaltigkeitslehre*. So use of the term '*Mannigfaltigkeit*', usually translated into English by 'manifold' or 'multiplicity', cannot be univocal within the context of the problems addressed here.

Cantor used the terms '*Menge*', '*Mannigfaltigkeit*' and '*Inbegriff*' interchangeably (Grattan-Guinness 2000, Chapter 3). Right from the start Husserl, however, directly confronted the terminological difficulties that one faces in speaking of *Mannigfaltigkeiten*. In his earliest writings, he noted that in place of the word '*Vielheit*' the practically synonymous terms '*Mehrheit*', '*Inbegriff*', '*Aggregat*', '*Sammlung*', '*Menge*', etc. (variously translated by 'quantity', 'aggregate', 'plurality', 'totality', 'collection', 'set', 'multiplicity') had been used and he acknowledged the ambiguity thus introduced into reasoning. In the beginning of *Philosophy of Arithmetic*, he informed readers that, while recognizing the differences, he would not initially restrict himself to using any one of these terms exclusively. He said that he hoped thereby to be able to neutralize differences in meaning among them (Husserl 1891, 14, note). This same strategy had been adopted in *On the Concept of Number*. (Husserl 1887, 96)

Although Husserl did use the various terms for set interchangeably in the late 1880s, in those days he rarely used the more Cantorian term '*Mannigfaltigkeit*', which he only began to use more frequently in posthumous writings of the 1890s and when he particularly studied geometrical manifolds (Husserl 1983). By that time, the manifolds of the *Mannigfaltigkeitslehre* that Husserl himself was developing, and that he ultimately considered to be the highest expression of pure logic, were quite different from Cantor's *Mannigfaltigkeiten* (Hill and Rosado Haddock 2000, Chapters 7, 8, 9).[2] From that time on, he strove to distinguish sets from manifolds, multiplicities, totalities, aggregates, and so on.

In what follows I translate '*Vielheit*' as 'multiplicity'; '*Allheit*' as 'totality'; '*Inbegriff*' as 'aggregate'; '*Menge*' as 'set'; '*Mannigfaltigkeit*' as 'manifold'. I translate '*Widerspruch*' by 'contradiction' and '*Widersinnigkeit*' by contradiction in terms. The translations appearing in the text are often my own and I have often modified published translations to make the terminology consistent.

---

[2] As is completely clear to anyone who has read Cantor's writings.

Sets in *On the Concept of Number* and *Philosophy of Arithmetic*

In the *Philosophy of Arithmetic*, Husserl endeavored to instill rigor into the rather inchoate set-theoretical foundations being proposed for numbers (Hill and Rosado Haddock 2000, Chapters 7, 8). In *Formal and Transcendental Logic*, he even explicitly characterized the *Philosophy of Arithmetic* as having represented an initial attempt on his part "to obtain clarity regarding the original genuine meaning of the fundamental concepts of the theory of sets and cardinal numbers"(*"Klarheit über den ursprungssechten Sinn der Grundbegriffe der Mengen- und Anzahlenlehre zu gewinnen"*). (Husserl 1929, §27a; also §24 and note)

In both *On the Concept of Number* and *Philosophy of Arithmetic*, Husserl in fact tried to make set theory the basis of arithmetic. Just as cardinal numbers relate to sets (*Mengen*), so ordinals relate to series that are themselves ordered sets (*Mengen*), he maintained in the introduction to *Philosophy of Arithmetic* (Husserl 1891, 10-13), chapter 1 of which opens with the assertion that the analysis of the concept of number presupposes that of the concept of multiplicity (*Vielheit*). There cannot be any doubt, Husserl went on to affirm there, as in *On the Concept of Number*, that the concrete phenomena that form the basis for the abstraction of the concepts in question are aggregates (*Inbegriffe*), multiplicities (*Vielheiten*) of determinate objects. Everyone, he stressed, knew what was meant by this, for despite difficulties experienced in analyzing it, the concept of *Vielheit* itself was perfectly precise and the range of its extension exactly delimited. It might therefore be considered as a given, he maintained, even though we might still be in the dark about the essence and formation of the concept itself. (Husserl 1891, 14-15; Husserl 1887, 96-97, 111)

Husserl then turned to study the experience of concrete *Inbegriffe* in order to consider how both the more indeterminate universal concept of *Vielheit* and the determinate number concepts were to be abstracted from them. Crucial to that account was his conviction that any talk of such *Inbegriffe* necessarily involved "collective combination", the combination of the individual contents into a whole, into a unity containing the individual contents as parts. Though the combination involved might be very loose, there *was*, the faithful student of Brentano then held, a particular sort of unification there that would also have to have been noticed as such since the concept of *Inbegriffs* (*der Vielheit*) could never have arisen otherwise. The concept of *Vielheit*, he explained, in fact arose through reflection upon the special way in which the contents were united together in each concrete *Inbegriff*, a way analogous to the way in which the concept of any other kind of whole arose through reflection upon the mode of combination peculiar to it. (Husserl 1891, 18-20; Husserl 1887, 97-98)

In *On the Concept of Number*, Husserl characterized the distinctive abstractive process that yielded the concept of set in this way:

> It is easy to characterize the abstraction which must be exercised upon a concretely given *Vielheit* in order to attain to the number concepts under which it falls. One considers each of the particular objects merely insofar as it is a something or a one herewith fixing the collective combination; and in this manner there is obtained the corresponding general *Vielheitsform*, one and one and ... and one, with which a number name is associated. In this process there is total abstraction from the specific characteristics of the particular objects.... To abstract from something merely means to pay no special attention to it. Thus in our case at hand, no special interest is directed upon the particularities of content in the separate individuals. (Husserl 1887, 116-17)

In *Philosophy of Arithmetic*, Husserl studied this same process in greater detail (chapter 4, especially), especially underscoring the uniqueness of the abstraction process yielding the number concept. There, number was characterized as the general form of a *Vielheit* under which the *Inbegriff* of objects *a*, *b*, *c* fell. He wrote,

> It is thus clear that this *Inbegriff* (this *Vielheit, Menge*, or whatever one may want to call it) makes up the subject of the number statement. Considered formally, the number and the concrete set (*Menge*) are related to one another as concept and object. *The number is not consequently assigned to the concept of the objects counted, but to their Inbegriff.* (Husserl 1891, 165-66)

In Chapter XI of *Philosophy of Arithmetic*, Husserl took up what he called a particularly remarkable way of extending the original concept of *Menge* or *Vielheit* that by its very nature reached beyond the necessary bounds of human cognition, thus winning for itself an essentially new content. Signs, symbolic presentation *à la* Brentano, he explained, might aid the mind in reasoning in regions of thought beyond what could be known through direct cognitive processes like perception or intuition; the repeated application of operations permitting the collecting together of a multitude of objects one after the other into a set could take the place of the direct cognitive grasp of *Mengen* with hundreds, thousands or millions of members, and this was a way of actually representing collections in an ideal sense and essentially unproblematic from a logical point of view.

The logical problems connected with infinite *Inbegriffen, Mengen, Vielheiten* were, however, he maintained, of a completely different order, since the very principle by which they were formed or symbolized itself immediately made collecting all of their members together, one by one, a logical impossibility. By no extension of our cognitive faculties could one conceivably cognitively grasp or even successively collect such sets. With them one had reached the limits of idealization. As examples of infinite sets, he proposed the extensions of most general concepts, the set of the numbers of the symbolically extended number series, the set of points on a line and of the boundaries of a continuum.

With infinite *Menge*, Husserl explained, one was in possession of a clear principle by which to transform any already formed concept of a certain given species into a new concept plainly distinct from the previous one. And this could be done over and over in such a way as to be certain a priori of never coming back to the original concept and to previously generated concepts. Repeated application of this process would yield successive presentations of continually expanding sets, and if the generating principle was really determinate, then it was determined a priori whether or not any given object could belong to the concept of the expanding set of concepts. In the case of the concept of the infinite *Menge* of numbers, one began with a direct presentation. The other natural numbers could be reached by repeated additions of 1, and nothing prevented one from advancing indefinitely in this way to new cardinals; we have a method for adding one to a previously given number, an operation necessarily generating one new number after another without return and limitation, each new number being determined by the process.

It is easy, Husserl added, to see why mathematicians had tried to transpose the concept of quantity onto such constructions, which were, however, of an essentially different logical nature. In the usual cases, the process by which the sets were generated was finite, there was always a last stage, and it was sometimes possible actually to bring the process to a halt and also to construct the corresponding set. But, this was quite absurd in the case of infinite sets. The process used to generate them was non-terminating, and the idea of a last stage, of a last member of the set meaningless. And this constituted an essential logical difference.

However, Husserl saw that, despite the absurdity of the idea, analogies fostered a tendency to transpose the idea of constructing a corresponding collection for infinite sets, thereby creating what he called a kind of "imaginary" concept whose anti-logical nature was harmless in everyday contexts precisely because its inherent contradictoriness was never obvious in life. This was, Husserl explained, the case when "All S" was treated as a closed set. However, he warned, the situation changes when this imaginary construct was actually carried over into reasoning and influenced judgments. It was clear, he concluded, that from a strictly logical point of view we must not ascribe anything more to the concept of infinite sets than is actually logically permissible, and above all not the absurd idea of constructing the actual set. (Husserl 1891, 218-21)

Husserl's Growing Doubts about Sets

Husserl quickly came to judge his first attempts to clarify the true meaning of the fundamental concepts of the theory of sets and cardinal numbers to have been a failure. While working on the logic of mathematical thought and mathematical calculation, he explained in introductions to the *Logical Investigations*, he had encountered disturbing problems while studying the logic of formal arithmetic and the theory of *Mannigfaltigkeiten*. The first volume of the *Logical Investigations*, he explained in the foreword to the first edition, had arisen out of reflections upon certain implacable problems that had constantly hindered, and ultimately interrupted, the progress of many years of work devoted to achieving a philosophical clarification of pure mathematics. In a later preface to the book, he confessed to having been "unsettled–even tormented" by doubts about the psychological analysis of *Mengen* from the very start. (Husserl 1900-01, 41-43; Husserl 1975, 16-17)

There was certainly some truth, Husserl concluded, in the idea that the idea of *Menge* arose out of collective combination. The *Kollektivum* was no material unity forming in the content of the items collected, he had initially reasoned, so the concept of collection had to arise through psychological reflection, in Brentano's sense, upon the act of collecting. But Husserl came to believe that the concept of number then had to be something basically different from the concept of collecting, which was all that could result through reflection on acts (Husserl 1975, 34-35). Such doubts eventually undermined his confidence in Brentano's theories, as well as those of Weierstrass and Cantor. (Hill and Rosado Haddock 2000)

Husserl's antipathy towards a calculus of classes is evident in articles published during the 1890s (Husserl 1994, 92-114, 115-30, 135-38, 199, 443-51), where he sought to show "that the *total* formal basis upon which the class calculus rests is valid for the relationships between conceptual objects", and that one could solve logical problems without "the detour through classes", which he considered to be "totally superfluous". (*Ibid.*, 109, 123)

Husserl's chief target in those articles was Ernst Schröder, whom he credited with having provided "the best present form of the extensional calculus" (Husserl 1994, 52-91, 123, 421-41). Of particular interest here are Husserl's comments on Schröder's attempt to show that bringing all possible objects of thought into a class gives rise to contradictions. Schröder's argument, Husserl wrote in an article of 1891, might appear astonishing at first glance, but was actually sophistical. Husserl was, though, tellingly prepared to concede that:

> in the case where we simultaneously have, besides certain classes, also classes *of* those classes, the calculus may not be blindly applied. In the sense of the calculus of sets as such, any set ceases to have the status of a set as soon as it is considered as an element of another set; and this latter in turn has the status of a set only in relation to its primary and authentic elements, but not in relation to whatever elements *of* those elements there may be. If one does not keep this in mind, then actual errors in inference can arise. (Husserl 1994, 84-85)

In Halle, Husserl had been sort of a victim *avant la lettre* of the crisis in foundations that broke out once Russell began advertising the famous contradiction about the set of all sets that are not members of themselves that he discovered while studying Cantor's theories.[3] Once appointed to

---

[3] Hill and Rosado Haddock 2000, Chapters 7 and 8. Cantor 1991, 387-485. Russell derived "his" paradox while studying Cantor's proof that there can be no greatest cardinal number Russell 1903, §§100, 344, 500; Russell 1959, 58-61; Grattan-Guinness 1978, 1980.

Göttingen in 1901, Husserl was drawn into mathematical discussions with mathematicians who had known the contradiction for some years before Russell publicized his finding.[4] However, no account of Husserl's reaction to the set-theoretical paradoxes has yet to be fully pieced together, and it is only in unpublished notes on set theory (Husserl Ms A 1 35),[5] that we find him directly grappling with the questions raised by them. Though the ideas expressed there are not developed into a systematic whole, they yield clues into his views on matters not explicitly addressed in his published writings.

Husserl Develops His Own *Mannigfaltigkeitslehre*

Even as he was cogitating on set theory, whether Cantor's, Frege's, or Schröder's, Husserl was at work developing his own theory of *Mannigfaltigkeiten*. Between 1886 and 1893, he related in a 1901 letter, he had busied himself with the theory of geometry, formal arithmetic and the theory of manifolds, at times exclusively devoting himself to this (Husserl 1983, 396). In prefaces to the *Logical Investigations*, he specifically alluded to having been troubled by the *Mannigfaltigkeitslehre*. (Husserl 1900-01, 41; Husserl 1975, 35)

---

[4] In a November 7, 1903 postcard, Hilbert informed Frege that Russell's paradox was already known to them in Göttingen and that he himself had found other even more convincing contradictions as long as four or five years ago. Hilbert further informs Frege that "Dr. Zermelo discovered it three or four years ago after I had communicated my examples to him" (Frege 1980a, 51). Volker Peckhaus and R. Kahle have tried to elucidate Hilbert's remark. Zermelo, they explain, came to Göttingen in 1897 to *Habilitieren*. Under the influence of Hilbert he shifted his interests to set theory and foundations and became Hilbert's collaborator in the philosophy of mathematics, a first member of Hilbert's school before it was established". The winter term of 1900/01 already found him giving a lecture course on set theory in Göttingen (Peckhaus and Kahle, 3). A record of Zermelo's early discovery of the famous contradiction is to be found known in a note of April 16, 1902 conveyed to Husserl about the comments about the contradictions involving sets of sets discussed his review of Schröder discussed above, (Husserl 1994, 442; Rang and Thomas 1981; Peckhaus and Kahle, 3-4). When asked in 1936 about Hilbert's comment to Frege, Zermelo answered that there had been much discussion of the set-theoretical paradoxes in Hilbert's circle around 1900 and that at that time he had given precise form to the antimony of the great cardinal number later named after Russell (of the set of all sets that do not contain themselves). By the time Russell's work appeared, Zermelo affirmed, that it was already familiar to them (Peckhaus and Kahle, 4, n. 7). Peckhaus and Kahle consider that it can be assumed that Hilbert formulated his paradox during exchanges with Georg Cantor documented in their 1897-1900 correspondence where one finds Hilbert concerned with what was later called "Cantor's paradox", the paradox of the greatest cardinal or of the set of all cardinals, which "served only as a paradigmatic example for other inconsistent multiplicities" and according to Peckhaus' and Kahle's analysis was closely related (*Ibid.*, 6, 14).

[5] Husserl Ms A 1 35. The dates found in these notes are 1912, 1918, 1920, 1926, and 1927. They have principally been known only through Rosado Haddock 1973. Professor Lohmar and his colleagues at the Husserl Archives in Cologne kindly sent copies of them to the Husserl Archives in Paris where Messieurs Courtine and Pernet made it possible for me to study them. S. Luft and B. Goossens of the Husserl Archives in Leuven helped me by finding the answers to some questions that I had. Pages 27a-28b of Husserl's notes on set theory are an almost word for word translation of passages from Russell 1911.

Of the 1890s, Husserl wrote that he had seen about him only ambiguously defined problems and profoundly unclear theories, that disappointed by the theories of those to whom he owed most of his intellectual training and sick of all the confusion, he had abandoned work on the second volume of the *Philosophy of Arithmetic* about the logic of the deductive sciences and set out in search of his own answers, engaging in very general reflections that took him beyond the confines of the mathematical realm towards a universal theory of formal deductive systems. Traditional logic, he complained, should have made the rational nature of the deductive sciences, their formal unity and symbolic method, transparent and easily understandable, but the study of the actual deductive sciences left all that problematic and obscure. Husserl saw the mathematical theory of manifolds as a realization of the idea of a science of possible deductive systems, but viewed it, and all of modern formal analysis as but a partial realization of his own ideal of a science of possible deductive systems. (Husserl 1900-01, 41-42; *Prolegomena*, §70; Husserl 1975, 16-17, 35; Husserl 1994, 490-91)

In notes on sets and manifolds dating from 1891/92, one finds Husserl asking what in definition of the concept of *Menge*, of *Mannigfaltigkeit* in the broadest sense of the word, leads to the concept of *Mannigfaltigkeit* in the narrower sense. He begins to answer his question by noting that by *Mannigfaltigkeit* Cantor merely meant an aggregate of any elements combined into a whole and cites the passage of the *Mannigfaltigkeitslehre* where Cantor wrote: "By *Mannigfaltigkeit* or *Menge*, I understand generally any many that can be thought of as one, i.e., any aggregate of determinate elements which can be bound together into a whole by a law...". (Husserl 1983, 95)

Husserl goes on to note that Cantor's concept does not correspond to Riemann's and other related ones in the theory of geometry.[6] For these, Husserl stresses, a *Mannigfaltigkeit* is an aggregate of elements that are not just combined into a whole, but are ordered and continuously interdependent (*Ibid.*, 95-96). Husserl defined order as "a concatenation that has the special property that each member possesses an unambiguous position in the narrow sense of the word in relation to any arbitrary one, i.e., can therefore be unequivocally characterized by the mere form of the direct or indirect connection with the last one" (*Ibid.*, 93). A *Mannigfaltigkeit* is not an aggregate of elements without relations, he underscored in 1892. It is precisely the relations that are essential and serve to distinguish it from a mere aggregate (*Ibid.*, 410).

In the *Prolegomena to a Pure Logic*, after his decade of lonely, intellectual struggle, Husserl unveiled his own theory of manifolds as the highest task of formal logic (Husserl 1900-01, *Prolegomena*, §§69-70). It

---

[6] Husserl extensively discussed and criticized the shortcomings of the theory of manifolds in geometry, including Riemann's theories, in a 1889/1890 course on issues in the philosophy of mathematics. (Husserl 1983, 312-347)

is also discussed in *Ideas I* (Husserl 1913, §§71-72). In *Formal and Transcendental Logic*, Husserl expressed his satisfaction with the theory as given decades earlier (Husserl 1929, §33).

However, while holding fast to the theory of manifolds as outlined in his published works, Husserl further refined and explained his theory in a particularly clear and explicit way in lecture courses only published decades after his death. In *Introduction to Logic and Theory of Knowledge* (Husserl 1906-07, §§18-19) and *Logik und allgemeine Wissenschaftstheorie* (Husserl 1917/18, Chapter 11), Husserl explained that he had come to discern a certain natural order in formal logic and to broaden its domain to include two levels above traditional formal logic in such a way as to account for the progress made by modern mathematics and, most particularly, the progress represented by the theory of manifolds. He considered the detection of these three levels of formal logic to be of prime importance for the understanding of logic and philosophy.

According to this new conception of formal logic, the logic of subject and predicate propositions and states of affairs, which deals with what might be stated about objects in general from a possible perspective, is found on the lowest level. The purely logical disciplines of the two higher levels of pure logic still deal with individual things, but these are no longer empirical or material entities. They are removed from acts, subjects, or empirical persons of actual reality. For example, in the set theory of the second level, it is not a matter of predicating something of the members, but of sets overall, having any members whatsoever.

On the second level, it is no longer to be a question of objects as such about which one might predicate something, but of objective constructions of a higher kind determined in purely formal terms and dealing with objects in an indeterminate, general way. It is to be a matter of forms of judgments, and forms of their constituents, forms of deduction, forms of demonstration, sets and relationships between sets, combinations, orders, quantities, objects in general, etc. Husserl placed the fundamental concepts of mathematics on this second level, which he conceived of as an expanded, completely developed, analytics. This is where one is to find the theory of cardinal numbers, the theory of ordinals.

On the second level, one investigates what is valid for higher-order objects. One proceeds in a purely formal manner since every single concept used is analytic. Here one calculates, reasons deductively, with concepts and propositions. Signs and rules of calculation suffice because each procedure is purely logical. One manipulates signs for which rules having such and such a form are valid, signs which like chess pieces acquire their meaning in the game through the rules of the game. One may proceed mechanically in this way and the result will prove accurate and justified. This is an enormous help in reasoning, for it is incomparably easier to think with signs and one is thereby freed from the equivocation and ambiguity that comes with using words. Moreover, the process itself calls for the maximum of rigor.

By abstracting further, one reaches Husserl's third level, that of the theory of possible theories, the theory of manifolds. Husserl described manifolds as pure forms of possible theories which, like molds, remain totally undetermined as to their content, but to which thought must necessarily conform in order to be thought and known in a theoretical manner. On this third level, we have a new discipline and a new method constituting a new kind of mathematics, the most universal one of all. Here formal logic deals with whole systems of propositions making up possible deductive theories. It is now a matter of theorizing about possible fields of knowledge conceived of in a general, undetermined way and purely and simply determined by the fact that they are in conformity with a theory having such a form, i.e., determined by the fact that its objects stand in certain relations that are themselves subject to certain fundamental laws of such and such determined form.

By using axioms of such and such a form, theories of such and such a form may be developed. These objects are exclusively determined by the form of the interconnections assigned to them. These interconnections themselves are just as little determined in terms of content as are the objects. Only their form determines them by virtue of the form of the elementary laws admitted as valid for these interconnections, laws that also determine the theory to be constructed, the form of the theories.

Through axiom forms, we define a multiplicity of anything whatsoever in an indeterminate, general way. After formalization words are completely empty signs that only have the purely formal sense laid down for them by the axiom forms. One then speaks of a certain something that must by definition stand in a certain relationship to something else in the defining manifold. We have concepts defined by pure, formal concepts, purely set forth, defined, as a formal possibility. Only a form is defined, but whether axioms as truths have existence in any objective real or ideal spheres corresponding to the prescribed form remains open. The form exists insofar as it is correctly defined, insofar as the axiom forms are ordered in such a way as to contain no formal contradictions, no violation of analytic principles, i.e., in the sense of formal analytical consistency. On the basis of the definition of the manifold, we can deduce conclusions, construct proofs, and it is then certain a priori that anything obtained in this way will correspond to something in our theory.

The general theory of the manifolds, or science of theory forms, is a field of free, creative investigation made possible once the form of the mathematical system was emancipated from its content. Once one discovers that deductions, series of deductions, continue be meaningful and to remain valid when one assigns another meaning to the symbols, one is then free to liberate the mathematical system, which can henceforth be considered as being the mathematics of a domain in general, conceived in a general and indeterminate manner. No longer restricted to operating in terms of a particular field of knowledge, we are free to reason completely on the level of pure forms. Operating within this sphere of pure forms, we can vary the systems in different ways. Nothing more need be presupposed than the fact that the objects figuring in it are such that, for them, a certain connective supplies new objects and does so in such a way that the form determined is assuredly valid for them. One finds ways of constructing an infinite number of forms of possible disciplines. And all that is of inexhaustible practical interest, Husserl maintained.[7] Now let us turn to our Fregean problems.

Problem One: Reference to Non-existent and Impossible Objects

Gottlob Frege insisted that "we have to throw aside proper names that do not designate or name an object" (Frege 1979, 122), that "in science and wherever we are concerned about truth... we also attach a reference to proper names and concept-words; and if through some oversight, say, we fail to do this, then we are making a mistake that can easily vitiate our thinking" (*Ibid.*, 118). He considered that he had shown that for certain proofs the whole cogency of the proof stands or falls on the question whether a combination of signs–e.g. $\sqrt{-1}$ has a meaning or not (*Ibid.*, 123). In "On Formal Theories of Arithmetic", he argued that unless an equation contained only positive numbers, it no more had a meaning than the position of chess pieces expressed a truth and he condemned the theory by which one might but set down rules by which one passed from given equations to new ones in the way one moved chess pieces. A proposition might very well be non-contradictory, he stressed, without being true. (Frege 1885, 112-21)

Frege *could* not in fact accept combinations of sign that do not designate an object because his logic was actually designed in such a way as not to be able to cope them. "Only in the case of objects can there be any question of identity (equality)", he held (Frege 1979, 182). And he put identity statements at the very heart of his project to provide foundations for the theorems of arithmetic (Frege 1884, x, §§55-67, 106).

It was his need for objects that had induced Frege to introduce the classes, extensions (problem two, below) that he eventually considered to be the cause of Russell's paradox (problem three, below). Consider his anguish as he wrote to Russell:

> I myself was long reluctant to recognize... classes; but I saw no other possibility of placing arithmetic on a logical foundation. But the question is, How do we apprehend logical objects? And I have found no other answer to it than this, We apprehend them as extensions of concepts.... I have always been aware that

---

[7] Husserl also studied manifolds at length in his unpublished notes on set theory. See, for example, Ms A 1 35 12b; 13b; 37a-b; 48; 51b; 52b.

there are difficulties connected with this, and your discovery of the contradiction has added to them; but what other way is there? (Frege 1979, 140-41)

In his appendix on the paradox, Frege wrote that he should gladly have dispensed with the law about extensions if he had known any substitute for it and that even then he did not know how arithmetic could be scientifically established, numbers could be apprehended as logical objects without it (Frege 1980b, 214). So with Richard Heck we may well suggest that we shall "not fully understand Frege's philosophy until we understand the enormous significance the question how we apprehend logical objects... had for him...". (Heck 1995, 287)

Husserl also anguished over the logical issues surrounding combinations of signs that do not and cannot refer to objects. Such questions were a major factor in his decision to abandon work on the second volume of the *Philosophy of Arithmetic*, where he had planned to deal with the fractions, negative, and irrational numbers that he included under the heading of the imaginary (Hill 1991, Chapter 5 §4). Through an easy verification, he observed in 1890, one might convince oneself of the correctness of any sentence deduced by means of negative, irrational and imaginary ("impossible") numbers and after innumerable such experiences naturally come to "trust in the unrestricted applicability of these modes of procedure, expanding and refining them more and more". He complained about all the mental energy wasted that had been wasted in connection with these numbers and expressed his opinion that arithmetic would have progressed much more quickly and securely if there had there been more clarity and insight about the logic involved. (Husserl 1994, 48-49)

In the foreword to the *Logical Investigations*, Husserl specifically alluded to having been troubled by the *Mannigfaltigkeitslehre*, with its expansion into special forms of numbers and extensions. The fact, he explained, that one could obviously generalize, produce variations of formal arithmetic which could lead outside the quantitative domain without essentially altering formal arithmetic's theoretical nature and calculational methods had brought him to realize that there was more to the mathematical or formal sciences, or the mathematical method of calculation than would ever be captured in purely quantitative analyses. (Husserl 1900-01, 41-43; see also Husserl 1975, 35)

In talks that he gave before the Göttingen Mathematical Society in 1901, Husserl explained that questions regarding imaginary numbers had come up in mathematical contexts in which formalization yielded constructions which arithmetically speaking were nonsense, but which could, nevertheless, be used in calculations. It was apparent, he noted, that when formal reasoning was carried out mechanically as if these symbols had meaning, if the ordinary rules were observed, and the results did not contain any imaginary components, then these symbols might be legitimately used. And this could be empirically verified. However, this fact raised significant questions and he named these three: (1) Under what conditions can one freely operate within a formally defined deductive system with concepts which according to the definition of the system are imaginary and have no real meaning? (2) When can one be sure of the validity of one's reasoning, that the conclusions arrived at have been correctly derived from the axioms one has, when one has appealed imaginary concepts? And (3) To what extent is it permissible to enlarge a well-defined deductive system to make a new one that contains the old one as a part? (Husserl 1891, 432-33; Husserl 1929, §31; Schuhmann and Schuhmann 2001)

A letter to Carl Stumpf from the early 1890s affords insight into how Husserl tackled the questions. In trying to come to a clear understanding as to how operating with contradictory concepts could lead to correct theorems, Husserl says that he had found that in the case of imaginary numbers like $\sqrt{2}$ and $\sqrt{-1}$, it was not a matter of the "possibility" or "impossibility" of concepts, for through the calculation itself and its rules, as defined for those fictive numbers, the impossible fell away, and a genuine equation remained. One could calculate again with the same signs, but referring to valid concepts, and the result was again correct. Even if one mistakenly imagined that what was contradictory existed, or held the most absurd theories about the content of the corresponding concepts of number, Husserl realized, the calculation remained correct if it followed the rules. So, he concluded that this must be a result of the signs and their rules (Husserl 1994, 15-16). In an 1891 letter to Frege, Husserl wrote that he did not have a clear idea of how Frege would justify the imaginary in arithmetic since in "On Formal Theories of Arithmetic" he had rejected the path that Husserl himself had found after much searching. (Frege 1980a , 65)

Quite unlike Frege, Husserl concluded that formal constraints banning reference to non-existent and impossible objects[8] unduly restrict us in our theoretical, deductive work. For Husserl, though no object could correspond

---

[8] See Husserl 1983, 244-49; Husserl 1891, 432. In *On the Concept of Number*, Husserl expressed his conviction that, as Karl Weierstrass had famously taught, all the more complicated and artificial forms of numbers had their origin and basis in the concept of positive whole numbers and their interrelations and were derivable from them in a strictly logical way. However, Husserl never again endorsed that thesis in the confident way he did in *On the Concept of Number* (Husserl 1887, 94-95). Most conspicuously apparent in the *Philosophy of Arithmetic* is Husserl's tergiversation regarding Weierstrass' teaching. He even warns readers that, although cardinal numbers in a certain way seem to be the basic numbers involved in arithmetic because the signs for them figure in expressions for positive, negative, rational, irrational, real, imaginary, alternative, ideal numbers, quaternions etc., the analyses of the second volume would perhaps show that thesis to be untenable (Husserl 1891, 11-12). In the April 1891 preface to the book, he went further to state that the analyses of the second volume would actually show that in no way does a single kind of concept, whether that of cardinal or ordinal numbers, form the basis of general arithmetic (*Ibid.*, 7). And before that book even made its way into print, he confessed to Carl Stumpf that through no manner of cunning could negative, rational, irrational and the various kinds of complex numbers be derived from the concept of the cardinal number. (Husserl 1994, 13)

to what was a contradiction in terms (*Widersinnigkeit*), a contradiction in terms nonetheless genuinely had a coherent meaning and could be determined to be true or false (Husserl 1900-01, LI IV). Unlike Frege again, Husserl found an answer that ultimately satisfied him. In *Logical Investigations*, he expressed his conviction that his theory of complete manifolds was the key to the only possible solution to the as yet unclarified problem as to how in the realm of numbers, impossible, non-existent, meaningless concepts might be dealt with as real ones (*Ibid., Prolegomena* §70). In *Ideas*, he wrote that his chief purpose in developing the theory of manifolds had been to find a theoretical solution to the problem of imaginary quantities. (Husserl 1913, §72 and n.)

Understanding the nature of theory forms, Husserl explained in several texts, shows how reference to impossible objects can be justified. According to his theory of manifolds, one could operate freely within a manifold with imaginary concepts and be sure that what one deduced was correct when the axiomatic system completely and unequivocally determined the body of all the configurations possible in a domain by a purely analytical procedure. It was the completeness of the axiomatic system that gave one the right to operate in that free way. A domain was complete, according to Husserl's theory, when each grammatically constructed proposition exclusively using the language of this domain was, from the outset, determined to be true or false in virtue of the axioms, i.e., necessarily followed from the axioms (in which case it is true) or did not (in which case it is false). In that case, calculating with expressions without reference could never lead to contradictions (Husserl 1900-01, *Prolegomena*, §70; Husserl 1906/07; Husserl 1917/18, §56; Husserl 1929, §31; Husserl 1891, 441). In *Ideas I*, Husserl especially defined complete manifolds as having the "distinctive feature that a finite number of concepts and propositions–to be drawn as occasion requires from the essential nature of the domain under consideration–determines completely and unambiguously on the lines of pure logical necessity the totality of all possible formations in the domain, so that in principle, therefore, nothing further remains open within it". In such complete manifolds, he stressed, "the concepts true and formal implication of the axioms are equivalent". (Husserl 1913, §§71-72)[9]

In *Logik und allgemeine Wissenschaftslehre*, Husserl explains that one finds that there may be two valid discipline forms standing in relation to one another in such a way that the axiom system of one may be a formal limitation of that of the other. It then becomes plain that everything that can be deduced in the narrower axiom system is included in what can be deduced in the expanded system. Now, all the theorems deducible in the expanded system must exclusively contain concepts that are either valid in terms of the narrower one, and thus not imaginary, or they must contain

---

[9] Husserl's views on completeness and the imaginary are treated in greater depth in chapters 6 and 7.

concepts that are imaginary. Thus it is that when one compares cardinal arithmetic and ordinal arithmetic, where the minuend may be greater than the subtrahend, and their respective discipline forms, besides theorems including only non-imaginary numbers having real meaning, one finds formulas and theorems that also include negative numbers, which are imaginary in terms of the more restrictive axioms of the arithmetic of cardinal numbers. (Husserl 1917/18, §56)

For example, Husserl explained, in the arithmetic of cardinal numbers, there are no negative numbers, for the meaning of the axioms is so restrictive as to make subtracting 4 from 3 nonsense. Fractions are meaningless there. So are irrational numbers, $\sqrt{-1}$, and so on. Yet in practice, all the calculations of the arithmetic of cardinal numbers can be carried out as if the rules governing the operations were unrestrictedly valid and meaningful. One can disregard the limitations imposed in a narrower domain of deduction and act as if the axiom system were a more extended one. (*Ibid.*)

We cannot arbitrarily expand the concept of cardinal number, Husserl explained in the notes for his Göttingen talks on imaginary numbers. But we can abandon it and define a new, pure formal concept of positive whole number with the formal system of definitions and operations valid for cardinal numbers. And, as set out in our definition, this formal concept of positive numbers can be expanded by new definitions while remaining free of contradiction (Husserl 1891, 435). Fractions do not acquire any genuine meaning through our holding onto the concept of cardinal number and assuming that units are divisible, he theorized, but rather through our abandonment of the concept of cardinal number and our reliance on a new concept, that of divisible quantities. That leads to a system that partially coincides with that of cardinal numbers, but part of which is larger, −meaning that it includes additional basic elements and axioms. And so in this way, with each new quantity, one also changes arithmetics. The different arithmetics do not have parts in common. They have totally different domains, but have an analogous structure. They have forms of operation that are in part alike, but different concepts of operation (*Ibid.*, 436).

So, Husserl found, it was formal constraints requiring that one not resort to any meaningless expression, to any meaningless imaginary concept that were restricting us in our theoretical, deductive work. But what is marvelous, Husserl believed, is that resorting to the infinity of pure forms and transformations of forms frees us from such conditions and at the same time explains to us why having used imaginaries, what is senseless, must lead, not to senseless, but to true results. (Husserl 1917/18, §57) [10]

---

[10] As different as Husserl's *Mannigfaltigkeitslehre* ultimately was from Cantor's, in it we still hear echoes of Cantor famous declaration that mathematics is "entirely free in her development and only bound to the obvious consideration that her concepts both be not

Problem Two: Concerning Frege's Use of Extensions

Upon learning of the paradox derivable in *Basic Laws*, Frege immediately spotted the source of the problem in Basic Law V, his law about extensions. He recognized its "lack of self-evidence" and confessed that he should have "gladly dispensed" with it if he had known of any substitute, but did not know how arithmetic could be scientifically established, how numbers might be apprehended as logical objects unless one could pass, at least conditionally, from a concept to its extension. He said that he found it consoling that anyone else who had used extensions of concepts, classes, sets in proofs was in the same position that he was. (Frege 1980b, 214)

In *Basic Laws*, Frege had accorded extensions "great fundamental importance", but pinpointed them as the place where any decision about defects or errors in his logic would ultimately be made (Frege 1893, vii, ix-x). He once alluded to having been "led by a certain necessity" to introduce them (Frege 1894, 197) and another time to having been "constrained to overcome" his "resistance" to them (Frege 1980a, 191). He ultimately concluded that his law about extensions had indeed undermined his work, that it was to blame for the collapse of his system, that the expression 'extension of concept' easily got one into a "morass", led "into a thicket of contradictions". (Frege 1980a, 55, 130-32; Frege 1979, 269-70)

As discussed in Chapter 9, Frege had gingerly introduced extensions into *The Foundations of Arithmetic* in order to circumvent two very specific problems: 1) the nonsensical conclusions to which his theories could lead; and 2) the sterility or unproductiveness to which they could lead (Frege 1884, §§66-69 and note; §107). In *Rethinking Identity and Metaphysics* (Hill 1997, 61-68), I duplicated Frege's reasoning, but used a more graphic example:

Suppose that it had finally been determined to be true that the man who fired the shots from behind the grassy knoll was identical with the man who killed John Kennedy. According to Frege's reasoning, his definition of identity would then afford us a means of identifying the man who fired the shots from behind the grassy knoll again in those cases in which he is referred to as the man who killed John Kennedy. Frege recognized, however, that this means would not provide for all the cases. For example, it could not decide for us whether Lee Harvey Oswald was the man who fired the shots from behind the grassy knoll. Although it might be perfectly nonsensical

---

self-contradictory and stand in ordered relationships fixed through definitions to previously formed, already existing and proven concepts". "In particular", Cantor continued, "with the introduction of new numbers she is only duty bound to give definitions of them that bestow upon them such determinacy and, if need be, such a relationship to the older numbers, that in given cases they may be distinctly differentiated from one another. As soon as a number meets these conditions, it can and must be considered as existent and real in mathematics". This is why, Cantor held, "one is to regard the rational, irrational, and complex numbers as altogether just as existent as the finite positive whole numbers". (Cantor 1883, 181-82)

to confuse Oswald with the man who fired the shots from behind grassy knoll, this would not be, according to Frege's reasoning, owing to his definition. For it says nothing as to whether the statement: 'the man who fired the shots from behind the grassy knoll is identical with Lee Harvey Oswald' should be affirmed or denied, except for the one case in which Oswald is given in the form of 'the man who killed John Kennedy'. Only if we could lay it down that if Oswald was not the man who killed John Kennedy (still parroting Frege's reasoning), could our statement be denied, while if he was that man, our original definition would decide whether it is to be affirmed or denied. But then, Frege saw, we have obviously come around in a circle. For in order to make use of this definition we should have to know already in every case whether the statement 'Lee Harvey Oswald is identical with the man who killed John Kennedy' was to be affirmed or denied.

Conscious that identities play such an important role in so many fields because we can recognize something as the same again even though it is given in a different way, Frege saw that, as it stood, his definition of number was "unproductive" because it afforded no means of recognizing that object as the same again when given differently; it presupposed that an object could only be given in one single way. All identities, he realized, would then amount simply to the principle "so obvious and sterile as not to be worth stating" "that whatever is given in the same way is to be reckoned as the same and one could not draw from it any conclusion which was not the same as one of the premises". (Frege 1884, §§66-67; §107)

Returning to the Kennedy example, suppose that we finally have incontrovertible evidence that a certain Mr. Knoll shot at Kennedy from behind the grassy knoll and those shots killed him. 'Mr. Knoll is identical with Kennedy's assassin' is then a true statement and we should then be able substitute Mr. Knoll's name every time we find a reference to Kennedy's assassin. However, our identity statement is so blind as concerns other alternatives and, in this case Kennedy's assassin has so often been identified with Oswald, that substitution in most contexts would yield sheer nonsense.

In his irresponsible review of the *Philosophy of Arithmetic*, Frege chastised Husserl for not having used the term 'extension of a concept'. Any view, Frege complained, according to which a statement of number is not a statement about a concept or about the extension of a concept is naive for "when one first reflects on number, one is led by a certain necessity to such a conception" (Frege 1894, 197, 201-02). Nonetheless, Husserl never seems to have been tempted by extensions. In *Formal and Transcendental Logic*, we still find him calling extensional logic naive, risky, doubtful and its fundamental unclarity the source of many a contradiction requiring every kind of artfulness to make it safe for use in reasoning. (Husserl 1929, §§23b, 26c)

In *Philosophy of Arithmetic*, Husserl wrote that he did not see how Frege's method might signify an enrichment of logic. What Frege's method actually enabled one to define, Husserl found, was the extension of the concepts. All Frege's "definitions become correct propositions if one substitutes extensions for the defining concepts, but of course become completely obvious and worthless propositions", Husserl deemed (Husserl 1891, 122). For him, Frege's theory showed that:

> the "direction of the straight line a is the extension of the concept 'parallel to straight line a'". By extension of a concept, though, one means the aggregate of the objects falling under it. The direction of the straight lines a would thus be the aggregate of straight lines parallel to a. Likewise, it showed that "the shape of the triangle d is the extension of the concept 'similar to triangle' d", meaning the aggregate of all the triangles similar to d. And "the number belonging to concept F" is, therefore, it too defined in this manner as the concept "numerically equal to concept F". In other words: the concept of this number is the whole of the concepts numerically equal to F, therefore a whole of infinitely many "equivalent" sets. (Husserl 1891, 122)

According to Frege's definition, Husserl observes, 'number of Jupiter's moons' would accordingly mean "having the same number as the concept Jupiter's moons", or more clearly expressed "having the same number as the aggregate of Jupiter's moons". Thus one obtains concepts having the same extensions, but not the same intension. The latter concept is identical with the concept "any set whatsoever from the equivalence class determined by the aggregate of Jupiter's moons". All these sets also fall under the number four. That different concepts are there requires no proof, Husserl considers. (*Ibid.*, 122 n. 1)

Foreshadowing ideas about replacing a calculus of classes with a calculus of concepts that he ultimately espoused, Husserl added that Frege himself seems to have sensed the doubtfulness of his definition since in a note he says of it that he believes that instead of "extension of concept one could simply say 'concept'" (*Ibid.*).[11] By 1890, Husserl may have actually been permanently inoculated against recourse to extensions. Several articles published during the 1890s find Husserl laying bare "the follies of extensional logic" (Husserl 1994, 199), which he would replace by a calculus of concepts. (*Ibid.*, 92-114, 115-20, 121-30, 135-38, 443-51)

Husserl wrote quite a bit about how extensions, objects, totalities, aggregates, sets, and manifolds stand in relationship one to the other in his unpublished notes on set theory. There his objections to extensions and his espousal of manifolds turn on his convictions that objects stand

---

[11] In the section of the *Philosophy of Arithmetic* called "Frege's Attempt" (Husserl 1891, 118-23), on the page just following the ones Husserl retracted in *Logical Investigations*. Husserl only ever retracted three pages of his criticism of Frege, not eight as a misprint in the English edition indicates. Close examination of Husserl's criticism of Frege's use of the principle of substitutivity of identicals is also a criticism of Frege's recourse to classes, and one that is especially rich in implications for the way that Frege's insights became integrated into analytic philosophy. Unfortunately, the implications of those arguments of Husserl are too rich and far reaching to be integrated into this chapter and have been examined elsewhere. See Hill and Rosado Haddock 2000, Chapters 1 and 5, and Hill 1997, 1991.

in relations with respect to certain properties and that among the former are those belonging to the essence (Ms A 1 35, 10a).[12] Husserl's manifolds are not aggregates of elements without relations. It is precisely the relations that are essential and serve to distinguish a manifold it from a mere aggregate. As explained above, Husserl saw manifolds as aggregates of elements that are not just combined into a whole, but are continuously interdependent and ordered so that each member possesses an unambiguous position in relation to any other one (Husserl 1983, 93, 95-96, 410). The properties and relations set out in the axioms of a complete manifold in Husserl's sense determine objects unequivocally, bring information and logically eliminate nonsensical conclusions, something that Frege hoped to do within his theory by resorting to extensions.

Problem Three: The Contradictions of Set Theory

Of course, the really big thorn in Frege's side turned out to be the contradiction derivable in the system of *Basic Laws*. Husserl characterized what he referred to as "Russell's paradox" as follows:

> All sets fall into such that contain themselves as an element, and such that do not do this. What is the set of sets that does not contain itself as an element to be viewed as being? If it does not contain itself as an element, then it is not the set of *all* sets that do not contain themselves. If it, however, does contain itself, then it is among the sets of all sets that do not contain themselves as element, one that contains itself as an element. (Ms A 1 35, 26a)

Of what he referred to as "Zermelo's paradox", Husserl wrote:

> Zermelo argues: A set M that contains each of its partial sets as elements is an inconsistent set. 1) We consider those partial sets that do not contain themselves as elements. 2) In their entirety these form a set Mo that is contained in M. 3) Mo is thus an element of M. 4) Mo is not an element of Mo. Proof: were Mo an element of Mo, then it would contain a partial set of M (namely Mo) that contains itself as element. However, Mo is to contain *ex definitione* partial sets of M that do not contain themselves as elements. 5) Thus Mo, since it is not an element of Mo, is a partial set of M, which does not contain itself as element. But all such sets *are ex definitione* contained in the concept of Mo, thus in opposition to 4. But Mo is an element of Mo. We come to a direct contradiction. If it essentially belongs to the concept of set that (without contradiction) no set can contain itself as an element, then Mo and M are identically the same set, and we show that the whole reasoning was untenable. (Ms A 1 35, 24a)

---

[12] See for example Ms A 1 35, 35a, 40b, 43b, 48, 53a, 62a-b, 63a. Much of what Husserl says is reminiscent of Cantor's comment in his review of *Foundations of Arithmetic* (Cantor 1932, 440-41) that it was unfortunate that Frege had taken extensions of concepts as the foundation of the number concept and that it was a reversal of the proper order to base it on the concept of the extension of a concept because the extension of a concept was generally something quantitatively completely undetermined, and that for such quantitative determination, the concept of number would have to have been given from somewhere else.

The telling phrase here is: "If it essentially belongs to the concept of set that (without contradiction)...". Given Husserl's conviction that logical, mathematical laws are laws of essence, it is not surprising to find him arguing over and over that the set-theoretical paradoxes must involve some violation of the essence of set.

His notes on set theory record his reflections about just what the essence, the concept, of set entails. It is part of the idea (*Idee*) of the set to be a unit, a whole comprising certain members as parts, but doing so in such a way that, vis-à-vis its members, it is something new which is first formed by them. All mathematico-logical operations performable with sets, he wrote, turn on the idea that sets can be looked upon as kinds of wholes, as new units, formations that are something new vis-à-vis their original members, so that out of these formations new units can then again be formed. The unity of a system is something new vis-à-vis the elements systematized. It would be a contradiction in terms for the system's unity itself to be able to be one among the elements of the same system, upon those the system bases itself. But system units can themselves again be systematized and then ground higher forms of system. They then, however, bring elements into a new system whole (Ms A 1 35, 20b). Wherever mathematicians speak of sets, Husserl maintained, if the concept is to be a mathematical one, they must have a set essence in view, and whatever sets may have as an essence, it is expressed with a relation that belongs to the essence, i.e., the relation between sets themselves and elements of a set. (*Ibid.*, 12b)

If the formal logical construction "set of all sets which do not contain themselves as parts" is considered, it may not be presupposed that they already exist, i.e. that the extension of the conception produces a totality without further ado (*Ibid.*, 43a). From the fact that one can speak of all sets, it does not follow that the totality of sets can in return be looked upon as a set (*Ibid.*, 20b). Mere talk like "a set which contains each of its partial sets as an element" guarantees nothing. One can initially, upon occasion, avoid talking about sets in general. But then one is bound to logic, which requires that sets, genuine manifolds must come under the formal rules for wholes and that consistency be there. (*Ibid.*, 21)

"It belongs essentially to the concept of set that (without contradiction) no set can contain itself as an element", Husserl reiterated constantly (*Ibid.*, 24a). An essence relation, i.e., the relation between sets themselves and elements of a set, makes it impossible for the members of the relation to be identical. Hence a set that contains itself as a element would be a contradiction in terms (*Widersinnigkeit*) (*Ibid.*, 12b). A whole cannot be its own part. Just as it is contradictory for a whole to be its own part at the same time, so it is contradictory for a set for it to be its own member (*Ibid.*, 20b). To the objection that there is no set that contains itself as an element, he maintained that one need merely respond that that is a contradiction in terms (*Widersinnigkeit*). (*Ibid.*, 17a)

Husserl thus repeatedly relegated the set-theoretical paradoxes to the category of contradictions in terms (*Widersinnigkeiten*), as defined in the Fourth Logical Investigation where it is the job of logical laws to guard against formal or analytical contradictions in terms, formal absurdity, by dictating what it is that objects require to be consistent in purely formal terms (Husserl 1900-01, LI IV Introduction; §12). If one is clear and distinct with respect to meaning, Husserl suggested in his notes, one readily sees the contradiction in terms (*Widersinnigkeit*) involved in the set-theoretical paradoxes. So, the solution to the paradoxes would then lie in demonstrating the shift of meaning that makes it that one is not immediately aware of the contradiction in terms and that once one perceives it, one cannot indicate wherein it lies (Ms A 1 35, 12a). This is studied more extensively in Chapter 12.

All sets, one finds Husserl reasoning in his notes, fall into one of two classes. Either they fall into class A of sets that contain themselves as an element, or they fall into class B of sets that do not. That being so, he asked to which of these two classes the set PM of all sets that do not contain themselves as an element belongs? By the law of the excluded middle, it would have to be one of the two. However, he observed, one can show that neither or both of the possibilities must be valid. For were PM to belong to class A, that would mean that it contained itself and that would be contradictory by definition. Were it fall into class B, PM would then not be the set of all sets of all Bs. In actual fact, it proceeds from the paradox, if no conceptual shift is demonstrable, that a set of kind A or a set of kind B must be a contradiction in terms. Then the classification is a contradiction in terms as well. He then looks at the alternative that both may not be the case and asks whether PM is not a contradiction in terms. (*Ibid.*, 17a)

Nonetheless, although Husserl's notes on set theory record many a reflection on what the concept of set entails, he also expresses his conviction there that we do not yet by any means have the real and genuine concept of set that the mathesis and logic need. The paradoxes, he declared, only demonstrate that a general logic of sets in general, of totalities is still lacking. He stressed that in his logic courses he had constantly from the beginning said that totality and set should not be identified and that this identification must be partly responsible for the paradoxes of set theory. (*Ibid.*, 43a, 69a)[13]

Now Husserl's theory of *Mannigfaltigkeiten* was precisely a theory about how to engage in pure a priori analyses of essence. His notes on set theory show him suggesting reforming the mathematical theory of manifolds by consciously transforming it into a transcendental theory of manifolds that consciously captures the formal essence of a genuine,

---

[13] Husserl devotes many a page of his notes on set theory to distinguishing between, multiplicities, aggregates, totalities, sets and manifolds. See, for example, his Ms A 1 35, 13a; 20a; 36a; 43a; 44a; 47a; 67b.

constructible totality, consciously analyzes what belongs to the essence of a concept defining a totality, what belongs to the essence of an axiom and axiom system that as such establishes the univocity and construction and only establishes the meaning of what is constructed, which then formally weighs the conditions of the possibility of such a totality and by this means derives the system of possible totality forms, or manifold forms, so that it inquires into the possible forms of construction principles and into the formal system in which they themselves can be constructed as forms (*Ibid.*, 38). Each field of an exact science, he maintained, is a constructible totality. For that field, "axioms" are valid in which construction principles for the totality must lie and they make a systematic thinking and relationships of genuine arguments and genuine inferences, apodictic conclusions possible (*Ibid.*, 45a). The validity and non-validity of all derivable concepts are to be decided in accordance with the axioms and they are to delimit the talk of a manifold of "existing" formations. (*Ibid.*, 13a)

Of course, the kinds of contradictions in terms flagged above would no longer have any place in such a theory, for which consistency and completeness were paramount. As someone who frequented Hilbert's school in Göttingen, and who upon several occasions pointed to the kinship existing between his own manifolds and Hilbert's axiomatic systems (Husserl 1913, §72; Husserl 1929, §31), Husserl seems to have concluded that once one indeed had the real and genuine idea of set needed, one could proceed with the axiomatization of set theory in a way compatible with his own theories about sets and manifolds. Needless to say, he did not believe that Frege's system of axioms, with its law about extensions, could embody the real and genuine idea of set needed.

Conclusion

Much more can be said about the significance of Husserl's theories about sets and manifolds than can be said here. Here I have set out the main steps in the development of those ideas and I have tried to put some flesh on Husserl's little explored theory of manifolds by showing how it developed in connection with some specific problems that Husserl perceived in set theories espoused by his contemporaries, –including his own. My discussion focused on three Fregean problems that Husserl explicitly addressed. I defined certain issues involved and provided information about Husserl"s theory of manifolds that provides clues as to how it might confront them. May what I have written here help bring Husserl's ideas about sets and manifolds out of the realm of abstract theorizing and prompt further exploration of this philosophical territory, which is as uncharted as it is rich in philosophical implications needing to be drawn and to be made known.

References

Cantor, Georg 1883. *Grundlagen einer allgemeinen Mannigfaltigkeitslehre. Ein mathematisch-philosophischer Versuch in der Lehre des Unendlichen*, Leipzig: Teubner, Cited as appears in *Gesammelte Abhandlungen*, Ernst Zermelo (ed.), Berlin: Springer, 1932, 165-246.
Cantor, Georg 1932. *Gesammelte Abhandlungen*, Ernst Zermelo (ed.), Berlin: Springer.
Cantor, Georg 1991. *Briefe*, H. Meschkowski and W. Nilson (eds.), Berlin: Springer.
Frege, Gottlob 1884. *The Foundations of Arithmetic*, Oxford: Blackwell, 1986.
Frege, Gottlob 1885. "On Formal Theories of Arithmetic", in his *Collected Papers on Mathematics, Logic, and Philosophy,* Brian McGuinness (ed.) Oxford: Blackwell, 1984, 112-21.
Frege, Gottlob 1893. *Basic Laws of Arithmetic*, Berkeley: University of California Press.
Frege, Gottlob 1894. "Review of E. G. Husserl's Philosophy of Arithmetic", in his *Collected Papers on Mathematics, Logic, and Philosophy*, 195-209, Brian McGuinness (ed.), Oxford: Blackwell, 1984.
Frege, Gottlob 1979. *Posthumous Writings*, H. Hermes et. al. (eds.), Oxford: Blackwell.
Frege, Gottlob 1980a. *Philosophical and Mathematical Correspondence,* Oxford: Blackwell.
Frege, Gottlob 1980b. *Translations from the Philosophical Writings*, 3rd ed., Oxford: Blackwell, 1952.
Gerlach, H. and Sepp, H. 1994 (eds.). *Husserl in Halle,* Bern: P. Lang.
Grattan-Guinness, Ivor 1978. "How Russell Discovered His Paradox", *Historia Mathematica* 5, 127-37.
Grattan-Guinness, Ivor 1980. "Georg Cantor's Influence on Bertrand Russell", *History and Philosophy of Logic* 1, 61-93.
Grattan-Guinness, Ivor 2000. *The Search for Mathematical Roots, Logics, Set Theories and the Foundations of Mathematics from Cantor through Gödel*, Princeton: Princeton University Press.
Heck, Richard 1995. "The Development of Arithmetic in Grundgesetze", in *Frege's Philosophy of Mathematics,* William Demopoulos (ed.), Cambridge MA: Harvard University Press, 257-94.
Hill, Claire Ortiz 1991. *Word and Object in Husserl, Frege and Russell, the Roots of Twentieth Century Philosophy,* Athens: Ohio University Press.
Hill, Claire Ortiz 1997. *Rethinking Identity and Metaphysics, On the Foundations of Analytic Philosophy,* New Haven: Yale University Press.
Hill, Claire Ortiz and G. E. Rosado Haddock 2000. *Husserl or Frege? Meaning, Objectivity and Mathematics*, La Salle IL: Open Court.
Husserl, Edmund 1887. "On the Concept of Number", *Husserl: Shorter Works*, P. McCormick and F. Elliston (eds.), Notre Dame: University

of Notre Dame Press, 1981, 92-119. Also published in Husserliana XII and in the English translation of it by Dallas Willard.

Husserl, Edmund 1891. *Philosophie der Arithmetik, mit Ergänzenden Texten (1890-1901)*, Husserliana XII, The Hague: Martinus Nijhoff, 1970. In English, *Philosophy of Arithmetic, Psychological and Logical Investigations with Supplementary Texts from 1887-1901*, translated by Dallas Willard, Dordrecht: Kluwer, 2003. The English edition contains page references to the Husserliana volume.

Husserl, Edmund 1900-01. *Logical Investigations*, New York: Humanities Press, 1970.

Husserl, Edmund 1906/07, *Einleitung in die Logik und Erkenntnistheorie, Vorlesungen 1906/07*, Husserliana XXIV, Ullrich Melle (ed.), Dordrecht: Martinus Nijhoff, 1984, translation by Claire Ortiz Hill, *Introduction to Logic and Theory of Knowledge*, Dordrecht: Springer, 2008. The English edition contains page references to the Husserliana volume.

Husserl, Edmund 1913. *Ideas, General Introduction to Pure Phenomenology*, New York: Colliers, 1962.

Husserl, Edmund 1917/18. *Logik und allgemeine Wissenschaftstheorie*, Husserliana XXX, Dordrecht: Kluwer, 1996.

Husserl, Edmund 1929. *Formal and Transcendental Logic*, The Hague: Martinus Nijhoff, 1969.

Husserl, Edmund 1913. *Introduction to the Logical Investigations, A Draft of a Preface to the Logical Investigations (1913)*, The Hague: Martinus Nijhoff, 1975.

Husserl, Edmund 1983. *Studien zur Arithmetik und Geometrie, Texte aus dem Nachlass (1886-1901)*, Husserliana XXI, The Hague: Martinus Nijhoff.

Husserl, Edmund 1994. *Early Writings in the Philosophy of Logic and Mathematics*, Dordrecht: Kluwer.

Husserl, Edmund. Ms A 1 35. Manuscript on Set Theory, available in the Husserl Archives in Leuven, Cologne and Paris.

Peckhaus, Volker, and R. Kahle 2000/2001. "Hilbert's Paradox", *Report No. 38, 2000/2001*, Institut Mittag-Leffler, The Royal Swedish Academy of Sciences. Published on the Internet.

Rang, B. and W. Thomas 1981. "Zermelo's Discovery of Russell's Paradox", *Historia Mathematica* 8, 16-22.

Russell, Betrand 1903. *Principles of Mathematics*, London: Norton.

Russell, Betrand 1913. "The Philosophical Implications of Mathematical Logic", *The Monist* 22, 481-93 and in *Essays in Analysis*, London: Allen & Unwin, 1973, 284-94. Partially translated in Husserl Ms A I 35.

Russell, Betrand 1959. *My Philosophical Development*, London: Unwin.

Schuhmann Elisabeth and Karl 2001. "Husserl: Manuskripte zu seinem Göttinger Doppelvortrag von 1901", *Husserl Studies* 17, 87-123.

11

Claire Ortiz Hill

ON FUNDAMENTAL DIFFERENCES BETWEEN DEPENDENT AND INDEPENDENT MEANINGS

Are there, as Gottlob Frege, Bertrand Russell and Edmund Husserl all concluded, fundamental differences between dependent meanings and independent meanings that lie concealed behind inconspicuous grammatical distinctions and ultimately prove inviolable because they are "founded deep in the nature of things" (Frege 1891, 41) in such a way that contradictions, paradoxes, antinomies, fallacies, nonsense, confusion, absurdity inevitably result when they are not respected?

In what follows, I argue that the actual existence of such differences is one of the major lessons to be learned from the logic and philosophy that grew out of Frege's and Russell's work. I study how issues involving such differences made their way into Fregeo-Russello-Quineo logic (FRQL). In particular, I study how the manner in which Frege placed the need to lay hold of *self-subsistent*, independent, objects at the heart of his logical project caused a string of closely interrelated problems that he and Russell had to face, the signal one being the famous contradiction about the class of all classes that are not members of themselves. I take this to be evidence that the fundamental differences between dependent and independent meanings are indeed, as Frege considered them to be, of the highest importance, inviolable and founded deep in the nature of things.

Frege on Functions, Concepts, Objects and the Deep Nature of Things

Frege studied the differences between dependent and independent meanings in "Function and Concept" (Frege 1891) and "On Concept and Object" (Frege 1892). The two parts into which mathematical expressions can be split are dissimilar, he explained, for the argument is a whole complete in itself, but the function is not. By itself, a function must be called incomplete, in need of supplementation, or 'unfilled'.[1] An argument must go together with the function to make up a complete whole (Frege 1891, 24-25, 31-32). He further observed that functions do, and at times must, take other functions as their arguments. But, he insisted, in that case, they are fundamentally different from functions whose arguments are objects and cannot be anything else; for just as functions differ fundamentally from objects, so functions whose arguments are and must be functions differ fundamentally from functions whose arguments are objects and cannot be anything else. He called the former second-level functions and the latter first-level functions. He stressed that the difference between first-level and second level functions was not made "arbitrarily, but founded deep in the nature of things". (*Ibid.*, 38, 41)

Like mathematical equations or inequalities, Frege also explained, statements in general can be split up into a part complete in itself and another part in need of supplementation, or unfilled. As an example, he proposed the sentence 'Caesar conquered Gaul', which he divided into 'Caesar' and 'conquered Gaul', where, 'Caesar' is the argument and the second, 'unfilled' part 'conquered Gaul' is the function. It contains an empty place and only when that place is filled with a proper name, or an expression replacing a proper name, can a complete sense appear. While an equation is a reversible relation, an object's falling under a concept is an irreversible relation. (*Ibid.*, 31-32)

Frege saw a close connection between functions and concepts, which he principally defined in terms of certain very specific logical properties. Like functions, he maintained, concepts need completion supplementation. Concept-words contain a gap that is intended to receive a proper name. They are 'unfilled'. They have to have something to fall under them and therefore cannot exist on their own. The sentence 'the morning star is a planet' is composed of a proper name, 'the morning star', and a concept-word 'planet'. Concepts are predicative, the meanings of grammatical predicates. When we say "This leaf is green", we are saying that something falls under a concept meant by the grammatical predicate. (Frege 1891, 30, 32; Frege 1892, 43-44, 48, 50)

For Frege, the predicable nature of concepts is conferred upon them by the particular kind of incompleteness and dependency they exhibit with respect to objects. He considered this to be the principal difference between them and objects, which he defined as "anything that is not a function, so that an expression for it does not bring any empty place" (Frege 1891, 32). He held that an object was something that could be the meaning of the subject, but could never be the total meaning of the

---

[1] In Frege's special vocabulary, functions, concepts, predicates were *ungesättigt*, which has usually been translated as 'unsaturated'. As I explained in chapter 11, I prefer to translate *ungesättigt* as 'unfilled', which I find clearer in meaning and more natural.

predicate. Hence, the names of objects, proper names, could not be used as grammatical predicates. For example, in the sentence 'the morning star is Venus', the word 'Venus' could never be a proper predicate. Its meaning could never occur as a concept, but only as an object, something which he believed involved admitting a very important distinction between what can only occur as an object and everything else. (Frege 1892, 43-44, 48, 50)

In "On Concept and Object", Frege warned that anyone who thought that there was no need to take into account anything as unmanageable as a concept by regarding an object's falling under a concept as a relation in which the same thing could at times occur as an object and at others as a concept had only shifted the problem, but not avoided it. Not all parts of a thought, he reminded, can be complete. At least one of them must in some way be 'unfilled' or predicative, or else they could not stick together. For example, in the sentence 'the number 2 falls under the concept *prime number*', the sense of the phrase 'the number 2' does not hold together with the sense of the expression 'the concept *prime number*' unless a link is added, for example, in the form of the words 'fall under' whose sense is unfilled making them capable of binding. It is only then that we have a complete thought (*Ibid.*, 54). It was not false to say concerning an object what he was saying concerning a concept, it was impossible, senseless to do so. Frege believed that with his distinction between concept and object he had gotten hold of a distinction of the highest importance (*Ibid.*, 50). In the introduction to *The Foundations of Arithmetic*, he had written that in his investigations he had kept to the fundamental principle of never losing sight of the distinction between concept and object and that it was "a mere illusion to suppose that a concept can be made an object without altering it". (Frege 1884, X)

Frege was, however, aware that with concepts we come up against an obstacle analogous to the one encountered with functions and he believed that thorough examination would show that the obstacle was "essential and founded in the nature of our language". While explaining what he was presenting as inviolable differences, Frege was in fact conscious that "Language has means of presenting now one, now another, part of the thought as the subject". Though he firmly believed that concepts were fundamentally different from objects and that by virtue of its predicative nature a concept could not play the role of a grammatical subject, he found that in "logical discussions one quite often needs to say something about a concept and to express this in the form usual for such predication". When "we say that the concept *horse* is not a concept, whereas, e.g., the city of Berlin is a city, and the volcano Vesuvius is a volcano", he famously wrote in "On Concept and Object", "we are confronted by an awkwardness of language, which... cannot be avoided". "Language is here in a predicament that justifies the departure from custom", he realized. Owing to its predicative nature, the concept as such cannot play the part of the subject, so "it must be converted into an object, or, more precisely, an object must go proxy for it". He concluded that "the behaviour of the concept is essentially predicative, even where something is being said about it; consequently it can be replaced there only by another concept, never by an object. Thus, what is being said concerning a concept does not suit an object". (Frege 1892, 46, 49, 54-55)

Another Hierarchical Structure

Logical form is as maddeningly impalpable as it is consequential and real. Talk of it is as frustratingly abstract as it is important and necessary. This makes it immensely hard to find analogies apt enough to illustrate the issues. In this section, I embark on a quest to make the specific points about the inviolability of logical form that Frege made in "Function and Concept" and "On Concept and Object" more perceptible by comparing the hierarchical structure he wrote about there with another one, in this case, the hierarchical structure appealed to in the usual biological classification of living beings into Kingdoms, Phyla, Classes, Orders, Families, Genera, and Species. In later sections, I complement this effort to make the internal logical structure and logical disposition of parts of FRQL more tangible by studying some of the consequences that that violation of logical form has actually had.

According to my dictionary, a species consists of individual organisms that resemble one another and have the very concrete property of being able to breed among themselves, but not with members of another species. *Homo sapiens* is defined as: "Modern man, the single surviving species of the genus *Homo* and of the primate family *Hominidae*, to which it belongs". An individual human being x, then, belongs to the species *Homo sapiens*, which belongs to the genus *Homo*, which belongs to the family of *Hominids*, which belongs to the order of *Primates*, which belongs to the class of *Mammals*, which belongs to the phylum of *Chordates*, which belongs to the *Animal Kingdom*.

It is clear that in this hierarchy, the species *Homo sapiens*, the genus *Homo*, the family *Hominids*, the class *Mammals*, the phylum *Chordates*, and the *Animal Kingdom* itself are all abstract entities (and ones very useful for science). The only individuals figuring in this entire hierarchy are the human beings belonging to the lowest type. In logical parlance, the words for them refer to objects. The linguistic expression for the species *Homo sapiens* refers to a first-level concept and the linguistic expression for the genus *Homo* refers to a second level concept, and so on. If the members of a species die out, the name for the species becomes an empty shell. So it is true that,

$$(\forall x)\ x \text{ is a human being} .\supset. x \in \textit{homo sapiens}.$$

But,

Because,
$(\forall y)$ y is *homo sapiens* $.\supset.$ $y \notin$ *homo sapiens*,

y is *homo sapiens* $.\supset.$ $y \in$ *Homo*.

It is clear that the abstract entity species *homo sapiens* cannot be a member of itself and that it is something radically different from a human being. It is plainly not an organism—say a featherless biped with a heart and kidneys—that is capable of breeding with human beings. So, the following chain of reasoning is wrong:

$(\forall x)$ $x \in$ F (organisms capable of breeding with human beings), and
$(\forall x)$ $x \in$ G (the abstract entity *homo sapiens*),
Therefore, being capable of breeding with human beings = being the abstract entity *homo sapiens*.

Back to Frege

Getting back to Frege, how did independency and dependency become an issue in FRQL? The first thing to say in response to that question is that they were an issue there from the very start. Frege developed his logical symbolism in such a way that the reference of words became "at every point... the essential thing for science". He insisted that "in science and wherever we are concerned about truth... we also attach a reference to proper names and concept-words; and if.... we fail to do this, then we are making a mistake that can easily vitiate our thinking". (Frege 1979, 118, 122, 123)

In *Foundations*, Frege maintained that to obtain the concept of number, it was a matter of fixing the sense of an identity (Frege 1884, x, §§62-70, 106) and that only in the case of objects could there be any question of identity (Frege 1979, 182, 120). Frege had to have objects, because that was the way he designed his logic. So it is not at all surprising to find him eventually emphasizing that the "prime problem of arithmetic may be taken to be the problem: How do we apprehend logical objects, in particular numbers? What justifies us in recognizing numbers as objects?" (Frege 1903, 224)

The importance that Frege accorded to apprehending numbers as logical objects has been recognized and studied. Richard Heck has even gone so far as to say that we shall "not fully understand Frege's philosophy until we understand the enormous significance the question how we apprehend logical objects... had for him" (Heck 1993, 286-87; Demopoulos 1995, 10). However, it is most vital to realize in addition that the logical objects that Frege needed to apprehend were not just any logical objects. They were *independent, self-subsistent* logical objects.[2]

---

[2] J. L. Austin chose to translate '*selbständiger*' as 'self-subsistent'. I use the more common translation 'independent', which better serves the purposes of this chapter.

Part IV (§§55-86) of Frege's five part *Foundations* is devoted to analyzing the concepts of arithmetic. The first portion (§§55-61) of it is entitled "Every individual number is an independent (*selbständiger*) object" and is entirely devoted to affirming the independency of numbers. It was clear to him, Frege explains there, "that the number studied by arithmetic must be conceived not as a dependent attribute, but substantivally", a distinction that he saw as corresponding to that between 'blue' and 'the color of the sky' (§106 and note). "Precisely because it forms only an element in what is asserted", Frege reasons, "the individual number shows itself for what it is, an independent object." He considered that the presence of the definite article 'the' in expressions like 'the number 1' served "to class it as an object" and that in arithmetic "this independence comes out at every turn, as for example in the identity $1 + 1 = 2$" (§57, §106). He argued that since it was a matter of arriving at a concept of number usable for scientific purposes, "we should not, therefore, be deterred by the fact that in the language of everyday life number appears also in attributive constructions" for that "can always be got around". Ever ready to "break the domination of the word over the human spirit" (Frege 1879, 7), Frege proposed that a statement like:

> the proposition "Jupiter has four moons" can be converted into "the number of Jupiter's moons is four". Here the word "is" is not to be taken as the copula, as in the proposition "the sky is blue". This is shown by the fact that we can say: "the number of Jupiter's moons is the number four, or 4". Here "is" has the sense of "is identical with" or "is the same as". So that what we have is an identity, stating that the expression "the number of Jupiter's moons" signifies the same object as the word "four". (Frege 1884, §57)

This affirmation of the independency of numbers is followed by a lengthy response (§§58-60) to the possible objection that we are incapable of forming of the object we are calling Four, or of the Number of Jupiter's moons, any sort of presentation in the sense of a mental picture that would make it something independent. That is not the fault of the independence that he has attributed to the number, he reassures readers, for we can have no mental picture of number as either an independent object or as a property in an external thing, because it is not in fact either anything sensible or a property of an external thing. He considers that the fact that we are unable to form any such mental picture of its content is

> therefore no reason for denying all meaning to a word, or for excluding it from our vocabulary. We are indeed only imposed on by the opposite view because we will, when asking for the meaning of a word, consider it in isolation, which leads us to accept a presentation as the meaning.... But we ought always to keep before our eyes a complete proposition. Only in a proposition have the words really a meaning.... It is enough if the proposition taken as a whole has a sense; it is this that confers on its parts also their content.... (Frege 1884, §60)

Frege concludes with the important proviso that the independence that he is "claiming for number is not to be taken to mean that a number word signifies something when removed from the context of a proposition, but only to preclude the use of such words as predicates or attributes, which appreciably alters their meaning" (§60). He then proceeds to show that to "obtain the concept of Number, we must fix the sense of a numerical identity". (§§62-69)

As his definition of identity, Frege gives Leibniz' definition: "Things are the same as each other, of which one can be substituted for the other without loss of truth". Spotting yet another opportunity to break the domination of the word over the human spirit, Frege added that it was "not of any importance" whether we talk of complete agreement in all respects, or only agreement in this respect or that, for "we can adopt a form of expression such that this distinction vanishes. For example, instead of 'the segments are identical in length', we can say 'the length of the segments is identical', or 'the same', and instead of 'the surfaces are identical in colour', 'the colour of the surfaces is identical'". (§65)

Now this particular recipe for expunging the differences between complete agreement in all respects and agreement in this respect or that changes a dependent predicative structure into an object. Frege has changed statements of sameness of concrete properties predicated of objects into statements affirming the identity of objects, in this case surfaces and lengths. However, he was perfectly aware that many of the inferences that could be made by appealing to such a principle would lead to evidently false and absurd conclusions. So, after devoting several pages of *Foundations* to ferreting out some very basic problems that he saw sticking to his theory and unable to silence questions, doubts and suspicions about the sterility and nonsensical consequences that he saw proceeding from it, he acknowledged that he could not by these methods obtain any satisfactory concept of number. To forestall the problems that he foresaw would vitiate his theories, Frege settled for the definition: "The Number which belongs to the concept F is the extension of the concept 'concept equal to the concept F'". (§§66-68, 107)

Although Frege ended the *Foundations* claiming that he was not attaching any decisive importance to bringing in extensions (§107), he did not propose any alternative, and by the time it came to actually proving his theory of number in the *Basic Laws of Arithmetic*, extensions had taken on "great fundamental importance". Their introduction, he now maintained was "an important advance which makes for far greater flexibility". (Frege 1893, ix-x)

So it is that even as he wrote of the inviolability of differences between dependent and independent meanings in "Function and Concept" and "On Concept and Object", Frege was developing the logical system of *Basic Laws of Arithmetic* that violated them. In *Basic Laws I*, he argued that the generality of an identity could always be transformed into an identity of courses of values and conversely, an identity of courses of values may always be transformed into the generality of an identity. To lay hold of the extension of concepts that his theories needed, Frege finally proposed transforming a sentence in which mutual subordination is asserted of concepts into a sentence expressing an identity. Since only objects could figure in identity statements, he knew he had to find a way of correlating objects and concepts in a way which correlated mutually subordinate concepts with the same object. He decided to achieve this by translating language asserting mutual subordination into statements of the form 'the extension of the concept X is the same as the extension of the concept Y in which the descriptions would then be regarded as proper names as indicated by the presence of the definite article (Frege 1979, 181-82). This meant that if it is true that (x) $F(x) = G(x)$, then those two functions have the same extension and functions having the same extension are identical. (Frege 1893, §§9, 21; Frege 1903; Frege, 1884, §§66-67; also §107)

So, if all readers of philosophical papers are human beings and F is the property of being a reader of this paper and G the property of being a human being, then $(\forall x)$ $F(x) \supset G(x)$ is a true statement. However, since there are countless human beings who are not and never will be readers of this paper, $F = G$ is manifestly false, as is $(\forall x)$ $G(x) \supset F(x)$. Nonetheless, Basic Law V willing, if two concepts have the same extension those two concepts are equal: $(\forall x)$ Fx & $Gx \supset F = G$.

But, by permitting such a transformation, Frege realized that he was conceding that such proper names have meanings. He never believed that any proof could be supplied that would sanction such a transformation. So, he devised Basic Law V to mandate the view of identity, equality and substitutivity that his system required. By transforming "a sentence in which mutual subordination is asserted of concepts into a sentence expressing an equality", Basic Law V would permit logicians to pass from a concept to its extension, a transformation which, Frege held, could "only occur by concepts being correlated with objects in such a way that concepts which are mutually subordinate are correlated with the same object". (Frege 1979, 181-82; Frege 1980b, 159-60)

Consequences of Mixing Dependent and Independent Meanings

Frege insisted upon differences between dependent and independent meanings in "Function and Concept" and "On Concept and Object", but in those articles, he was not explicit about the actual consequences of violating such differences. Nonetheless, since he himself ultimately did mix dependent and independent meanings, we are in a position to study some of the actual consequences of his having done so.

So, the next step of our quest to apprehend the logical structure of dependent and independent meanings consists of describing a string of very interrelated consequences that Frege's attempt to transform dependent meanings into independent meanings by creating a law that permitted one to pass from a concept to its extension actually brought in its wake. The list is not exhaustive.

What, then, can actually happen when that hierarchical structure "founded deep in the nature of things" that assigns different roles to dependent and independent meanings is broken and dependent meanings come to be in places for independent meanings?

Contradictions. Of "Russell's" contradiction about the set of all sets that are not members of themselves, Frege wrote in 1912 that his first aim had been to obtain objects out of concepts. So, he overcame his resistance and admitted the passage from concepts to their extensions. However, when extensions of concepts are introduced, he acknowledged, Russell's contradiction arises. (Jourdain 1912, 191, n. 29)

Of that contradiction, Russell had written to Frege that he had found Frege's idea that a function could also constitute the indefinite element dubious because it led to the following contradiction:

> Let w be the predicate of being a predicate which cannot be predicated of itself. Can w be predicated of itself? From either answer follows its contradictory. We must therefore conclude that w is not a predicate. Likewise, there is no class (as a whole) of those classes which as wholes, are not members of themselves. From this I conclude that in certain circumstances a set does not form a whole. (Frege 1980a, 130-31)

Faithful to his convictions about the fundamental differences between predicates and objects, Frege responded that 'A predicate is predicated of itself' did not seem exact to him and that one should rather say 'A concept is predicated of its own extension'. A predicate, he explained, was as a rule a first-level function which required an object as argument and which could not therefore have itself as argument (subject). (*Ibid.*, 132-33)

Frege tested the validity of the chain of inferences leading up to the contradiction and concluded that it indicated that the transformation effected by Basic Law V was not always permissible, that the law was false and his explanations did not suffice to secure a reference for his combinations of signs in all cases (*Ibid.*, 132). "I do not see how arithmetic can be scientifically established; how numbers can be apprehended as logical objects, and brought under review; unless we are permitted—at least conditionally—to pass from a concept to its extension", Frege agonized in his study of the contradiction in his 1903 appendix to *Basic Laws II*. (Frege 1903, 214)

Russell's publicizing of the contradiction sent up a series of red flags signaling problems in the logical system of *Basic Laws of Arithmetic I* and initiated a flurry of activity on his part that culminated in the writing of *Principia Mathematica*. He acknowledged that a very large part of the labor involved in the writing of that book had been expended dealing with and avoiding the contradictions that had infected logic and set theory, that the book's system had been specially framed to solve the paradoxes that had troubled students of symbolic logic and set theory and to avoid the contradictions and detect the precise fallacy that had given rise to them. (Russell 1927, vii; 1)

The contradictions of set theory, Russell came to realize, drew attention to important fallacies that arise from not recognizing that there are different kinds of symbols and different kinds of relations between a symbol and what it symbolizes. He said that his struggle with the contradictions derivable in Frege's logic taught him that if a word or phrase that is devoid of meaning when separated from its context is wrongly assumed to have independent meaning, paradoxes and contradictions are apt to result. (Russell 1956, 206; Russell 1973, 165)

From the contradictions, Russell said he learned that the meanings of words are of different types, that an attribute expressed by an adjective is of a different type from the objects to which it can be attributed. He came to realize that a predicate, by which he meant a word used to designate a quality such as red, white, square, or round, could never occur except as a predicate, and that statements in which a predicate seems to be the subject must be transformed into ones in which what is predicated is predicated and if this is not done, such sentences are liable to produce contradictions.

As an example, Russell suggested the word 'Socrates', which means a certain man, the word 'mortal', which means a certain quality, and the sentence 'Socrates is mortal' which means a certain fact. The three sorts of meaning involved are entirely distinct, Russell pointed out, and "you will get into the most hopeless contradictions if you think the word 'meaning' has the same meaning in each of these three cases". (Russell 1956, 186-87, 205-06).

Breakdown of logical structure. Language, not only everyday language, but also philosophical and logical languages, are composed of what is talked about and what is said about it. A statement about x is generally analyzable into two components, x and something said about x. Only the former names an independent entity with an independent meaning of its own.

Putting the predicative part, what is said about x, on the same footing as what it is talking about robs discourse of its way of talking *about* things (and about itself). When principles of extensionality assume that a function can only enter into a proposition through its values (Russell 1927, xxix and note, 401), or that a class is completely determined by its elements, as in Gödel's famous incompleteness proof, discourse is flattened down to first-order particulars. The basic logical relief, the essential features of which Frege outlined in "Function and Concept" and "On Concept and Object", is lost.

Blurring distinctions between dependent and independent meanings by allowing a concept-word to be transformed into a proper name and to come to figure on the wrong tier in the hierarchy of meaning breaks the logical structure that Frege was so intent upon preserving and opens the way to confusion. Once logical structure is broken and meaning categories are violated, coherency of meaning is threatened and absurdity can be the result. Confusion, is bound to arise if, Frege warned at the end of his life, as a result of its transformation into a proper name, a concept-word comes to be in a place for which it is unsuited (Frege 1980a, 54-55). When two words have two different types of meanings, Russell once warned, the relations of those words to what they stand for are also of different types and the failure to realize this is "a very potent source of error and confusion in philosophy". (Russell 1956, 133)

Since concepts and objects belong to different logical types, it should not be surprising to find that concepts exhibit strange behavior when treated as objects. Dependent meanings strenuously resist being identified with the objects to which they are inextricably linked, or with any fictional objects like classes. When one damages the original logical structure and one breaks the rules that it dictates by trying to subject dependent meanings to the same logical rules that self-subsistent objects obey, then dependent meanings rebel.

It is this logical relief that Frege defended when he explained in his 1902 correspondence with Russell that the proposition '3 is a prime number' can be analyzed into two essentially different parts: '3' and 'is a prime number'. The first part, Frege explained, is complete in itself, but the second stands in need of completion. It makes sense to fit together the complete part with the part of the proposition in need of completion, it makes no sense to fit together the two complete parts. They will not hold together. The "difference between the signs must correspond to a difference in the realm of meanings.... The analysis of the proposition corresponds to an analysis of the thought, and this in turn to something in the realm of meanings and I should like to call this a primitive logical fact". (Frege 1980a, 141-42)

Experience taught Russell not to affirm, as he once had in *The Principles of Mathematics* that predicates are individuals (Russell 1903, §499). It no longer seemed possible to him that a statement about x was generally analyzable into two parts, x and something said about x, both of which named independent entities with independent meanings of their own (Russell 1973, 137). What is predicative, he realized, suggests a structure, and no significant symbol can symbolize it in isolation. When dependent meanings come to be in the places for independent meanings, the fundamental logical structure grounded in the differences between the two is broken. For example, predicative structure—that is the relationship of dependence of what is predicated on something independent —breaks down.

Russell came to stress how important it was "to realize how much of what there is in phraseology is of the nature of incomplete symbols". "There are a great many sorts of incomplete symbols in logic, and they are sources of a great deal of confusion and false philosophy, because people get misled by grammar...". Incomplete symbols, he began arguing, have absolutely no meaning whatsoever in isolation, but only became significant as parts of appropriate propositions. (Russell 1956, 253-54; Russell 1973, 170)

For example, for Russell, symbols for functions did not have a meaning in themselves, but required some supplementation in order to acquire a complete meaning (Russell 1973, 225). For him, a function was not a definite object, but essentially a mere ambiguity awaiting determination. For it to occur significantly, it had to receive the necessary determination. It had to occur in such a way that the ambiguity disappeared and a wholly unambiguous statement resulted (*Ibid.*, 229-30). He called a propositional function standing on its own "a mere schema, a mere shell, an empty receptacle for meaning, not something already significant". (Russell 1919, 157)

So, let us compare, imperfectly of course, dependent meanings to a wineglass. Unless a wineglass contains wine it is just an empty receptacle. The wine and the glass go together to make the glass of wine. The role of the wineglass is such that it only can be replaced by another receptacle, never by wine. By its very nature, the receptacle, the wineglass, as such cannot play the part of wine. It cannot be a glass of wine and it obviously cannot be converted into wine. The wine is just as obviously dependent on a receptacle. Without it, it would spill all over the place.

Things come to be in places unsuited to them. Flattening logical structure smoothes the way for things to come into places not intended for them. When differences between independent and dependent meanings are ignored, one easily loses sight of the essential differences between concept and object lying concealed behind inconspicuous grammatical distinctions in our word languages, and dependent meanings easily slip into places unsuited to them. A concept-word can slip into a place suited to a proper name, a function into a place suited to an argument, or a set can seem to be a member of itself: both $(w \in w) \wedge \sim (w \in w)$.

When abstract logical entities like functions or concepts come to be represented by letters of alphabets or other symbols, it becomes very easy to think that one can govern them by rules of one's own creation. Led by superficial outward appearances, that is, by the linguistic or syntactic form, one is easily misled by symbolic look-alikes to the point of losing sight of the original logical structure and the rules that it dictated—until the abstract entities rebel....

As Russell once stressed, one

can never place a particular in the sort of place where a universal ought to be, and vice versa. If I say '*a* is not *b*', or if I say '*a* is *b*', that implies that *a* and *b* are of the same logical type. When I say of a universal that it exists, I should be meaning it in a different sense from that in which one says that particulars exist. E.g. you might say 'Colours exist in the spectrum between blue and yellow'. That would be a perfectly respectable statement, the colours being taken as universal. You mean simply that the propositional function '*x* is a colour between blue and yellow' is one which is capable of truth. But the *x* which occurs there is not a particular, it is a universal. (Russell 1956, 258)

Say we have two medicine bottles F and G. They could, among numerous other possibilities, be filled with the same medicine, different medicines, placebos, poison, generic medicines having the same chemical compositions, but different names and appearances, or any combination of these possibilities. According to principles of extensionality like Basic Law V, if the contents of the bottles are the same, if every pill in bottle F is the same as every pill in bottle G, it does not matter how the bottles are labeled.

Perhaps, in the case of generic medicines, for example, a bottle of medicine may conceivably be said to be completely determined by the chemical composition of its contents: $(\forall x)\ Fx = Gx \supset F = G$. But, what happens if the bottles are mislabeled or the medicine is misbottled and someone receives the wrong medication? For, it is not at all true that if two medicine bottles are identical or labeled identically, then every pill in one is the same as every pill in the other: $F = G \supset \forall x\ Fx = Gx$. This example is inspired by the death of a three year old boy in a Paris hospital who died after having received the wrong medication.

Modes of inference threatened. Writing of what he called the "downright catastrophic effect" of the contradiction derivable in Frege's system, David Hilbert once explained:

> In their joy over the new and rich results, mathematicians apparently had not examined critically enough whether the modes of inference employed were admissible; for purely through the ways in which notions were formed and modes of inference used—ways that in time had become customary—contradictions appeared.... In particular a contradiction discovered by Zermelo and Russell had, when it became known, a downright catastrophic effect in the world of mathematics .... The reaction was so violent that the commonest and most fruitful notions and the very simplest and most important modes of inference in mathematics were threatened and their use was to be prohibited.... Just think: in mathematics, this paragon of reliability and truth, the very notions and inferences, as everyone learns, teaches and uses them, lead to absurdities. (Hilbert 1925, 375)

In this statement, Hilbert does not specify which important modes of inference were endangered. However, Analytic philosophers have produced an abundant literature on the failures of the principle of substitutivity of identicals and the principle of existential quantification in FRQL, where those two principles of inference work together work hand in hand to license inference between sentences and make up a basic part of the deductive apparatus that yields logical truths.

Dependent meanings in fact have the power to bring about the failure of those two most basic and important modes of inference. For one thing, dependent meanings cause inference failures when they are expected to behave like objects in identity statements. If only in the case of objects can there be a question of identity, then one commits logical sin when one identifies an object with anything that is not an object, or which refers only indirectly to an object, for example, a predicate, a concept, a class, a property, a function.

So, it stands to reason that failures of substitutivity will result if symbols standing for dependent meanings are expected to obey the same formal rules of identity that symbols for complete meanings do. It stands to reason that if one makes the mistake of identifying one type of entity with an entity of an entirely different type (i.e., a concept with an object, a function with an argument, a class with a particular, etc.), substitutivity is going to fail. If, as Frege stressed at the end of his life, concept-words and proper names must occupy essentially different places, and it is obvious that a proper name will not fit into the place intended for a concept-word (Frege 1980a, 54-55), and if, as he wrote to Heinrich Liebmann in 1900, there is a "radical difference between concepts and objects, which is such that a concept can never stand for an object or an object for a concept" (*Ibid.*, 92), then the principle of substitutivity of identicals is bound to fail when one puts one in the place intended for the other.

Substitutivity presumes identity and only in the case of objects can there be a question of identity. No identity without entity. A singular term that fails to name cannot meaningfully figure in an identity statement. So, failures of substitutivity are helpful in spotting confusions caused by identifying things of different logical types. They are often the first sign that something has gone wrong, that you do not have a firm grasp on the objects designated by the signs on either side of your equals sign, or that they are not objects at all.

Moreover, an identity statement is reversible, while the relation of an object's falling under a concept is an irreversible relation. It does not follow from the fact that x is a Y that Y is an x (or for that matter that y is an X). So, when talk of classes and their extensions make it look as if the irreversible relation of an object's falling under a concept can be turned into the reversible relation of identity, the door is opened to illegitimate inferences. The hierarchical structure established by the fundamental differences between predicates and objects ordinarily protects against invalid inference, so that when those differences are not respected, problems with inference are liable to result.

Like the principle of substitutivity of identicals, the principle of existential quantification requires a firm grasp on the objects that one is substituting, identifying and quantifying over. So, it is undermined by the kind of irreferentiality that is present where dependent and independent meanings have been confused. For many analytic philosophers the objects referred to in a theory are not the things the singular terms name. They are the values of the variables of quantification. Thus, it is essential for those philosophers to be able to move from a statement in which something is affirmed of an object to a quantified statement that affirms that there is an object x such that.... This move is licensed by the principle of existential generalization. Since the basic idea underlying this theory of quantification is "that whatever is true of the object named is true of something", existential generalization clearly loses its justification when the singular term fails to name (Quine 1961, 144-46, Quine 1960, §35). So, by obstructing the substitutivity of identicals and foiling attempts at quantification, confusing dependent and independent meanings brings two basic, simple, common, important modes of inference to grief. (Hill 1997, 17-18).

Pseudo-objects are created. When asked late in life about the causes of the paradoxes of set theory, Frege replied that it was language's propensity to undermine the reliability of thinking by forming apparent proper names to which no objects correspond that had allowed concept-words to be transformed into proper names and so come to be in places unsuited to them that had "dealt the death blow" to his set theory. (Frege 1980a, 55)

He described the "essence of the procedure" that leads into "a thicket of contradictions" as follows: The objects that fall under F are regarded as a whole, as an object, and designated by the name 'set of Fs'. This is inadmissible, he explained, because of the essential difference between concept and object, which is covered up in our word languages. Such a transformation of a concept into an object is inadmissible, because the set formed only seems to be an object, while in truth there is no such object at all. Experience, he said, had shown him "how easily this can get one into a morass". He even suggested that one "must set up a warning sign visible from afar: let no one imagine that he can transform a concept into an object". (*Ibid.*, 54-55)

To illustrate what he called the fatal tendency of language to form proper names to which no objects correspond, Frege gave the formation of a proper name after the pattern of 'the extension of the concept a'. In the case of 'the extension of the concept star', 'the concept *star*' is a proper name, while 'concept *star*' is a designation of a concept and could not be more different from a proper name. However, though there is no object for which this phrase could be a linguistically appropriate designation, because of the definite article, it appears to designate an object. Frege warned that the difficulties which this idiosyncrasy of language entangles us in are incalculable and threaten to undermine the reliability of thinking. (Frege 1979, 269-270; Frege 1980a, 54-55)

As for Russell, he said that his struggle with the contradictions derivable in Frege's logic had taught him that if a word or a phrase that is devoid of meaning when separated from its context is wrongly assumed to have an independent meaning, false abstractions, pseudo-objects, and paradoxes and contradictions are apt to result. (Russell 1973, 165)

Russell had originally believed that:

> When we say that a number of objects all have a certain property, we naturally suppose that the property is a definite object, which can be considered apart from any of all of the objects, which have, or may be supposed to have, the property in question. We also naturally suppose that the objects which have the property form a *class*, and that the class is in some sense a new single entity, distinct, in general, from each member of the class. (Russell 1973, 163-64)

However, the contradiction about the classes that are not members of themselves showed Russell that classes must be something radically different from individuals (Russell 1956, 81). He came to believe that if one assumes that the class is an entity, one cannot escape the contradiction about the class of classes that are not members of themselves (Russell 1973, 171). He decided that classes could not be independent entities. The symbol for a class, he argued, is an incomplete symbol. It does not actually stand for part of the propositions in which it occurs. (Russell 1956, 262)

Russell was therefore led to distinguish between classes and particulars, to maintain that a class consisting of two particulars is not itself in turn a fresh particular, "that in the sense in which there are particulars, in that sense it is not true to say there are classes". There "are particulars, but when one comes on to classes, and the classes of classes, and classes of classes of classes, one is talking of logical fictions .... It is not significant to say 'There are such things', in the same sense of the words 'there are' in which you can say 'There are particulars'". Regarding them as such leads inescapably to the contradiction about the class of classes that are not members of themselves (Russell 1956, 260-65; Russell 1973, 163-64). As he explained,

> if you think for a moment that classes are things in the same sense in which things are things, you will then have to say that the class consisting of all the things in the world is itself a thing in the world, and that therefore this class is a member of itself. Certainly you would have thought that it was clear that the class consisting of all the classes in the world is itself a class.... If there is any sense in asking whether a class is a member of itself or not, then certainly in all the cases of the ordinary classes of everyday life you find that a class is not a member of itself. (Russell 1956, 261)

Type ambiguities creep into reasoning unnoticed. Early in his search for ways to evade (his choice of verb) the problem of the contradiction about the class of all classes that are not members of themselves, Russell thought that "the key to the whole mystery" would be found by inventing (his choice of verb) a hierarchy of types. So, he invented a hierarchy of classes according to which the first type of classes would be composed of classes made up entirely of particulars, the second type composed of classes whose members are classes of the first type, the third type composed of classes whose members are classes of the second type, and so on. (Russell 1903, §§104-105; Russell 1956, 264)

By forbidding certain contradiction generating inferences, Russell explained, his hierarchy of types was to perform "the single, though essential, service of justifying us in refraining from entering on trains of reasoning which lead to contradictory conclusions. The justification is that what seem to be propositions are really nonsense" (Russell 1927, 24). It had become clear to him that the contradiction about the classes that are not members of themselves could only be avoided by realizing that no class either is or is not a member of itself, that the entire question as to whether a class is or is not a member of itself is nonsense. (Russell 1956, 261-62)

The hierarchy of types, Russell was satisfied, yields a series of types, such that, in all cases where a paradox might have emerged, a difference of type renders the paradoxical statement meaningless. The types obtained are mutually exclusive, making reflexive fallacies impossible and the notion of a class being a member of itself meaningless. It is never to be possible for a class of one type either to be or not to be identical with a class of another type. One can therefore lay it down that a totality of any sort cannot be a member of itself, that the totality of classes in the world cannot be a class in the same sense in which they are. (Russell 1903, §104; Russell 1956, 75-80, 264; Russell 1973, 201)

Russell's theory of types can be, and has been, seen from many angles, including as an attempt to restore the logical structure created by dependency and independency that was vitiated by Basic Law V. When philosophers are freed to put their symbols to wrong uses, type ambiguities creep into their reasoning. Although Russell said that he *invented* his hierarchy of types, the type differences he was defending with it were ones that Frege had judged inviolable and founded deep in the nature of things even as he authorized their violation. Russell was trying to put back the hierarchical structure broken when principles of extensionality open the door to a confusion of types by blurring distinctions between dependent and independent meanings, when principles of extensionality say that a concept-word can be transformed into a proper name and so occupy places that both he and Frege considered unsuited to it, when principles of extensionality allowed the normally predicative W be asserted of itself in statements of the form 'w is a w'....

Russell believed that the theory of types he had invented led "both to the avoidance of contradictions, and to the detection of the precise fallacy which has given rise to them" and that no solution to the contradictions was technically possible without that theory (Russell 1927, 1, 24), but he finally realized that it could not be "the key to the whole mystery". Deeper problems caused the old contradiction to break out afresh and he realized that "further subtleties would be needed to solve them". (Russell 1919, 135; Russell, 1956, 333)

To cope with these new outbreaks of the "old contradiction", Russell introduced the axiom of reducibility, which he leaned on at every crucial point in his definition of classes in *Principia Mathematica* (Russell 1973, 250; Russell 1927, 58-59, 75-81). He considered that many of the proofs of *Principia* "become fallacious when the axiom of reducibility is not assumed, and in some cases new proofs can only be obtained with considerable labour" (Russell 1927, xliii). Yet in spite of all that might be achieved by means of the axiom, Russell expressed reservations about it reminiscent of those Frege had expressed regarding Basic Law V. Russell deemed the axiom "only convenient, not necessary", and even called it "a dubious assumption" and a "defect" (Russell 1919, 192-93). "This axiom," he admitted, "has a purely pragmatic justification: it leads to the desired results, and to no others. But clearly it is not the sort of axiom with which we can rest content". (Russell 1927, xiv)

In the second edition of *Principia Mathematica*, Russell assumed "that a function can only enter into a proposition through its values", an assumption that he deemed "amounts to saying that mathematics is essentially extensional rather than intensional". Appealing to reasoning like that that Frege had used to justify his use of extensions, Russell believed that his use of the assumption could "be validated by definition, even if the assumption is not universally true". He considered that: "This assumption is fundamental.... It has its difficulties, but for the moment we ignore them. It takes the place, not quite adequately of the axiom of reducibility" (*Ibid.*, xxix and note, 401). Basic Law V flattens natural hierarchical logical structure. It is re-flattened when, as became common in FRQL, one dictates that a function only figures in its values, as Russell did this in this new rendition of Frege's axiom of extensionality.

Conclusion

Here, I have examined the development of FRQL from the angle of the differences between dependent and independent meanings that Frege argued in "Function and Concept" and "On Concept and Object" were inviolable and "founded deep in the nature of things". I have tried to show how a string of problems associated with FRQL turns on issues of dependency and independency.

By studying some of the consequences of the failure to respect such differences, I have sought to shed additional light on the deep, unsolved problems connected with Frege's efforts to apprehend self-sufficient logical objects. I have endeavored to show that when fundamental differences between dependent meanings (i.e., concepts, universals, functions, predicates, attributes, descriptions, classes, propositional functions, intensions) and independent meanings (i.e., objects, arguments, subjects, substantives, extensions, extents, graphs) that lie concealed behind inconspicuous grammatical distinctions are violated, various kinds of contradictions, paradoxes, antinomies, fallacies, ambiguity, confusion, nonsense, absurdity do result. Frege, Russell and Husserl, the inventors of 20th century western philosophy, all came to that conclusion.

If, to take up Hilbert's point brought up above, one's logical system is generating contradictions purely through the customary ways in which notions were formed and modes of inference used, it is not logical to see that as posing any particular threat to those notions and inferences themselves. And, it most certainly does not indicate that their use should be prohibited. There is no reason to blame the modes of inference, to think that they might be inadmissible. Though the use of those notions and inferences may have led to the detection of the contradictions, they were not the cause of them. Those basic logical notions and modes of inference were just doing what they were supposed to do: telling you whether or not you could do what you were trying to do. They were saying that a logic that brings about their failure by violating basic principles of meaning is bound to be illogical.

Viewed from the angle of the differences between dependent and independent meanings, Russell's contradiction is just faithfully telling us that: the set X of x's is not a member of what it is a set of; what is predicated of an object is of a different logical type from the object itself; a concept is not an object; a function is not an argument; what is dependent is not independent... In short, logic is doing what logic is supposed to do.

Likewise, there is no reason why an inconsistent theory about the foundations of arithmetic, a theory that violates basic principles of meaning, should have any lasting "downright catastrophic effect in the world of mathematics". If that theory is producing contradictions, if it leads to the failure of the very simplest and most important modes of inference in mathematics as everyone learns, teaches and uses them, it is only reasonable to conclude that there is something wrong with the theory. If, as Husserl lamented in *Formal and Transcendental Logic*, extensions generate contradictions requiring every kind of artful device to make them safe for use in mathematical reasoning (Husserl 1929, 74, 76, 83), there is no reason at all why that paragon of reliability and truth that is mathematics should "crumble" as a consequence, as Russell once feared it might (Russell 1903, § 489). And, it has not.

In such a case, however, one should seriously ask whether one's logical point of view is really the right one. Frege and Russell did that, and so did Alfred North Whitehead and Ludwig Wittgenstein. Try as they might, they never really weeded out the problems that they faced, problems which have yet to be uprooted. Philosophers went on to integrate the logic of *Principia Mathematica* and related systems in their manner into mainstream Analytic philosophy. FRQL has gone on to be become "classical" logic, and trying to solve the problems it generates has become the stuff of logic and much of philosophy in our times.

Frege's contradiction generating, structure breaking axiom of extensionality has yet to be extirpated from FRQL. It remains there to this day. Philosophers keep on doctoring the symptoms of the problems without attacking their causes, or even really identifying them. One popular activity that has suited the spirit of the times has been to disguise defects by performing ingenious, expert plastic surgery on chunks of Frege's logical project. For instance, the major principle governing the selection of the papers anthologized in William Demopoulos' *Frege's Philosophy of Mathematics* was evidence of "a sympathetic, if not uncritical, reconstruction, evaluation, or extension of one or another facet of Frege's thought", and worthwhile papers not satisfying that criterion were not included in the collection (Demopoulos 1995, x). Close examination of the book shows that, perfectly lucid about shocking consequences of Frege's reasoning up to that point, those sympathetic minds jump into Frege's reasoning only after he started to try to fix his theories with extensions in *Foundations*. (Hill 2002; ex. Boolos 1986/1987; 1987; 1993; Dummett 1996)

In the anthology, Demopoulos lauds Michael Dummett for having set an intellectual standard to which most philosophers of his generation aspire, but fails to note that, while calling Frege "the best philosopher of mathematics" in the preface to *Frege: Philosophy of Mathematics*, Dummett writes that the reason why Frege's work in the philosophy of mathematics has been "dismissed as a total failure" is probably that his work "does not prompt any further line of investigation in mathematical logic" and "does not even appear to promise a hopeful basis for a sustainable general philosophy of mathematics". The "evidences of the blindness and lack of generosity which were such marked features of Frege's work after 1891 combine", Dummett considers, "with his great blunder in falling into the contradiction to suggest that he cannot have much to teach us". (Dummett 1991, xi-xiii)

Dependent meanings are intensional, and that one fact, as philosophers like Quine, the premier American philosopher of the latter half of the 20th century, would be the first to admit, is the key to opening of a big can of worms for modern logic, semantics and analytic philosophy. He considered it smart to flee from the problems. He called intensions "creatures of darkness" (Quine 1956, 185) and advised philosophers to run away from them. One of the seven chapters of his classic work *Word and Object* is called "Flight from Intension" (Quine 1960, 191-232). Was he afraid that confronting the hard questions raised by dependent meanings within the context of FRQL would have a damning effect on it?

As an epigram to *Word and Object*, Quine quotes, in German, Otto Neurath's statement that we are all sailors who must make alterations in their ship in the open sea, without ever being able to take it apart in dock and construct it anew. Assuming that this boat is what I have been calling FRQL, Analytic philosophers have indeed failed to bring it into dock, take it apart and construct it anew. They have not wanted to be cast out of the paradise that they believe Frege created for them.

Is FRQL really going to be left in the middle of the ocean to spring leaks for all time? All considered it is perfectly amazing that philosophers have managed to keep it afloat as long as they have. Is it not about time for those still sailing on FRQL to stop trying to repair it in the middle of the ocean, and to let it be brought in into port where it can be made seaworthy and rebuilt from the bottom up using the best parts? What is really keeping philosophers from doing this? From a logical point of view, that is the only course of action with a future. Should it not be the eternal delight of philosophers to confront challenging problems rather than fleeing them?

References

Boolos, George 1986/1987. "Saving Frege from Contradiction", in Demopoulos (ed.), 438-53.
Boolos, George, 1987. "The Consistency of Frege's Foundations of Arithmetic", in Demopoulos (ed.), 211-34.
Boolos, George, 1993. "Whence the Contradiction?", in Schirn (ed.), 234-52.
Demopoulos, William (ed.) 1995. *Frege's Philosophy of Mathematics*, Cambridge MA: Harvard University Press.
Dummett, Michael 1991. *Frege: Philosophy of Mathematics*, London: Duckworth.
Dummett, Michael 1996. "Reply to Boolos", in Matthias Schirn (ed.), *Frege: Importance and Legacy*, Berlin: de Gruyter, 253-60.
Frege, Gottlob 1879, "Begriffsschrift, a formula language, modeled upon that of arithmetic, for pure thought," in van Heijenoort (ed.), 5-82.
Frege, Gottlob 1884. *The Foundations of Arithmetic*, Oxford: Blackwell, 2nd rev. ed., 1986.
Frege, Gottlob 1891. "Function and Concept", in Frege 1980b, 21-41.
Frege, Gottlob 1892. "On Concept and Object", in Frege 1980b, 42-55.
Frege, Gottlob 1893. *Basic Laws of Arithmetic*. Berkeley: University of California Press, 1964.
Frege, Gottlob 1903. "Frege on Russell's Paradox", in Frege 1980b, 214-24.
Frege, Gottlob 1979. *Posthumous Writings*, Oxford: Basil Blackwell.
Frege, Gottlob 1980a. *Philosophical and Mathematical Correspondence*, Oxford: Blackwell.
Frege, Gottlob 1980b. *Translations from the Philosophical Writings*, Peter Geach and Max Black (eds.), Oxford: Blackwell, 3rd ed., (1952).

Heck, Richard 1993. "The Development of Arithmetic in Frege's *Grundgesetze der Arithmetik*", in Demopoulos (ed.), 1995, 257-94.
Hill, Claire Ortiz 2002. "Review of W. Demopoulos' *Frege's Philosophy of Mathematics* and W. W. Tait's *Early Analytic Philosophy*", *Synthese* 133, 441-52.
Hilbert, David 1925. "On the Infinite", in *From Frege to Godel. A Source Book in Mathematical Logic 1879-1931*, Jean van Heijenoort (ed.), Cambridge MA.: Harvard University Press, 1967, 367-92.
Husserl, Edmund 1929. *Formal and Transcendental Logic*, The Hague: Martinus Nijhoff, 1969.
Jourdain, Philip 1912. "The Development of the Theories of Mathematical Logic and the Principles of Mathematics", *The Quarterly of Pure and Applied Mathematics* 48, 219-315.
Quine, Willard van Orman 1956. "Quantifiers and Propositional Attitudes", *Journal of Philosophy* 53 (1956), 177-87.
Quine, Willard van Orman 1960. *Word and Object*, Cambridge MA: The M.I.T. Press.
Quine, Willard van Orman 1961. *From a Logical Point of View*, New York: Harper's Row (1953).
Russell, Bertrand, 1903. *Principles of Mathematics*, London: Norton.
Russell, Bertrand, 1919. *Introduction to Mathematical Philosophy*, London: Allen & Unwin.
Russell, Bertrand, 1927. *Principia Mathematica to *56*, Cambridge UK: Cambridge University Press, 2nd ed. (1964).
Russell, Bertrand, 1956. *Logic and Knowledge, Essays 1901-1950*, R. C. Marsh (ed.), London: Allen &Unwin.
Russell, Bertrand 1973. *Essays on Analysis*, D. Lackey (ed.), London: Allen & Unwin.

12

Claire Ortiz Hill

INCOMPLETE SYMBOLS, DEPENDENT MEANINGS, AND PARADOX

The subject of Husserl's Fourth Logical Investigation, namely the fundamental distinction between independent and dependent meanings that lies concealed behind inconspicuous grammatical distinctions like those between syncategorematic and categorematic expressions, complete and incomplete expressions, is a topic of prime importance for the understanding of major issues in 20th century western philosophy.

Most philosophers would, however, view that as a most exaggerated claim. For the full implications of Husserl's distinction have yet to be drawn and confronted. Wishing to redress that situation, I systematically studied a number of the issues in *Rethinking Identity and Metaphysics, On the Foundations of Analytic Philosophy* (Hill 1997) and in chapters 9 and 10 here. However, I did not explicitly integrate *Husserl's* ideas into what I had to say there.

I undertake to do that here. I do so in the spirit of J. N. Mohanty's interesting and perspicacious observation of many, many years ago that Husserl anticipated many recent investigations in logic and semantics in recent times and what "is more, a true understanding and appraisal of his logical studies is not possible except in light of the corresponding modern investigations" (Mohanty 1976, VII). By drawing connections between Husserl's, Frege's and Russell's theories about incomplete symbols, dependent meanings and the causes of the set-theoretical paradoxes, I hope to lay more groundwork for understanding and appraising Husserl's logical investigations and the contribution that they must eventually make to modern logic and semantics. To this end, I here integrate Husserl's unpublished, and little known, remarks on Frege's "Function and Concept" (Frege 1891) and "On Concept and Object" (Frege 1892), and on Russell's paradox[1] into ideas developed in the Fourth Logical Investigation.

---

[1] Husserl owned copies of Frege's articles "Function and Concept" (Frege 1891) and "On Concept and Object" (Frege 1892). He marked them and commented on them in ways

Terminological Considerations

Of course, when trying to understand and appraise the significance of Husserl's writings for modern philosophy of logic and philosophy mathematics, readers of English must still navigate their way around a number of terminological obstacles, because very much of what Husserl had to say to them still lies concealed behind misleading translations.[2]

In this case, since the Fourth Logical Investigation is devoted to the study of a distinction between kinds of meaning, it is here necessary to a special look at the issues surrounding words like: '*Sinn*', '*sinnvoll*', '*sinnlos*', '*Unsinn*', '*unsinnig*', '*Widersinn*', '*widersinnig*', '*Bedeutung*', '*bedeutsam*', '*bedeutend*', '*Bedeutsamkeit*', '*Bedeutungslosigkeit*' and '*bedeutungslos*'. Since Frege's theory of meaning is under study here too, it is also necessary to keep in mind: 1) that he had already made his famous distinction between *Sinn* and *Bedeutung* in all the writings discussed here, and 2) that Husserl made a point in the First Logical Investigation of saying that he himself used the words '*Sinn*' and '*Bedeutung*' as synonyms and thought it dubious to differentiate between their meanings by using '*Sinn*' for meaning in his own sense and '*Bedeutung*' for the objects expressed as Frege had proposed (Husserl 1900-01 I, §15). On Husserl's personal copy of "Function and Concept", he marked the passage about distinguishing between sense and meaning and underlined the words '*Sinn*' and '*Bedeutung*'. (Frege 1891, 29)

Husserl's choice not to distinguish between sense and meaning simplifies matters for us here. For in English we can speak of sense, nonsense, and senselessness. We do not, however, say senseful or unsenseful, and there is no reason to start doing that here when we can use the words 'meaningful' and 'meaningless'. Harder problems, though, surround the translation of the words '*Widersinn*' and '*widersinnig*', which play a central role in the arguments of this chapter. Although Husserl used these words in a perfectly normal way, they do not translate as neatly into English as one may wish. The German word '*wider*' means against, counter, contrary to, in opposition to. So a very literal translation of these words might be 'countersense' and 'countersensical' (which might not be English words at all). Some have chosen to translate them thus; others have

---

revealing for the purposes of this chapter. I examined Husserl's personal copies of these articles in November 2000 at the Husserl Archives in Leuven, Belgium. Husserl also wrote about 100 pages on set theory. They have principally been known only through Guillermo E. Rosado Haddock's doctoral thesis (Rosado Haddock 1973). Professor Dieter Lohmar and his colleagues at the Husserl Archives in Cologne kindly sent photocopies of them to the Husserl Archives in Paris where Messieurs Jean-François Courtine and Alain Pernet made it possible for me to study them in March 2000. In addition Sebastian Luft and Bernard Goossens of the Husserl Archives in Leuven helped me by finding the answers to some questions that I had. Interestingly, pages 27a-28b of Husserl's notes on set theory are an almost word for word translation of passages from Russell 1911.

[2] I have, of course, at times modified the published translations of Husserl's and Frege's works to bring them into line with these terminological considerations

chosen 'absurdity' and 'absurd'. Husserl himself used '*Absurdität*' and '*absurd*' as synonymous with '*Widersinn*' and '*Widersinnig*'. (ex. Husserl 1900-01 I, §19; IV, Introduction, §12)

'*Widersinn*' and '*widersinnig*' may, however, be understood in the sense of paradox or contradiction and paradoxical, contradictory, illogical, which better suits the purposes of this chapter. In that case, these words fall into the family of '*widersprechen*' (to contradict), '*Widerspruch*' (contradiction), and '*widersprechend*' and '*widerspruchsvoll*', two common German words meaning contradictory. Given the importance of these terms to my main argument, I have chosen to leave '*Widersinnigkeit*', '*Widersinn*', and '*widersinnig*' in German.

Moving on, the kinds of meanings studied in the Fourth Logical Investigation lie hidden behind grammatical distinctions, like those between syncategorematic and categorematic expressions, *geschlossenen* and *ungeschlossenen* expressions. Now, J. N. Findlay used the words 'closed' and 'unclosed' to translate '*geschlossenen*' and '*ungeschlossenen*' (Husserl 1900-01 IV, Intro.), a choice that really leaves readers in the dark about exactly what basic distinction Husserl had in mind. Since '*geschlossen*' and '*ungeschlossen*' also mean complete and incomplete in English, by translating them in that way we can, however, connect Husserl's ideas with those of his contemporaries. Frege often used the term '*abgeschlossen*' (ex. Frege 1891, 25, 31, 32; Frege 1892, 54, 55) to describe what was complete as opposed to what was incomplete (*unvollständig*), in need of completion (*ergänzungsbedürftig*) to make up a complete whole (*ein vollständiges Ganzes*) (ex. Frege 1891, 24, 25, 27, 31, 32). And this is perfectly consistent with choices made by Husserl (ex. Husserl 1900-01 IV, §§4,10), who actually wanted to make much the same point as Russell later would when he wrote of the "great many sorts of incomplete symbols in logic" that "are sources of a great deal of confusion and false philosophy because people get misled by grammar". (Russell 1918, 253)

To describe what was complete as opposed to what was incomplete, Frege usually used his own words '*gesättigt*' and '*ungesättigt*', which have always been translated by 'saturated' and 'unsaturated' (ex. Frege 1892, 38, 47 n., 54-55). In Frege's special vocabulary, functions, concepts, predicates were *ungesättigt*. Here I translate *ungesättigt* as 'unfilled,' which I find clearer in meaning and more natural. According to my *Random House College Dictionary*, 'to saturate' means: 1. to cause (a substance) to unite with the greatest possible amount of another substance through solution, chemical combination, or the like; 2. to charge to the utmost, as with magnetism; 3. to soak, impregnate, or imbue thoroughly or completely; 4. to destroy (a target) completely with many bombs or missiles; 5. to furnish (a market) with goods to the point of oversupply. None of these meanings speak to Frege's concerns.

Related to the German word for satiation, '*gesättigt*' and '*ungesättigt*' might however, just as well have been translated by 'filled' or 'unfilled', which more neatly fit in with Husserl's discourse in terms of meaning fulfillment, filling up (*Ausfüllung*) (ex. Husserl 1900-01 IV, §10; compare Frege 1891, 25) or *Erfüllung* (Husserl 1900-01 IV, §9), which Frege translators have translated by 'realized' (ex. Frege 1892, 49, 50), but which could just as well have been translated by 'fulfilled', which facilitates comparisons with Husserl's ideas.[3]

The Primitive, Essential Distinction between Independent and Dependent Meanings

Since the distinction between categorematic and syncategorematic expressions is a grammatical one, Husserl realized that it might at first sight seem that mere grammatical considerations lie behind it and that the structure of the expressions had no relation to the structure of their meaning. In that case, the syncategorematic words going into the making of the expressions would then always be completely meaningless (*bedeutungslos*), and only the whole expression would have meaning (*nur dem gesamten Ausdruck kommt wahrhaft eine Bedeutung zu*). (Husserl 1900-01 IV, §§4, 5)[4]

Husserl, however, considered that not to be the case. He rather saw the completeness (*Vollständigkeit*) and incompleteness (*Unvollständigkeit*) of expressions as being emblematic of a more primitive, essential distinction between independent and dependent meanings. One must not merely distinguish between categorematic and syncategorematic expressions, he taught, but also between categorematic and syncategorematic meanings. He reasoned that only meaningful (*bedeutsame*) signs were referred to as expressions and that it was superficial to put syncategorematic parts of expressions on the same level as other, generally entirely meaningless (*bedeutungslosen*) parts of expressions like letters of the alphabet, sounds, syllables, prefixes, or suffixes that may, for example, only be part of the sensory apparatus of the expression. (*Ibid.* IV, §§4, 5, 7)

---

[3] Less essential to this particular discussion are Peter Simons' important comments in that classic work on Husserl's theory of wholes and parts, *Parts and Moments, Studies in Logic and Formal Ontology,* about the translation of the terms '*selbständig*' and '*unselbständig*', '*abhängig*' and '*unabhängig*' (Smith ed., 1982, 262-63). In the introduction to his translation of *Logical Investigations*, J. N. Findlay explained to readers that he had translated '*unselbständig*' by 'non-independent', "since the English word 'dependent' has less negativity and more relativity than '*unselbständig*'" and that he "had traffic with 'self-sufficient' for *selbständig*, instead of 'independent'" (Husserl 1900-01, 39). Simons, however, reasons that since Husserl "makes no attempt to distinguish the senses in any explicit way, he understands the two German words as synonymous, and simply uses the commoner one". Simons concludes that "it appears unnecessary to use Findlay's 'non-independent', but to render the most common German word by the most common English word, 'dependent'" (Smith ed., 1982, 263), which is what I have done in this chapter.

[4] In this discussion, Husserl explicitly takes issue with both Bernard Bolzano's theory that every word has its own meaning and Anton Marty's theory that syncategorematic means the same as *mitbedeutend*, i.e., "having no meaning of its own". (Husserl 1900-01 IV, §4).

We are right, Husserl pointed out, to say that words like 'but' or 'father's' have a meaning, but the same cannot be said of 'bi'. All three stand in need of completion, but the need of completion (*Ergänzungsbedürftigkeit*) is essentially different in each case. In the one case, it only concerns the expression; in the other, it not merely concerns the expression, but above all the thought. Whenever syncategorematic expressions function in a normal manner, i.e. whenever they appear in connection with an independent, complete (*abgeschlossenen*) expression, Husserl maintained, they always have a determinate meaning relation to the whole thought, are bearers of meaning for a certain dependent component of the thought and in this way contribute to the expression. (*Ibid.* IV, §§4, 5)

So, Husserl concluded, if one wants to demarcate the ambiguity surrounding the completeness and incompleteness of expressions and determine both its problematical significance and the inner reasons why certain expressions can stand on their own as discourse complete in itself (*als abgeschlossene Reden*) and others cannot, one must go back to the realm of meanings and detect the need for completion (*Ergänzungsbedürftigkeit*) inherent in certain "dependent" meanings. The laws governing this dependency are grounded *a priori* in the nature of the corresponding contents. Each kind of dependency is accompanied by a law stipulating that a content of the relevant kind, say a, can only be in connection with a whole $W(a, b...n)$, where the signs $b...n$ stand for specific kinds of contents. Clarifying what lies concealed behind such grammatical distinctions thus leads to an application of the general ontological distinction made in the Third Logical Investigation between dependent and independent objects to the realm of meaning. (*Ibid.* IV Introduction; §7)

The Laws of Meaning

It was further Husserl's conviction that the primitive, essential distinction between dependent and independent meanings formed the necessary basis for discovering the essential categories of meaning in which were grounded a variety of essential laws of meaning whose business it was to distinguish sense (*Sinn*) from nonsense (*Unsinn*) by determining the *a priori* forms in accordance with which the meanings of the different meaning categories might combine into *One* meaning instead of producing chaotic nonsense (*Unsinn*). (*Ibid.* IV, Introduction)

According to Husserl, any instance of dependent meaning was accompanied by an essential law that governed its need for completion (*Ergänzungsbedürftigkeit*) by other meanings and established the ways in which they might be connected together that ruled out other remaining possible combinations that would yield just a jumble of meanings instead of One meaning. The impossibility of combining meanings in certain ways was not merely subjective, Husserl insisted. It did not merely lie in our actual incapacity to achieve unity. It was objective, ideal and grounded in the "nature", the pure essence of the realm of meaning. Meanings, he repeated over and over, were governed by *a priori* laws that regulated the ways in which they might be combined with new meanings, in which they might fit together and constitute meaningful (*sinnvolle*), coherent meanings. (*Ibid.* IV, §10)

Mere combinations of words like 'a round or', 'king but or', or 'a man and is' are nonsensical, meaningless (*unsinnig, sinnlos*), Husserl emphasized. In such expressions, each word has a meaning, but their meanings do not combine to give a coherent meaning to the whole expression. They are meaningless, utterly incomprehensible. It is completely obvious that so combined no meaning exists, or can possibly exist, for them. On no account can they refer to any object. Moreover, not only is there not any question of reference to objects, but there is not any question of truth either. They break the laws about what can be meaningful. Meaning itself is missing. (*Ibid.* IV, §12)

To make his point Husserl took the coherent, meaningful expression 'this tree is green' and proposed formalizing the given meaning, the independent logical statement, to obtain the corresponding pure meaning form: 'this S is p'. It is clear that formalized in that way it can be interpreted in infinitely many ways, he explained. The statement 'this tree is green' can of course be transformed. We can put any noun or noun phrase in the place of 'S' and any adjective in the place of 'p'. We can say 'this gold', 'this algebraic number', 'this blue crow', etc. 'is green' and we will again obtain a coherent, meaningful (*sinnvolle*) meaning and an independent sentence of the form indicated. Such free exchange of expressions within a given category may yield false, dumb, or funny meanings, but it will necessarily yield coherent meanings. (*Ibid.* IV, §10)

We are not, though, absolutely free in how we go about this, but are bound by strict limitations. Not just any meaning can be substituted for S or for p. Once meaning categories are violated, the coherency of the meaning is lost. As examples of the kind of transgression that he had in mind, Husserl gave: 'this reckless is green'; 'more intensive is round'; 'this house is equal'. We can, he noted, substitute 'horse' for 'similar' in the relational form 'a is similar to b', but we thus still obtain only a series of words, in which each word as such indeed has a sense, refers to a complete combination of meanings (*vollständigen Sinneszusammenhang*), but we do not have a coherent, complete meaning (*geschlossenen Sinn*). (*Ibid.*)

According to Husserl, the job of a science of meanings would be to construct meanings in accordance with essential laws, to discover the laws of combining meanings and transforming them and trace them back to a minimal number of independently elementary laws. It would be necessary first to identify the primitive meaning formations and investigate their inner structure in order to identify the pure meaning categories, which define the sense and extension of what is indeterminate (or what is precisely analogous in mathematics, the variable) in the laws, he maintained. (*Ibid.* IV, §§10, 12, 13)

Changes in Meaning

On his personal copy of Frege's "On Concept and Object", Husserl marked the sentence that reads: "Language has means of presenting now one, now another, part of the thought as the subject". And he tellingly underlined the word 'language', as if to underscore the fact that one could do with language things that logic did not allow (Frege 1892, 49). Section 11 of the Fourth Logical Investigation is, in fact, devoted to elucidating this very matter. In it, Husserl studies the possibility of being led astray by the fact that meanings of each category, and even syncategorematic forms like 'and', may figure in the subject position otherwise reserved for substantival meanings. Taking a closer look, he noted, one sees that in the meaning modification process it happens that a meaning of another syntactical form, say adjectival or even just a mere form, may simply be transplanted into the subject position, as occurs, for example, in sentences of the kind: "'if' is a particle', "'and' is a dependent meaning'. The words are definitely in the subject position, but their meanings are not the same as they normally are.

That each word and each expression can generally be put into the subject position is not in itself surprising, Husserl confirms. What we need to look at, though, is not the composition of the words, but that of the meanings. For from a logical point of view, all change in meaning is to be judged logically abnormal. However, although logic's concern with identical, coherent meanings demands invariability of the meaning function, certain changes in meaning belong even to the normal grammatical state of each language. Within the context of what is being said, the modified meaning is always easily understandable, and if the reasons behind the modification are of sweeping enough generality, if they are rooted, for example, in the general nature of the expression as such, or even in the pure essence of the meaning realm in itself, then the types of abnormality concerned recur everywhere, and what is logically abnormal then appears to be grammatically sanctioned. (Husserl 1900-01 IV, §11)

Turning to consider the *suppositio materialis* of the Scholastics, Husserl further noted that each expression, whether its normal meaning is categorematic or syncategorematic, can name itself. If we say "the Earth is round' is a statement', it is not the meaning of the statement that is in the subject position, but the statement as such; judgment is passed not on the fact that the Earth is round, but on the statement sentence, and this sentence is functioning abnormally as its own name. If we say "'and' is a conjunction', we have not brought the meaning normally corresponding to the name into the subject position, but rather the independent meaning of the word 'and'. Relative to this abnormal meaning, 'and' is actually not a syncategorematic expression, but a categorematic one. It names itself as a word. Something precisely analogous occurs, Husserl explained, when we say: "and', 'but', 'bigger' are dependent meanings'. As a rule, we say: the meanings of the words 'and', 'but' and 'bigger' are dependent. Likewise, in the expression: "man', 'table', 'horse' are concepts of things'. According to Husserl, it was not the concepts themselves that were here functioning in the subject position, but rather representations of these concepts (*Vorstellungen dieser Begriffe*). (*Ibid.*)

In this section Husserl also discusses adjectives, which are normally predestined to play a predicative and attributive role. The adjective 'green', Husserl explains, functions normally with its "original", unmodified meaning in the sentence 'This tree is green' and remains unchanged in itself (*in sich selbst*) when we say 'this green tree'. Husserl maintains that this way of changing syntactical form as opposed to syntactical matter, which, for example, takes place when a nominative meaning functioning as a subject changes to function as a object, or when a sentence functioning as the antecedent changes to function as the conclusion, is to be established first of all and is a main theme of the description of the permanent structures of the realm of meaning.

However, Husserl points out, what is adjectival can undergo still another modification in which the adjective not merely functions attributively in an nominative whole, but is itself used as a noun as in 'Green is a color' or 'Greenness is a difference in coloring'. In that case, 'green' used in the original way and 'green' used as a subject obviously share an identical, common, abstract "core" (*Kern*), which has in each case taken on different forms that are to be distinguished from their syntactical form. If the modification in the core form of the adjectival core content (of the core itself) has produced syntactical matter of the nominative type, then this noun enters into all the syntactical functions, even those that according to formal rules require nouns as syntactical matter. Further details on this matter, Husserl then concludes, belong to a systematic construction of his theory of meaning. (*Ibid.*)

Logical Laws as Opposed to Laws of Meaning

Laws of meaning, according to Husserl, are not themselves logical laws, but rather supply pure logic with the possible forms of meaning whose formal truth or falsehood and reference to objects is determined by logical laws. Whereas laws of meaning serve to distinguish sense (*Sinn*) from nonsense (*Unsinn*) by providing pure logic with possible meaning forms, *a priori* forms of complex coherent, meaningful meanings, it is the job of logical laws to guard against formal or analytical *Widersinn*, formal absurdity, by dictating what it is that objects require to be consistent in purely formal terms. While laws of meaning distinguish what is *sinnvoll* from what is *sinnlos*, logical laws distinguish what is *sinnvoll* from what is *widersinnig*. (*Ibid.* IV, Introduction; §12)

Husserl was adamant about the differences between *Sinn*, *Unsinn* and *Widersinn*. What is *widersinnig*, he complained, is too often wrongly spoken of as being *sinnlos*. However, what is *sinnlos, unsinnig*, he insisted, has no meaning whatsoever and cannot have any. In contrast, what is *widersinnig* genuinely has a coherent meaning and can be determined to be true or false. What is *widersinnig* rightly belongs in the realm of the meaningful, constitutes a partial domain of what is *sinnvoll*. However, in contrast to what is *sinnvoll*, no object can correspond to what is *widersinnig*. The meaning is there, but no existing object can correspond to the existing meaning.[5]

Into the category of *Widersinn*, Husserl placed expressions like 'wooden iron', 'round square', 'all squares have five corners'. These are as respectable names and sentences as any, he maintained. They have meaning, but no object, no thing or fact such as is described by such expressions exists or can exist (*Ibid.*). In his unpublished notes on set theory, Husserl jotted down the following comments on the sentences: 'The present emperor of France is blond' and 'The present emperor of France is not blond'. Both, he noted, are objectless (*gegenstandslos*). 'The present emperor is blond' implies that France presently has a blond emperor and she has no emperor at all. The sentence is not valid, because it is objectless. A sentence is not valid, because it is objectless, in actual fact or owing to a contradiction (*Widerspruchs*) in the subject term. A sentence is not valid because it asserts something of its object that is not attributable to it. The sentence is *widersinnig* with regard to what is predicated of it. A sentence is not valid because it affirms something of its object, having valid existence, or recognized as valid, that is not generally attributable to objects of this sort, or not in fact attributable to this object itself. (Husserl Ms A 1 35, 19a-b)

Husserl, *Widersinn*, and the Set-theoretical Paradoxes

Husserl never published his thoughts on the set-theoretical paradoxes that were the bane of Frege and a thorn in the side of Russell and of so many of their colleagues. Husserl did, however, leave behind a hundred pages of notes on set theory and, as has just been indicated, important points made there are of one piece with what he wrote in the Fourth Logical Investigation. Moreover, as will be seen below, they tie in with conclusions that Russell and Frege themselves came to regarding the causes of the set-theoretical paradoxes.

---

[5] Husserl further distinguishes between material synthetic *Widersinn*, which is true or false on factual grounds, ex. 'A square is round' is a false proposition of pure geometry; and analytic, formal *Widersinn*, which is non-factual, concerns the pure realm of meanings, the objective validity of meanings prior to any material, factual considerations.

Remember that in 1902 Russell wrote to Frege of a contradiction derivable within the system of the *Basic Laws of Arithmetic* (Frege 1893). In his letter, Russell informed Frege that his idea that a function could also constitute the indefinite element was dubious because it led to the following contradiction:

> Let *w* be the predicate of being a predicate which cannot be predicated of itself. Can *w* be predicated of itself? From either answer follows its contradictory. We must therefore conclude that *w* is not a predicate. Likewise, there is no class (as a whole) of those classes which as wholes, as not member of themselves. From this I conclude that in certain circumstances a set does not form a whole. (Frege 1980, 131)

In his notes on the set-theoretical paradoxes, Husserl plainly relegated them into the category of *Widersinn*. He affirms this over and over. For him, a set which contains itself as an element is *widersinnig* (Husserl Ms A 1 35, 12b). To the objection that there is no set that contains itself as an element, he maintained that one need merely respond that that is *widersinnig* (*Ibid.*, 17a). In the case of the set-theoretical paradoxes, he wrote that if one is clear and distinct with respect to meaning (*Sinn*), one readily sees the *Widersinn*. The solution to the paradoxes lies in demonstrating the shift of meaning (*Sinnverschiebung*) that makes it that one is not immediately aware of the contradiction (*Widersinn*) and that once one perceives it one cannot indicate wherein it lies. (*Ibid.*, 12a)

All sets, Husserl reasoned in these notes, fall into one of two classes. Either they fall into class A of sets that contain themselves as an element, or they fall into class B of sets that do not. That being so, he asked to which of these two classes did the set PM of all sets that do not contain themselves as an element belong? By the law of the excluded middle, it would have to be one of the two. However, Husserl observed, one can show that neither or both of the possibilities must be valid. For, were PM to belong to class A, that would mean that it contained itself and that would be contradictory by definition. Were it fall into class B, PM would then not be the set of all sets of all Bs. (*Ibid.*, 17a)

In actual fact, Husserl further noted, it proceeds from the paradox, if no conceptual shift is demonstrable, that a set of kind A or a set of kind B must be a *Widersinn*. Then the classification is *widersinnig* as well. He then looks at the alternative that both may not be the case and asks whether PM is not *widersinnig*. Since, he adds in parenthesis, 'member' has a completely secure meaning (*Sinn*), we have two cases: either concept A is *widersinnig*, or concept PM. Being mathematical, he adds having closed the parenthesis, the sense (*Sinn*) of the classification should be a classification of all *possible* sets in general, independently of questions of real existence having to do with real things of life instead of possible things generally, possible counted things, collected things. (*Ibid.*, 17a)

Three pages later, we find Husserl asking whether a set can contain itself as a partial set. If by partial set, we understand any set which only contains members of M (as its members), he reasoned, then each set has itself as a partial set. However, if by partial set, we understand only any set that only contains members of M among its members, but does not contain at least one member of M among its elements, then no set contains itself as a partial set. To his next question as to whether a set contains itself as a member, he answers that that could only naturally be valid of sets of sets. (*Ibid.*, 20a)

It is part of the idea (*Idee*) of the set to be a unit, he explains, as it were a whole comprising certain members as parts, but doing so in such a way that, vis-à-vis its members, it is something new which is first formed by them. A whole cannot be its own part. Just as it is contradictory (*widersprechend*) for a whole to be its own part at the same time, so it is contradictory (*widersprechend*) for a set for it to be its own member. From the fact that I can speak of all sets, it does not follow that the totality of sets can in return be looked upon as a set. All mathematico-logical operations performable with sets turn on the idea that sets can be looked upon as kinds of wholes, as new units, formations that are something new vis-à-vis their original members, so that out of these formations new units can then again be formed. (*Ibid.*, 20b)

With respect to the set-theoretical paradoxes one can therefore say, Husserl affirms, that wherever mathematicians speak of sets, if the concept is to be a mathematical one, they must have a set essence in view. And whatever sets may then have as an essence, it is expressed with a relation that belongs to the essence, i.e., the relation between sets themselves and elements of a set. An essence relation makes it impossible for the members of the relation to be identical. Thus a set which contains itself as a element is *widersinnig*. (*Ibid.*, 12b)

Husserl's Reactions to Frege's Articles on Functions, Concepts, and Objects

For Husserl, we have seen, the study of syncategorematic and categorematic expressions, complete and incomplete expressions drew attention to the differences between dependent and independent meanings as well as to the differences between laws of meaning that serve to distinguish meaningfulness from meaninglessness and logical laws that that serve to distinguish meaningfulness from *Widersinn*. In "Function and Concept" (Frege 1891) and "On Concept and Object" (Frege 1892), Frege too studied problems surrounding the completeness or incompleteness of expressions, meaningfulness, *Widersinnigkeit*, truth, falsehood and objective reference. Like Husserl, he affirmed that the distinctions to be drawn there were deeply grounded in the nature of the matter and of language. (Frege 1891, 41; Frege 1892, 55)

To my knowledge, Husserl never referred to "On Concept and Object" in print. In a 1903 review, he did, however, fleetingly refer to

"On Function and Concept" as a work by the "ingenious mathematician G. Frege... which unhappily has not found the attention that it deserves from professional logicians" (Husserl 1903, 247). Husserl did, though, possess offprints of both articles and by marking passages and writing his comments in the margins, he left a valuable record of his keen interest in the ideas expressed there. Interestingly, he mainly, and almost exclusively, marked the very passages that deal with themes that went into the making of the Fourth Logical Investigation. Moreover, he signaled points of weakness in Frege's reasoning that both he and Russell ultimately concluded were the cause of the paradoxes derivable within Frege's system.

The matter of the completeness or incompleteness of expressions figures prominently in both of Frege's articles. "Statements in general", he informed readers in "Function and Object", "just like equations or inequalities or expressions in Analysis, can be imagined to be split up into two parts; one complete in itself (*abgeschlossen*), and the other in need of supplementation (*ergänzungsbedürftig*) or 'unsaturated'". In a passage that Husserl marked, Frege chose 'Caesar conquered Gaul' as an example. The second part, 'conquered Gaul', he explained, "is 'unsaturated'–it contains an empty place; only when this place is filled up (*ausgefüllt*) with a proper name, or with an expression that replaces a proper name, does a complete sense (*abgeschlossener Sinn*) appear". (Frege 1891, 31)

According to Frege, the predicable nature of concepts was conferred upon them by a particular kind of incompleteness and dependency that they exhibited with regard to objects. For him concepts stood in need of supplementation and of completion, and he always cited their fundamentally incomplete nature as being what constituted the principal difference between them and objects. In Frege's special vocabulary "a concept is unsaturated in that it requires something to fall under it; hence it cannot exist on its own" (Frege 1980, 101). A concept word "contains a gap which is intended to receive a proper name". (*Ibid.*, 55)

Not all of the parts of a thought may be complete (*absgeschlossen*), Frege reminded readers in "On Concept and Object". At least one must be in some way 'unsaturated' or predicative, or they would not stick together. The sense of the phrase 'the number 2', he explained, does not stick to that of the expression 'the concept *prime number*' without something binding them. In the sentence 'the number 2 falls under the concept *prime number*', this bond is contained in the words 'falls under', which need to be completed in two ways–by a subject and an accusative. It is only because their sense is 'unsaturated' in that way, Frege stresses, that they are capable of binding. Only when they have been supplemented in this twofold respect, he stressed, do we have a complete sense (*abgeschlossenen Sinn*), do we have a thought. (Frege 1892, 54; also 46-48, 50; Frege 1891, 24, 25, 31; Frege 1979, 119-20; Frege 1980, 101, 141)

In these two articles, Frege was also particularly intent upon drawing analogies between concepts and what mathematicians call functions. In "Function and Concept", he explained that in a sentence like 'Caesar conquered Gaul', he called the meaning (*Bedeutung*) of the unsaturated portion the 'function' and Caesar the argument (Frege 1891, 31). On his personal copy of the article, Husserl underlined the word 'meaning' in that passage. On the page before, he had marked where Frege had written: "We thus see how closely that which is called a concept in logic is connected with what we call a function. Indeed, we may say at once: a concept is a function whose value is always a truth-value". (*Ibid.*, 30)

In answer to the question as to what, once he had admitted objects without restriction as arguments and values of functions, it was that he was calling an object, Frege answered that an "object is anything that is not a function, so that an expression for it does not bring any empty place" (*Ibid.*, 32). Husserl underlined the word 'object' in that answer. On one of the first pages of "Function and Concept", he had written that the object of a concept would be something hard to express, but is easily represented in arithmetical notation. (*Ibid.*, 24)

On his copy of "Function and Concept", Husserl also marked where Frege had written:

> I am concerned to show that the argument does not belong with a function, but goes together with the function to make up a complete whole (*ein vollständiges Ganzes*); for a function by itself must be called incomplete (*unvollständig*), in need of supplementation (*ergänzungsbedürftig*), or 'unsaturated'. And in this respect functions differ fundamentally from numbers. (*Ibid.*)

On his copy of "On Concept and Object", Husserl marked the statement that a "concept is the meaning (*Bedeutung*) of a predicate; an object is something that can never be the total meaning (*Bedeutung*) of a predicate, but can be the meaning (*Bedeutung*) of the subject" (Frege 1892, 48), as well as Frege's declaration that the concept must always be distinguished from the object, even in cases like that of the concept word 'Venus', where just one object falls under the concept. 'Venus', Frege had stressed, could never be a proper predicate, although it can form part of a predicate. "The meaning of this word is thus something that can never occur as a concept, but only as an object" (*Ibid.*, 44). Husserl further marked the paragraph in which Frege says that with his distinction between concept and object he had got hold of a distinction of the highest importance. (*Ibid.*, 54)

A link suggests itself between what Frege wrote about the incompleteness of concepts and functions and the completeness of objects in these articles and what Husserl would have to say about meaningfulness, meaninglessness, *Widersinn*, truth, falsehood and having an objective reference in the Fourth Logical Investigation. For Husserl marked the place where Frege wrote that if we substitute 'Julius Caesar' for the proper name formed by the first six words of the sentence 'the concept square root of 4 is 'fulfilled' (*erfüllt*), we get a sentence that has a sense (*Sinn*) but is false; for that fulfillment (*Erfülltsein*)... can truthfully only be said of a quite special kind of objects, viz. Such as can be designated by proper names of the form 'the concept F'. (*Ibid.*, 50)

Frege had just written of the sentence 'there is Julius Caesar' that it is neither true nor false, but senseless (*sinnlos*), whereas the sentence 'there is a man whose name is Julius Caesar' has a sense (*einen Sinn hat*)". (*Ibid.*) Husserl marked Frege's conclusion that what the example showed

holds good generally; the behaviour of the concept is essentially predicative, even where something is being said about it; consequently it can be replaced there only by another concept, never by an object. Thus what is being said concerning a concept does not suit an object. (*Ibid.*)

In several places, Husserl showed himself at variance with Frege about matters of meaninglessness and truth value. For example, next to Frege's comment that for any argument x for which '$x + 1$' were meaningless (*Bedeutungslos*), the function $x + 1 = 10$ would have no value and thus no truth value either, so that the concept: 'what gives the result 10 when increased by 1' would have no sharp boundaries, Husserl penned the objection that the function does not, however, do this to concepts first, but delimits their logical use. The call to work with possible concepts is generally justified, he acknowledged, but impossible concepts are also concepts. There we find the ambiguity again, Husserl continued writing. Having a value, he protested, surely does not mean having a truth value. Each function has *eo ipso* a value, or this not being the case, then value and truth value collapse together (*Ibid.*, 33). Alongside Frege's statement "that we give the name 'the value of a function for an argument' to the result of completing the function with the argument", Husserl had commented that the definition was unclear and that value should surely be distinguished from truth value (*Ibid.*, 25). And next to Frege's statement that the value of our function was a truth value, Husserl had noted that this was again not completely clear, –that each value of a function was necessarily true or false, existing or not existing (*Ibid.*, 29). A few pages later he asked in the margins whether or not he could not speak of the truth or falsehood, meaning the existence of the capital of the number 4, whether he could not say that there was no capital of the number 4, as it appears absurd (*absurde*) concepts would not reckoned as concepts. (*Ibid.*, 31-32)

Frege's Linguistic Predicament

In "Function and Concept" and "On Concept and Object", Frege also, and most importantly, struggled with what he eventually came to call the "fatal tendency of language" to allow a concept word to be transformed into a proper name and so to come to be in a place for which it was unsuited (Frege 1979, 269-70; Frege 1980, 54-55). The essential differences between concepts and objects are, we find him explaining in letters, "covered up in our word languages" (Frege 1980, 55) where "the two merge into each other" (*Ibid.*, 100) and "the sharpness of the difference is somewhat blurred, in that what were originally proper names (e.g., 'moon'), can become concept words, and what were originally concept words (e.g., 'God') can become proper names". (*Ibid.*, 92)

By "a kind of necessity of language", he confessed in "On Concept and Object", his expressions, taken literally, sometimes missed his thought, so that he mentioned an object, when what he intended was a concept (Frege 1892, 54). Indeed, although he firmly believed that concepts were fundamentally different from objects, by the early 1890s he had found that in "logical discussions one quite often needs to say something about a concept and to express this in the form usual for such predication" (*Ibid.*, 46). When "we say that the concept *horse* is not a concept, whereas, e.g., the city of Berlin is a city, and the volcano Vesuvius is a volcano", he famously wrote in "On Concept and Object", "we are confronted by an awkwardness of language, which... cannot be avoided". (*Ibid.*)

"Language is here in a predicament that justifies the departure from custom", he concluded. Owing to its predicative nature, the concept as such cannot play the part of the subject, so "it must be converted into an object, or, more precisely, an object must go proxy for it" (*Ibid.*). In a note to that claim, Frege wrote:

> A similar thing happens when we say as regards the sentence 'this rose is red': The grammatical predicate 'is red' belongs to the subject 'this rose'. Here the words 'The grammatical predicate "is red"' are not a grammatical predicate, but a subject. By the very act of explicitly calling it a predicate, we deprive it of this property. (*Ibid.*, 46 n.†)

Husserl marked this note. As we have seen, he marked Frege's sentence: "Language has means of presenting now one, now another, part of the thought as the subject" and underlined the word 'language'. (*Ibid.*, 49)

Frege was additionally aware that "over the question what it is that is called a function in Analysis, we come up against the same obstacle; and on thorough examination one will find that the obstacle is grounded in the matter itself and in the nature of our language" (*in der Sache selbst und in der Natur unserer Sprache begründet*) (*Ibid.*, 55). Functions do, he found, and sometimes have to, take other functions as their arguments. When that is the case, however, they are then fundamentally different from functions whose arguments are objects and cannot be anything else, he insisted. Just "as functions are fundamentally different from objects, so also functions whose arguments are and must be functions are fundamentally different from functions whose arguments are objects and cannot be anything else"(Frege 1891, 38). Frege called the latter first-level functions and the former second-level functions and distinguished between first-level and second-level concepts in the same way (*Ibid.*). "Function and Concept" closes with the affirmation that the difference between first-level and second level functions discussed there "is not made arbitrarily, but is deeply grounded in the nature of the matter" (*nicht willkürlich gemacht, sondern in der Natur der Sache tief begründet*). (*Ibid.*, 41)

On his copy of "On Concept and Object", Husserl marked the sentence: "The relation of an object to a first-level concept that it falls under is different from the (admittedly similar) relation of a first-level to a second-level concept" (Frege 1892, 50). He also marked the line where Frege had written in the paragraph before: "I do not want to say it is false to say concerning an object what is said here concerning a concept; I want to say that is impossible, it is senseless (*sinnlos*) to do so" (*Ibid.*). Here Husserl tellingly underlined the word '*sinnlos*' on his personal copy.

During the 1890s, Frege convinced himself that he could overcome this linguistic predicament by creating a "logical law", his Basic Law V governing the behavior of graphs of functions and extensions of concepts. As he explained in a sentence that Husserl marked in "Function and Concept": "Graphs of functions are objects, whereas functions themselves are not.... Extensions of concepts likewise are objects, although concepts themselves are not" (Frege 1891, 32). Frege knew well that what he wished to sanction through Basic Law V was "forbidden by the basic difference between first and second level relations" (Frege 1979, 182), but he temporarily convinced himself that, though an actual proof could "scarcely be furnished" and "an unprovable law" would have to be assumed, a transformation might "take place, in which concepts correspond to extensions of concepts..." (*Ibid.*). "Of course", he knew by 1906, "it isn't as self-evident as one would wish for a law of logic. And if it was possible for there to be doubts previously, these doubts have been reinforced by the shock the law has sustained from Russell's paradox" (*Ibid.*). (It is worth noting here that Husserl put a question mark where Frege had written in "Function and Concept" that the possibility of regarding an equality holding generally between values of functions as an equality is not demonstrable and must be taken to be a fundamental law of logic. (Frege 1891, 26))

Frege's Initial Reactions to Russell's Paradox

Russell's paradox about the class of all classes that are not members of themselves ultimately led both Frege and Russell to many of Husserl's conclusions about what can and cannot be done in logic. For understanding the logical behavior of incomplete symbols and dependent meanings came to play a preeminent role in the efforts of those two makers of modern logic and semantics to grasp the causes of the set-theoretical paradoxes and to avoid them. (Hill 1997, 20, 22, 40-41, 69-70, 77-82, 87-90, 91-110, 145-47, 149-51)

When first confronted with Russell's paradox in 1902, Frege responded that it seemed to indicate that his Basic Law V was false and that his explanations did not suffice to secure a meaning for his combinations of signs in all cases (Frege 1980, 132). True to his convictions that the fundamental differences between predicates and objects are inviolate and grounded in the deep nature of the matter, another of his first reactions upon receiving the bad news had been to write to Russell that his expression 'A predicate is predicated of itself' did not seem exact to him (Frege) and that he would rather say 'A concept is predicated of its own extension'. A predicate, Frege explained to Russell, was as a rule a first-level function required an object as argument and could not therefore have itself as argument (subject) (*Ibid.*, 132-33). The problem with the proposition 'A function never takes the place of a subject', he told Russell, had already been dealt with in 'On Concept and Object'. It was "only an apparent one occasioned by the inexactness of the linguistic expression; for the words 'function' and 'concept' should properly speaking be rejected. Logically, they should be the names of second-level functions; but they present themselves linguistically as names of first-level functions. It is therefore not surprising that we run into difficulties in using them". (*Ibid.*, 141)

To express ourselves precisely, Frege further informed Russell, our only choice would be to talk about words or signs. The proposition '3 is a prime number', he told him, could be analyzed into two essentially different parts: '3' and 'is a prime number'. The first part was complete (*abgeschlossen*) in itself, but the second stood in need of completion (*ergänzungsbedürftig*), and the same was to be said for the proposition '4 is a square number'. It makes sense, Frege maintained,

> to fit together (*sinnvoll zusammenfügen*) the complete (*abgeschlossenen*) part of the first proposition with that part of the second proposition which is in need of completion (*ergänzungsbedürftigen*) (that the proposition is false is a different matter); but it makes no sense to fit together (*sinnvoll zusammenfügen*) the two complete (*abgeschlossenen*) parts; they will not hold together; and it makes just as little sense (*ebensowenig sinnvoll*) to put 'is a square number' in place of '3' in the first proposition". (*Ibid.*, 142)

"This difference between the signs", Frege continued,

> must correspond to a difference in the realm of meanings; although it is not possible to speak of it without turning what is in need of completion (*Ergänzungsbedürftige*) into something complete (*Abgeschlossenes*) and thus falsifying the real situation. We already do this when we speak of 'the meaning

of "is a square number"'. Yet the words 'is a square number' are not meaningless (*bedeutungslos*). The analysis of the proposition corresponds to an analysis of the thought, and this in turn to something in the realm of meanings. (*Ibid.*, 141-42)

In the appendix to the 1903 volume of *Basic Laws*, Frege still concluded that there was "nothing left but to regard extensions of concepts or classes, as objects in the full and proper sense of the word" (Frege 1903, 217) and that there was "nothing 'unsaturated' or predicative about classes that would characterize them as functions, concepts, or relations. What we usually consider as a name of a class... has rather the nature of a proper name; it cannot occur predicatively, but *can* occur as the grammatical subject of a singular proposition. (*Ibid.*, 215)

Frege's Ultimate Conclusions

It was ultimately this very propensity of language to undermine the reliability of thinking by forming apparent proper names to which no objects correspond that Frege blamed for having "dealt the death blow" to his set theory (Frege 1979, 269). When asked shortly before his death to write on the paradoxes of set theory in order to provide further justification for his belief that it was untenable, he replied that the paradoxes of set theory arise because a concept is connected with something that is called the set which appears to be determined by the concept–and determined as an object (Frege 1980, 54). "The essence of the procedure which leads to the thicket of contradictions", he said, could be summed up as follows:

> The objects that fall under *F* are regarded as a whole, as an object and designated by the name 'set of *F*s'. This is inadmissible because of the essential difference between concept and object, which is indeed quite covered up in our word languages.... Because of its need for completion (*Ergänzungsbedürftigkeit*), (unsaturatedness, predicative nature), a concept word is unsaturated, i.e., it contains a gap which is intended to receive a proper name.... Through such saturation or completion (*Ergänzung*) there arises a proposition whose subject is the proper name and whose predicate is the concept word.... In such a proposition, concept word and proper name occupy essentially different places, and it is obvious that a proper name will not fit into the placed intended for the concept word. Confusion is bound to arise if a concept word, as a result of its transformation into a proper name comes to be in a place for which it is unsuited. (*Ibid.*, 55)

"One feature of language that threatens to undermine the reliability of thinking", he wrote that same year, "is its tendency to form proper names to which no objects correspond". The paradoxes of set theory arise, he explained then, because a concept is connected with something called a set, which appears to be determined as an object. One thus thinks of the objects falling under the concept as combined into a whole, which is construed as an object and designated by a proper name. Such an expression appears to designate an object, but there is no object for which this phrase could be a linguistically appropriate designation, he came to realize. The expression is surely a designation of a concept and thus could not be more different from a proper name (Frege 1979, 269-70). The "difficulties which this idiosyncrasy of language entangles us in are incalculable", Frege warned then (*Ibid.*, 270). Experience had shown him "how easily this can get one into a morass" (Frege 1980, 55). He even went so far as to suggest that one "must set up a warning sign visible from afar: let no one imagine that he can transform a concept into an object". (Frege 1979, 55)[6]

Russell on Incomplete Symbols and the Set-theoretical Paradoxes

As for Russell, one of the things that his struggle with the paradox derivable within Frege's system taught him was that it

> is important, if you want to understand the analysis of the world, or the analysis of facts, or if you want to have any idea what there really is in the world, to realize how much of what there is in phraseology is of the nature of incomplete symbols. (Russell 1918, 253)

In particular, he learned through it that if a word or phrase that is devoid of meaning when separated from its context is wrongly assumed to have an independent meaning, we then get what may be called false abstractions, pseudo-objects, and paradoxes and contradictions are apt to result (Russell 1906c, 165). Indeed, Russell came to believe that "if we assume, as Frege does, that the class is an entity, we cannot well escape the contradiction about the class of classes which are not members of themselves" (*Ibid.*, 171), and that it resulted from that contradiction that it "must under all circumstances be meaningless (not false) to suppose a class a member of itself or not a member of itself...". (Russell 1919, 185)

Russell saw that

> When we say that a number of objects all have a certain property, we naturally suppose that the property is a definite object, which can be considered apart from any or all of the objects which have, or may be supposed to have the property in question. We also naturally suppose that the objects which have the property form a *class*, and that the class is in some sense a new single entity, distinct, in general, from each member of the class. Both these natural suppositions can be proved, by arguments so short and simple that they scarcely admit a possibility of error, to be at any rate not *universally* true" (Russell 1906b, 163-64).[7]

---

[6] Peter Simons studies incomplete meanings and symbols in Frege's latest published writings, particularly his *Logical Investigations*, in an article entitled "Unsaturatedness" (Simons 1981). Husserl's *Logical Investigations* are discussed there too. The points he makes are indirectly related to the points I make here.

[7] It is worthwhile noting here that the only two sentences that Husserl marked in all of his copy of Russell's *Principles of Mathematics* (Russell 1903) were from §104 of Chapter X entitled "The Contradiction". They were: 1) "In such cases there is only a class as many, not a class as one. We took it as axiomatic that the class as one is to be found wherever

So Russell came to teach that "there are no such things as classes and relations and functions as entities, and that the habit of talking of them is merely a convenient abbreviation" (Russell 1906b, 145). Symbols for functions did not have a meaning in themselves, he saw, but required some supplementation in order to acquire a complete meaning (Russell 1910, 225); such incomplete symbols were entirely devoid of meaning by themselves and only became significant as parts of appropriate propositions (Russell 1906c, 170). Standing alone, a propositional function was "a mere schema, a mere shell, an empty receptacle for meaning, not something already significant" (Russell 1919, 157). A function was not a definite object, but essentially a mere ambiguity awaiting determination. For it to occur significantly, it had to receive the necessary determination, had to occur in such a way that the ambiguity disappeared, and a wholly unambiguous statement resulted. (Russell 1910, 229-30)

Once Russell had concluded that it was "natural to suppose that classes are merely linguistic or symbolic abbreviations", the important thing for him to do became "merely to provide a mode of interpreting the ordinary statements about classes without assuming that classes are entities" (Russell 1906a, 200). So he set to work to devise a way to escape the paradoxes "by the fact that classes are now not single entities... and are only parts of significant phrases, without being themselves significant in isolation". (*Ibid.*, 210)

Analogies that he was able to draw between the logical behavior of descriptions and classes as incomplete symbols and the success that he had with his 1905 theory of definite descriptions suggested to him a way in which classes could be analyzed away in much the same way as descriptions could. Thus he developed a procedure for analyzing away classes modeled after his way of handling descriptions as incomplete symbols with his theory of definite descriptions. He judged this move to be a major breakthrough in his efforts to unlock the mystery of contradictions arising from treating incomplete symbols as if they stood for objects.

Analyzed in his new way,

> all the formal properties that you desire of classes, all their formal uses in mathematics, can be obtained without supposing... that a proposition in which symbolically a class occurs, does in fact contain a constituent corresponding to that symbol, and when rightly analysed that symbol will disappear, in the same sort of way as descriptions disappear when the propositions are rightly analysed in which they occur. (Russell 1919, 266)

---

there is a class as many; but this axiom need not be universally admitted, and appears to have been the source of the contradiction"; and 2) "In this view, a class as many may be a logical subject, but in propositions of a different kind from those in which its terms are subjects". Examined at the Husserl Archives in Leuven in November 2000.

This adaptation of his theory of descriptions to the theory of classes was Russell's way of sweeping away some of his most onerous problems, but the whole story is much too long to tell here. I endeavored to tell it in *Rethinking Identity and Metaphysics, On the Foundations of Analytic Philosophy* (Hill 1997). It also figures in Chapter 9.

Conclusion

In this essay I have discussed some of the main themes of Husserl's Fourth Logical Investigation and shown that important lessons that Frege and Russell learned from the contradictions derivable within system of the *Basic Laws of Arithmetic* turned out to be compatible with Husserl's ideas. All three actually came to many of the same conclusions about the inviolability of the laws governing the use of complete and incomplete expressions, dependent and independent meanings.

Of the three, Frege was the first to draw attention to the problems in "Function and Concept" and "On Concept and Object". However, silencing his doubts, he took logical law into his own hands; even as he published those two articles, he was developing the logical system of *Basic Laws of Arithmetic* that violated the very logical principles that he was proclaiming inviolable in his articles. By assuming "an unprovable law", he tried to force symbols standing for incomplete meanings to obey the same formal rules of identity that symbols for complete meanings do. By so mandating that a function could take another function as an object, however, he broke the law and suffered the contradiction about the class of all classes that are not members of themselves as a consequence. At least that is the way he saw it. Something deeper and logically prior to Basic Law V was at work that caused its failure. And what was it if not an *a priori* law of the kind Husserl wrote about in the Fourth Logical Investigation?

For the sake of those who feel queasy about finding Russell and Husserl discussed in the same place, or about any talk of self-evident *a priori* laws, it is important to note here that in Russell's article on the philosophical implications of mathematical logic that is found translated in Husserl's notes on set theory, Russell affirmed "that there is *a priori* and universal knowledge" (Russell 1911, 292). He wrote there that "all knowledge which is obtained by reasoning, needs logical principles which are *a priori* and universal" (*Ibid.*). He further declared that "it is necessary that there should be self-evident logical truths" and that these were "the truths which are the premises of pure mathematics as well as of the deductive elements in every demonstration on any subject whatever". (*Ibid.*) Russell wrote,

> Logic and mathematics force us, then to admit a kind of realism in the scholastic sense, that is to say, to admit that there is a world of universals and of truths which do not bear directly on such and such a particular existence.... We have

> immediate knowledge of an indefinite number of propositions about universals: this is an ultimate fact.... (*Ibid.*, 293)

Here I have tried to lay the foundations for the further discussion of what I have called a topic of prime importance for the understanding of some major issues in 20th century western philosophy. Of course really big questions that remain are: Were Husserl, Frege, and Russell, who were after all as important philosophers as any that the 20th century ever produced, right about these inviolable laws whose existence they believed that they had discerned? How responsible really was Frege's violation of the "deep nature of the matter" for the paradoxes? And most importantly, what does that have to do with us now?

I am of the conviction that a close and thorough investigation of the matter, of the kind I undertook in *Rethinking Identity and Metaphysics* (Hill 1997) and investigations into the connections between the set-theoretical paradoxes and the paradoxes of modal and intensional logics carried out in Chapter 9 of this book on Husserl's philosophy of logic and mathematics will show that it is precisely the kind of errors about incomplete symbols and dependent meanings that Husserl, Frege and Russell discerned that have been generating the problems with substitutivity, identity, reference, existential generalization and so on that have done so much to heat up many philosophical discussions since their time.

Modern logic has yet to draw the full implications of Frege's and Russell's struggle with the paradoxes derivable in Frege's system has for the "classical", Fregeo-Russello-Quineo logic (FRQL) that was defended with such passion by so many during the 20th century. The way that that logic took hold in English-speaking countries and the virulent attitude that its proponents displayed towards modal and intensional logics effectively dissimulated many of the most important lessons that the set-theoretical paradoxes hold for philosophy. For intensions are dependent meanings and that one fact is the key to the opening of a great big can of worms for modern logic, semantics and analytic philosophy.

References

Frege, Gottlob 1891. "Function and Concept", *Translations from the Philosophical Writings of Gottlob Frege*, 3rd ed. Oxford: Basil Blackwell, 1980 (1952), 21-41.

Frege, Gottlob 1892. "On Concept and Object", *Translations from the Philosophical Writings of Gottlob Frege*, 3rd ed. Oxford: Basil Blackwell, 1980 (1952), 42-55.

Frege, Gottlob 1893. *Basic Laws of Arithmetic*, Berkeley CA: University of California Press, 1964.

Frege, Gottlob 1903. "Frege on Russell's Paradox", in *Translations from the Philosophical Writings of Gottlob Frege*, 3rd ed., Oxford: Basil Blackwell, 1980 (1952), 214-24.

Frege, Gottlob 1979. *Posthumous Writings*, Oxford: Basil Blackwell.

Frege, Gottlob 1980. *Philosophical and Mathematical Correspondence*, Oxford: Basil Blackwell.

Hill, Claire Ortiz 1997. *Rethinking Identity and Metaphysics, On the Foundations of Analytic Philosophy*, New Haven CT: Yale University Press.

Husserl, Edmund 1900-01. *Logical Investigations*, New York: Humanities Press, 1970.

Husserl, Edmund 1902. "Memorandum of a Verbal Communication from Zermelo to Husserl", *Early Writings in the Philosophy of Logic and Mathematics*, translated by Dallas Willard, Dordrecht: Kluwer, 442.

Husserl, Edmund 1903. "A Report on German Writings in Logic from the Years 1895-1899, Third Article", *Early Writings in the Philosophy of Logic and Mathematics*, translated by Dallas Willard. Dordrecht: Kluwer, 246-59.

Husserl, Edmund Ms A I 35. Untitled, undated manuscript on set theory available at the Husserl Archives in Cologne, Leuven, and Paris.

Mohanty, J. N. 1964. *Edmund Husserl's Theory of Meaning*, 3rd ed., The Hague: Martinus Nijhoff, 1976.

Rang, B. and W. Thomas 1981. "Zermelo's Discovery of Russell's Paradox", *Historia Mathematica* 8, no. 1, 15-22.

Rosado Haddock, Guillermo Ernesto 1973. "Edmund Husserl's Philosophie der Logik und Mathematik im Lichte der gegenwärtigen Logik und Grundlagenforschung", Doctoral Thesis, Bonn: Rheinischen Friedrich-Wilhelms-Universität.

Russell, Bertrand 1903. *Principles of Mathematics*, London: Norton.

Russell, Bertrand 1905. "On Denoting", *Logic and Knowledge, Essays 1901-1950*, R. C. Marsh (ed.), London: Allen &Unwin, 1956, 41-56.

Russell, Bertrand 1906a. "On 'Insolubilia and their Solution by Symbolic Logic", *Essays in Analysis*, London: Allen &Unwin, 1973, 190-214.

Russell, Bertrand 1906b. "On Some Difficulties in the Theory of Transfinite Numbers and Order Types", *Essays in Analysis*, London: Allen & Unwin, 1973, 135-64.

Russell, Bertrand 1906c. "On the Substitutional Theory of Classes and Relations", *Essays in Analysis*, London: Allen & Unwin, 1973, 165-89.

Russell, Bertrand 1910. "The Theory of Logical Types", *Essays in Analysis*, London: Allen &Unwin, 1973, 215-52.

Russell, Bertrand 1911. "L'importance philosophique de la logistique", *Revue de Métaphysique et de Morale* 19, 281-91, published in English as "The Philosophical Implications of Mathematical Logic", *The Monist* 22 (Oct. 1913): 481-93 and in *Essays in Analysis*. London: Allen & Unwin, 1973, 284-94. Partially translated in Husserl Ms A I 35.

Russell, Bertrand 1918. "The Philosophy of Logical Atomism", *Logic and Knowledge, Essays 1901-1950*, R. C. Marsh (ed.), London: Allen & Unwin, 1956, 177-281.

Russell, Bertrand 1919. *Introduction to Mathematical Philosophy*, London: Allen &Unwin, 1956.

Simons, Peter 1981. "Unsaturatedness", *Grazer Philosophische Studien* 14, 73-96.

Smith, Barry (ed.) 1982. *Parts and Moments, Studies in Logic and Formal Ontology*, Munich: Philosophia Verlag.

13

Jairo José da Silva

HUSSERL'S PHENOMENOLOGY AND WEYL'S PREDICATIVISM

I

In 1918 Hermann Weyl (1885-1955), one of the most influential mathematicians of last century, published a small book called *The Continuum* (Weyl 1918a) in which he carries out a partial reconstruction of classical real analysis in accordance with the Poincaré-Russell vicious-circle principle but avoiding type-theory. Weyl's aim, from a strictly mathematical perspective, was to establish sound *arithmetical* foundations for analysis.

In the preface to *The Continuum*, Weyl explicitly states his agreement with Husserl's *Logical Investigations* (Husserl 1900-01) with respect to the epistemological aspects of logic and also mentions Husserl's *Ideas I* (Husserl 1913). These references to the founder of phenomenology in a mostly technical work on the foundations of mathematics are less surprising if we bear in mind that Weyl had actually been Husserl's student in Göttingen, beginning in the period 1904-05, when Husserl was lecturing on the phenomenology of time. Moreover, the experience of time was for Weyl the paradigm of the continuum that mathematical analysis strives to capture conceptually and express as faithfully as it can (if, as Weyl believed unavoidable, only in its formal aspects).

Husserl is also frequently mentioned and quoted in Weyl's *Philosophy of Mathematics and Natural Science* (Weyl 1949) and in an autobiographical lecture given in 1954 at the University of Lausanne Weyl comments on the impact Husserl had on him at the time he wrote *The Continuum*. Furthermore, in *Space, Time and Matter*, a book Weyl published at about the same time as *The Continuum*, he points to his inquiries into space as a good example of the kind of analysis of being aspired to by phenomenology.

All this attests to his lifelong interest in Husserl's philosophy and his agreement with substantial portions of it. As a matter of fact, Weyl did not see his efforts in *The Continuum* as directed exclusively towards technical mathematical problems but also, and to no lesser extent, as a contribuition to the clarification of the epistemological problem of the relation between *intuition* and *conceptual knowledge* (Weyl 1918a, preface), a problem, as is well known, central to both the *Logical Investigations* and *Ideas I*. The question then naturally arises as to the extent to which Husserl's thought influenced Weyl in *The Continuum* and shaped his approach to the foundations of analysis. This is the question I want to address in this chapter.[1]

It is clear enough, however, that Weyl cannot be regarded as anything like a disciple of Husserl's. His philosophy of mathematics differs from Husserl's in many points, some of which will be discussed in this chapter, and Husserl was by no means the only philosopher to have influenced him. Around the time Weyl came into contact with Husserl's ideas he was also introduced to Fichte's philosophy, from which he seems also to have extracted important ideas, such as that of an interplay between the *I* and the *not-I* in the constitution of an *objective* reality[2]. Weyl's "constructivism", which, it should be stressed, must not be confused with any form of what is traditionally known by this name in mathematics, was probably influenced by Husserl as to its intuitive foundations and by Fichte as to the idea that intuitions must be confronted with opposing forces to acquire objective validity (which, for Weyl, can only be granted to its formal aspects, those only that mathematics can express).

But Weyl refused Fichte's stubborn apriorism. For him, our rational reconstruction of the given (that which is offered to *me* in *my* intuition) must take into consideration, besides the other and his or her intuitions, also the findings of science. Weyl's contructivism is not a priori. The reconstruction of the given is also a task for mathematics, which accomplishes it by recasting the given (the intuitive data) in *symbolic* form. The idea that mathematical theories provide a *symbolic* representation of the given, which must, like any scientific theory, pass the test of experience is clearly detected in *The Continuum*, where Weyl gives us precisely a "symbolic reconstruction" of the continuum of intuition, "which must establish its own reasonableness... in the same way as a physical theory" (Weyl 1918a, Chapter 2 §6). Such a reconstruction, which strives to be as faithful as possible, is, nonetheless, unavoidably imperfect, given the unbridgeable gap between the intuitive continuum and the world of mathematical concepts. However, Weyl believes, it at least avoids the gross distortions of *classical* analysis, at whose core lies a contradiction, a *continuum* constituted of *individual* points. Weyl's arithmetical continuum is not, in fact, the genuine article. The intuitive continuum, unlike the

---

[1] See Weyl's letter of August 13, 1918, pp. 392-93.

[2] Although the constitution of objectivity in intersubjectivity is also, of course, a Husserlian theme (see, in particular, his *Cartesian Meditations* (Husserl 1931)). Weyl seems to have interpreted phenomenology as a reflection of the *I* upon itself.

arithmetical one, is not made of points and those we mark on the former in our effort to capture it in mathematical concepts are *imposed* upon it by an external sieve of arithmetically definable numbers; what passes through it is lost for science.

As a mathematician, Weyl could not rest content with simply describing what he found in his intuitions, as a descriptive phenomenologist would. Husserl, despite his detailed analyses of intuitive time, did not have a mathematical theory of it, since he only wanted to investigate the genesis of intuitive time in intentional subjectivity, not "reconstruct" it in symbolic terms and reshape it in mathematical garb. In short, Weyl is *not* giving us a phenomenology of the intuitive continuum, but taking a step (or several steps) beyond the mere phenomenological description, trying to capture the intuitively given in *objectively valid mathematical language and concepts*. This is not to dismiss the importance of the phenomenological description, which will always remain the *necessary beginning*, but is the mathematical complement to the phenomenologist's work.

Weyl's task goes beyond the "covering" of the intuitive continuum with the arithmetic continuum, it also includes the constitution of the arithmetic continuum itself out of an intuitively given basis, the *discrete* sequence of the positive integers, the natural numbers–it is important to note that, for Weyl, the *whole* sequence of natural numbers is intuitively given. The foundational role played by intuition in Weyl's mathematical analysis is, then, twofold, it provides the fundamental stuff mathematics is called to describe and the basic domain of mathematical entities, including its language, with which to describe it.

Despite the naturalness of approaching Weyl's mathematical analysis in phenomenological terms, given his relation to Husserl's philosophy at the time he wrote *The Continuum*, we must always keep in mind, as I have already noted, that Weyl was not in the least concerned with a faithful interpretation of Husserl's thought and that, moreover, he was at that time already in disagreement with many fundamental tenets of phenomenology. Let me take a moment to point out some of these points of disagreement.

In *Space, Time and Matter*, Weyl expressed his doubts concerning the possibility of a *direct* intuition of the essence of *space*, unaided by the findings and creations of mathematical and physical theories of space.[3] This amounts, I believe, to a clear disqualification of the effectiveness of the process of intuition of essences, at least with respect the essence of space, within the scope of phenomenological *epoché*, which, as we know, requires the abandonment of "scientific prejudices" in the search for the *eidos* of space. He, however, never manifested the same concerns as to the immediate intuition of the *whole* sequence of natural numbers and

---

[3] For Weyl's views on geometry, see Chapter 2.

its essential features, including the principle of induction, even though the idea that we can have *immediate* access to an infinite realm of ideal entities is, of course, quite un-Husserlian. For Husserl, ideal objects are necessarily non-independent entities which depend on certain constitutive acts being carried out before manifesting themselves in consciousness. This idea of a *direct* intuitive access to the realm of *natural numbers* probably came from Henri Poincaré. Nonetheless, I believe that the fundamental role Weyl gives to a broad form of intuition in his new approach to the arithmetization of analysis, the fact that in his analysis, sets and functions are not given independent existence, but are constituted, and his theory of meaning for judgments can all plausibly be credited to the influence of Husserl.

Besides Husserl's notion of *epoché*, which Weyl criticizes for imposing the suspension of scientific knowledge in the disclosure of the essence of space, Weyl also criticizes what he sees as a confinement of the Husserlian pure *I*. For Weyl, any individual *I* must encounter other *I*s in order to constitute an *objective* reality: objective reality is what emerges from the conflict of opposing subjectivities. We know that Husserl eventually had to face the problem of the intentional constitution of objective reality in transcendal subjectivity, but at the time Weyl met him and studied his works phenomenology could easily be seen as a form of solipsism. However, *this* criticism is expressed only much later (for instance, in "Insight and Reflection", Weyl 1955).

Fichte and Husserl were probably the two major influences on Weyl's epistemological ideas at the time he wrote *The Continuum* (Kant had already been left behind, defeated by the non-Euclidean geometries) and, in the end, Fichte seems to have been the stronger voice. In "Insight and Reflection", Weyl says:

> In the antithesis of construction and phenomenology, my sympathies lie entirely on his [Fichte's] side; yet how a constructive procedure which finally leads to the symbolic representation of the world, not *a priori*, but rather with continual reference to experience, can really be carried out is best shown by physics. (Weyl 1955, 215)

Weyl believed that the symbolic recasting of the given, carried out by physics and mathematics could be best seen in terms of a sort of Fichtean dialectics rather than by the concentrated efforts of an isolated pure *I* within the scope of phenomenological *epoché*. How clear those ideas were by the time he wrote *The Continuum* is hard to determine. But, be that as it may, what we find in *The Continuum* is a clearly phenomenological description of the intuition of the continuum, exemplified by the flux of *immanent* time (we must remember that Weyl had followed Husserl's lectures on precisely this topic), together with (and here is where he believes to be abandoning Husserl) an effort to capture this intuition in mathematical terms. The entire mathematical theory, however, is conditioned by the previous phenomenological clarification of the intuitive continuum, which requires that "points" in the arithmetical continuum (which do not correspond to anything intuitively given) be constituted out of *given* natural numbers by superposing layers of intentional structures (to which certain arithmetical "constructions" in the domain of natural numbers correspond). The question Weyl is asking is this: how can we obtain from the *given* natural numbers, by mathematical constructions corresponding to intentional experiences, a mathematical structure that comes as close as possible to "capturing" our experience of the intuitive continuum mathematically so as to grant it *objective* status?

Weyl was aware, as was Husserl, of the gap between the description of the intuitive continuum in terms of *necessarily* vague and imprecise morphological concepts and the casting of the intuitively given into exact mathematical concepts (for an explanation of Husserl's distinction between *morphological* and *ideal* concepts see especially *Ideas I* (§74). In Chapter 2, §6 of *The Continuum*, together with an unmistakably phenomenological description of the intentional constitution of phenomenal time (which essentially reproduces Husserl's analysis as given in *Ideas I* (§§81-82) and the commitment to developing a mathematical theory of the continuum as close as possible to the intuitively given continuum, which is primarily displayed in phenomenal time (as opposed to objective, clock-measured time), Weyl manifests his awareness that this intuition, nonetheless, escapes the sharp boundaries of the mathematical concepts. The continuum, as experienced in intuition, does *not* have the structure of a *Weylsches Zahlsystem*, which is only a "symbolic reconstruction" of it that must, like a physical theory, pass the test of *experience*.

It is interesting to notice that Weyl apparently saw the relation of the mathematical theory of the continuum to the intuition of the continuum as being the same as the relation between physical theories and perceptual experience, which indicates that, for Weyl, intuition was, indeed, a form of perception. As we shall see below, Husserl's notion of intuition is precisely that, a generalization of the notion of perception. And the fact that Weyl also thought of intuition as a *seeing* and a *seeing that* is consistent with the hypothesis of a Husserlian influence.

Moreover, the strict adherence of Weyl's arithmetical theory to what had been disclosed in the phenomenological analysis of the intuitive continuum seems to conform to what Husserl calls the "principle of all principles" of phenomenology (which, we should note, Weyl quotes in "Insight and Reflection"):

> whatever presents itself in 'intuition' in primordial form (as it were in its bodily reality), is simply to be accepted as it gives itself out to be, though only within the limits in which it then presents itself. (*Ideas I*, §24)

This distinctive feature of *The Continuum* is, I believe, a strong enough indication that phenomenological ideas can help us achieve a deeper insight into the details of the theory presented in that beautiful book.

Weyl was driven towards the philosophy of Husserl after a brief flirtation with what he calls "conventionalism", a position that in his view involved an effort to clarify Zermelo's notion of *definite Klassenaussage* without presupposing the concept of natural number. A *definite Klassenaussage* was supposed to be a unary predicate derived from the basic relation of set membership by a few logical operations *finitely* iterated. Definition by recursion was, of course, excluded. The concept of natural number was not to be presupposed at all, and was expected to be introduced formally. (Weyl 1918a, chapter 1 §8)

This approach lead Weyl into a hopeless dead end from which, he says, he was rescued by a "general philosophy" (*allgemeine philosophische Erkenntnisse*) arrived at after turning his back on conventionalism. I claim that this "general philosophy" was very close to Husserl's philosophy, if not phenomenology itself. Weyl does not mention Husserl explicitly at this point, but there are a number of reasons supporting this identification. Certainly the most important is the substitution of a form of intuitionism for the conventionalist approach he had just given up. In *The Continuum*, Weyl embraces a notion of intuition as a form of givenness that plays a fundamental role in the ontology and epistemology of that work that is arguably derived from Husserl.

Weyl acknowledges the fact that his views in *The Continuum* bring him close to Poincaré, but Weyl's *explicit* rejection of the rest of Poincaré's philosophy suffices to rule out Poincaré's views as constituting that "general philosophy" that pushed Weyl into embracing a theory of knowledge founded on a notion of intuition. In what follows, I shall try to substantiate these claims, but before going into details I should like to make clear what precisely the essential features are of the phenomenological notion of intuition to which I shall be referring.[4]

First of all, this brand of intuition has nothing to do with a view of intuitions as mere sensations of a strictly subjective nature with no real objective reference, no matter how strong the sensations are. An intuitive experience is *not* a revelation of a hitherto hidden reality to a *passive* consciousness, a sort of mundane annunciation in which one is impregnated with truth without really knowing how. For Husserl, the concept of intuition is required on both the ontological and epistemological levels in order to ground the concepts of truth and knowledge, *and it is nothing more than a generalization of the notion of perception*. In my view, it is precisely this concept of intuition that was inherited by Weyl, and it is precisely this foundational role that the concept plays in the system of *The Continuum*.

Critics of intuitionistic approaches to knowledge tend to dismiss them as obscure and contaminated by a tinge of irrationalism. They believe that nothing can be rationally accepted as true independently of a justification

---

[4] See Levinas 1984 or Sokolowski 1994 for a more detailed analysis of the Husserlian notion of intuition.

that is essentially argumentative. Husserl—and also Weyl—sees the vicious-circle and infinite regress such a position leads to and accepts some sort of *immediate*, i.e. non-argumentative, justification as a primitive and founding form of rational justification. For Weyl, this was precisely the way out of his prior conventionalism.

Intuitions are those experiences in which consciousness is directed at an object (or objectivity, although the neologism "objectuality" would be preferable) that is not merely represented but actually given "in person". By "object" ("objectivity", "objectuality") Husserl understands just about anything that can be conceived as a bearer of attributions: physical, ideal or generic objects, states-of-affairs, judgments, various kinds of syntheses, nominalized attributes, and what have you. At the most basic level, intuition is nothing but sense perception, but even at this level it already requires some constitutive activity on the part of the perceiver in which an *object* is constituted as an identifiable unity out of a mass of sense data. Nonetheless, the *activity* of perceiving, in its traditional sense, is not yet directed by the cognitive interest that presides over the constitution of higher level objectivities such as those Husserl calls syntactic objectivities, i.e. ideal, non-independent objectivities constituted in the process of thinking. In order to be intuited (i.e. given "in person" to consciousness) these syntactic objectivities demand a series of constitutive experiences, abstraction, reflection, ideation, identification, and others that together articulate the intentional life of consciousness, in which those objectivities emerge in a way strictly similar to objects of perception. Hence, the intuition of higher-order objectivities is no more mysterious than sense perception, and denying the validity of the former as an object-giving experience is tantamount of denying that of the latter.

It is now clear that the intuition of any objectivity, including truth as the objective correlate of a synthesis between a state-of-affairs as intuited and the content of a judgment, demands a complex constitutive experience of consciousness and cannot be equated with a happy accident in which something is simply revealed. This elaborate, and original, notion of intuition is fundamental to the whole of Husserlian ontology and epistemology, in the realist as well as the transcendental idealist periods. Weyl, I believe, inherits this notion and gives it a twofold interpretation with respect to the realm of mathematical analysis, *realist* with respect to a basic class of objectivities (the natural numbers given in a sort of *immediate, unfounded* intuition, together with some basic truths about them) and *idealist* with respect to all higher-level objectivities constitutited on the basis of what is immediately given in intuition. In other words, for Weyl, natural numbers and relations among them are *independently* existing objectivities, whereas objectivities founded upon this given are not. As is clear, and as his criticism of Husserlian "solipsism" would require, Weyl does not import Husserl's notion of *epoché*, which would, of course, require the suspension of all ontological commitments.

In view of this, Weyl's conversion to Brouwerian intuitionism shortly after the publication of *The Continuum* comes as no surprise. Weyl was, I believe, probably unsatisfied with this double interpretation he gave to the notion of intuition in *The Continuum* and in the end favored the idealistic reading to the point of identifying it with Brouwerian constructivism.

For Weyl, intuition is, as it is for Husserl, characterized in most general terms as the experience, not necessarily passive, in which objectivities present themselves *in person*, not indirectly through the intermediary of mere intentions or signs. In Husserl, the intuition of higher order objectivities is, as he says, *founded* on lower level objectivities in the sense that these are required as a basis on which constitutive experiences are performed; for instance, the intuition of non-independent abstract moments such as color or shape by the action of abstraction on objects given in perception. The relation of foundation obviously requires the existence of basic elements given in some elementary and unfounded form of immediate intuition. For Husserl, this basic domain is constituted by the primordial data of passive perceptual awareness.

In Weyl we also find a clear delimitation of two classes of objectivities. On the one hand, we have the *closed* totality of the natural numbers (the basic objectivities) ordered according to the successor relation (the basic states-of-affairs) in which the elementary axioms of arithmetic, including the principle of complete induction, together with the principle of definition by recursion, (the basic facts) hold. On the other hand, an *open* domain of higher level objectivities constituted by cognitively oriented consciousness as *non-independent* entities founded on objectivities of the first realm. The former domain is presented in *immediate* intuition; the latter in *mediate*, or founded, intuition.

In Weyl the above distinction is marked by many features of an ontological, logical and epistemological nature. On the epistemological side, Weyl takes the primitive realm of objectivities as the exclusive focus of cognitive interest and he assumes that it alone constitutes the object of knowledge. The realm of higher order objectivities is constituted in the course of carrying out investigations focused on this privileged realm and is not itself an object of knowledge (obviously, this is enough to block ramification in the theory).

These higher order objectivities stand as noematic poles, i.e. objective counterparts of subjective, intentional experiences, which cannot exist independently of their noetic counterparts, i.e. the experiences in which they are constituted. This idealistic reading of the notion of constitution gives the system of *The Continuum* its constructive flavor, which, together with Weyl's implicit characterization of the class of meaningful assertions, imposes certain logical restrictions on the system.

According to Husserl no judgment can be true if its content is not fulfilled by intuition, that is, if what it expresses is not given *as expressed* in an intuitive experience. Our cognitive activity is, of course, directed at the pursuit of truth and hence our cognitively oriented activity of judging must be forbidden to bring forth judgments that are seen *a priori* as incapable of being *ideally* adequately fulfilled by intuition. This is, in short, the fundamental principle behind Husserl's theory of meaning, in accordance with which, Husserl distinguishes meaning with respect to *form* from meaning with respect to *matter*, the former being determined by *a priori* rules of compatibility of *logical types*, the latter by *a priori* rules of compatibility of *contents*. (Husserl 1900-01, LI IV §§10, 12, 14; LI VI ch.4)

Weyl takes up this notion of meaning with respect to matter and the associated notions of *material* sense and *material* counter-sense as central to his theory of meaning for declarative assertions. In Weyl, as in Husserl, if a judgment is meaningful with respect to matter, then what it expresses can *in principle* be an object of intuition, which is only another way of saying that we can see *a priori* that its content does not bring together things that are *conceptually* incompatible. Weyl gives us, in the first paragraph of *The Continuum*, an example of what he considers a meaningless statement: saying that an ethical value is green. It is meaningless because an ethical value cannot possibly be an object of visual perception and hence what the judgment expresses cannot be given in an appropriate experience of intuition (perception, in this case). This can be seen *a priori* simply by insight into the essential (in)compatibility of the contents of the terms of the judgment. Therefore, according to Weyl, attributes and objects cannot be put together *arbitrarily* as unified contents of meaningful judgments.

This conception of meaning showed Weyl the way out of the paradoxes that plagued the foundations of mathematics at that time, although, as a matter of fact, Weyl gives us only a very sketchy analysis of Grelling's paradox (Is "heterological" heterological?) before concluding that the question it poses is devoid of sense.

Nonetheless, it is not difficult to envisage what would in general be Weyl's solution for the self-referential paradoxes. Given any predicate-name P and any object-name a there is no *formal* restriction in forming the judgment Pa. Weyl thinks that the formal correctness of this judgment, is enough to persuade anyone willing to equate sense with formal sense to think that Pa expresses a proper judgment to which a state-of-affairs does or does not correspond (*tertium non datur*). In certain cases, this leads to paradoxes such as Grelling's or Russell's, which theories that possess only a purely formal theory of meaning regard as insoluble contradictions. Weyl's solution, or rather dissolution, of paradoxes of this sort requires a restriction of scope of predicates (and relations). Not an *ad hoc* restriction arbitrarily devised to circumvent problematic attributions of the property in question but, rather, one determined by a clear-cut notion of *material* sense. For Weyl, extensions of properties or relations admit only those objects for which the attribution of the property or relation in question constitutes proper judgments, i.e. those denoting *possible* states-of-affairs, the insertion of the state-of-affairs into the realm of the experienciable or intuitable being the determining criterion.[5] The contradictions exposed by the paradoxes are symptoms that the limits of (material) sense have been surpassed and, consequently, that the classical logical principles (such as *tertium non datur*) have ceased to be valid. The validity of classical principles of reasoning, for Weyl, demands not only *formal*, but also *material* meaningfulness. What I want to stress here is that Weyl's theory of (material) meaning is teleologically oriented by the *ideal* of the adequate fulfilling of intentions expressed by judgments by appropriate intuitions. Here, I believe, Weyl's theory of meaning is clearly indebted to Husserl's.

Now, does the understanding that a judgment is meaningful if it is in principle experienceable (i.e. provided the intuitive filling of its intentional content is in principle possible) mean that it has an intrinsic truth value, either the true or the false, independently of our being *effectively* capable of going through those intuitive experiences? For Weyl, an assertion has attached to it a unique truth value, either truth or falsehood, if we can go through the complex experiences by which the facts of reality are experienced as either fulfilling or frustrating the propositional content expressed in the assertion. Whenever I say that S is p, and p is an attribute that is not incongruous with S on conceptual grounds, then S is p is a meaningful assertion. Now, S is p has a truth value attached to it if, and only if, either the objectivity $S(p)$, i.e. S *as determined by p*, or S *(not-p)*, belongs to the field of objectivities that can be presented to me in an intuitive experience. If we take this "we can" in the widest possible sense and understand that *we can* experience whatever is not *a priori seen* as non-experienciable, then the content of any meaningful assertion can be experienced and meaningful assertions are precisely those with true-false polarity.

However, we may want to give this "can" a narrower interpretation, closer to what is actually or potentially feasible. Obviously, the validity of *tertium non datur* will depend upon how we interpret this modality, upon how powerful we are in bringing about intuitive experiences, or even upon to whom we are refer by "we", our limited selves or some idealized version of ourselves. In *Formal and Transcendental Logic* (Husserl 1929) Husserl gave an overview of the presuppositions involved in the acceptance of *tertium non datur* and other principles of logic.[6]

Let us consider how Weyl dealt with these questions. For him, the basic domain of the natural numbers is a *closed* domain existing independently, and hence, given enough time, perhaps even an *infinite* amount of time (or else, super-tasks being admitted)–which is, of course, an idealization

---

[5] 'Experienciable' or 'intuitable', in this context does not mean that which is or can be *effectively* included in our field of experiences or intuitions. These expressions only denote that which is not excluded *a priori* from being included therein on the grounds of conceptual incongruity. A state-of-affairs is in principle intuitable if it is not non-intuitable, conceptual incongruity being the criterion of non-intuitability.

–any numerical assertion can eventually be decided. But the same is not true with respect to the derived domains of objectivities, constituted in the course of our investigations, and hence *open* and in constant determination. Our efforts to decide assertions involving free variables ranging over them will be frustrated no matter how much time we have, even if it is an infinite amount of time. The point is that while we are busy carrying out our investigations, new derived objectivities are constantly popping up, and so assertions with free variables ranging over derived domains cannot be decided once and for all. The derived domains of objectivities are seen, so to speak, as open fields of ever-renewed possibilities. The conclusion is that numerical assertions, i.e. those whose variables are exclusively restricted to the domain of natural numbers have a truth value attached to them whereas "transfinite" assertions, i.e. those whose variables may range over derived domains of objectivities, and so incapable of being in general fulfilled by adequate intuition, do not. Here Weyl has a choice to make, either to allow transfinite assertions and give up classical logical principles, or else to preserve those principles and ban transfinite assertions. Weyl takes the second alternative and imposes a restriction on which expressions are allowed in carrying out our investigations focused on the basic domain, namely, only those expressions whose variables range exclusively over that basic domain. Weyl calls this the *narrower procedure*. His justification for it is as follows:

> To decide the validity of finite judgments, i.e. those in conformity with the narrower procedure, we have only to take under consideration the objects belonging to the basic domain, whereas for 'transfinite' judgments we must also survey, besides those the totality of derived properties and relations too. (Weyl 1918a, ch. 1 §6)

Weyl is basically telling us is that the basic and the higher order domains have different ontological status, which entails that whereas the possibility of an intuitive fulfilling is granted for any numerical assertion by the *facts* subsisting independently in the basic domain completely determined in itself once and for all, the possibility of intuitively filling assertions quantifying over the totality of higher order non-independent objectivities cannot be guaranteed *a priori*. Hence transfinite assertions, which are not in conformity with the "narrower procedure", have no intrinsic truth value and cannot appear in our cognitively oriented investigations directed at the basic domain.

Now we can see how much Weyl's conversion to that "general philosophy" centered on the notion of intuition is responsible not only for his theory of meaning, and consequently his answer to the antinomies, viz., they are mere consequences of attributing meaning to meaningless expressions, but also for his theory of truth and, consequently, for the logical restrictions imposed on the system of *The Continuum*. Moreover, given his particular interpretation of the intuition of higher order objectivities, the ontology of *The Continuum* is also a consequence of this "intuitionist" turn.

The distinction between a realm of objects that is *given* prior to the activity of judging (the pre-predicative realm of being) and an open domain in constant determination of non-independent objectivities constituted in the activity of judging echoes Husserl's distinction between *substrative* and *syntactical*[6] categories of objects as expressed in *Ideas I* §11. There Husserl distinguishes between syntactical substrata of judgments, which we can call terms, and syntactical forms and, correlatively, between substrata, that is, objects to which syntactical substrata refer, and syntactical objectivities, i.e. objects "which are mere correlates of the functions of thought". A syntactical objectivity can, of course, also be a substratum in a higher order level judgment, but according to Husserl we must admit, on pain of an infinite regress, that there are substrata that are syntactically simple. These are what Husserl calls the *ultimate* substrata, which are, of course, in Weyl's mathematical analysis, the natural numbers.

The point to be noticed here is that syntactical objectivities are constituted *as intentional objects in the activity of judging*. Their constitution as intuited objects will demand, of course, the experiences by which the intention of the judgment is fulfilled. By judging, we are either rendering an object more explicitly by attributing certain predicates to it, or relating it to other objects by means of external relations. In either case, syntactical objectivities are posited: qualities, qualitatively determined objects, relations, pluralities, etc. In *Experience and Judgment* (Husserl 1939a), this theme is extensively analyzed, but it is already clearly stated in *Ideas I*, and in *Logical Investigations* similar distinctions are drawn, for instance in the VIth Investigation, between real (which in that context just means perceived by the senses) and ideal (or categorial) objectivities.

All these characteristic features of Husserl's ontology are present in Weyl's *The Continuum*, namely, a distinction between independent and non-independent categories of objectivities, a notion of foundation as an ontological relation between objectivities of different categories, and the view that non-independent objectivities are posited in the activity of judging. All this, together with the distinction between a merely intentional objectivity and an intuited one and the role given to intuitive experiences with respect to the notions of meaning, truth and knowledge, which are implicit in *The Continuum*, strengthen my interpretation that it was indeed Weyl's conversion to Husserl's "general philosophy" that

---

[6] Husserl uses the term 'syntactical' in order to emphasize the parallel between ontological categories and categories of meaning. The distinction between substrative and syntactical categories of objects corresponds to the distinction between elements (*Stoffen*) or syntactical substrata and syntactical forms in the formal theory of meaning. This does *not* mean that syntactical objectivities are 'mere' linguistic structures.

determined most of the epistemological, logical and ontological aspects of his approach to the foundations of analysis.

One might say that the distinction that Weyl draws between basic elements and ideal objects comes from Hilbert and owes nothing to Husserl's ontology. The reason I do not accept that view rests on the fact that, for Weyl, ideal elements are not purely *formal* entities, but *intuitable* ones. Hence, contrary to Hilbert's ideal statements, Weyl's assertions involving ideal elements (provided there is no quantification over them) also have intuitive content.

Weyl's allegiance to Husserl's notion of truth as itself an objectivity that must be given either in immediate intuition, in the case of the basic truths of arithmetic, or in founded intuition, in the case of the rest of the truths of mathematical analysis, together with his idealistic interpretation of the intuition of higher-order objectivities, impose, as already observed, a choice upon Weyl: either adopt the so-called narrower procedure or give up classical logic. Extending the validity of *tertium non datur* to general assertions about the basic as well as the derived domains of objectivities would require a notion of truth that would be independent of intuitability conditions, i.e. a notion inconsistent with the one implicit in *The Continuum* and which is, as I have been arguing, derived from Husserl's epistemology. The fact that Weyl chose to accept the restrictions imposed by the narrower procedure shows that he was not willing to abandon either classical logical laws or the constructive notion of truth that his particular interpretation of Husserl's notion of intuition imposed on him.

The idealistic reading of Husserl to which Weyl is committed is clearly stated in the second chapter §6 of *The Continuum*: "Being is and can only be given as the intentional content of the mental experiences of a pure I that bestows meaning (*Dasein nur gegeben ist und gegeben sein kann als intentionaler Inhalt der Bewußtseinserlebnisse eines reinen, sinngebenden Ich*)". This is probably the clearest sign of Weyl's allegiance in *The Continuum* to Husserl's phenomenological standpoint.

Nonetheless, and I must insist on this, despite its debt to phenomenological ideas, the system of *The Continuum* cannot be seen as a prototype of how the whole of mathematics should be developed from the phenomenological perspective. The phenomenology of knowledge takes seriously the dialectics of *presence* and *absence* in the dynamics of knowledge and mathematics too, as seen by phenomenology, is permeated by it. Mathematics is a form of knowledge of *actual* being *as well as possible* realms of being. The mathematics of presence, so to speak, does not demand exclusivity and, consequently, phenomenology does not rule out purely formal, non-constructive, intuitively empty forms of mathematics.[7]

Husserl's notion of intuition, moreover, is not a matter of all or nothing, an objectivity being or not being intuited, with no alternative in-

---

[6] See Chapter 3.

between. Intuitions can present degrees of clearness in a continuous scale ranging from complete absence to full presence. Analogously, the degree of fulfillment of empty intentions by corresponding intuitions can also vary continuously in a similar way. Hence, knowledge is not always certain and unrevisable. It can come in partial forms and be perfected—provided we accept, of course, Husserl's characterization of knowledge as the fulfillment of empty intentions by intuitions and *certain* knowledge as the fulfillment of intentions by *adequate* intuitions.

Mathematical knowledge is no exception, it can also range from a *proper* form of knowledge in which merely distinct judgments are always also rendered clear by means of appropriate intuitions—as is probably the case in the realms of constructive mathematics—to merely symbolic reasoning of the type carried out in the more formal provinces of mathematics. Husserl never, as far as I know, either in print or in lectures, advocated a restriction of mathematical knowledge within strictly intuitive boundaries. Quite the opposite, he appreciated the relevance of the study of formal non-interpreted systems and their metamathematics (which he saw as belonging to a province of *logic*, that of *formal ontology*, see *Formal and Transcendental Logic*). So, Husserl would not oppose the development of formal or axiomatic theories of mathematical analysis.

And neither would the more traditional Weyl, who was responsible for the development of many mathematical theories that are not given the intuitionist treatment of *The Continuum*. Whether these merely formal theories can actually *describe* some intuitively given realm of being is an altogether different matter. Husserl appreciated the accomplishments of formal mathematics in describing *possible* realms of being, but would certainly refrain from accepting most of them as expressing *knowledge* about *actual* realms of being. I believe that these distinctions are useful in reconciling Husserl's epistemology, which is cast in terms of presence and absence (of intuitions), with his acceptance of mathematical practice as he found it.[8]

In general, Weyl would probably agree with Husserl, but in *The Continuum* he was after a theory founded on presence. The constructive character of the arithmetical theory of the continuum in *The Continuum* was determined by Weyl's commitment to carrying out a description (as faithfully as possible) in mathematical terms, of an intuitively given continuum. From a *strictly formal* perspective, Weyl was probably quite satisfied with classical theories of mathematical analysis, but from this perspective the epistemological goal he imposed upon himself in *The Continuum* was unattainable. Weyl's adherence to his goals conditions a constructive development of mathematical analysis in order to reflect

---

[8] Although Husserl was always aware, and would become even more so towards the end of his life, of the danger posed by "formalist alienation". (See Chapter 17)

the intentional constitution of the continuum given in intuition, that of phenomenal, immanent time that emanates from consciousness.

Weyl's *predicative* analysis is the result of the following presuppositions. First, real numbers are the mathematical counterparts of temporal instants. Second, temporal instants do not exist in themselves, i.e. independently. Consequently, the only admissible real numbers are those that are "constructed", i.e. definable by arithmetical relations. Predicative mathematics in general is not, as I believe my comments above show, a necessary outcome of a Husserlian philosophy of mathematics, although, I believe, Weyl's predicative analysis can be seen as following from a particular reading of Husserlian constitutive phenomenology coupled with a strict adherence to the Husserlian epistemological principle of the primacy of intuition in the grounding of *knowledge*.

Weyl's "constitution" of real numbers as intentional objects from the intuitively given domain of natural numbers harbors a peculiar interpretation of what the intuition of a real number might amount to that deserves some attention. I want to stress that Weyl's views on this question cannot be seen as faithfully Husserlian and that alternative interpretations of the notion of intuition may either restrict even more the system of *The Continuum* or, on the contrary, expand it into less constructive versions.

For Weyl, to intuit a real number is basically to define it by means of arithmetical relations. A similar assertion also holds for numerical functions. Using Husserlian terminology, a specific real (or more to the point, irrational) number stands as the objective focus of intentional experience(s) in which it appears as an identifiable unity and of which the defining relations(s) constitute the characteristic sense (noematic sense). It is the defining relation that is responsible for the intuitive experience of picking *this* (and not another) number which then comes into being as having the property the relation expresses. The number, nonetheless, can also be singled out by any other extensionally equivalent relation, in which case it is the *same* number but with another sense, as if seen from a different perspective. This is the principle of extensionality, which says that despite the fact that the intuition of a real number requires a subjective constitutive experience, what is given in intuition stands as a fully objective, although non-independent, entity. This is enough, I believe, to free Weyl's constructivism from readings of a psychologist orientation.

The requirement that defining relations should be *arithmetical* is, of course, a consequence of Weyl's assumption that the domain of natural numbers is given in immediate intuition. As already pointed out, such an assumption does not square with Husserl's views on ideal objects. So, the theory of *The Continuum*, despite inviting an analysis in terms of phenomenological notions, cannot be seen as a natural mathematical development of what might be called the Husserlian ontology and epistemology of being and truth *in presence*.

From the noetic (i.e. subjective) side, the defining relations can be seen as prescriptions to be followed in order to bring real numbers into consciousness. If we choose to see them in this way we must cope with an idealization that more orthodox phenomenologists would find difficult to accept, viz. that an *actually infinite* sequence of steps can be carried out by consciousness in the task of intuiting, in an *adequate* way, an infinite set. A more conventional reading of Husserl's ideas on the intuition of sets seems to indicate that infinite sets can at best be only *inadequately* (i.e. partially) intuited.

Another feature of Weyl's theory that may seem at odds with phenomenological ideas is his refusal to iterate the constitutive process in order to obtain a hierarchy of higher order numbers. After all, Husserl himself did accept this as a legitimate procedure. But, contradictory as it may seen, I believe that it was precisely Weyl's strict compliance with Husserl's requirement of fidelity to our primordial intuitions that determined such an outcome. Instants in the continuous flux of time are not, in fact, distributed in a nested hierarchy. This, together with more technical reasons such as the need for some Russell-type axiom of reducibility, seem to have been responsible for the absence of stratification in Weyl's theory.

II

Before closing, I want to give an overview of the reconstruction of classical real analysis carried on in *The Continuum* from the perspective of the interpretation sketched here.[9]

Immediate, unfounded intuition forces upon Weyl the full complete domain of natural numbers ordered according to the successor relation, the basic elementary axioms of the arithmetic of natural numbers, the principle of definition by recursion and the principle of complete induction.

In the sense that not only objects (the natural numbers), but also a basic relation among them and basic axioms and rules of reasoning are given, one can say that intuition gives Weyl a privileged language with which to describe and reason about the domain of natural numbers.

Weyl's adherence to a view of mathematics based on intuition is responsible for his complete disregard of formalization in *The Continuum*. Proofs are not seen as rigorous substitutes for vague intuitions, but as chains of intuitive experiences constituting a complex intuitive experience. Proofs are not mere symbolic manipulations within a system, but intuitive presentations of truth. Mathematical formal rigor cannot replace intuition, which must remain the ultimate ground of justification. (Weyl 1918a, ch. 1 §3)

---

[9] For a detailed presentation of the system of *The Continuum* from a strictly mathematical perspective, see Feferman 1988.

Given the basic domain of objects and facts and the validity of certain forms of reasoning, such as the principle of complete induction, Weyl specifies, then, how to derive through a family of principles of definition complex derived relations by mens of which to express possible state-of-affairs in the basic domain.

Of course, one cannot expect a theory of the continuum, i.e. classical analysis, if natural numbers are the only available objects. So Weyl has to have ways of obtaining complex objects such as rational and irrational numbers, sets, and functions. In the classical construction of real numbers (rationals or irrationals), they are usually introduced as sets of pairs of natural numbers (for the rationals) or cuts, sets of rational numbers (for the irrationals).

Hence sets are the objects that Weyl has to obtain. As they are not given in the original intuition they can only be constituted as founded by an object-positing act of consciousness. We may call such an act *collecting*. Naive phenomenological analysis of this act tells us that: first, collecting consciousness has to collect something; second, the collection has to stand as an identifiable unity; and, third, and more importantly, the collecting has to be guided by some principle of choice.

The only primitive objects Weyl has are the natural numbers and the only language where collecting-guiding rules can be stated is that determined by the original intuition. Therefore, for Weyl, the only sets admitted are sets of (n-tuple of) natural numbers that are *definable* by (primitive or derived n-ary) relations in the language.

From a strictly mathematical point of view, the condition of definability is justified by the fact that, as Weyl sees it, ideal elements (such as sets) are introduced in mathematics in such a way that assertions involving them can always be translated into assertions involving certain relations only among "real" elements (Weyl 1918a, ch. 1 §8). It is important to notice that sets are not identified with the defining relations but extensionally determined by them. Weyl explicitly states the principle of comprehension: to any n-ary relation *corresponds* a n-ary set (a set of n-tuples of natural numbers) and the principle of extensionality: two sets are equal if, and only if, they have the same elements, independent of whether they are or are not determined by the same relation.

It is then clear that Weyl is not looking for any *nominalistic* reduction. These principles bear witness to the fact that sets are irreducible objects, although in the words of Husserl, founded, *non-independent ones,* whose *fully granted existence* is nonetheless, now in Weyl's words, *not in themselves.* (Weyl 1918a, ch. 1 §1)

Nor is Weyl simply presupposing their existence in the manner of *conventionalism*, or *admitting* their independent existence, as in ontological *Platonism*. Sets, as ideal objects, are *given by consciousness.* As a result, Weyl was very critical of Cantorian set theory. What distinguishes Weyl from other constructivists such as Kant or Brouwer is that the "constructions" involved are not supposed to be carried out in *finitely* many steps. The arithmetical relation defining sets of natural numbers are recipes for actually obtaining them intuitively, only not necessarily in a finite amount of time.

In analysis based on set theory, a numerical function is nothing but a particular kind of set of pairs of numbers. In an analysis founded upon careful phenomenological inquiry, such as Weyl's, functions are ideal objects of an altogether different kind: objective (noematic) poles corresponding to a peculiar subjective (noetic) act. We may call such an act *correlating,* and since it does not, if taken in its purity, involve collecting, the object it posits, a function, cannot be reduced to a set. Moreover functions also have to be *explicitly defined* by binary relations, and again this is so because correlation always has to be guided by a principle. Here Weyl takes the concept of function back to a point prior to Dirichlet, whose notion he considered "vague". (Weyl 1918a, ch. 1 §4)

It is interesting to note that Weyl consistently claimed that it is not evident that an infinite set must contain a denumerable subset, foreseeing a result proved by Paul J. Cohen only in 1963 (Weyl 1918a, ch. 2 §5). Weyl refused to accept the *absolute* point of view of classical analysis where sets and functions that cannot be characterized are quite legitimate. *Ideal existence demands consciousness.*

The condition of definability in a well-defined language implies that an object of higher type not only has a name in the language, or some loose characterization that does not enable us to tell it apart from other objects of a certain class, but a *corporeality* given by a syntactic structure by means of which it is perfectly identifiable. This does not mean that a higher order object is *only* a syntactic structure. Objects of higher types are non-independent ideal objects *corresponding to* the linguistic expressions that define them. Linguistic expressibility grants objective availability; language brings objects from within individual consciousness to collective consciousness. This is reminiscent of Husserl's claim that language provides a body for ideal objects (Husserl 1939b). Incidentally, in this respect Weyl differs sharply from constructivists of Brouwer's school, for whom language is at best only a vehicle for communication.

Since ideal objects such as real numbers and functions do not have an *independent* life, quantification over them is not allowed. Existence is determined by definition, which, as Weyl sees it, is a recipe for intuition to be brought about: by defining higher order objects appropriately we bring them to *distinct* consciousness, opening concomitantly the possibility for *clear* consciousness, i.e. intuition. Higher order entities that are not individually defined do not exist (unlike natural numbers, which exist without being individually identified). Of course, Weyl could very well admit quantification over, say, real numbers already defined, but considerations of a technical and *epistemological* nature rule this out. To cope with such a restriction Weyl's analysis has to be confined to the theory of *continuous* functions.

In *The Continuum*, immanent time is the standard model of an intuitively given continuum and Weyl's scrutiny of it reproduces and refers to Husserl's analysis as given in *Ideas I* §§81-82. If it is this intuitive continuum that analysis has to model then, concludes Weyl, classical analysis is unacceptable. The very notion of a point in the continuum lacks intuitive support. So, if real numbers are seen as formal counterparts of instants, they cannot be given independent existence.

I believe that we can conclude that Husserl's phenomenology does indeed constitute an important and perhaps essential component of the philosophical environment in which Weyl carries out his reconstruction of classical mathematical analysis, not only by providing its fundamental notion, that of a general form of perception, but also by shaping its technical features. Later in life, as I have already pointed out, Weyl became critical of Husserl and, under the influence of Brouwer, abandoned the approach to the foundation of mathematics pursued in *The Continuum* in favor of a more strictly constructivist position like Brouwer's intuitionism. However, this too was also eventually abandoned. But the ability to change his mind when he saw it necessary was only one of Weyl's many virtues.

References

Feferman, Solomon 1988. "Weyl Vindicated: 'The Continuum' 70 Years Later'", *Atti del Congresso Temi e Prospettive della Logica e della Filosofia della Scienza Contemporanee* vol.1, Bologna: CLUEB, 59-93.

Husserl, Edmund 1900-01. *Logische Untersuchungen*, Halle: Niemeyer, in English, *Logical Investigations*, New York: Humanities Press, 1970.

Husserl, Edmund 1913. *Ideen zu einer reinen Phänomenologie und phänomenologischen Philosophie, I,* Husserliana II, The Hague: Martinus Nijhoff, 1950, in English, *Ideas, General Introduction to Pure Phenomenology*, translated by W. R. Boyce Gibson New York: Collier, 1962 (1931).

Husserl, Edmund 1929. *Formale und transzendentale Logik*, Husserliana XVII, 2nd ed., The Hague: Martinus Nijhoff, 1974, in English *Formal and Transcendental Logic*, The Hague: Martinus Nijhoff, 1969.

Husserl, Edmund 1931. *Cartesian Meditations: An Introduction to Phenomenology*, The Hague: Martinus Nijhoff, 1973, translation of *Cartesianische Meditationen* in Husserliana I.

Husserl, Edmund 1939. *Erfahrung und Urteil*, Claassen & Goverts, Hamburg, 1948, in English, *Experience and Judgment*, London: Routledge and Kegan Paul, 1973.

Husserl, Edmund 1939b, "Die Frage nach dem Ursprung der Geometrie als intentional-historiches Problem", *Revue internationale de philosophie* 1, 1939, 98-109.

Levinas, Emmanuel 1984. *Théorie de l'intuition dans la phénoménologie de Husserl*, 5th ed., Paris: J. Vrin.
Sokolowski, Robert 1994. "Le concept Husserlien d'intuition catégoriale", *Études Phénoménologiques* 19, 39-61.
Weyl, Hermann 1918a. *Das Kontinuum, Kritische Untersuchungen über die Grundlagen der Analysis*, in English, *The Continuum*, New York: Chelsea, 1973.
Weyl, Hermann 1918b. *Raum, Zeit, Materie*, Berlin: Springer, 7th ed., 1988, in English, *Space, Time and Matter*, New York: Dover, 4th ed. 1952.
Weyl, Hermann 1955. "Erkenntnis und Besinnung (Ein Lebensrückblick)", *Studia Philosophica* 15, in English, "Insight and Reflection", *Mind and Nature, Selected Writings on Philosophy, Mathematics and Physics*, Princeton NJ: Princeton University Press, 2009.
Weyl. Hermann 1949. *Philosophy of Mathematics and Natural Science*, New York: Atheneum, 1963.
Weyl, Hermann 1994. "Der circulus vitiosus in der heutigen Begründung der Analysis", *Jahresbericht der Deutschen Mathematikervereinigung* 28, in English, "The *Circulus Vitiosus* in the Current Foundation of Analysis", in *The Continuum: A Critical Examination of the Foundation of Analysis*, translated by S. Pollard and T. Bole, New York: Dover.

14

Jairo José da Silva

HUSSERL AND THE PRINCIPLE OF BIVALENCE

Every distinct judgment is *in itself* either true or false, *tertium non datur*. This is the principle of bivalence, a basic principle of truth-logic. Its most striking feature is to give *any* proper judgment a *definite* truth value independently of verification. Given any judgment p, either p is true or p is false, even if the subject has not experienced either the adequation or the conflict of the content of p with the facts of reality. Does this not contradict the most basic tenet of Husserl's epistemology, that no judgment can be *declared* true (or false) if not actually experienced as true (or false)? Does the validity of bivalence not then harbor presuppositions of some sort?

I want to argue here that the first question must be answered in the negative by pointing out that the concept of *intrinsic* truth (which can but only *in principle* be verified) does not coincide with that of *verifiable* truth (which can be *effectively* verified) and that the second can only admit a positive answer if these presuppositions are properly understood as tied up to the intentional constitution of a unitary field of experiences possible in principle (and correlatively of a "world", which although experienceable in principle, may not be so in fact). In addition, in the spirit of *Experience and Judgment* (Husserl 1939), I bring out the hidden presuppositions that, by acting at the level of passive receptivity, are at the root of the presuppositions, effective at the judicative level, embodied in the principle of bivalence in its two versions, objective and subjective. From this perspective, I close with a critical analysis of views on the matter put forward by some philosophers who have approached it, Michael Dummett and Willard van Orman Quine in particular.

The project of delimiting the field and scope of logic, which Husserl launched with the "Prolegomena" to his *Logical Investigations* (Husserl 1900-01)–written before 1896, published in 1900–was only completed with the publication of *Formal and Transcendental Logic* (Husserl 1929). As Husserl argued in this work, formal logic contains two distinct but intimately related domains–*apophansis*, the theory of predicative judgments (taken as ideal entities, not mere noetic "occurrences"), and *formal ontology*, the a priori theory of objects simply as such–each containing essentially three levels: morphology at the base and, at the top, what we would call today metalogic, the theory of formal theories, on the apophantic side, and the theory of their objective correlates, formal domains, on the ontological side.

But, for Husserl, this was only the beginning. A philosophical grounding of logic was also required: a complete clarification of the origins of the predicative judgment in pre-predicative experience and the life-world (*Lebenswelt*) that it presupposes; the intentional constitution, on the basis of this *original* life-world, of a higher-level realm of experience and its objective correlate, the "world", whose *active* apprehension and *predicative determination* is the goal of a cognitively motivated I, and the (usually hidden) presuppositions and idealizations such a constitution involves. This part of the project was essentially carried out in the posthumous *Experience and Judgment* (Husserl 1939), although some of the problems it touches upon had already been addressed in *Formal and Transcendental Logic*. Such a radical investigation could not spare logical principles themselves, whose claim to universal validity, particularly in the scientific realm, also had to be clarified, grounded and ultimately justified. To criticize the principles of logic, a task Husserl imposed upon himself, is not, for him, to doubt them or deny their validity, but "to uncover the hidden presuppositions implicit in them". (Husserl 1929, §87)

I shall concentrate here exclusively on the principle of bivalence (or *tertium non datur*), valid in what Husserl called truth-logic, the domain of apophansis that considers judgments not only as objective senses, but as claims to truth.[1] The task I impose upon myself here is to render explicit Husserl's criticism, in the sense just established, of this principle, the presuppositions it harbors and the nature of these presuppositions.

Tertium non datur states, in its *objective* version, that any proper judgment has a definite truth value–either the true or the false– regardless of whether we are able, *now or ever*, to decide which. Husserl himself

---

[1] In truth-logic "the judgments are thought of from the very beginning not as mere judgments, but as judgments pervaded by a dominant *cognitional striving*, as meanings that have to become *fulfilled*, that are not objects by themselves, like the data arising from mere distinctness, but passages to the 'truths' themselves that are to be attained". (Husserl 1929, §19)

noticed how baffling this principle is.² Why should one of the two truth values, the true or the false, invariably stick by right to any judgment, taken in a proper sense, when truth and falsity are not constitutive features of judgments *qua* judgments? No wonder intuitionists reject the principle. But, for Husserl, the validity of bivalence is not at stake.³ The philosopher's task, he thought, is not one of reforming logic, but rather of clarifying its meaning, and, ultimately, uncovering the sources of validity of its principles.⁴

Whenever Husserl directed his attention to the sciences, empirical and a priori, he accepted them as creations in no need of reform, only clarification.⁵ In his last published work, *The Crisis of European Science and Transcendental Phenomenology* (Husserl 1936), for example, although detecting the alienation of modern science vis-à-vis the *Lebenswelt*, Husserl never questioned its technical accomplishments, asking only that its methods be philosophically clarified. In his philosophy of geometry, also an exercise in clarification and understanding,⁶ Husserl showed how, from immediate sensorial perceptions, intentionally motivated psycho-physical functions (esthetic acts) and higher level intentional categorial acts constitute a "world", physical space, that although extrapolating substantially the limits of *immediate* sensorial perception is *in fact* the world in which we live and which Euclidean geometry correctly describes (as far as our perception can tell, but which can also be described by non-Euclidean variants of geometry, if scientific conveniences, not only immediate sensorial perception, are also taken into consideration– provided these geometries are *locally* Euclidean, so that perception and science are not in conflict).

This much is worth remembering: Physical space is not simply sensorial space, in any of its guises (visual, kinesthetic, etc.); empirical reality is not only what meets the senses.⁷ "Words" are intentional constructs that harbor idealizations, opening up to internal and external horizons, regions maybe not yet trod upon, but that cannot be put *a priori* out of reach of *adequate* intuitive experiencing (the important word here

---

² "*In itself every judgment is decided*; its predicate truth, or its predicate falsity, '*belongs*' to its essence–though... it is not a constituent mark of any judgment as a judgment. This is very remarkable". (Husserl 1929, §79)
³ "The following pertains *a priori* to every proposition: each one is true or false". (Husserl 1939 §73)
⁴ Of course, this poses a problem of interpretation: How are we to harmonize such a task with Husserl's epistemology, centered as it is on a notion of truth based on intuitive fulfillment, which seems to be incompatible with *tertium non datur*? This paper intends to offer a solution to this dilemma.
⁵ See Husserl 1939 §10.
⁶ Husserl's philosophy of geometry is essentially contained in the second part of *Studien zur Arithmetik und Geometrie* (Husserl 1983). See my "Husserl on Geometry and Spatial Representation", originally published in *Axiomathes* 22(1):5-30, 2012, anthologized in this book (Chapter 2)
⁷ See Husserl 1939, §10.

is "a priori"). Empirical reality, but also the "worlds" of science are always constituted as *a priori capable of being experienced; otherwise, they would not be of interest for science*. This is a fact that logic, the theory of science, cannot ignore, and the principle of bivalence is how logic takes this into account. Bivalence, we could say, is a logically valid instrument for "peeking" into the recesses of scientific domains even before we have any idea of how actually to get there.

In Husserl, as we very well know, truth cannot be conceived independently of facts and experience. To state the truth of p is to state that the judicative content of p, that which p asserts to be the case, the situation or state-of-affairs depicted by p, is *indeed* the case, i.e. it is a *fact* of reality, and so *can* be the content of an intuitive act of presentation. This much is indisputable. The *nature* of this modality, however, demands clarification.

Bivalence tells us that, given any *proper* judgment p, whose content pictures a *possible* situation in a "world" W (p states a *possibility* as far as W is concerned, or better, as far as W is *conceived*), *either p* is true *or p* is false, in which case *not-p* is true, that is, *either* the situation depicted by p or that depicted by *not-p* is a *fact* of W. This implies that either an *experience* of *harmony* or one of *conflict* of the situation depicted by p vis-à-vis the facts of W is an evidential experience *possible* to the judging subject (the subject is not *a priori* doomed to be forever in the dark as to this particular aspect of W–but, *nota bene*, eternal ignorance with respect to a particular aspect of the "world", although never a matter of *principle*, may very well be a matter of *fact*).

There are so many undefined terms in these formulations that it is not possible to know what they really express. But before any clear sense can be given to them, many questions must be answered, here are some: when does a judgment depict a *possible* fact of W? What are *proper* judgments anyway? What are the criteria of *possibility* for evidential experiences of a judging subject in general? Are these questions to be answered a priori, both objectively and subjectively, that is independently of an investigation of W or the cognitive possibilities of the subject? If a priori, do they require any *presupposition* concerning W or the subject (or both), and if so, what is the nature of these presuppositions? Can they be factual *hypotheses* that may *actually* be false?

The clarification of the principle of bivalence must then occur on two fronts simultaneously; objectively, vis-à-vis the "world" W, and subjectively, vis-à-vis the judging subject. Accordingly, Husserl recognizes two versions of the principle (notice, they are versions of the *same* principle, not two different, although closely related principles), one objective, another subjective. Objectively, the principle asserts that any proper or *well-formed* judgment (this is already a first clarification of the sense of a proper judgment) is *in-itself* true or *in-itself* false; subjectively, that any well-formed judgment can be verified by a knowing subject in general. By passing from the objective to the subjective version of the principle, Husserl brings to the fore the most basic presupposition regarding bivalence. It is, in fact, by *presupposing* that well-formed judgments are *verifiable* that we can assert, says Husserl, that they possess intrinsic truth values, i.e. that they are *in-themselves* either true or false. But to what extent, if at all, does this presupposition involve our cognitive abilities, understood as what we, knowing subjects, are *actually capable* of doing, i.e. the evidential experiences that we are actually capable of *implementing*?

As I believe Husserl makes clear, this presupposition does not, in any robust sense, involve our cognitive abilities, our capability of *establishing*, if we cared so to do, *either* the veracity *or* the falsity of any proper judgment. The presupposition of *decidability* behind bivalence has *nothing to do* with the *hypothesis* that any well-posed question can be *effectively* answered (hypothesis so eloquently expressed in Hilbert's epistemological optimism by his "*Wir müssen wissen, wir werden wissen*", "We must know, we shall know") (Hilbert 1930, 1165). As so often, if not always in science, *idealizations* are involved. The "can" of these presuppositions is an *ideal* "can". By saying that we *can* verify any proper judgment we are simply saying that this verification is not *a priori* (emphasis on "a priori") excluded from the field of our evidential experiences, that this evidencing experience, even perhaps beyond our present possibilities, can conceivably be thought of as actualizable in an *open* future (emphasis on "open").

The actual verification may be only an ideal, but, even so, it still plays an important *practical* role in the dynamics of knowledge: it pushes us into actually fulfilling it, not unlike the carrot hanging in front of the ass' nose. Idealized experiences of verification are *representations* in which cognitive consciousness depicts the perfect realization of its goal, the *actual living* experience of truth. In metaphorical terms, ideal experiences organize the field of actualizable evidential experiences as perspective points organize the pictorial space. Although belonging to this field only as limit points–like vanishing points with regard to the pictorial space–they bring some perspective into it, so to speak. The idealization of truth experiences, moreover, goes together with idealizations concerning the "world" W the subject strives to know (and experience "in the flesh") that I shall uncover soon.

If we interpret Husserl, as we do, as grounding the truth-in-itself of judgments on verifiability, and interpret *this* notion in terms of the *actual* existence of decision procedures, then bivalence immediately loses its character of a logical *principle*, for the actual existence of implementable decision procedures cannot be granted a priori (and principles that are not a priori are not principles). But Husserl never considered trimming mathematics along intuitionist lines, and never suggested restricting bivalence to a proper subclass of the class of well-formed judgments.[8]

We seem to have only two choices in this matter, if we interpret the condition of verifiability in terms of actualizable procedures of verification, either to abandon Husserl's analysis of the ultimate grounding of the principle of bivalence, or embrace intuitionist theses concerning the a posteriori character of logic and the invalidity in general of bivalence. But, in this case, there is a *tertium*. We can simply not give the notion of verifiability involved in the subjective version of bivalence the sense of *effective* verifiability.[9]

Obviously, if the principle of bivalence is to have *unconditional* validity, it cannot depend on matters of fact. The validity of any principle can only be a matter of principle, and then a problem for transcendental philosophy to investigate. And this means that we must investigate the intentional life of a transcendental I, in particular in the role of predetermining the field of evidential experiences available *in principle* to her or him (that is, determining *a priori* which evidential experiences he or she can *reasonably* expect to live through). The validity of bivalence will be shown to depend essentially on how this domain and, correlatively, the "world" the subject strives to know are constituted.

In order to answer the questions raised above, let us begin with the characterization of proper judgments, those only that admit intrinsic truth values. One thing, however, must be absolutely clear, I insist: the sources of validity of any logical principle, which we as philosophers must uncover and whose uncovering amounts to a *justification* of the principle (for we then know why it is valid and the preconditions of its validity), cannot depend on matters of fact or hypotheses; otherwise the principle would not be a *principle* at all, enjoying *a priori* validity, but, at best, a factual *law*.

---

[8] In his introduction to Husserl's "Origin of Geometry", Jacques Derrida has an interesting discussion about how Husserl's notion of decidability, which is closely related to his idea of a *complete*-or *definite*-theory, need not to be read in strict effective terms. According to Derrida, the notion of decidability, correctly understood, must be conceived in terms of the horizon of science in general, and mathematics in particular, thus saving Husserl's idea of a nomological mathematical theory from the limitations imposed by Gödel's theorem. Derrida correctly notices that "[g]eometrical determinability in the broad sense [as opposed to decidability in strict sense, that is *effective* decidability–*my note*] would only be the regional and abstract form of an infinite determinability of being in general, which Husserl so often called the ultimate horizon for every theoretical attitude and for all philosophy". (Derrida, n. 51, 55)

[9] We can read the statement that, for any proper judgment, *there is* a decision procedure as a *mathematical* claim of existence, understood as Husserl understands *any* such claim: "All mathematical propositions of existence have this modified sense... not simply a 'there is' but rather: *it is possible a priori that there is*.... All existential judgments of mathematics, as *a priori* existential judgments, are in truth judgments of existence about possibilities ...". (Husserl 1939, §96) For Husserl, *a priori* factual judgments are not really judgments concerning *facts*, but a priori *possibilities*.

I have already advanced that proper judgments are the *meaningful* judgments,[10] and judgments are meaningful, i.e. capable of playing a representational role vis-à-vis the "world" to which they refer, if and only if they are both *syntactically* and *semantically* meaningful. Syntactically meaningful judgments are those that conform to a priori rules of combination of logical types. For example, in syntactically correct judgments predicate terms apply to subject terms, determining them, not the contrary. If "*S*" denotes the subject and "*p*" the predicate *S is p* is syntactically correct, but *p is S* is not.[11]

But syntactic sense, although necessary, is not a sufficient condition for meaningfulness. 'Jealousy is green', although perhaps expressive in a metaphorical sense, is not a proper judgment. It does have syntactic sense, but lacks semantic sense: greenness can only be predicated of spatial, concrete objects, not feelings. Or, in other words, "green" is a concept restricted to the ontological domain of *bodies*; to physical, not psychic nature. So, judgments have semantic sense if, and only if, the terms occurring in them denote things having something to do with one another, that are *ontologically compatible*. When both conditions are satisfied, judgments are meaningful, and we can say they represent *a priori possibilities* in the domain, the "world" to which they refer.[12]

Of course, the notion of possibility involved here is a very weak one, weaker than even logical possibility. Let us examine a particularly problematic example. Consider the judgment '465 is a prime number'. It has both syntactic and semantic senses. So, it is meaningful and represents a possible fact of the numerical domain. However, 465 is not a prime number, and more importantly, it is *necessarily* not a prime number. So, how can a necessarily false judgment represent a possible fact? By saying that meaningful judgments represent possible facts of the "world", we are then saying no more than that the situations they represent *cannot be ruled out a priori as being facts* considering *only* ontological compatibilities and incompatibilities or, equivalently, the *semantic* compatibility of the terms occurring in them. In our example, 465 occurs *only* as a particular token of the type *number*, without any

---

[10] Since meaningful judgments are those that possess well-determined judicative contents by means of which states-of-affairs are intended, we can assume that they correspond to those judgments that Husserl called *distinct*, as opposed to indistinct or obscure, not completely articulated judgments. Of course, they can still be unclear if not clarified by adequate intuition.

[11] The task of revealing the rules for the lawful combination of logical or syntactic types so as to avoid non-sense befalls upon apophantic morphology.

[12] "*[A] priori...* means *by reason of their validity, preceding all factuality*, all determinations arising from experience. Every actuality given in experience, and judged by the thinking founded on experience, is subject, insofar as the correctness of such judgments is concerned, to the unconditional norm that it must first comply with the *a priori* 'conditions of possible experience' and the possible thinking of such experience: that is, with the conditions of its pure possibility, its representability and positability as the objectivity of a uniform identical sense". (Husserl 1939, §90)

further specification, and obviously the type *number* is compatible with the subtype *prime number*. In short, meaningful judgments can represent even logically impossible situations, a fact that shows that logical possibility is a more stringent condition than meaningful representability.

Now, judging is always judging about something, a "world", and there must be a close connection between situations that can be represented in this "world" and how it is intentionally conceived.[13] The criteria of meaningfulness are then related to how a world is conceived or intentionally constituted, syntactic meaningfulness to how *any* world is conceived, and semantic meaningfulness to how *a particular* world is. A judging subject can only judge meaningfully if he or she knows what he or she is talking about, that is, the *meaning* intentionally attributed to the world about which she or he makes a judgment, and only such a subject can be a competent user of the language in which judgments concerning this world are expressed. Consequently, meaningfulness is accessible to competent users of this language and only to them.[14]

In the sense that, for Husserl, logic is always world-logic, logical principles and "world-making" are necessarily correlated. But we must be careful so as not to give this expression an idealistic reading in a metaphysical sense: "world-making" means only the *intentional* constitution of a "world".

It now seems that some of our questions have been answered: a judgment depicts a *possible* fact of W whenever it is meaningful, and meaningful judgments are those that are syntactically and semantically meaningful. On the subjective side, meaningfulness of judgments serves as a criterion of *a priori possibility of verification* by a judging subject in general. Evidential experiences are *possible in principle* (maybe not effectively, or even logically possible) whenever they are expressed by meaningful judgments. Therefore, any situation in W that can be possibly conceived (but that may not necessarily be a *fact* of W) is in principle capable of being verified by the subject.

To give meaningful judgments the way we characterized them the power of determining situations amenable *in principle* to verification is a way—maybe the only way—of characterizing verifiability independently of factual presuppositions or hypotheses concerning the cognitive powers of judging subjects, that is, presuppositions of a factual nature. This is so because, as I have already emphasized, questions of principle must be settled a priori. Hence, in this context, verifiability can only mean verifiability in principle, maybe not in actual fact.

---

[13] Remember that, for Husserl, logic is always world-logic; so, it is basically the sense with which "worlds" are constituted which determine the preconditions of sensefulness for predication concerning them.

[14] Parrots can "enunciate" correctly constructed phrases, but they do not "speak" meaningfully.

Now, objectively, the principle of bivalence guarantees that, for any judgment *p*, *either p* is true *or p* is false; that is, either the situation depicted by *p* is a *fact* of W, or the situation depicted by *not-p* is.[15] This means that bivalence presupposes W to be *ontologically complete*. *No* situation is *objectively* indeterminate in it; it is always a matter of *fact* that either *p* or *not-p*. This gives W the sense that Husserl recognizes as proper to any "world" intentionally constituted and offered to scientific investigation under the rule of law provided by logic: a world standing out there, objectively given, already completely determined in-itself. And, from the subjective side, since any situation can be verified, a "world" infinitely determinable by gradually developed means of cognition, capable at the very most of being completely mastered by science. This much is, for Husserl, *a fundamental a priori presupposition of science* (which, remember, is an ever open *communal* activity cutting through space and time, objectified in language and culture).[16]

The principle of bivalence, not being a matter of factual presupposition or a hypothesis, is then, as it must be, a matter of principle, closely tied to the intentional constitution of an *objective*, although *not necessarily metaphysically real* domain of cognition,[17] a "world" out there, offered to me and anyone else who happens to take a theoretical stand with respect to it, now and in the open infinite future.

One question, however, cannot easily be put to rest: In what sense can a judgment be proven to be true (in the sense of true in-itself) without being actually offered to an evidential experience of confirmation? Is this not in contradiction with the notion of truth as the *living experience* of the adequation between representation and fact?

The obvious answer is that truth-in-itself does not coincide with the living experience of truth.[18] The former is only an idealization of the

---

[15] The experience of a "negative" fact, that is, that *not-p* is the case, is not simply the *failure* to experience that *p* is the case, but the experience that *p* is *not* the case, a positive rather than a negative experience. The attempt, so to speak, to make *p* conform to the facts must, in the experience of *not-p*, meet with *irresistible resistance* on the part of the facts. *Frustration* must result, not simply a lack of satisfaction. (See Husserl 1900-01, Logical Investigation VI, chapter 4)

[16] "The possibility of sciences depends entirely upon this certainty that their provinces exist in truth, and that, concerning their provinces, theoretical truths-in-themselves exist, as actualizable by following explorable and gradually actualizable ways of cognition". (Husserl 1929, §80)

[17] In his *Studien zur Arithmetik und Geometrie* (Husserl 1983, 266), Husserl asks whether our representation of space has "metaphysical value" (*metaphysischen Wert*), that is, whether something *real* (*ein Wirkliches*) in a transcendent sense corresponds to it. I use the expression "metaphysically real" with the same meaning.

[18] In *Formal and Transcendental Logic* §76, Husserl says something relevant for our considerations here, namely, that logical principles *explicate* the concept of truth axiomatically. So, since he does *not* think we should give up bivalence, there must be a sense of truth this principle helps to explicate. Indeed, this is, objectively, the notion of truth as being objectively existing in itself and, subjectively, that of the *progressive and unobstructed givenness* of being. Consider this quote: "logic, by its relation to a real world, presupposes not only a real world's being-in-itself, but also the possibility, existing 'in

latter. A judgment that is true in-itself, but not yet effectively experienced as true, depicts a *fact* of a "world" *that is supposed to stand out there completely determined in itself* (the ontological completeness of the "world" that, remember, may not be *metaphysically* real); a fact, therefore, waiting to be intuitively experienced. To say that *p* is true in-itself because we have a proof of *p* that essentially involves the principle of bivalence is not to say that the objective fact depicted by *p* is at our disposal to experience *effectively*, but that the *situation*, the state-of-affairs depicted by *not-p cannot, as a proven fact,* be experienced as a fact of reality. In more precise terms, truths in-themselves express *anticipations of experience* in the overall context of the intentional positing of a "world". The principle of bivalence, in a few words, simply states that any well-formed judgment expresses a truth epiphany *in principle* available to us. Can a logical *principle* accomplish more than this?

Before proceeding, it is important to offer some extra textual evidence that the criticism, or rather the philosophical clarification of the principle of bivalence that I presented here coincides with Husserl's explicitly stated views on the matter. The best place to find them is in §90 of *Formal and Transcendental Logic*. Although Husserl does not believe that the principle of bivalence has unconditional validity, what he has in mind has nothing to do with the intuitionist criticism of it. For him, the scope of validity of bivalence is much wider than intuitionists are willing to grant.

Let us consider an example that Husserl explicitly gives of an assertion that he thinks is, "in its senselessness", "exalted above truth and falsity": "the sum of the angles of a triangle is equal to the color red". Although syntactically meaningful, this assertion is not *semantically* meaningful; being colored is a property that does not *in principle* fit idealities such as geometrical figures, being restricted to the realm of real, concrete bodies. "And yet", as he also explicitly says, "every declarative sentence that fulfills *only* [*my emphasis*] the condition for unitary *purely grammatical* sense (every unity of a sentence that is at all understandable) is also *thinkable as a judgment*–a judgment in the widest sense". In other words, *syntactic* meaningfulness is a *necessary and sufficient* condition for declarative sentences to express judgments; *obscure, indistinct, improper* judgments, but judgments nonetheless. Hence, the principle of bivalence is not valid for the totality of judgments, "for all judgments that are 'senseless' in respect to content violate this law". (Husserl 1929, §90)

For bivalence to be valid, it is also necessary that judgments be *distinct*, that is, that their "cores are congruous in respect of sense".

---

itself, of acquiring cognition of a world as genuine knowledge, genuine science, either empirically or a priori. This implies: Just as the realities belonging to the world are what they are, in and for themselves, so also they are substrates for truths that are validated in themselves–'truths in themselves'". (Husserl 1929, §92b)

Summing up, the principle of bivalence, for Husserl, applies to *all and only* the assertions that are meaningful with respect to both form and content–grammatically and semantically meaningful–for these only are the "judgments that fulfill the conditions for unitary sensefulness". *Husserl has no further restrictions for the validity of* tertium non datur. For judgments that have both syntactic and semantic senses "it is given *a priori*, by virtue of their genesis, that they relate to a unitary experiential basis. Precisely because of this, it is true of every such judgment, in relation to such a basis, either that it can be brought to an adequation and, with the carrying out of the adequation, either the judgment explicates and apprehends categorially what is given in harmonious experience, or else that it leads to the negative of adequation: it predicates something that, *according to the sense* [*my emphasis*], indeed belongs to this sphere of experience; but what it predicates conflicts with something experienced". (*Ibid.*)

This quote also makes it clear that, for Husserl, judgments with syntactic and semantic sense, proper judgments that is, whether they be either true or false, *predicate something that is in accordance with the sense belonging to the unitary experiential basis to which they refer and upon which they are grounded* (and, as we know, sense always emanates from transcendental subjectivity), and, as a consequence, that they can, *in principle*, be verified. So, a "unitary basis of experience", besides a "harmonious unity of possible experience", i.e., a domain in which "syntactic cores" can co-exist in harmony, is, most importantly, also one in which any situation that can be represented, i.e. that does not bring disharmony into the domain, is decided in it, or, as I put it, is a domain intentionally constituted as *ontologically complete*.

I shall now conclude with a few critical remarks on the treatment of this problem by some other authors.

Although he rejects Brouwer's criticism of bivalence on the charge that it is directed towards a subjective, instead of objective notion of truth,[19] Alonzo Church (Church, 75-78) takes a typical formal-mathematical, non-philosophical approach to the problem. For him, there are as many formal-logical systems as we care to define consistently; bivalence being a valid principle in some, not necessarily in others. Being non-philosophical this perspective is alien to our interests here.

Agreeing with Michael Dummett, for whom "bivalence is the hallmark of realism" (a claim we shall examine shortly), Willard van Orman Quine, as a good American pragmatist, accepts bivalence, but at a price. Considering that presupposing the bivalence of well-formed expressions involving imprecise terms, such as "there is an odd number of blades of grass in Harvard Yard at the dawn of Commencement Day, 1903" hurts

---

[19] *Effectively* verifiable truth and truth *in-itself*, which can, but only *in principle*, be verified and presuppose the ontological completeness of the "world", are, in a sense, a subjective and an objective conception of truth, respectively.

our common sense, for we hesitate to accept that there is a fact of the matter concerning this, but also that, without bivalence, our theory of reality would be much less elegant and much more complicated, Quine proposes a tradeoff: for the sake of theoretical simplicity, accept bivalence and keep your qualms to yourself. (Quine, 90-95)

From Husserl's more elaborate philosophical perspective, it is easy to see the confusions that Quine gets himself involved in. What he in fact proposes is that, in order to validate bivalence, we *presuppose* the ontological completeness of reality with respect to the categories of the theoretical apparatus we want to squeeze it into, giving, however, this presupposition a naturalistic, empiricist interpretation. The reality Quine has in mind is Nature as conceived in empiricism, a being *in-itself*, not Nature as Husserl's analyses showed always to be the case in science, *intentionally constituted* Nature, a being *for-ourselves*, bearing our stamp. As a consequence, the presupposition of ontological completeness enters into the "negotiation" as a philosophically ungrounded and unjustified presupposition regarding something upon which we have no bearing, not as Husserl wanted it, as a *transcendental* presupposition concerning the intentional constitution of reality. The contrast between Husserl and Quine here highlights the misery of naturalism and empiricism vis-à-vis a truly philosophical conception of reality.

Let us conclude with Michael Dummett. In *Truth and Other Enigmas*, he claims, as already noted, that "bivalence is the hallmark of realism" (Dummett, 145-65), understanding by "realism", of course, ontological and epistemological realism, for which there is an *independently* existing reality out there, *completely determined* in-itself. There are two hidden presuppositions here: that metaphysically real, ontologically independent realities are *necessarily* ontologically complete; and conversely, that *only* such realities are so. As I have shown above, both are false. The first, as I have argued, is a metaphysical presupposition that—as even Quine's examples show—some would decline. The second is just plainly not the case, for, as I have also shown, metaphysically *unreal*, but objectively real, ontologically dependent realities can very well be *intentionally constituted as* ontologically complete, enjoying the fullness of being and metaphorically "out there" now and forever, for us all to investigate and, at the *very most*, know *completely*. So, bivalence is *not* really the hallmark of realism, either ontological or epistemological. From all that has been said, we can also, with no effort, see that the validity of bivalence in no way presupposes, as a matter of *fact*, that any well-posed problem can be solved.

As we see, bivalence, as a formal logical principle, is rooted in the theoretically oriented intentional life of a knowing subject, who *posits* her or his "world" as an objectively existing (but not necessarily *metaphysically independent*), ontologically complete domain in which no situation that can be represented properly is *a priori* incapable of being verified. In fact, as *Experience and Judgment* (Chapter I, part I §19) tells us, the roots of bivalence go as deep as the pre-predicative life of a subject immersed in his or her life-world, who lives in it, is drawn to it, is interested in it, building expectations and anticipations that will, in the sequence of apprehensions and explications, be either satisfied or frustrated, but that, or so the subject *takes for granted*, even if not *consciously*, can go on *unobstructed* as far as his or her interest persists, even, in the extreme case, forever. We can say that bivalence, in its two versions, objective and subjective, is the counterpart in the sphere of active judging of the subject's passive *belief* that no particular aspect of his life-world is inaccessible to him. This is a general phenomenon: what is logical, properly speaking, is rooted in the pre-logical structure that the subject builds into his relation with the life-world by simply *being immersed in it* as a living consciousness.

References

Church, Alonzo 1928. "On the Law of Excluded-Middle", *Bulletin of the American Mathematical Society* 34 1, 75-78.

Derrida, Jacques 1962. *Edmund Husserl's 'Origin of Geometry': An Introduction*, Lincoln NE: The University of Nebraska Press, 1989, translation of his *Introduction à "L'Origine de la géométrie" de Husserl*, Paris: PUF.

Dummett, Michael 1978, *Truth and Other Enigmas*, Cambridge MA: Harvard University Press.

Hilbert, David 1930, "Logic and the Knowledge of Nature", in Ewald (ed.), 1157-65.

Husserl, Edmund 1900-01. *Logische Untersuchungen*, Halle: Niemeyer, in English, *Logical Investigations*, New York: Humanities Press, 1970.

Husserl, Edmund 1913. *Ideen zu einer reinen Phänomenologie und phänomenologischen Philosophie, I,* Husserliana II, The Hague: Martinus Nijhoff, 1950. In English, *Ideas, General Introduction to Pure Phenomenology*, translated by W. R. Boyce Gibson, New York: Collier, 1962 (1931).

Husserl, Edmund 1929. *Formale und transzendentale Logik*, The Hague: Martinus Nijhoff, in English *Formal and Transcendental Logic*, The Hague: Martinus Nijhoff, 1969.

Husserl, Edmund 1936. *The Crisis of European Sciences and Transcendental Phenomenology: An Introduction to Phenomenological Philosophy*, Northwestern University Press, Evanston IL, 1970, translation of his *Die Krisis der europäischen Wissenschaften und die transzendentale Phänomenologie*, Husserliana VI, The Hague: Martinus Nijhoff, 1954.

Husserl, Edmund 1939. *Erfahrung und Urteil*, Claassen & Goverts, Hamburg, 1948, in English, *Experience and Judgment*, London: Routledge and Kegan Paul, 1973.

Husserl Edmund 1983. *Studien zur Arithmetik und Geometrie (1886-1901)*, Husserliana XXI, The Hague: Martinus Nijhoff.
Quine, Willard 1981. "What Price Bivalence", *The Journal of Philosophy* 78 2, 90-95.

15

Claire Ortiz Hill

GEORG CANTOR'S PARADISE, METAPHYSICS AND HUSSERLIAN LOGIC

Introduction

Twentieth century schools of thought as diverse as those associated with Friedrich Nietzsche, Karl Marx, Bertrand Russell, the Vienna Circle, Martin Heidegger, Rudolf Carnap, Jean-Paul Sartre, Willard Van Orman Quine, certain partisans of secular political systems, and so on, all strove to shut the doors to metaphysical inquiry. This profound philosophical antagonism towards metaphysics acted on several fronts to discredit, undermine, proscribe, kill a wide range metaphysical notions associated, rightly or wrongly, with the follies and excesses of 19th century idealism, whether transcendental, subjective, absolute, or religious, associated with the names of people like Immanuel Kant, G. F. W. Hegel, F. H. Bradley, not to mention Jesus Christ.

One of the principal strategies adopted by Bertrand Russell, Rudolf Carnap, Willard Van Orman Quine, and like-minded philosophers was to create an inhospitable climate for metaphysical thought through a transformation of logic. Set-theoretical notions appeared as promising instruments for achieving their ends, and through *Principia Mathematica* (Russell 1927) and related systems, which did so much to determine the course of modern logic, basic ideas of set theory came to play a key role in laying the logical foundations for analytic philosophy's well known scheme to overcome metaphysics.

However, there is another story to be told with a different ending. For, while it is well known that work on set theory had a role to play in shaping the destiny of analytic philosophy, Edmund Husserl, the founder of the rival phenomenological school, is not generally perceived as having considered set theory tempting at all, and the tale of his encounter with it decades before the logic shaped by *Principia Mathematica* and kindred systems took hold is barely known.

Adepts of the analytic and the Husserlian schools have seemed to almost everybody to have been at odds from almost the very beginning, and irreconcilable differences of logic and metaphysics have most often been cited as the cause of their estrangement. Yet, however incompatible analytic philosophy and phenomenology turned out to be, variations on Georg Cantor's set theory played a preeminent role in developing logic and in redrawing the boundaries between it and metaphysics in both schools. So, along with Bernard Bolzano, Franz Brentano, and Hermann Lotze, Cantor deserves to be ranked as one of the progenitors of both analytic philosophy and phenomenology. Here, I seek to add new dimensions to standard discussions by taking readers back to the place where the two logical roads diverged and affording them a look down the one less traveled by, the Husserlian one not taken by mainstream logicians in the 20th century.

On the Roots of the 20th Century's Strongly Anti-Metaphysical Animus

A rediscovery of metaphysics took place during the final years of the 19th century. Emblematic of those playing roles in rehabilitating the respectability of metaphysical inquiry was Hermann Lotze. In *A Critical Account of the Philosophy of Lotze, the Doctrine of Thought* published in 1895, Henry Jones of the University of Glasgow explained Lotze's "power" over his age as having sprung from the fact that he dealt with problems that it saw as vital and solved them in a way which, on the whole, accorded with its convictions. For Jones, Lotze's relation to Hegel, on one hand, and to the pure Naturalists, on the other, and his attempt to correct the errors of both by recourse to the teachings of Kant, made Lotze "a most interesting figure in the history of philosophic thought" (Jones, 32). He seemed, Jones wrote,

> to have restored to us possessions which Kant and his immediate followers had made insecure, and which the Materialists and the Pessimists had rendered untenable. In the service of these convictions, he has, at least for the time, stemmed the tide of Idealism and given pause to that ambitious Monism which seemed to have confused the old boundaries of thought, mingling together nature and spirit, good and evil, things and thought, the human and the divine. In the same interest he has also "stayed the Bacchic dance of the Materialists," who had occupied the place left vacant by the spent Idealism. So that it is no matter for surprise that some... should consider that they owe it to Lotze... that, after the reign of chaos, there is once more "a firmament in the midst of the waters, dividing the waters from the waters". (*Ibid.*, 3-4)

"The yearning for the real, or at least the palpable and the particular, under whose impulse the thought of Lotze's day threw itself upon the natural world of perceptible facts and events, and which seemed to be the direct and necessary consequence of confining the German people to the thin Hegelian diet of abstract and ambitious Idealism, had complete possession of Lotze", according to Jones (*Ibid.*, 31-32). Hegelian Idealism, Jones explains, seemed to Lotze "to reduce the world to a 'solemn shadow-land' of general conceptions, to convert the infinite variety of its chances and changes into a system of logical notions at once empty and ruled by necessity.... to be an attempt to establish a universal mechanism, which was not the less fixed and relentless because it was called 'spiritual'". (*Ibid.*, 10)

Jones considered that "Lotze possibly divined a truth which is ever becoming clearer, that there is a close affinity between natural science and Idealism, that modern science when it understands itself is idealistic in temper and tendency..." (*Ibid.*, 8). "It is not Idealism with its spiritual construction of the world that is at war with the inner spirit of science," Jones maintained, "but the scepticism which.... conceals its true nature under the names of Dualism and Agnosticism". (*Ibid.*)

Much the same understanding of Lotze and his times is conveyed in a 1902 Paris doctoral thesis on Lotze's metaphysics by Henri Schoen, who explained there how Lotze had inspired courage in worried and tormented consciences, how he had been able to communicate faith in the triumph of a spiritualistic conception of the world to young people whose confidence had been shaken by the ineffectiveness of idealism and the successes of materialism. To those impressed by positivism, he had showed an exact method starting from observation and not a priori reasoning. He had taught a generation disgusted with abstractions to start from given facts and their relations among themselves and to them. Schoen saw his generation as being disgusted with materialism, with vague and confused aspirations, but disposed to accept a metaphysics that was not in contradiction with its scientific views. He explains how he had been guided and had tried to guide his students through the philosophical and psychological crisis of German metaphysics, of how he had felt that it was impossible to go back to the old dogmatism, but also recognized the inadequacy of pure reason, of how, eclipsed by idealism, Kant's realism was to wreak vengeance in the modern metaphysics of the end of the 19th century, whose goal it was to develop the seeds of realism contained in Kant's doctrine, and not the idealism found there as well.

For Schoen, the true method fell midway between pure criticism and absolute idealism. The old extreme skepticism and naive dogmatism had both become untenable. There was an equal balance to be maintained between the realm of the ideal and reality, between the things of the suprasensible world and the things of the real world. After having been at the point of doubting the future of spiritualistic metaphysics, he said that he came to understand that genuine criticism ultimately gives back more than it takes away. Far from attacking metaphysics, most genuine scholars recognize the mystery where their science is obliged to stop. Schoen expressed complete confidence in the future of metaphysics. (Schoen, 8-9, 18, 22-23)

Since much discourse in recent times seems wrongly to assume that the late 19th century could not see beyond the metaphysical dictates of a very conventional form of Christianity, it is important to stress that many were eagerly casting off its shackles and that, once liberated, some of them were engaging in behavior deemed irrational, superstitious, and unsavory by scientifically minded thinkers. While the end of the 19th century did witness attempts to rehabilitate the respectability of metaphysical inquiry and to situate it centrally on the philosophical agenda alongside rigorous, rational, scientific thinking, there was another, more deeply disturbing, dimension in the turn towards metaphysics at that time.

A chapter of Nicholas Goodrick-Clarke's study of the occult roots of Nazism is devoted to the modern German occult revival from 1880 to 1910. Though modern occultism was represented by many varied forms, he explains, its function appeared relatively uniform.

> Behind the mantic systems of astrology, phrenology and palmistry, no less the doctrines of theosophy, the quasi sciences of 'dynamosophy', animal magnetism and hypnotism, and a textual antiquarianism concerning the esoteric literature of traditional cabbalists, Rosicrucians, and alchemists, there lay a strong desire to reconcile the findings of modern natural science with a religious view that could restore man to a position of centrality and dignity in the universe. Occult science tended to stress man's intimate and meaningful relationship with the cosmos in terms of 'revealed' correspondences between the microcosm and macrocosm, and strove to counter materialist science, with its emphasis upon tangible and measurable phenomena and its neglect of invisible qualities respecting the spirit and the emotions. These new 'metaphysical' sciences gave individuals a holistic view of themselves and the world in which they lived. This view conferred both a sense of participation in a total meaningful order and, through divination, a means of planning one's affairs in accordance with this order. (Goodrick-Clarke, 29)

So it is that while some took up arms against mystical and idealistic philosophies, others revolted against the various forms of positivism, empiricism, naturalism and materialism that they felt modernity was foisting upon them. Much like the end of the 20th century, the end of the 19th century witnessed growing participation in cults, spiritism, Satanism, the occult, magic, witchcraft, and so on, and this surely did its share to fan antagonism toward metaphysics, or even fear of it, and to inflame desires to defeat it.

Georg Cantor and the Metaphysical Sciences

In his quirky way, Georg Cantor was part of the movement to reconcile the findings of modern natural science with metaphysical views. An avowed adversary of psychologism, positivism, the new empiricism, sensualism, skepticism, Kantianism, naturalism and related trends, he was an enthusiastic metaphysician. In 1894, he wrote to the French mathematician Charles Hermite that "in the realm of the spirit" mathematics had no longer been "the essential love of his soul" for more than twenty years. Metaphysics and theology, he "openly confessed", had so taken possession of his soul as to leave him relatively little time for his "first flame" (Cantor 1991, 350). In 1884, he had expressed his high regard for metaphysics and his belief in a close alliance between metaphysics and mathematics. He expressed gratitude for having been paid the honor of having had philosophical, and even metaphysical, worth accorded to his writings. (Cantor 1884, 83-84)

Cantor considered that his theories showed the way to a new, abstract realm of ideal mathematical objects that could not be directly perceived or intuited (Cantor 1883, 207 n. 6, 7, 8; Cantor 1887/8, 418 n. 1) and that the transfinite realm he was exploring "presented a rich, ever growing field of ideal research" (Cantor 1887/8, 406). For example, in 1895, we find him writing to Hermite that: "the reality and absolute uniformity of the whole numbers seems to be *much stronger* than that of the world of senses. That this is so has a single and quite simple ground, namely the whole numbers both separately and in their actual infinite totality exist in that highest kind of reality as eternal ideas in the Divine Intellect". (cited Hallett, 149)

Cantor wanted to provide his numbers with adequate metaphysical foundations and he filled his writings about set theory with metaphysical reflections aimed at explaining and justifying his novel ideas to a readership chary of such talk (ex. Cantor 1991, 100, 113, 118, 178, 199, 227). In a 1890 letter to Giuseppe Veronese, Cantor wrote that contradictions found in his theories were but apparent and that one must distinguish between the numbers that we can to grasp in our limited ways and "numbers as they are *in and for themselves, and in and for the Absolute intelligence*", each of which "is a simple concept and a unity, just as much a unity as one itself. Taken absolutely", he told him, "the smaller numbers are only *virtually* contained in the bigger ones. They are, taken absolutely, all independent one from the other, all equally good and all equally necessary metaphysically". (*Ibid.*, 326)

Cantor was also explicit about the precise nature of the metaphysical foundations that he envisioned for his numbers. He stressed that the "certainly realist, at the same time, however, no less than idealist foundations" of his reflections were essentially in agreement with the basic principles of Platonism (Cantor 1883, 181, 206 n. 6)."My idealism", he wrote to Paul Tannery, "is related to the Aristotelian-Platonic kind, which as you know is at the same time a form of *realism*. I am just as much a *realist* as an *idealist*" (Cantor 1991, 323). To Giuseppe Peano, Cantor once wrote: "I conceive of numbers as 'forms' or 'species' (general concepts) of sets. In essentials this is the conception of the ancient geometry of Plato, Aristotle, Euclid etc". (*Ibid.*, 365). By manifold or a set, he said he generally meant "any Many which can be thought of as a One, any totality of determinate objects which can be united by a law into a whole" and thus was defining something related to the Platonic *eidos* or *idea* and to what Plato called a *mikton* (Cantor, 1883, 204 note 1). He believed that the whole real numbers were "related to the *arithmoi noetoi* or *eidetikoi* of Plato with which they probably even fully coincide" and that his transfinite numbers were but a special form of these *eidetikoi*. (Cantor 1887/8, 420; Cantor 1884, 84)

"No one shall be able to drive us from the paradise that Cantor created for us", David Hilbert is famous for having declared (Hilbert, 1925, 376), but Cantor's colleagues did not find everything about that paradise tempting. Gösta Mittag-Leffler warned Cantor in 1883 that his work would be much more easily appreciated in the mathematical world "without the philosophical and historical explanations" (Cantor 1991, 118). In 1885, he warned him that his new terminology and philosophical way of expressing himself might be so frightening to mathematicians as to seriously damage his reputation among them. (*Ibid.*, 241)

A chapter of Joseph Dauben's *Georg Cantor, His Mathematics and Philosophy of the Infinite* is devoted to studying Cantor's personality. In it, Dauben cites the 1862 letter in which Cantor wrote to his father: "my soul, my entire being lives in my calling; whatever one wants and is able to do, whatever it is toward which an unknown, secret voice calls him, *that* he will carry through to success!" (Dauben, 277). For Dauben,

> There can be no mistake about Cantor's identification of his mathematics with some greater absolute unity in God. This also paralleled his identification of transfinite set theory with divine inspiration.... Cantor... told Mittag-Leffler that his transfinite numbers had been communicated to him from a "more powerful energy"; that he was only the means by which set theory might be made known.... The religious dimension which Cantor attributed to the *Transfinitum* should not be discounted as merely an aberration. Nor should it be forgotten or separated from his life as a mathematician. The theological side of Cantor's set theory... is... essential for the full understanding of his theory and the development he gave it. Cantor believed that God endowed the transfinite numbers with a reality making them very special.... He felt a duty to keep on, in the face of all adversity, to bring the insights he had been given as God's messenger to mathematicians everywhere. (*Ibid.*, 290-91)

The fact of the matter is that Cantor's views were more that just metaphysical, religious, or even mystical. The still unpublished letters of his letter books from the 1880s and 1890s reveal a robust interest in the occult, something of which Gottlob Frege may have had an inkling, for in a posthumously published draft of a review of Cantor's *Contributions to the Theory of the Transfinite* (Frege 1890-1892, 68-71), Frege refers mockingly to "magical effects" , the pronouncing of a "magic incantation", "supernatural powers", the possession of "miraculous powers ...not far removed from the Almighty", and even alludes someone who "hears an inner voice whispering". Interestingly, some of these very criticisms figure in Frege's irresponsible review (Frege 1894) of Husserl's *Philosophy of Arithmetic*, a review that has been lent far more credence than it deserves. (Hill and Rosado Haddock, Chapter 6)

Trouble in Cantor's "Paradise"

In 1886, Edmund Husserl arrived at the University of Halle to prepare his *Habilitationsschrift* called *On the Concept of Number* (Husserl 1887). Cantor served on Husserl's *Habilitation* committee and approved the mathematical portion of the work (Gerlach and Sepp). He took a liking to his younger colleague and was very supportive of him. Husserl's wife remembered that "Cantor, the greatest mathematician since Gauss, the creator of set theory (a new and very fruitful branch of mathematics)" loved her husband tenderly (*liebte H. zärtlich*). "They were alike in many ways," she commented, "but with otherwise great dissimilarity. The Cantors' house was like home..." (Malvine Husserl, §E). The same letter books that reveal Cantor's involvement in the occult also find him multiplying efforts to find Husserl an official professorship. (Cantor, 1884-1896; Cantor 1991)

Husserl remained in Halle for the next 15 years, enough time to have more than just a taste of Cantor's ideas at the very time he was creating his paradise (Hill and Rosado Haddock, Chapters 7, 8). During those years, Cantor was coping with the antinomies of set theory through, for example, his correspondence with David Hilbert (Cantor 1991, 387-485; Dauben, 240-70). As Joseph Dauben has commented, it is "not a little ironic" that the first mathematician to discover the antinomies of set theory was Cantor himself, who had anticipated the problem and by 1895 was already "trying to remedy the paradoxes with a minimum of damage to his system of transfinite numbers" (Dauben, 241). Using his diagonalization proof of 1891 (*Ibid.*, 165-68), Cantor "could argue that the set of all sets had to give rise to a set of larger cardinality; the set of all its subsets. But since this set had to be a member of the set of all sets, the paradoxical conclusion was inevitable that a set of lower cardinality actually contained a set of higher cardinality". (*Ibid.*, 242)

So it was that Husserl had a foretaste of the crisis in foundations that broke out once Cantor's proof by diagonal argument that there is no greatest cardinal number opened Bertrand Russell's eyes to the contradiction of the set of all sets that are not members of themselves and he began advertising that finding (Russell 1903, §§100, 344, 500; Russell 1959, 58-61; Grattan-Guinness 1978, 1980, 2000).

The first four chapters of Husserl's *On the Concept of Number* would go into the making of his 1891 *Philosophy of Arithmetic*, a book characterized by him in *Formal and Transcendental Logic* as having been an initial attempt on his part "to obtain clarity regarding the original genuine meaning of the fundamental concepts of the theory of sets and cardinal numbers" ("*Klarheit über den ursprungssechten Sinn der Grundbegriffe der Mengen- und Anzahlenlehre zu gewinnen*") (Husserl 1929, §27a; also §24 and note). However, from the very beginning Husserl displayed a critical attitude towards aspects of set theory. Chapter XI of *Philosophy of Arithmetic* contains a discussion of the logical problems of infinite sets (Husserl 1891, 230-34). Husserl's deep reservations about a calculus of classes is much in evidence in articles published during the 1890s (Husserl 1994, 92-114, 115-30, 135-38, 199, 443-51), in which he sought to show that the total formal basis upon which the class calculus rests was valid for the relationships between conceptual objects, and that one could solve logical problems without the detour through classes, which he considered to be totally superfluous. He was advocating a calculus of concepts (*Ibid.*, 109, 123). His chief target in those days was Ernst Schröder. An article dated 1891 finds him arguing that Schröder's attempt to show that bringing all possible objects of thought into a class gives rise to contradictions. (*Ibid.*, 84-85)

Husserl also quickly adopted a critical attitude towards his own attempts to clarify the true meaning of the fundamental concepts of the theory of sets and cardinal numbers using the empiricistic approach that he had learned from Franz Brentano, whose austere ideal of a strict, philosophic science was most nearly realized in the exact natural sciences. Brentano's clear, rigorous, insightful, objective, precise philosophical analyses had convinced Husserl that philosophy was a serious discipline that could to be dealt with in the spirit of the strictest science, but Husserl admitted to having been disturbed, even tormented, by doubts about a Brentanian analysis of sets from the very beginning. He finally concluded that Brentano's methods left him "in the lurch", that "once one had passed from the psychological connections of thinking, to the logical unity of the thought-content (the unity of theory) no true continuity and unity could be established". (Husserl 1900-01, 42; Husserl 1913, 20, 34-35; Husserl 1919, 343-45; Hill 1998)

Husserl's search for solutions finally led him to espouse idealistic metaphysical views that Brentano, bound as he was to the naturalistic tradition, considered odious. Husserl began to accord idealist systems the highest value, to see them as shedding light on totally new, radical dimensions of philosophical problems. The ultimate and highest goals of philosophy, he came to believe, are only opened up when the philosophical method that those particular systems call for is clarified and developed (Husserl 1919, 345). Every possible effort had been made, he came to state unequivocally of his *Logical Investigations*, "to dispose the reader to the recognition of this ideal sphere of being and knowledge ... to side with 'the ideal in this truly Platonistic sense', 'to declare oneself for idealism' with the author". (Husserl 1913, 20)

Husserl endeavored to explain to Brentano in 1905 that the

> empirical sciences–natural sciences", "–are sciences of 'matters of fact'.... Pure Mathematics, the whole sphere of the genuine Apriori in general, is free of all matter of fact suppositions.... We stand not within the realm of nature, but within that of Ideas, not within the realm of empirical. . . generalities, but within that of the ideal, apodictic, general system of laws, not within the realm of causality, but within that of rationality.... Pure logical, mathematical laws are laws of essence.... (Husserl 1905)

In so writing, however, Husserl strove to make it understood that he was "far from any mystico-metaphysical exploitation of 'Ideas', ideal possibilities and such" of the kind Brentano so despised. (Husserl 1905, 39; Hill 1998)

The sciences are in need of metaphysical foundations, Husserl would teach students. But he strove to make it perfectly clear to them that by that he "meant anything but a dialectical spinning of the concrete results of these sciences out of some abstract conceptual mysticism". Rather, he explained, he had in mind something "much more modest and fruitful, a level-headed clarification and testing of those general presuppositions that the factual sciences make about actual being, and, in more far-reaching scientific work and the recuperation of the most mature, recent knowledge of real being, of its elemental principles, forms, and laws that the present state of the individual sciences permits". (Husserl 1896, 5)

Husserl on Metaphysics, Empiricism, and the Natural Sciences

Husserl saw the metaphysical needs of his time going unmet and gave this as an explanation as to why spiritism and the occult were thriving and superstition of every kind was spreading. As he saw it, metaphysics had come to be seen as a relic of scientifically backward times on a par with alchemy and astrology. He saw the fight against metaphysics and most of the chance contemptuous remarks against it as being directed at a kind of a hobgoblin (*eine Art Popanz*) that people had concocted. (Husserl 1902/03a, 232)

Husserl considered the overriding role and authoritative influence that the natural sciences had acquired in the lives of educated people in his times to be especially to blame for the prevailing contempt for metaphysics and its transformation into a hobgoblin. As he saw it, the natural sciences had taken abundant revenge for the injustice that they had had to suffer from the pseudo-scientific natural philosophy of the Romantics, but in speaking of metaphysics, natural scientists were still thinking of a kind of philosophizing that was up to the old tricks of the

Hegelian school (*Ibid.*, 232-33). Resorting to colorful language, Husserl explained to students that when,

> after the collapse of idealistic philosophy in the middle of the 19th century, the great awkward lull set in, when the philosophical race of Titans of Romanticism, who trained themselves to be able to storm the Mount Olympus of philosophy with their dialectical tricks, were flung down into the dark Tartarus of dissension and unclarity, and uneasy (*katzenjämmerliche*) disenchantment, even disillusionment, followed the earlier exuberance, then sounded ever louder the call back to Kant, the great theorist of knowledge, who had set limits on the presumptuousness of an uncritical metaphysics and established critique of knowledge as the true foundations for philosophy. (*Ibid.*, 229)

Husserl was particularly concerned about the extent to which the hard questions about the objectivity of knowledge raised in the wake of Kant's work could determine one's entire conception of being in the world. The pressing problems of metaphysics, pressing too from the standpoint of the most exact scientific thinking, which in no way coincided with what Kant had in mind in his philosophical context, were no longer being distinguished from problems of the theory of knowledge. Since the collapse of idealistic philosophy, the rise of Romanticism, with its extravagant promises and flaunting of the requirements of rigorous science, and the revival of Kantianism, for which metaphysics an a priori science of concepts was impossible, the term "metaphysics" had taken on ominous overtones and people preferred to avoid using it. (Husserl 1902/03a, 9; 232; Husserl 1902/03b, 13)

Husserl, however, believed that science needed metaphysical foundations, that above and beyond the relative sciences of Being, there had to be a definitive science of Being to explore what had to be considered real in the final and ultimate sense. He called for a science of metaphysics to study problems lying beyond empirical investigation, to engage in this exploration of what is *realiter* in the ultimate and absolute sense, and so provide ultimate and deepest knowledge of reality. He believed that such a science of metaphysics was possible, justifiable, and that human beings were ultimately capable of attaining knowledge of reality. (Husserl 1902/03a, 232, 233, 252; Husserl 1906/07, §§ 20, 21)

Husserl defined metaphysics as the science of absolute Being in contrast to the individual sciences, which he saw as merely sciences of Being in the relative, provisional sense sufficient for practical orientation in the phenomenal world and for the practical mastery of nature (Husserl 1902/03b, 12-13). He proposed to have metaphysics understood in a broad sense as radical ontology, as the radical science of Being, the science of Being in the absolute sense, instead of the science of Being in the empirical sense, which we think we know so well, but which upon closer inspection at times turns out to be deceptive and an illusion. (Husserl 1906/07, §20)

In certain respects, Husserl pointed out, each empirical science is a science of what is real. It deals with real things... with their real becoming, their real relations, etc. Each such science is, then, in its own way, an ontology. And, since each empirical science explores a special sphere of real Being, the whole of empirical science appears to exploit the sum total of reality and to satisfy all epistemological interests regarding reality in a manner commensurate to one of the states of development of these sciences. (*Ibid.*)

It is certain, though, he argued, that knowledge of the world of the natural sciences, even the most highly developed ones, is not definitive knowledge of reality. Through them, a highly worthwhile goal, namely the practical mastery of nature, is attained, a far-reaching orienting of empirical reality, the possibility of formulating laws by which we exactly foresee and foretell the course of empirical processes, redirect the course of the processes but, Husserl stressed, with them we are not yet in possession of definitive knowledge, of ultimate, conclusive knowledge of the essence of nature. And advances made in the natural sciences alter nothing in this. One undoubtedly arrives at worthwhile results, but the lack of that critical insight into the meaning of the fundamental concepts and fundamental principles makes it impossible to be clear about what is thereby ultimately achieved and consequently about the sense in which one may claim to take the results as expressions of ultimate Being. (*Ibid.*)

Wherever it is a question of reality, in life and in all empirical sciences, he explained, we apply certain concepts like thing, real property, real relation, state, process, coming into being and passing away, cause and effect, space and time, that seem to belong necessarily to the idea of a reality. Whether or not all these concepts are actually intrinsic to the idea of reality, there are surely such concepts, the basic categories, in which what is real as such is to be understood in terms of its essence. Thus, investigations must be possible that simply reflect everything without which reality in general cannot be conceived. For Husserl, this was where the idea of a metaphysical a priori ontology came in. (*Ibid.*, §21)

It is certain, Husserl considered, that a most universal concept of what is real in general, of the particularities grounded in the essence of what is real, can and must be delineated. Concepts like that of an individual real thing, like Being for itself, or thing in the broadest sense, real property in the broadest sense, real relation, time, cause, and effect, are surely necessary thoughts concerning possible reality and require a study of the analysis of essence and of essential laws. There must therefore be, he concluded, a science of real Being as such in the most universal universality, and this a priori metaphysics would be the necessary foundation for empirically based metaphysics, which not only claims to know what lies in the idea of reality in general, but claims to know what is now actually actual, first of all universally as inquiry into what is general, but merely actual determinations (elements, properties, laws) of actual reality, and then claims to determine what is definitively real in a particular way in the actual sphere of Being in order to be able to understand definitively what is *realiter* there. (*Ibid.*)

As Husserl saw it, such a science of metaphysics is so necessary for science that even natural scientists can not do without it (Husserl 1902/03a, 233). The empirical sciences are not, he pointed out, creations of a purely theoretical mind, are not based on absolutely scrupulously lain foundations in accordance with a rigorous logical method. They are subject to principles that govern thinking and research in the natural sciences, that make natural science in general possible, and that consequently cannot in turn be searched for again by thinking and research in the natural sciences. Even the most highly developed, most exact natural sciences, also uncritically use concepts and presuppositions originating in a prescientific understanding of the world (Husserl 1906/07, §20). As soon as they begin reflecting on the principles of their science, he considered, they fall into metaphysics, though they most certainly do not want to call it by that forbidden name. (Husserl 1902/03a, 233)

Husserl on Metaphysics and Logic

Husserl may have been turned in the direction of metaphysical idealism by Cantor's experiments in the abstract realm of ideal mathematical objects, but more likely than not Cantor's excesses came as a shock to Brentano's disciple. Be that as it may, it was Hermann Lotze's interpretation of Plato's doctrine of Ideas, Husserl always maintained, that turned him away from empirical psychology and gave him the idea "to transfer all of the mathematical and a major part of the traditionally logical world into the realm of the ideal". His own concepts of "Ideal" significations and "Ideal" contents of presentations and judgments, Husserl stressed, originally came from Lotze, who was already writing about truths in themselves (Husserl 1913, 36; Husserl 1994, 201; Lotze 1888, Chapter II).

Husserl also gave Lotze credit for the theory that pure arithmetic was basically no more than a branch of logic that had undergone independent development and had developed very early through independent treatment (ex. Husserl 1896, 241; Husserl 1902/03b, 19, 249; Husserl 1906/07, §15). Husserl entreated his students not to be "scared" ("*Ich bitte Sie nicht zu erschrecken!*") by that thought (Husserl 1902/03b, 34) and "to accustom themselves to the initially strange view of Lotze that arithmetic is only a relatively independent, and from time immemorial, particularly highly developed piece of logic". (Husserl 1896, 271-72)

Via Lotze, Husserl also came to understand what he had initially thought of as the "metaphysical abstrusities" and "naive", "curious conceptions" of Bernard Bolzano. In the light of Lotze's ideas, Bolzano's theory that propositions were objects which nonetheless had no existence now seemed quite intelligible to Husserl, who "with one stroke" realized that Bolzano had not hypostasized presentations and propositions in themselves, for they could be seen as enjoying the ideal existence or validity characteristic of universals of objects that are universals and, therefore, that kind of being that is established in the existence proofs of mathematics. Husserl saw that what he had thought of as "mythical entities, suspended between being and non-being" were actually

> to be understood as what is designated in ordinary discourse --which always objectifies the Ideal–as the "sense" ("*Sinn*") of a statement. It is that which is explained as one and the same where, for example, different persons are said to have asserted the same thing. Or, again, it is what, in science, is simply called a theorem, e.g., the theorem about the sum of the angles in a triangle, which no one would think of taking to be someone's lived experience of judging. And it further became clear... that this identical sense could be nothing other than the universal, the species, which belongs to a certain *Moment* present in all actual assertions with the same sense, and which makes possible the identification just mentioned, even where the descriptive content of the individual experiences (*Erlebnisse*) of asserting varies considerably in other respects. (Husserl 1994, 201)

Husserl now saw the parts of Bolzano's *Wissenschaftslehre* on presentations and propositions in themselves as being an initial attempt to provide a unified presentation of the domain of pure ideal doctrines and as already providing a complete plan of a pure logic (Husserl 1994, 201-02; Husserl 1913, 36-38, 46-49; Husserl, 1908/09, 241). Hidden in that book was something that Husserl now saw as "one of the most momentous logical insights" that the "core content of any normative and practical logic consists in propositions that do *not* deal with acts of thought, but rather with those Ideas instanced in certain of their Moments" (Husserl 1994, 209), a key thesis of Lotze (Lotze 1888, Chapter II).

In his logic courses, Husserl would "reiterate and emphatically stress" that the ideal entities so unpalatable to traditional, and to empiricistic logic especially, and so consistently ignored by his contemporaries were not artificial inventions, but were given beforehand by the meaning of the universal discourse of propositions and truths indispensable in all the sciences. That indubitable fact, he emphasized, had to be the starting point of all logic. No more is to be meant by this ideality, he maintained, than that it is a matter of a kind of possible objects of knowledge, of objects whose particular characteristics can, and in scientific investigation must, be determined, while they are just not objects in the sense of real objects. (Husserl, 1908/09, 45, 47)

The continual talk of propositions, of true and false, Husserl taught, never at any time means what is reproduced in repeated stating, understanding, believing, seeing, but rather something identical and atemporal in contrast to it (*Ibid.*, 45). Science, he maintained, was a system of ideal meanings that unite into a meaning unit. The theory of gravitation, the system of analytic mechanics, the mechanical theory of heat, the theory of metric or projective geometry were all systematic units composed of ideal material, out of what we called meanings. And located in this ideal material were truth and falsehood, what science makes into an objective, supraindividual validity unit logically apprehending and exhausting a sphere of objectivity. (Husserl, 1906/07, §12)

Reality as objectivity, Husserl taught, comes under all forms and laws belonging to the essence of objectivity in general, and the theory of each real objectivity necessarily comes under the laws belonging to the theory in general of any objectivity whatsoever. Consequently, formal logic would be the science of this first a priori. On the other hand, the a priori belonging to the idea of reality as such would come under consideration. The body of truths relating to the essential categories of reality (thing, property, real relation between things, real whole, real part, cause and effect, real genus and species etc.) is a foundation and prerequisite for any further knowledge of reality. Taking into consideration the entire sphere of the sciences of reality, it is a necessary, common resource and "science theoretical" with respect to them. (*Ibid.*, §23).

The realm of truth, Husserl sought to impress upon students, is no disorderly hodgepodge. Truths are connected in systematic ways, are governed by consistent laws and theories, and so the inquiry into truth and its exposition must be systematic. The systematic representation of knowledge must to a certain degree reflect the systematic representation grounded in the things themselves. All invention and discovery involves formal patterns without which there is no testing of given propositions and proofs, no methodical construction of new proofs, no methodical building of theories and whole systems. No blind omnipotent power has heaped together some pile of propositions P, Q, R, strung them together with a proposition S, and then organized the human mind in such a way that the knowledge of the truth of P unfailingly, or in certain normal circumstances, must entail knowledge of S. Not blind chance, but the reason and order of governing laws reigns in argumentation. (Husserl 1896, 9, 13, 16-17)

Real things have their logical form insofar as they become the objects of statements formed in one way or another, and we can reflect upon what is attributable to objects in general in virtue of this form. But, this is basically only of interest because we aspire to knowledge of reality, and knowledge of form is naturally of extraordinary methodological significance for knowledge of things. Forms are precisely forms of actual and possible things, and without them, nothing becomes a thing knowable to us. Accordingly, through logic one could comprehend everything *a priori* belonging to the possibility of knowledge of reality in general, or if one likes, by logic understand theory of science, but not theory of science in general, rather theory of reality science in general. Then logic would embrace a dual a priori, one of pure form and one of content determined by form. (Husserl 1906/07, §23).

Sets, *Mannigfaltigkeiten*, and Analyses of Essence

Husserl's earliest attempts to obtain clarity regarding the meaning of the concepts of the theory of sets and cardinal number left him tormented by questions about the incredibly strange realms of actual consciousness and of pure logic. Disillusioned with the theories of those to whom he owed most of his intellectual training, he came to believe that what had been given as analyses of immanent consciousness had to be seen as pure a priori analysis of essence and this opened up the immense fields of the givens of consciousness as fields for "ontological" investigations". (Husserl, 1900-01, 41-43; Husserl 1903, 201-02; Husserl 1913, 17, 42; Husserl 1994, 490-91)

For Husserl, concepts and principles were purely logical which, abstracting from all subject matter, and owing to this abstraction, namely owing to their fully indeterminate universality with respect to content, relate to every possible field of knowledge, to every possible science, namely as concerns its theoretical content. As examples of concepts having such most universal significance, he gave object, characteristic, relation, whole, part, multiplicity, unit, cardinal number, order, universal, and particular. (Husserl 1906/07, §22)

He defined pure logic as formal, analytic logic, or the science of what is analytically knowable in general, a category into which we find him at various times putting all of the purely analytical theories of mathematics, algebra, arithmetic, number theory, the pure theory of cardinal numbers, the pure theory of ordinal numbers, set theory, *Cantorian* sets, the theory of manifolds in the broadest sense, the pure mathematical theory of probability, traditional syllogistics, the entire area of formal theories. He defended analytic logic against charges of being a "useless" spinning out of "sterile" formalizations, charges that he considered revelatory of considerable philosophical deficiency, a lack of understanding of crucial basic issues, and a disgraceful ignorance of the essence of modern mathematics and of the extraordinary significance that the scientifically rigorous, theoretical exploration of forms of pure deduction had acquired for the perfection and most rigorous grounding of the systems of pure mathematics in his day. (Husserl, 1908/09, 39, 244, 263-65; Husserl 1913, 28; Husserl 1994, 250, 490-91)

Whereas all of natural science is an a posteriori discipline grounded in experience with its actual occurrences, the world of the mathematical and purely logical is a world of ideal objects, a world of "concepts", Husserl argued. Pure mathematics, pure arithmetic, pure logic are a priori disciplines entirely grounded in conceptual essentialities. There, all truth is nothing other than the analysis of essences or concepts. With them, we are just not in psychology, in any sphere of empiricism and probability. The number series is a world of its own kind of ideal, not real, objects. The number 2 is not an object of perception and experience. Two apples come into being and pass away, have a place and time, but if they are eaten up, the number 2 is not eaten up. The number series of pure arithmetic has not suddenly then acquired a hole, as if we were to have to count 1, 3, 4. (Husserl 1906/07, §13c)

Pure arithmetic, Husserl pursued his reasoning along the same lines, explores what is grounded in the essence of number. It has nothing at all to do with nature. It is not concerned with things, physical things, souls, real occurrences of a physical or mental nature, does not acquire its universal propositions by perception and empirical generalizations on the basis of the perception and the substantiation of the resulting individual judgments. One does not state $a + 1 = 1 + a$ as a hypothesis that has to be established as true in further experience or else inductively in keeping with the methods of the natural sciences. Rather mathematicians start with $a + 1 = 1 + a$ as something unconditionally valid and certain, for it is obviously part of the meaning of the term "cardinal number" that each thing can be increased by one. To say that a cardinal number cannot be increased amounts to not knowing what one is talking about. It amounts to being in conflict with the meaning of "cardinal number". (*Ibid.*)

For Husserl, arithmetical laws, genuine axioms develop directly in the self-evidence of certainty. And this certainty and self-evidence carries over to all theses in deductive substantiation. All purely mathematical propositions, he reasoned, express something about the essence of what is mathematical, something about the meaning of what forms part of the same. Their denial is consequently an absurdity. In contrast, no proposition of the natural sciences, no proposition about real matters of fact is substantiated as being certain by self-evidence. Its denial never means an absurdity, a contradiction in terms. In denying the law of gravity, or the law of the parallelogram of forces, and the like, experimentation is cast to the wind, but this is not a matter of a contradiction in terms. Of two contradictory propositions, one is true and one false. That is generally to be looked upon as absolutely certain. Whoever denies this does not know what contradictory signifies, what true and false signify. One cannot deny this without casting the meaning of those words to the wind. The proposition is simply an "unfolding" of the intension of the "concepts". It is purely grounded in them. (*Ibid.*)

Appointed to the University of Göttingen in 1901, Husserl was quickly drawn into the discussions of the set-theoretical paradoxes in David Hilbert's circle of mathematicians (Peckhaus and Kahle; Husserl 1994, 442; Rang and Thomas 1981). Unpublished notes on set theory available at the Husserl Archives (Husserl, Ms A 1 35) show Husserl directly grappling with the questions raised by the set-theoretical paradoxes. In those notes, he recorded his ideas about just what the essence, the concept, of set entails. His reflections on sets and the set-theoretical paradoxes illustrate what he meant by analyses of essence.

Given his conviction that logical, mathematical laws are laws of essence, it is not surprising to find Husserl arguing over and over that the set-theoretical paradoxes must involve some violation of the essence of set. Wherever mathematicians speak of sets, he maintained, if the concept is to be a mathematical one, they must have a set essence in view, and whatever sets may have as an essence, it is expressed with a relation that belongs to the essence, i.e., the relation between sets themselves and elements of a set (*Ibid.*, 12b). It is part of the idea (*Idee*) of the set, he wrote, to be a unit, a whole comprising certain members as parts, but doing so in such a way that, vis-à-vis its members, it is something new which is first formed by them. All mathematico-logical operations performable with sets, he considered, turn on the idea that sets can be looked upon as kinds of wholes, as new units, formations that are something new vis-à-vis their original members, so that out of these formations new units can then again be formed. The unity of a system is something new vis-à-vis the elements systematized. It would be a contradiction in terms for the system's unity itself to be able to be one among the elements of the same system, upon those the system bases itself. But system units can themselves again be systematized and then ground higher forms of system. They then, however, bring elements into a new system whole. (*Ibid.*, 20b)

The paradoxes, he declared, only demonstrate that a general logic of sets in general, of totalities, is still lacking. He stressed that in his logic courses he had constantly from the beginning said that totality and set should not be identified and that this identification must be partly responsible for the paradoxes of set theory. He expresses his conviction there that we do not yet by any means have the real and genuine concept of set that logic needs. (*Ibid.*, 43a, 69a)

"It belongs essentially to the concept of set that (without contradiction) no set can contain itself as an element", Husserl repeated over and over (*Ibid.*, 24a). An essence relation, i.e., the relation between sets themselves and elements of a set, makes it impossible for the members of the relation to be identical. Hence a set that contains itself as an element would be a contradiction in terms (*Widersinnigkeit*) (*Ibid.*, 12b). A whole cannot be its own part. Just as it is contradictory for a whole to be its own part at the same time, so it is contradictory for a set to be its own member (*Ibid.*, 20b). To the objection that there is no set that contains itself as an element, he maintained that one need merely respond that that is a contradiction in terms (*Widersinnigkeit*) (*Ibid.*, 17a). If one is clear and distinct with respect to meaning, Husserl suggested in his notes, one readily sees the contradiction in terms (*Widersinnigkeit*) involved in the set-theoretical paradoxes. So, the solution to the paradoxes would then lie in demonstrating the shift of meaning that makes it that one is not immediately aware of the contradiction in terms and that once one perceives it, one cannot indicate wherein it lies. (*Ibid.*, 12a)

Husserl's search for answers to questions raised during his time in Halle took him beyond the confines of the mathematical realm towards the development of the science of deductive systems in general that he called his *Mannigfaltigkeitslehre*, theory of manifolds, which he considered to be the highest task of formal logic. His manifolds were theory forms, logical molds totally undetermined as to their content and not bound to any possible concrete interpretation. His theory of manifolds was his project for limning the true and ultimate structure of reality, a technique for engaging in pure a priori analyses of essence through an austere scheme of axiomatization that knows no acts, subjects, or empirical persons or objects belonging to actual reality. It was a matter of theorizing about possible fields of knowledge conceived of in a general, undetermined way, simply determined by the fact that the objects stand in certain relations that are themselves subject to certain fundamental laws of such and such determined form, are exclusively determined by the form of the interconnections assigned to them that are themselves just as little determined in terms of content as are the objects. (Husserl 1913, 35; Husserl 1900-01, 41; *Prolegomena*, §§69-70; Husserl 1906/07, §§18-19; Husserl 1917/18, Chapter 11, Hill and Rosado Haddock, Chapters 9, 10)

In his notes on set theory, Husserl's choice of manifolds turns on his convictions that objects stand in relations with respect to certain properties and that among these relations are those belonging to the essence (Ms A 1 35, 10a; also 35a, 40b, 43b, 48, 53a, 62a-b, 63a). In notes dating from the early 1890s, we find him already pointing out that by *Mannigfaltigkeit* Cantor merely meant an aggregate of any elements combined into a whole and that that concept does not correspond to that of Bernard Riemann and other related ones in the theory of geometry, for which, Husserl stresses, a *Mannigfaltigkeit* is an aggregate of elements that are not just combined into a whole, but are ordered and continuously interdependent. He defines order as

> a concatenation that has the special property that each member possesses an unambiguous position, in the narrow sense of the word, in relation to any arbitrary one, i.e., can therefore be unequivocally characterized by the mere form of the direct or indirect connection with the last one. (Husserl 1983, 93, 95-96)

Manifolds, he stressed, are not mere aggregates of elements without relations. It is precisely the relations that are essential and distinguish them from mere aggregates. (*Ibid.*, 410)

Husserl's notes show him suggesting a reform of the mathematical theory of manifolds by consciously transforming it into a transcendental theory of manifolds that consciously captures the formal essence of a genuine, constructible totality, consciously analyzes what belongs to the essence of a concept defining a totality, what belongs to the essence of an axiom and axiom system that as such establishes the univocity and construction and only establishes the meaning of what is constructed, which then formally weighs the conditions of the possibility of such a totality, and by this means derives the system of possible totality forms, or manifold forms. (Husserl Ms A 1 35, 38)

Conclusion

Investigations at the intersection of Cantor's *Mannigfaltigkeiten* and 19th century attitudes towards metaphysics and logic raise many unanswered questions, many more than philosophers seem to realize, about the ultimate structure of reality. A good many facts about what logical experimentation carried out there over the past hundred years has to tell us about what Bertrand Russell once colorfully called the ultimate furniture of the universe remain to be divulged and analyzed. In particular, the full, long, story of set theory's role in shaping modern logic and in redrawing the boundaries between metaphysics and logic in both the analytical and the phenomenological traditions is yet to be told and its full implications to be drawn.

Ironically, a strong anti-metaphysical animus thoroughly incompatible with the intentions of the creator of set theory was one of the most striking features of the new logic that grew out of his pioneering work. That antipathy for metaphysics did not just require eschewing 19th century metaphysical idealism in its various guises, but extended to include any vestige of it. So it happened that in the campaign to rid philosophy of metaphysics once and for all, metaphysical considerations of very different kinds were confused which should never have been.

Spurned and cast out too were essences, universals, ideas, senses, meanings, concepts, attributes essential properties, modalities, propositions, intensions... anything hinting of the a priori ... and, *above all*, anything that failed to obey the rules of the strictly "extensional" Eden into which the logical establishment strove to lock logical reasoning during much of the 20th century.

The emblematic figure in this was Willard van Orman Quine, who fought to defend his sterile realm of strong extensional calculi at all costs and made exposing and bewailing any *soupçon* of connivance with metaphysics and warning of the temptations and dangers of modal and intensional logics principal planks of his philosophical program. He admonished philosophers to remain within the confines of his logical Eden, to flee creatures of darkness, and to stay away from what he called curiously idealistic ontologies that repudiated material objects. He conjured up nightmare visions of the ontological crisis that would ensue were logicians to disobey his strictures and begin a retreat back into what he called "the metaphysical jungle of Aristotelian essentialism" (for example, Quine 1947, 43, 47; Quine 1960; Quine 1976, 159-76; 177-84; Hill 1997, Chapter 11).

Intense anti-metaphysical sentiment fostered an uncritical climate that helped unfounded anti-metaphysical prejudices to hold sway for decades. It was long professionally necessary to philosophize within the power of them and few dared to contradict what seemed false in them. Besides, it was convenient to adopt a shallow approach that made it easy to dismiss annoying, embarrassing, and disturbing questions about blunt, crude, blind features of a modern anti-metaphysical logic by charging questioners with harking back to a loathsome metaphysics to be overcome. It was realized that the logic contained many unsolved difficulties, but Quine and his followers found it desirable for achieving their ends, elegant, and aesthetically pleasing. So they, so to speak, sewed fig leaves together and made coverings for themselves. They thought of themselves as sailors trying to fix their craft out at sea and unable to bring it into dock, dismantle it, and use the best bits to build it anew, or so suggests the Quine's well-known epigram to his classic work *Word and Object*.

Fortunately, a handful of philosophers braved the strictures and adopted a bolder approach toward limning the true and ultimate structure of reality. They set out to increase the depth, versatility, and utility of the standard languages and to create languages capable of investigating epistemic and deontic contexts and of analyzing the many non-extensional statements that figure significantly in the empirical sciences, law, medicine, ethics, politics, and ordinary philosophy, but were being sloughed off by the philosophical establishment because they complicated matters by not conforming to the rigid standards set for admission into a logical world as stark as Quine's.

For example, finding extensional functional calculi "inadequate for the dissection of most ordinary types of empirical statement", Ruth Barcan Marcus insisted that "modal logic was worthy of defense, for... useful in connection with many interesting and important questions, such as the analysis of causation, entailment, obligation, and belief statements, to name only a few" (Marcus, 5). Dagfinn Føllesdal once pointed out that if Quine's judgment were to prove conclusive this would have "disastrous consequences", among which would be that "any attempt to build up adequate theories of causation, counterfactuals, probability, preference, knowledge, belief, action, duty, responsibility, rightness, goodness, etc. must be given up" (Føllesdal, 179, 184). Far too prolific to do him any justice has been Jaakko Hintikka, whose recent work on independence-friendly logic, to take just one example of his contributions to the field, represents another significant scheme of his for remedying problems that were written right into the foundations of the ineffectual logic that so many desired so ardently to have and to hold on to. (ex. Hintikka 2002; Hintikka 2004a; Hintikka 2004b)

The work of such logicians proved to be particularly effective in exposing logical form and displaying the inner workings and shortcomings of strong extensional systems. So reasons for not confining reasoning into an extensional mold mounted as objections that had muddled the issues for decades proved untenable and the Quinean hegemony was undermined. Metaphysical considerations were increasingly invoked and a more hospitable environment was created for the logical phenomena that had been vilified as metaphysical abstrusities, curious conceptions, unintelligible, mythical entities suspended between being and non-being. It was shown that they were not just scientifically permissible, but germane, even vital to knowledge and science and that it was most wrongheaded and impetuous to try to cleanse reasoning of them. What had been shunned and excoriated was increasingly shown to be just what was needed to provide the unity and continuity necessary to reasoning in philosophy and in science in general... just what is needed to clear up reasoning, help remove ambiguity and imprecision, draw fine distinctions both germane and indispensable to many scientific undertakings... the very thing needed to bring clarity, simplicity, precision, even elegance, to reasoning. Not just ethics and religion depend on the logico-metaphysical considerations in question, but knowledge and science in all its forms do. (Hill 1997)

Above, I told the story of how, after struggling with doubts about his early empirico-naturalistic approach to logic, sets, and the foundations of arithmetic, Husserl came to realize that what he had once thought of as metaphysical abstrusities could be conceptually sundered from the odious, pernicious, invidious forms of metaphysics that he had wished to avoid, how he came to embrace the a priori, to espouse an idealistic ontology, to philosophize in a metaphysical jungle of essentialism, to reason with intensions, in short, to adopt an metaphysical and logical perspective fundamentally antithetical to the one that was later imposed by the analytic establishment.

Husserl's conclusions accord with the late 20th century findings of philosophers like Ruth Barcan Marcus, who scandalized Quine and Quineans by declaring that essentialist talk was commonplace in and out of philosophy, that it was frequently unproblematic, and that it was surely dubious whether it could be replaced by nonessentialist, less 'problematic' discourse (Marcus, 55). For example, she dared to maintain that:

> A sorting of attributes (or properties) as essential or inessential to an object or objects is not wholly a fabrication of metaphysicians. The distinction is frequently *used* by philosophers and nonphilosophers alike without untoward perplexity. Given their vocation, philosophers have also elaborated such use in prolix ways. Accordingly, to proclaim that any such classification of properties is "senseless" and "indefensible," and leads into a "metaphysical jungle of Aristotelian essentialism" is impetuous. It supposes that cases of use that appear coherent can be shown not to be so or, alternatively, that there is an analysis that dispels the distinction and does not rely on equally odious notions. (Marcus, 54)

Being a human being or being gold or is not accidental, she pointed out.

> No metaphysical mysteries. Such essences are dispositional properties of a very special kind: if an object had such a property and ceased to have it, it would have ceased to exist or it would have changed into something else. If by bombardment a sample of gold was transmuted into lead, its structure would have been so altered and the causal connections between its transient properties that previously obtained would so have changed, that we would not reidentify it as the same thing. (Marcus, 69)

So, by the end of the 20th century, the logical road most taken ended up leading some of its most lucid philosophical minds to confront metaphysical issues and imperatives and will most likely continue to do so. However, upon reflection, in retrospect, logic being what it is, it is logical that that would transpire. For, if metaphysics is to be understood as *metaphysica generalis*, ontology, or the science of the most general categories of being understood not only as categories of what there is, but also as logical categories, we are not talking about the idea that logical or grammatical categories are mirrored in the ways *in which we structure being*, but about the idea that logical or grammatical categories mirror the ways *in which being itself is structured*. In both cases, metaphysics would elucidate certain universally applicable concepts, but only if logical or grammatical categories are taken as mirroring the ways being itself is structured are we talking about metaphysics as First Philosophy. In the other case, we are ultimately talking about some form of psychology or some derivative of Kantian transcendental idealism.

Cantor, Frege, and Russell all three recognized that there was something about the structure of being itself that could not be manipulated at will, that logic must in some sense mirror the ways in which being is structured or it would turn out illogical. Cantor said that he had been logically compelled to introduce new, strange, number classes almost against his will, but did not see how he might proceed further with set theory and function theory without them. When asked about the causes of the paradoxes of set theory, Frege once answered that the essence of the procedure leading into a thicket of contradictions consisted in regarding the objects falling under F as a whole, as an object designated by the name 'set of Fs', 'extension of 'F', or 'class of Fs'. He alluded to having been in a certain way forced to introduce them almost against his will because he saw no other alternative. In 1911, in a text that Husserl copied directly into his notes on set theory, Russell wrote of how logic and mathematics forced one to admit that there is a world of universals and of truths which do not bear directly on any particular existence. He affirmed that it was an ultimate fact that we have immediate knowledge of propositions about universals, that there was a priori and universal knowledge, and that all knowledge obtained by reasoning needed a priori, universal logical principles (Russell 1911). This sense that there is something about reality that resists manipulation is surely much of what is behind the new rehabilitation of metaphysical ideas, which has never been a matter of a return to the stifling atmosphere of 19th century idealism, or the excesses of irrationality, that engendered such determination to put an end to metaphysics once and for all.

References

Cantor, Georg 1883. *Grundlagen einer allgemeinen Mannigfaltigkeitslehre. Ein mathematisch-philosophischer Versuch in der Lehre des Unendlichen*, Leipzig: Teubner, cited as appears in Cantor 1932, 165-246.

Cantor, Georg 1884. "Principien einer Theorie der Ordnungstypen" (dated November 6, 1884), first published by Ivor Grattan-Guinness in *Acta Mathematica* 124, 1970, 65-107.

Cantor, Georg 1887/8. "Mitteilungen zur Lehre vom Transfiniten", *Zeitschrift für Philosophie und philosophische Kritik* 91 (1887), 81 125; 92 (1888), 240-65, cited as appears in Cantor 1932, 378-439.

Cantor, Georg 1932. *Gesammelte Abhandlungen*, Ernst Zermelo (ed.), Berlin: Springer.

Cantor, Georg 1991. *Georg Cantor Briefe*, H. Meschkowski and W. Nilson (eds.) New York: Springer

Dauben, Joseph 1979. *Georg Cantor, His Mathematics and Philosophy of the Infinite*, Princeton: Princeton University Press.

Føllesdal, Dagfinn 1969. "Quine on Modality", in *Words and Objections. Essays on the Work of W. V. Quine*, Donald Davidson and Jaakko Hintikka (eds.) Boston: Reidel.

Frege, Gottlob 1894. "Review of E. G. Husserl's *Philosophy of Arithmetic*", in *Collected Papers on Mathematics, Logic and Philosophy*, Oxford: Blackwell, 1984, 195-209.

Frege, Gottlob 1890-1892. "Draft towards a Review of Cantor's *Gesammelte Abhandlungen zur Lehre vom Transfiniten*", in his *Posthumous Writings*, Oxford: Blackwell, 1979, 68-71.

Frege Gottlob 1980. *Philosophical and Mathematical Correspondence*, Oxford: Blackwell.

H. Gerlach and H. Sepp (eds.) 1994. *Husserl in Halle*, Bern: Peter Lang.

Goodrick-Clarke, Nicholas 1992. *The Occult Roots of Nazism, Secret Aryan Sects and their Influence on Nazi Ideology*, London: I. B. Tauris & Co. Ltd.

Grattan-Guinness, Ivor 1978. "How Russell Discovered His Paradox", *Historia Mathematica* 5, 127-37.

Grattan-Guinness, Ivor 1980. "Georg Cantor's Influence on Bertrand Russell", *History and Philosophy of Logic* 1, 61-93.

Grattan-Guinness, Ivor 2000. *The Search for Mathematical Roots 1870-1940, Logics, Set Theories and the Foundations of Mathematics from Cantor Through Russell to Gödel*, Princeton: Princeton University Press.

Hallett, Michael 1984. *Cantorian Set theory and Limitation of Size*, Oxford: Oxford University Press.

Hilbert, David 1925. "On the Infinite", in Jean van Heijenoort (ed.), *From Frege to Gödel, a Sourcebook in Mathematical Logic, 1879-1931*, Cambridge MA: Harvard University Press, 1967, 376-92.

Hill, Claire Ortiz 1997. *Rethinking Identity and Metaphysics, On the Foundations of Analytic Philosophy*, New Haven CT: Yale University Press.

Hill, Claire Ortiz 1998. "From Empirical Psychology to Phenomenology: Husserl on the Brentano Puzzle", in *The Brentano Puzzle*, Roberto Poli (ed.), Aldershot: Ashgate, 151-68.

Hill, Claire Ortiz and Guillermo E. Rosado Haddock 2000. *Husserl or Frege, Meaning, Objectivity, and Mathematics*, La Salle IL: Open Court.

Hintikka, Jaakko 2002. "Hyperclassical Logic (A.K.A. IF Logic) and its Implications for Logical Theory", *The Bulletin of Symbolic Logic* 8, 3, September, 404-23.

Hintikka, Jaakko 2004a. "Independence-friendly logic and axiomatic set theory", *Annals of Pure and Applied Logic* 126, 313-33.

Hintikka, Jaakko 2004b. "On Tarski's Assumptions", *Synthese* 142, 353-69.

Husserl, Edmund 1887. *On the Concept of Number*, published in Husserl 1891 and in the English translation.

Husserl, Edmund 1891. *Philosophie der Arithmetik, mit ergänzenden Texten,* Husserliana XII, The Hague: Martinus Nijhoff, 1970, in English, *Philosophy of Arithmetic, Psychological and Logical Investigations with Supplementary Texts from 1887-1901*, translated by Dallas Willard, Dordrecht: Kluwer, 2003.

Husserl, Edmund 1896. *Logik, Vorlesung 1896*, Dordrecht: Kluwer, 2001.

Husserl, Edmund 1900-01. *Logical Investigations*, New York: Humanities Press, 1970.

Husserl, Edmund 1902/03a. *Allgemeine Erkenntnistheorie, Vorlesung 1902/03*, Dordrecht: Kluwer, 2001.

Husserl, Edmund 1902/03b. *Logik, Vorlesung 1902/03*, Dordrecht: Kluwer, 2001.

Husserl, Edmund 1905. *Briefwechsel, Die Brentanoschule I*, Dordrecht: Kluwer, 1994.

Husserl, Edmund 1906/07. *Introduction to Logic and Theory of Knowledge 1906/07*, translated by Claire Ortiz Hill, Dordrecht: Springer, 2008.

Husserl, Edmund 1908/09. *Alte und Neue Logik, Vorlesung 1908/09.* Dordrecht: Kluwer, 2003.

Husserl, Edmund 1913. *Introduction to the Logical Investigations, a Draft of a Preface to the Logical Investigations (1913)*, The Hague: Martinus Nijhoff, 1975.

Husserl, Edmund 1917/18. *Logik und allgemeine Wissenschaftstheorie 1917/18.* Dordrecht: Kluwer, 1996.

Husserl, Edmund 1919. "Recollections of Franz Brentano", in *Husserl: Shorter Works*, McCormick, and F. Elliston (eds.), Notre Dame: University of Notre Dame Press, 1981, 342-49. (Also translated by Linda McAlister in her *The Philosophy of Brentano*, London: Duckworth, 1976, 47-55)

Husserl, Edmund 1929. *Formal and Transcendental Logic*, The Hague: Martinus Nijhoff, 1969.

Husserl, Edmund 1994. *Early Writings in the Philosophy of Logic and Mathematics*, translated by Dallas Willard, Dordrecht: Kluwer.

Husserl, Edmund 1983. *Studien zur Arithmetik und Geometrie, Texte aus dem Nachlass (1886-1901)*, The Hague: Martinus Nijhoff.

Husserl, Edmund Unpublished, *Ms A 1 35*, manuscript on set theory available in the Husserl Archives in Leuven, Cologne and Paris.

Husserl, Malvine 1988. "Skizze eines Lebensbildes von E. Husserl", *Husserl Studies* 5, 105-25.

Jones, Henry 1895. *A Critical Account of the Philosophy of Lotze, the Doctrine of Thought*, Glasgow: James Maclehouse and Sons.

Lotze, Hermann 1888, *Logic*, New York: Garland, 1980 (reprint of B. Bosanquet's translation of his *Logik*).

Marcus, Ruth Barcan, 1993. *Modalities*, New York: Oxford University Press.
Peckhaus, Volker and R. Kahle 2000/2001. "Hilbert's Paradox", *Report No. 38, 2000/2001*. Institut Mittag-Leffler, The Royal Swedish Academy of Sciences, published on the internet.
Quine, Willard 1947. "The Problem of Interpreting Modal Logic", *Journal of Symbolic Logic* 12, 2 (June), 43-48.
Quine, Willard 1960. *Word and Object*, Cambridge MA: M.I.T. Press.
Quine, Willard 1976. *Ways of Paradox*, Cambridge MA: Harvard University Press.
Rang, B. and W. Thomas 1981. "Zermelo's Discovery of Russell's Paradox'", *Historia Mathematica* 8, 16-22.
Russell, Bertrand 1903. *Principles of Mathematics*, New York: Norton.
Russell, Bertrand 1911. "The Philosophical Implications of Mathematical Logic". *The Monist* 22 (October 1913) 481-93 and in *Essays in Analysis*, London: Allen & Unwin, 1973, 284-94. Partially translated in Husserl Ms A I 35. Originally published in French in 1911.
Russell, Bertrand 1959. *My Philosophical Development*, London: Unwin, 1985.
Russell, Bertrand 1927. *Principia Mathematica*, Cambridge UK: Cambridge University Press, 1964.
Schoen, Henri 1902. *La Métaphysique de Hermann Lotze, ou la philosophie des actions et des réactions réciproques*, Paris: Librairie Fischbacher.

16

Jairo José da Silva

GÖDEL AND TRANSCENDENTAL PHENOMENOLOGY

I

Kurt Gödel believed that we could perceive concepts, among other abstract entities like sets, in a way not unlike the perception we have of physical objects (at one point he even believed that there might exist a physical organ involved in the perception of concepts). Like bodies in physical space concepts were believed to be perceived one aspect at a time, in a sequence of partial perspectives that hopefully would harmonize in a progressively clearer apprehension of the concept. Misperception of concepts was supposed to be as likely to happen as misperception of common objects, though both could be corrected by further and more careful observations. Concepts, Gödel believed, must exist as objective entities in order to make sense of our experience of the world.

Edmund Husserl, the founder of phenomenology, also believed that concepts, or for that matter just about anything abstract or concrete, could be perceived, and that the structure of experiences of perception of concepts and physical objects was basically the same. The original sin of naturalism, a point of view to be abandoned in order to gain admission to the realm of phenomenology, was, according to Husserl, its inability to conceive any other form of existence but that of natural objects existing in space and time and able to enter in causal interactions.[1]

This proximity of their points of view might be only a matter of coincidence had Gödel not been, as he was after 1959, an attentive reader of Husserl, whom he took as one of the three philosophers, together with Plato and Leibniz, most congenial to his own way of thinking. Hence the hypotheses that Gödel's views were actually influenced by Husserl *and*

---

[1] As one can see, *causal* theories of perception (and *causal* theories of knowledge) are not to be expected in phenomenology.

that Gödel's ideas on conceptual and set intuition could be given a phenomenological reading arise naturally and have been suggested by some philosophers like Charles Parsons and defended by others like Richard Tieszen.

In this paper I do not want to question the undeniable fact that Husserl had a strong influence on Gödel's central philosophical ideas, such as the possibility of a rigorous philosophy, the distinction between sciences of facts and sciences of essences, and the relevance of the method of intuition of essences, but what I *do* want to deny is that Gödel's philosophical views can properly be called phenomenological. I believe that Gödel basically tried to bring together his own Platonism with respect to the existence of concepts and sets and Husserl's method of essential intuition, interpreted as a way of approaching these objects, without paying enough attention to the fact that the inaugural act of the phenomenological attitude, the phenomenological reduction (*epoché*), was incompatible with the metaphysical theses he cherished.

In what follows, I shall first introduce some basic notions of transcendental phenomenology in order to clarify Husserl's position in the realism versus idealism debate. Second, I shall analyze Gödel's Platonism and related issues, both before and after he became acquainted with Husserl's philosophy, highlighting both the similarities and dissimilarities with Husserl's ideas, so as to support my thesis that Gödel's views, in particular his notion of intuition, are inconsistent with some fundamental aspects of transcendental phenomenology.

II

As Husserl explained in his preface to Boyce Gibson's translation of *Ideas I*, transcendental phenomenology is the descriptive science of the essential structures of "transcendental subjectivity". It is an "eidetic" (i.e. "directed upon the universal in its original intuitability") (Husserl 1913a, 5-6) *a priori* science, since its objects are not facts but essences, namely the essential structures of transcendental subjectivity, revealed in eidetic intuition.[2] Transcendental phenomenology is restricted to "the realm of essential structures of transcendental subjectivity immediately transparent to the mind". (*Ibid.*, 6)

Transcendental subjectivity and its experiences are given to me by means of an act of abstention (*epoché* or *phenomenological-transcendental reduction*) which, so to speak, "purifies" my mind and its experiences as those of a person living among others in the natural world in order to obtain the essence of mental life in general. As a prolegomenon to this

---

[2] A means by which we can, according to Husserl, obtain the universal by going through imaginable variations of an instance of it and preserving that which they have in common.

reduction I concentrate my attention exclusively on my mental experiences "setting aside all the psycho-physical questions which relate to man as a corporeal being" (*Ibid.*, 8), thus obtaining the realm of a purely descriptive *psychology*, which Husserl also once called "phenomenological psychology", an appellation he later rejected as inadequate. At this level, mine is still a human mind immersed in nature. Reduction proper begins when I refuse to make *any* existential claim with respect to the world that my mind experiences and simultaneously remove my mind from its natural setting. As a consequence,

> I am now no longer a human Ego in the universal, existentially posited world, but exclusively a subject *for* which this world has being, and purely, indeed, *as* that which appears to me, is present to me, and of which I am conscious in some way or other, so that the real being of the world thereby remains unconsidered, unquestioned, and its validity left out of account. (*Ibid.*)

This pure, denaturalized I that no longer inhabits any world, but, rather, is the *raison d'être* of the existence of any world at all is transcendental subjectivity (*Ibid.*, 11). This reduction introduces what Husserl calls the phenomenological-transcendental attitude. By taking it, we no longer posit a subjectivity that is only part of a world and a world whose being is independent of any subjectivity, rather, from the phenomenological-transcendental perspective any posited domain has being *only* as a correlate of the experiences of a pure I that has, and alone has, absolute existence for which evidence is immediately given. (See Husserl 1913a, §46.)

In transcendental phenomenology, there is no being that is not a correlate of experiences of the transcendental I that endows it with the "meaning" it has. The "meaning" of an object is what provides the referential link between the I and the object *as intended by the I*. But it is important to keep in mind that the object in question is not supposed to exist outside the "transcendental field", i.e. the domain of transcendentally reduced consciousness, its experiences, and the correlates of these experiences. In other words, the intended object is merely an *intentional* object, i.e. an object-for-the-I. The question as to whether any intentional object has *real* existence—understanding by this a form of existence conceived in the manner of the objects of the natural world, that is, consciousness-independent in a strong sense—is a question that makes *no sense* in the transcendental attitude. Since nothing can be without being an object *for* the pure I, there is no neutral standpoint, with the I on the one side and the world on the other, from which to consider the question. Since the question makes no sense, none of its possible answers makes sense either, which is already a clear indication that any imputation of idealism, or realism, in the traditional sense, to transcendental phenomenology is incorrect. Both the claims "this object exists independently of the experiences in which it is 'given' to me" and its negation are meaningless in transcendental phenomenology, for, observing the restrictions imposed by the *epoché*, I cannot *conceive* of what would make either claim true. The transcendental attitude requires a sustained refusal to consider these questions because they are senseless.

If, as I believe, this interpretation is correct, then how can Husserl say that, for instance, in an experience of perception of an object what is perceived is not a mental image of the object but rather the object *itself*, which exists out there as an object of the natural world, located in space and time and having the properties that it presents itself as possessing? Is he not presupposing the *real existence* of the percept while viewing my experience of perception as only the way in which the object is *given* to me? The answer is a resounding no. For, among other reasons, if he were presupposing this, transcendental phenomenology would be indiscernible from descriptive psychology, two areas of inquiry that Husserl insists upon keeping separate, although in constant dialogue with each other. See, for instance, his preface to *Ideas I* (Husserl 1913a, 7-9). Let us see what Husserl had to say about this:

> [S]tarting as men in our natural setting, the real object is the thing out there. We see it, we face it, we have turned our eyes towards it and fixed them upon it, and as we find it there in space over against us, so we describe it and make our statements concerning it.... If we now carry out the phenomenological reduction, every transcendent setting, that above all which is bound up with perception, receives its suspending bracket, which envelops all the derivative acts, every perceptual judgment with the valuations grounded in it, and eventually the judgment of value, and so forth. What it comes to is this: we suffer all these perceptions, judgments and so forth, but only as conditions that they be regarded and described as the essentialities which they are in themselves; if anything in them or in relation to them is presented as self-evident, that we establish and fortify. But we allow no judgment that makes any use of the affirmation that posits a "real" thing or "transcendent" nature as a whole, or "cooperates" in setting up these questions. As *phenomenologists* we avoid all such affirmations. But if we "do not place ourselves on their ground", do not "co-operate with them", we do not for that reason cast them away. They are still, and belong essentially to the phenomenon as a very part of it. Rather, we contemplate them ourselves; instead of working with them, we make them into objects; and we take the thesis of perception and its components also as constituent portion of the phenomenon.... The "bracketing" which perception has undergone prevents any judgment being passed on the perceived reality (i.e. any judgment that has its ground in the unmodified perception and therefore accepts its thesis as its own). But it does not hinder any judgment to the effect that perception is the consciousness *of* a real world (provided the thesis thereof is not set in action). (Husserl 1913a, §90)

I quote Husserl at length because in this important passage he clearly undermines the interpretation accepted by many analytical philosophers that, since he does not contest the independent existence of the world, there is an irreducible realist core in the phenomenological theory of perception. In fact, nothing of the sort is the case, since consciousness *of* a real world is a purely intentional experience confined to the transcendental field, it *does not* and it *cannot* support the *thesis* (already eliminated by the *epoché*) that there is a real world independent of consciousness that is given to me in perception. As a man in the world I see an object of the world with properties and relations typical of objects of the world. My *interest* is focused on the world. As a *descriptive psychologist* my interest turns to the mental experiences by means of which I perceive this object, my interest turns "inward". In either case my standpoint is still the *natural attitude*, in which the object-consciousness dichotomy still holds. As I perform the *epoché* and "ascend" to the phenomenological transcendental point of view, as a *phenomenologist,* my interest changes radically, it is now focused on the object purely as a "*unity of meaning*", all the properties that can be attributed to the object based on this specific experience of perception are constitutive of this meaning, *nothing* is subtracted from the object when we "ascend" to the transcendental point of view. Anything that I as a man in the world accept as evident with respect to the object of my perceptual experience, as a phenomenologist I also accept as constitutive of the unity of meaning "this object of perception", only now the "meaning" which the object has "emanates" from the I, not from the object. The world-consciousness dichotomy is abolished, and it is this that constitutes the Copernican Revolution introduced by the transcendental-phenomenological attitude.

To endow an object with meaning is what Husserl calls the constitution of the object. As Paul Ricoeur says, "*Sinngebend* and *Konstitution* are perfectly synonymous" (Husserl 1913b, 183 n. 1). Phenomenological constitution, as Husserl calls an entire domain of transcendental problems, takes "in all the conceivable objects we could ever meet with in experience, briefly the whole real world spread out before us together with all its categories of the object, and likewise all 'ideal' worlds [*including, of course, the world of mathematical objects*–my note], and makes these all intelligible as transcendental correlate" (his preface to Husserl 1913a, 13).

So, the "meaning" which an object has as *given* to me in my pre-phenomenological disposition is *preserved* by the reduction, but as the meaning *bestowed* upon the object by transcendental subjectivity. Transcendental constitution is not a production, a free creation, it cannot endow an object with a meaning it does not have. We can say, metaphorically, that transcendental subjectivity is not completely free. It cannot, for instance, constitute the unity of meaning "spatial object" as excluding the meaning "transcendent" (i.e. that which can present an indefinite array of aspects) for it is part of the "meaning", or the essence, of a spatial object to be transcendent. Just as we (or even God) cannot draw a square circle.

> Unities of meaning presuppose... a *sense-giving consciousness* which, on its side, is absolute and not dependent in its turn on sense bestowed on it from another source.... *An absolute reality is as valid as a round square*. Reality and world, here used, are just the titles for certain valid unities of meaning, namely, unities of "meaning" related to certain organizations of pure absolute consciousness which dispense meaning and show forth its validity in certain *essentially* fixed, specific, ways. (Husserl 1913a, §55)

In the natural attitude the existence of the world as an absolute reality is not questioned. But when we enter the proper realms of philosophy through its only gate, the phenomenological reduction, we must abandon this pre-philosophical naiveté, realism. Husserl notices that any form of *philosophical* realism (as opposed to the naive realism of the natural attitude) is as *absurd* as any form of idealism that stands in contrast to it. The absurdity arises, as already pointed out, from the senselessness of their opposing theses considered from the phenomenological transcendental perspective. In his preface to the first English edition of *Ideas I*, Husserl wrote,

> [N]ow as ever I hold every form of current philosophical realism to be in principle absurd, as no less every idealism to which in its own arguments that realism stands contrasted, and which in fact it refutes. (Husserl 1913a, 12)
>
> So long as it was only the psychological subjectivity that was recognized, and one sought to posit it as absolute, and to understand the world as its correlate, the result could only be an absurd Idealism, a psychological Idealism–the very type which the equally absurd realism has as its counterpart. (*Ibid.*, 15)

In Husserl's view, psychological subjectivity and any "world", real or ideal, are not, *as a matter of principle*, from the transcendental perspective, absolute. Therefore both psychological idealism (which asserts that psychological subjectivity is the only absolute) and realism (which asserts the absoluteness of the real world) are *in principle absurd*. Only transcendental subjectivity is absolute, this founding *evidence* is the basic thesis of transcendental idealism, the only possible philosophical perspective. In fact, as Husserl points out, phenomenological transcendental idealism is not, properly speaking, a thesis, a point of view among others, rather, it is the *only* perspective coherent with transcendental phenomenology (*Ibid.*, 14). "Transcendental phenomenology... is universal Idealism worked as a science". (*Ibid.*, 13)

Also:

> [P]henomenology is ... *transcendental idealism*, though in a fundamentally new sense. It is not so in the sense of a psychological idealism which wants to deduce a meaningful world from meaningless sense data. It is not an idealism of the Kantian type which believes that it can leave open the possibility of a world of things in themselves, if only as a limit concept. It is an idealism which is *nothing* more than an operation of making explicit my ego as a subject of possible knowledge.... This idealism is not formed by a chain of arguments and is not opposed to some "realism" in a dialectical confrontation. It is the *rendering explicit of the sense* of any type of being that I, the ego, can imagine... which means: to unveil in a systematic way the constituting intentionality itself. *The proof of this idealism is phenomenology itself.* Only those who misunderstand the deep sense of the intentional method or the sense of the transcendental reduction–or both–can want to separate phenomenology and transcendental idealism. (Husserl, 1992, 143-44)

From the point of view of transcendental idealism, any "reality" is relative to transcendental subjectivity, i.e. it is only a correlate. Therefore, any sense this reality has is bestowed on it by transcendental subjectivity (his preface to Husserl 1913a, 14).

> If anyone objects, with reference to these discussions of ours, that they transform the whole world into subjective illusion and throw themselves into the arms of an "idealism such as Berkeley's," we can only make answer that he has not grasped the *meaning* of these discussions. We subtract just as little from the plenitude of the world's Being, from the totality of all realities, as we do from the plenary geometrical Being of a square when we deny... that it is round. It is not that the real sensory world is "recast" or denied, but that an absurd interpretation of the same, which indeed contradicts its *own* mentally clarified meaning, is set aside. It springs from making the world absolute in a *philosophical* sense.... Absurdity first arises when one philosophizes and... fails to notice that the whole being of the world consists in a certain "meaning" which presupposes absolute consciousness as the field from which the meaning is derived. (Husserl 1913a, §55)

How, then, can Husserl say that "no ordinary 'realist' has ever been as realistic and as concrete as I, the phenomenological 'idealist' (a word which by the way I no longer use)"?[3] First, Husserl refuses to use the word 'idealism', not because he has given up transcendental idealism, but because this expression has created all sorts of misunderstandings which persist even today (see Husserl 1913a 12, for instance). Second, Husserl does *not* deny the existence of the world, or even doubt it (the *epoché* is *not* Cartesian doubt), he only denies its *absolute* existence (it is precisely absoluteness that is eliminated by the *epoché*).

The following is a pervasive interpretation of Husserl in analytical philosophy. The phenomenological reduction, it is said,

> is Husserl's way of describing the turning of attention away from both objects in the world and psychological activity to the mental contents which make possible the reference of each type of mental state to each type of object. (Dreyfus and Hall, 2)

This turning away of attention does not, supposedly, alter the ontological status of the world, which still unquestionably exists "in itself" out there. In this interpretation, the natural world is taken to be causally responsible for the basic matter (*hyle*), primarily sensations, which noesis unifies "into a set of appearances of one object"[4], and hyletic data, the "boundary conditions" or constraints that our "in-forming" (in the Aristotelian sense of endowing with form) of the world must respect.[5] This shows, so goes the interpretation, that we are not completely free to

---

[3] Letter from Husserl to Abbé Baudin in 1934, quoted in Iso Kern, *Husserl und Kant: Eine Untersuchung über Husserls Verhältnis zu Kant und zum Neukantianismus* (Phenomenologica 16), The Hague: Martinus Nijhoff, 1964, 276n. Reproduced in Dreyfus and Hall, 182.

[4] Dagfinn Føllesdal, "Husserl's Theory of Perception", in Dreyfus and Hall, 93.

[5] Dagfinn Føllesdal, "Brentano and Husserl", in Dreyfus and Hall, 40.

"mold" the world to our liking, which, it is claimed, shows that there is a clear anti-idealistic aspect in Husserl's thought. In short, still according to the "analytic" interpretation, Husserl's philosophy displays a combination of realism (with respect to the world), Platonism (with respect to *noemata*) and constructivism (with respect to *noeses* and the phenomenological constitution); Husserl is a realist in ontology and an idealist in epistemology.[6]

The suspension of the *thesis* (as opposed to the *pre-philosophical* unquestioned *certainty*) of the existence of the world, the exclusive absoluteness of transcendental subjectivity, the absurdity of realism and psychological idealism as *philosophical* theses, the unavoidability of *transcendental* idealism in the phenomenological attitude are all conspicuously missing in this interpretation. In fact, since it does not take the phenomenological reduction to its full extent, it is closer to being an account of genetic psychology rather than transcendental phenomenology.

The origin of such an interpretation lies, I believe, in the fact that the reception of Husserl's thought among analytical philosophers was mediated by Frege's ideas. The parallel usually drawn between Frege's notions of experience, sense and reference and Husserl's notions of noesis, noema and the intentional object, respectively, contaminated Husserl's thought with Frege's realism and induced a tension between realism and idealism in the pioneering analytical interpretations of Husserl's transcendental philosophy.

Fortunately, more recently we see a fair amount of work on Husserl that is more sensitive to the aspects of his thought that are less palatable to analytical philosophy. Michael Dummett, for instance, raises doubts about the effectiveness of sensations for guaranteeing objectual reference for perception and warding off the threat of "idealism" (Dummett 1993, 82-83). But still here the distinction between transcendental idealism proper and the (according to Husserl, absurd) denial of (or, at least, doubt about) the existence of the world is not clearly drawn. Herman Philipse, on the other hand, clearly recognizes Husserl's unconditional commitment to transcendental idealism, to the point of drawing him close to George Berkeley,[7] who, it must be admitted, Husserl himself saw as a forerunner of transcendental philosophy (Husserl 1913a, 15). Even hyletic data, according to Philipse, fail to point to an ego-independent world (as Føllesdal believes).[8] But it is Harrison Hall,[9] I believe, who presents the interpretation that best does justice to Husserl's position on the traditional debate realism versus idealism: Husserl cannot, *as a matter*

---

[6] This last aspect of the interpretation seems to be favored, in particular, by Føllesdal. See, for instance, his introduction to Gödel *1961/? (Gödel 1995, 372).

[7] "Transcendental Idealism", in Smith and Smith, 239-322.

[8] See "Husserl's Theory of Perception", in Dreyfus and Hall, 95.

[9] "Was Husserl a Realist or an Idealist?" in Dreyfus & Hall, 169-90.

of principle, side with either position. This is precisely the interpretation for which I have been arguing. We turn now to the analysis of Gödel's ideas on the nature of mathematical concepts and objects.

III

As mentioned earlier, Gödel only began to read Husserl in 1959, so we can say with certainty that whatever Gödel said before this date about the issues that concern us here, in particular that of perception of concepts and sets, was not directly influenced by Husserl's thought. Nonetheless, independently of any influence, some of Gödel's philosophical ideas, even before 1959, might seem to invite a phenomenological reinterpretation. In what follows, I shall analyze Gödel's Platonism and his views on mathematical intuition in order to determine to what extent he was influenced by Husserl after 1959 and to what extent, if at all, his ideas permit a phenomenological interpretation, despite any direct or indirect influence Husserl may have had on him.

In the Gibbs lecture of 1951 we read that mathematical "concepts form an objective reality of their own, which we cannot create or change, but only perceive and describe" (Gödel 1951, 30). Gödel aptly named this point of view "Platonism" and distinguished it from Aristotelian realism, for he did not think that concepts were only aspects of things. Of course, such a view was thought to be clearly opposed to any form of psychologism or conventionalism. Gödel believed, in fact, that mathematical concepts inhabit a third realm, outside the human mind and the world of things exterior to the mind, which was, nonetheless, accessible to the human intellect. However, the access to this third realm was in no way considered unproblematic. Gödel believed that we "perceive" concepts in a way that closely resembles the way we perceive the real world. In both cases we can expect only a limited and incomplete, and sometimes even deceptive, perception. This parallel between the "perception" of concepts and the perception of real objects, Gödel says, can be extended further. Certain methods typical of the physical sciences, such as inductive methods, Gödel thought, are banned from mathematics more from prejudice than anything else. The world of mathematics is as objective and mind-independent as the physical world, and hence there is no reason why it cannot be explored by similar methods (this is a striking claim, indeed, for if taken seriously it would amount to abandoning the ideal of science that has guided the development of mathematics since Euclid).

We can summarize Gödel's Platonism as presented in the Gibbs lecture in one sentence: mathematical concepts belong to a realm of their own that we do not create and cannot change but only describe and to which we have access in a way (some form of "intuition", which Gödel does not investigate) not unlike the way we perceive physical objects.

This "intuition" is obviously a notion that must be better understood before Gödel's view can be accepted as something more than a metaphor.

Husserl could have offered Gödel a way of understanding this relation of our consciousness to an objective realm of ideal (as opposed to real) entities in terms of the central notion of intentionality. However, within the context of transcendental phenomenology, the objectivity of the world of concepts could not be understood in naturalistic-like terms; rather, it had to be understood in terms of the "objectifying" functions of (individual and also intersubjective) transcendental consciousness. Consequently, in order to accommodate phenomenological reduction, the ontological status that Gödel granted to the third realm of mathematical concepts would have to be drastically revised. It would have to be redefined as an "ontological region" inhabited by "unities of meaning" conceivable only as correlates of intentional experiences. But nothing of this sort can be found in Gödel's writings published so far.

In "Russell's Mathematical Logic", published in 1944, Gödel wrote:

> classes and concepts may, however, also be conceived as real objects... existing independently of our definitions and constructions. It seems to me that the assumption of such objects is quite as legitimate as the assumption of physical bodies and there is quite as much reason to believe in their existence. They are in the same sense necessary to obtain a satisfactory system of mathematics as physical bodies are necessary for a satisfactory theory of our sense perceptions.... (Gödel 1990, 128)

In this passage, Gödel is not saying that classes and concepts *are* real objects existing independently of our definitions and constructions, but that they may be so *conceived*. One possible interpretation for this quotation would be that, for Gödel, concepts and classes were mere *cogitata*, useful, or perhaps indispensable for best organizing our experience, but about whose *real* existence we would suspend disbelief. This attitude does not necessarily coincide with an outright assertion of real existence. Incidentally, according to this interpretation, assertions of the same kind can be made with respect to physical bodies and the existence of the external world.

Although very convenient for the phenomenological reinterpretation of Gödel, for presumably it would make it possible to see concepts and classes as only intentional objects, i.e. "mere" correlates of mental experiences, such an interpretation has not in fact been put forward by phenomenologically oriented readers of Gödel. And the reason is that it is almost certainly wrong. Although Gödel on at least one occasion *did* consider the possibility that *nothing* might correspond *objectively* to experiences of mathematical intuition (1963 supplement to "What is Cantor's continuum problem?"), he never took this suggestion seriously, as far as I know. But a stronger objection to the claim that for Gödel classes and concepts were *cogitata* is that if this were the case the role of subjectivity in connection with perceptions of these objects would be much enhanced and *something like* a *proper* phenomenological inquiry into the structure of subjectivity, at least with respect to these experiences, would be urgently needed. But, as Wang tells us (Wang 1987, 122), even *after* becoming acquainted with Husserl's later philosophy, which he was not at the time he wrote "Russell's Mathematical Logic", Gödel "probably did not accept Husserl's emphasis on subjectivity".

This is a point that we need to discuss further. If the kind of existence Gödel claimed for classes and concepts were to be understood as "merely" intentional, it would be incomprehensible why Gödel not only never discussed the role played by transcendentally reduced consciousness in the constitution of these objects when phenomenology became available to him, but also why he seemingly never accepted or appreciated Husserl's extremely elaborate analysis of the transcendental Ego. I think that this is strong enough evidence for believing that Gödel always thought, even after 1959, that classes and concepts, albeit ideal entities, existed just like real objects, independently of our mental or intentional experiences. I believe that Gödel was a metaphysical realist who never questioned the real existence of the physical world and did not see any reason for not believing in the real existence of the world of mathematics as well, believing as he did that there were more similarities than differences between the two, and that mathematical intuition was the equivalent of sense perception.

Hao Wang, who was for many years close to Gödel, classifies him unreservedly as a realist and says that, according to Gödel, "a (mathematical) 'realist' is someone who considers mathematical objects to exist independently of our constructions and of having an intuition of them individually..." (Wang 1987, 296). Another piece of evidence for Gödel's metaphysical realism can be found in his belief in the existence of *abstract* impressions, which he saw as the equivalent of sense impressions, both being *immediately* given to us. He even speculated about the existence of some physical organ, "closely related to the neural center for language", for the handling of these abstract impressions (Wang 1974, 85). All this seems to point to the fact that Gödel really believed in the *independent* existence of abstract entities that were somehow responsible for these abstract impressions. Consider the following quotation from Wang:

> Gödel sometimes calls his position 'objectivism' instead of 'realism'. We are, he says, more certain that we have objectivity than that we have found the right objects. For example, we are certain that there are infinitely many prime numbers and that Fermat's conjecture is either true or false. G. infers from this fact *there must be objects* [my emphasis], but we are, he seems to say, less certain that the natural numbers are the 'right' objects. (Wang 1987, 285)

It seems to follow from this that for Gödel abstract objects such as numbers can be seen as mere conceptions that we may eventually abandon, but these conceptions must also *correspond* to something existing independently of any conception. With respect to sets in particular, Gödel presumably saw any of the available clarifications of the concept of set embodied in the different axiomatizations of it as a particular apprehension of a concept that lies out there, still only partially apprehended by any conception and hence not reducible to any of them. The belief that such a concept exists as an independent entity is what, I believe, gives Gödel the conviction that a definite and complete clarification that would make possible the decision of any question concerning the concept, such as the continuum hypothesis, *must* exist. Otherwise, if the notion that lies beyond all of its apprehensions were only a conception, Gödel's rationalist optimism would seem unfounded, for a conception could very well be incomplete and uncompletable, whereas an existing entity cannot.

So far, little has been said about mathematical intuition and the reason for this is that Gödel himself did not say much in order to clarify his views about intuition before the well-known addendum of 1963 to "What is Cantor's Continuum Problem?" Since by then Gödel was already familiar with Husserl's philosophy, we can be reasonably confident that it was precisely Husserl's elaborate analysis of the intentional experience of an object being "bodily" present, as opposed to being only represented to consciousness, which is, according to Husserl, what characterizes intuitions, and also the prominent role this notion plays in Husserl's epistemology, that led Gödel to present a somewhat more detailed analysis of his views on the matter.[10]

Consider the following well-known quotation:

> But, despite their remoteness from sense experience, we do have something like a perception also of the objects of set theory, as is seen from the fact that the axioms force themselves upon us as being true. I don't see any reason why we should have less confidence in this kind of perception, i.e. in mathematical intuition, than in sense perception.... It should be noted that mathematical intuition need not be conceived of as a faculty giving an *immediate* knowledge of the objects concerned. Rather it seems that, as in the case of physical experience, we *form* our ideas also of those objects on the basis of something else which *is* immediately given. Only this something else here is *not*, or not primarily, the sensations. That something besides the sensations actually is immediately given follows (independently of mathematics) from the fact that even our ideas referring to physical objects contain constituents qualitatively different from sensations or mere combinations of sensations, e.g., the idea of object itself, whereas, on the other hand, by our thinking we cannot create any qualitatively new elements, but only reproduce and combine those that are given. Evidently the "given" underlying mathematics is closely related to the abstract elements contained in our empirical ideas. It by no means follows, however, that the data of this second kind, because they cannot be associated with actions of certain things upon our sense organs, are something purely subjective, as Kant asserted. Rather they, too, may represent an aspect of objective reality, but, as opposed to the sensations, their presence in us may be due to another kind of relationship between ourselves and reality. (Gödel 1990, 268)

---

[10] Parsons 1995 gives us an accurate and detailed historical and critical analysis of the development of Gödel's Platonism and his not always clear and completely articulate views about mathematical intuition. Since my interest here lies exclusively in the relation of Gödel's ideas concerning these issues to Husserl's thought, I concentrate on the texts that I think illuminate this problem and refer the reader wanting a more elaborate historical analysis of the development of Gödel's philosophical ideas to Parsons' article.

Gödel's views on mathematical intuition as expressed in this text seem to be essentially the following:

1. Mathematical intuition is both *de re* and *de dicto*, we can perceive (i.e. intuit) both objects (like sets or concepts) and truths about these objects (the latter being apparently more fundamental than the former). It seems that the truth of the axioms is given *first*. The perception of the objects of which these axioms are true is somehow dependent on the perception of basic truths about them. In an earlier article (Gödel *1959/9), as pointed out by Parsons (Parsons 1995, 59), Gödel can also be interpreted as accepting the primacy of intuitions *de dicto* over intuitions *de re*.[11]

It is tempting to see here something similar to Husserl's idea that intentional objects are always presented to us by means of *noemata*, complex structures that include components (*noematic "Sinne"*) that, when linguistically expressed, take the form of assertions about the objects "framed" by the *noemata*. This is one aspect of Gödel's views on mathematical intuition that may have been influenced by Husserl.

2. Mathematical intuition can be viewed as strictly analogous to sense perception. The idea that we can perceive abstract objects in a way similar to the perception of real objects is, of course, an idea present in Husserl. But the fact that Gödel says the same thing *before* being acquainted with Husserl's philosophy shows that there is no direct influence of Husserl on him with respect to this particular point. Nonetheless, this aspect of transcendental phenomenology was certainly very attractive to Gödel.

3. Mathematical intuition (like sense perception) does not give us immediate or absolutely reliable access to the objects intuited; rather, it usually requires that we develop our ideas about them. Nonetheless, these ideas are *not* free creations. They must conform to certain *immediately* given *data*. In the case of sense perception the *data* are sense impressions. In the case of mathematical intuition they are *abstract* impressions, which always come associated with sense impressions but cannot be reduced to them. Abstract impressions are then *immediately given* elements from which we somehow obtain our ideas about mathematical objects like sets.

A number of assertions are made here. They all demand clarification. First, sense data contain abstract components. Second, we can *develop* certain ideas having these abstract impressions as boundary conditions. It is not clear if by "ideas" Gödel meant abstract *objects*, like concepts, or "meanings" referring to these objects (some analogue to Husserl's *noemata*). In any case, as we saw above, objects and "ideas" about them are not independent of one another. Incidentally, we notice here a constructive aspect of Gödel's notion of intuition: ideas about mathematical objects are *our* creations, although not completely free creations since they must

---

[11] This interpretation is in opposition to Føllesdal's introductory note to *1961/? (Gödel 1995, 370), for whom Gödel might have taken the intuition of objects as primordial.

conform to certain boundary conditions. Third, intuitions do not have an apodictic character, i.e. they are not absolutely reliable guides to truth. The ideas we develop can be restructured provided the boundary conditions are respected.

All these views have a distinctively Husserlian character. It is a commonplace that Husserl distinguishes between, for instance, the perception of *a* and *b* and the perception of *a and b*. In the second case, above the perception of the *objects* there is *also* the perception of the *copula*, which is an abstract component of the perception of a state-of-affairs. The copula, is, to use a Husserlian term, a *non-independent content* of the state-of-affairs, which means that it cannot be conceived as existing separately from the objects *a* and *b*, which are the *independent* contents of the state-of-affairs. Gödel seems to be saying something similar to this when he says that abstract impressions are *contained* in empirical perception.

But there is an important Husserlian distinction that Gödel does not make, or at least not explicitly: for Husserl there is no *real* difference between *a* and *b* and *a and b*, i.e. *two* objects and a *pair* of objects are not distinguishable as elements of the natural world. But they are perfectly distinguishable as *objects of perception,* the *objective* difference between them being, according to Husserl, the *objective correlate* of a "mental" experience (*Erlebnis*), that which makes two objects into a pair. To the higher level intentional experience that makes a pair out of two objects that are given in a lower level experience, there corresponds something objective which lies, with respect to these objects, at a higher level of abstractness. Moreover, the higher level *abstractum* is, to use a Husserlian notion, dependent on the lower level objects. This hierarchy is absent in Gödel's considerations. In his view, sense impressions and abstract impressions both seem to be *simple data*. Moreover, Gödel does not mention the role played by consciousness in the *constitution* of abstract impressions, or the complex objects of which they are components. Rather, he says that they are immediately *given*, as if our minds somehow only "captured" them without being actively involved in their constitution as objective entities. In fact, in Gödel's theory of mathematical intuition, abstract impressions play a role similar to the *hyle* in Husserl's transcendental theory of perception. Similar, but not identical, since for Husserl hyletic data are not constraints imposed on consciousness by a reality *independent* of it, as seems to be the case with Gödel's abstract impressions.

The final remark made by Gödel in the passage just quoted defies straightforward interpretation. What could possibly be meant by what Gödel called "another type of relationship" between ourselves and reality? Was Gödel postulating some sort of sixth sense for the perception of concepts, as might be suspected from his earlier talk of a physical organ involved in this task? Or was this cryptic "another type of relationship" meant to be understood as the intentional relationship Husserl describes?

Independently of what Gödel had in mind, to understand this relationship in terms of an intentional relationship as Husserl understands it would be, as should be evident by now, inconsistent with the ontological status Gödel conceded to the realm of concepts. On the other hand, I do not believe that a *proper* notion of intentional relationship was what he had in mind, for otherwise it would be inexplicable why he did not mention, or even hint at the *active* role transcendental consciousness (which is not to be confused with the mind) plays in the affair of constituting abstract impressions. In fact, Husserl explicitly says the opposite, that they are *not* something purely subjective. It seems that, for Gödel, abstract impressions are *caused* by independently existing entities (impressions are, after all, impressions *of something*). This "something" cannot affect our sense organs but they might somehow affect our brains. (Remember Gödel's talk of a physical organ for the handling of abstract impressions located somewhere in the brain. The fact that he even considered this possibility shows that intentional relations were not the type of non-sensorial relations he had in mind with respect to abstract impressions).

Finally, Gödel's insistence on giving abstract impressions "objective reality" and denying that they are "purely subjective" shows clearly that he was far from appreciating the commonplace in transcendental phenomenology that *objectivity must be grounded in subjectivity*. Missing this point is to miss the point of transcendental idealism, and consequently transcendental phenomenology, altogether.

Nothing in the quotation we are analyzing contradicts the following interpretation: the objects of mathematical intuition are *not* properly speaking "things out there", but rather are more or less faithful "sketches" of things out there, in the sense of being conceptions of and ideas about things out there that we develop on the basis of certain impressions we receive from them whose main task is to impose limits on our freedom to develop these conceptions and ideas. For Gödel, mathematical intuition does *not* provide a direct reference to independently existing objects; rather, it provides (provisional, uncertain, revisable) "knowledge" (we might also say a *representation*) of an objectively and *independently* existing "something" on the basis of abstract impressions "emanating" from it. Gödel seems to believe that in the course of science, through the shared effort of cooperating mathematicians, these partial sketches (or representations) will probably become more and more faithful and we can hope that they will eventually stabilize in a true and complete description of a transcendent reality.

It is not difficult to notice many phenomenological or phenomenologically inspired ideas at work here. To mention but a few: something like a generalization of the notion of *hyletic data* (the abstract impressions); the ideas that abstract objects can also be objects of a generalized form of perception; that "truths" known by intuition can be corrected, i.e. that intuitions are not always irrefutable; that the intuitively given is not necessarily immediately given; that the objectivity of objects and truths given in intuition is expressed in the intersubjectivity of a community of co-workers. So, I believe we can safely say that Gödel's notion of (mathematical) intuition does indeed owe a lot to Husserl's notion of intuition.

Nonetheless there are some *essential* aspects of Gödel's notion that do not square with the phenomenological notion. The most important is Gödel's apparent refusal to follow Husserl in confining intuitive experiences to the realm of mere *intentional* experiences. For Gödel, mathematical intuition points to "something" out there that could *not* have survived the phenomenological reduction, whereas for Husserl intuitions never transcend the intentional life of consciousness. Gödel's understanding of Husserl's transcendental phenomenology does not seem to give the notion of phenomenological reduction its due role, and so he cannot appreciate its elimination of the gap between being and consciousness. Phenomenological epistemology cannot be a theory of how we simply came to *know* a world that transcends the intentional realm. As Husserl says in "Philosophy as Rigorous Science":

> [I]f knowledge theory will nevertheless investigate the problems of the relationship between consciousness and being, it can have before its eyes only being as the correlate of consciousness, as something 'intended' after the manner of consciousness. (Husserl 1965, 89)

This is only a way of saying that *there is no being outside consciousness*. But Gödel's notion of intuition is, I believe, still committed to the world-consciousness and phenomena-noumena dichotomies (although not in a strictly Kantian sense).

The relation between Gödel's and Husserl's views can be better understood if we turn our attention now to a paper by Gödel called "The Modern Development of the Foundations of Mathematics in the Light of Philosophy" (Gödel 1995, 374-87), where Gödel is explicitly concerned with Husserlian philosophy, or rather methodology. For Gödel, phenomenology possesses a method for clarifying meanings that can precede the proper mathematical job of defining and proving. According to him:

> [C]larification of meaning consists in focusing more sharply on the concepts concerned by directing our attention in a certain way, namely, onto our acts in the use of these concepts, onto our powers in carrying out our acts, etc.... phenomenology... is [or in any case should be] a procedure or technique that should produce in us a new state of consciousness in which we describe in detail the basic concepts we use in our thought, or grasp other basic concepts hitherto unknown to us. (*Ibid.*, 383)

It is clear from this quotation that Gödel practically identifies the phenomenological method with phenomenological *reflection*, the turning of attention from the objects of consciousness to the experiences in which they are given to us, together with phenomenological *description*. There is no mention at all of the *epoché* or transcendental subjectivity. As viewed by Gödel, phenomenology is indistinguishable from phenomenological or descriptive *psychology*. He continues:

> [T]he whole phenomenological method, as I sketched it above, goes back in its [[central]] ideas to Kant, and what Husserl did was *merely that he first formulated it more precisely*, made it fully conscious and actually carried it out for particular domains [my emphasis]. (*Ibid.*, 385)

Here Gödel is clearly saying that the phenomenological method, as he understands it, is essentially Kantian and that Husserl *only* formulated it more precisely. It is illuminating for the question concerning Gödel's understanding of Husserl to see what Husserl himself considered as an essential aspect of the phenomenological method and its relation to Kant's philosophy. In *Erste Philosophie,* Husserl says that his method, the phenomenological-transcendental method, includes the search for the "origins" of objectivity in transcendental subjectivity, that is, the whole of "constitutive phenomenology" (Husserl 1923/24, 382-83),[12] which is, Husserl points out, completely alien to Kant's philosophy: "Of this method Kant and the entire neo-Kantianism that depends on him did not have the slightest idea" (*Ibid.*, 382). Gödel's claim that Husserl's method is essentially Kantian shows how little he appreciated Husserl's originality with respect to Kant, and that he did not see the fundamental differences between Kant's and Husserl's brands of transcendental idealism. For Kant, transcendental subjectivity is *part* of the world and is *affected* by it, whereas for Husserl it is neither one nor the other, rather it is the objective world, which must be constituted in it. Husserl criticizes Kant *precisely* for not drawing a clear distinction between empirical and transcendental subjectivities and, therefore, lapsing into a form of transcendental psychologism. According to Husserl, Kant ignored completely the new provinces opened up by the *epoché*. We can say of Gödel what Husserl says of Kant, that he confuses transcendental and empirical subjectivities.

Instead of idealism, though, Gödel embraces realism, from which perspective he criticizes Kant. According to Wang, "one aspect of Gödel's objectivism is to deny Kant's sharp separation between the thing-in-itself and our mind. In particular, he questions that 'synthesis' must be subjective" (Wang 1987, 228). In another passage, Wang says that, "Gödel's objectivism offers a way to bypass the problem of intersubjectivity that is so difficult for Husserl". (*Ibid.*, 219)

As we know, since for Husserl the object of intuition is only a *cogitatum*, he must solve this "difficult problem of intersubjectivity" in order to grant objectivity to objects of intuition. If, as Wang believes, Gödel's notion of objectivity by itself offers a way to bypass the problem of intersubjectivity, this seems to be only because it already implies that

---

[12] In Husserl 1990, 21-22.

objects of possible intuition are *not* mere *cogitata*, but rather independently existing objects.

According to Wang, "Husserl's considerations of *Wesenschau* can be borrowed to support G's belief in the objective existence of mathematical objects" (*Ibid.*, 304). But, nonetheless, Gödel did not "understand Husserl's apparent emphasis on subjectivity". Wang continues: "It is not clear to me whether Husserl only uses subjectivity as a tool to refine a realist position". In my opinion, it is quite clear, from the passages just quoted, that Wang fails to appreciate the scope and problems of transcendental phenomenology and the *central* position that *transcendental* subjectivity plays in it. His belief that Husserl's notion of *Wesenschau* can support Gödel's notion of mathematical (or more specifically, conceptual) intuition seems to derive from this failure. Were he able to understand the reasons for Husserl's (surely not only apparent) emphasis on subjectivity, he would also be able to see that Husserl and Gödel did not have the same notion of objectivity. For Gödel objectivity implied *reality*, primarily of *facts* and derivatively of *objects*, whereas for Husserl it had no such an implication (understanding by "reality" the quality of things whose existence is in no way dependent on being thought).

In *Reflections on Kurt Gödel*, Wang tells us that Gödel studied Husserl's "Philosophy as Rigorous Science" carefully. So, it would be interesting to see what Gödel might have learned from it and what he probably failed to learn. First, something Gödel did not seem to appreciate, although it is quite clearly stated by Husserl, that "being" is to be always understood as a correlate of consciousness. Though the notion of *epoché* is only introduced explicitly in 1913, it is already clearly at work in this essay and its main function is to eliminate the dichotomy between *being* and *consciousness*. Being has a sense only as being-for-consciousness, whereas consciousness is always consciousness-of-something.[13]

A note by Quentin Lauer in the English translation of "Philosophy as Rigorous Science" can stand as a possible answer, within the phenomenological perspective, to Gödel's (and Wang's) assumption that an objective realm of being can be *given* to conscious-ness without being *constituted* in it. Lauer writes:

> Since objectivity is a function of pure consciousness, the rôle of philosophy is to ground objectivity by investigating consciousness. To the very end Husserl will insist on the philosophical task as one of "self-knowledge". (Husserl 1965, 90 n.26)

Some of the things Husserl says in "Philosophy as Rigorous Science" certainly must have appealed to Gödel very much, for instance:

---

[13] See, for instance, the famous §49 of *Ideas I* where this idea is expressed in an exemplary fashion.

> Intuiting essences conceals no more difficulties or 'mystical' secrets than does perception.... [T] hat the 'essences' grasped in essential intuition permit, at least to a very great extent, of being fixed in definite concepts and thereby afford possibilities of definitive and in their own way absolutely valid objective statements, is evident to anyone free of prejudice. (Husserl 1965, 110-11)

Two ideas in "Philosophy as Rigorous Science" certainly impressed Gödel greatly. First, the idea that philosophy must be a science of essences (not of facts) and that essences can be perceived. Second, that scientific philosophy should be a cooperative effort cutting through space and time, not restricted to a *Weltanschauung*, i.e. a manner of thinking, valuing and acting in accordance with the spirit of the time. We can see these ideas supporting Gödel's essentialism and his understanding of what reflection on the foundations of mathematics (set theory, in particular) should be focused on. But Gödel failed to see that for Husserl the *only* rigorous philosophy is phenomenology, whose inaugural act is the reduction of all being to consciousness and believed that he could put Husserl's ideas to work for an epistemology of mathematics alien to the new Copernican Revolution brought about by the phenomenological *epoché*, precisely the *reduction* of the realm of being to the realm of consciousness.

References

Dreyfus, Hubert L. and Harrison Hall (eds.) 1982. *Husserl, Intentionality and Cognitive Science*, Cambridge MA: MIT Press.
Dummett, Michael 1993. *Origins of Analytical Philosophy*, Cambridge MA: Harvard University Press.
Gödel, Kurt 1951. "Some Basic Theorems on the Foundations of Mathematics and their Implications", in Gödel 1995, 304-23.
Gödel, Kurt 1961/?. "The Modern Development of the Foundations of Mathematics in the Light of Philosophy", in Gödel 1995, 375-87.
Gödel, Kurt 1990. *Collected Works* vol. II, S. Feferman et al. (eds.), Oxford: Oxford University Press.
Gödel, Kurt 1995. *Collected Works* vol. III, S. Feferman et al (eds.), Oxford: Oxford University Press.
Husserl, Edmund 1913a. *Ideas, General Introduction to Pure Phenomenology*, translated by W. R. Boyce Gibson New York: Collier.
Husserl, Edmund 1913b. *Idées directrices pour une phénoménologie*, translated by Paul Ricoeur, Paris: Gallimard.
Husserl, Edmund 1923/24. *Erste Philosophie (1923/24) Erste Teil: Kritische Ideengeschichte,* Rudolf Boehm (ed.), Husserliana VII, The Hague: Martinus Nijhoff, 1959.
Husserl, Edmund 1960. *Cartesian Meditations*, translated by Dorion Cairns, The Hague: Martinus Nijhoff.
Husserl, Edmund 1965. *Phenomenology and the Crisis of Philosophy: Philosophy as Rigorous Science and Philosophy and the Crisis of*

European Man, translated by Quentin Lauer, New York: Harper & Row.

Husserl, Edmund 1990. *Kant e l'idea della filosofia trascendentale,* translated by Claudio La Rocca, Milan: Mondadori.

Kern, Iso 1964. *Husserl und Kant: Eine Untersuchung über Husserls Verhältnis zu Kant und zum Neukantianismus,* The Hague: Martinus Nijhoff.

Parsons, Charles 1995. "Platonism and mathematical intuition in Kurt Gödel's thought", *The Bulletin of Symbolic Logic* 1, 44-74.

Smith, Barry and Smith, David Woodruff (eds.) 1995. *The Cambridge Companion to Husserl,* Cambridge: Cambridge University Press.

Tieszen, Richard 1992. "Kurt Gödel and Phenomenology", *Philosophy of Science* 59, 176-194.

Wang, Hao 1974. *From Mathematics to Philosophy,* London: Routledge & Kegan Paul.

Wang, Hao 1987. *Reflections on Kurt Gödel,* Cambridge, MA: MIT Press.

17

Jairo José da Silva

MATHEMATICS
AND THE CRISIS OF SCIENCE

Edmund Husserl's last book, *The Crisis of European Science and Transcendental Phenomenology*, was written during one of the bleakest periods of European history.[1] European culture, he rightly saw, was in crisis. But more than a political crisis, Husserl saw a cultural crisis, which he thought phenomenology could help to overcome by bringing about cultural renewal.

But, interestingly, Husserl chose to focus his criticism on physical science rather than other aspects of culture. This choice may seem curious, but Husserl believed that the modern science of Nature exemplified paradigmatically what he took to be wrong with modern culture in general. For the crisis of culture was, Husserl thought, essentially a crisis of *meaning*, or the lack thereof. So, he saw what he supposed to be a manifestation of this crisis in modern physical science (by which we usually mean science from the time of the scientific revolution of the 17th century on)—namely, in the fact that a gap exists between the practices of the modern science of Nature and the sources from which he believed those practices derive their meaning. He adopted this as a vantage point from which to conduct his diagnosis and to suggest the prophylaxis that he considered adequate for this supposed cultural malaise of modernity (perhaps the prominent position physical science enjoys in modern culture and the methodological paradigm it posits for other sciences also played a role in his choice of target).[2]

---

[1] According to Gérard Granel (Husserl 1936a, I-IX), the main text dates from 1935-36.
[2] Given the success of modern physics, Husserl had to separate clearly its *technical* accomplishments, which he thought must be preserved, from what he took for an "alienated" interpretation of its methods, which he thought must be overcome for the sake of cultural renewal. But, as I hope to establish here, Husserl's remedy for this 'alienation', namely, driving physical science back to the "sources from which meaning derives", rooting it firmly in the *Lebenswelt*, interferes with scientific methodology.

But how, according to Husserl, does this crisis manifest itself in science? Is it related to the exciting new developments that had occurred in physics and mathematics not long before the book was written?

In the beginning of the 20th century, mathematics was going, and had been going for some time already, through a period of crescent formalization,[3] leaving behind the supposed solid grounds of intuition[4] and giving free rein to pure, formal imagination.[5] Bernhard Riemann,[6] for instance, had created the theory of general abstract geometrical manifolds; Georg Cantor, the theory of transfinite numbers; William Rowan Hamilton, the theory of quaternions, and non-commutative algebras in general; Sophus Lie, the theory of transformation groups; Hermann Grassmann, the theory of extensions.[7] But none of these creations seemed problematic to Husserl. All the above theories were duly appreciated by him and are mentioned here only because Husserl explicitly praised them (see §§ 69-70 of the *Prolegomena* (Husserl 1900)), but, let us make this clear, not as *scientific* theories in the strictest sense, since these theories do not provide us with *knowledge* of *particular* types of objects, but as formal ontological theories, being as they are theories of purely formal manifolds, i.e. logical *forms* that could in principle, but maybe not actually (or maybe *not yet*), give form to materially determinate domains. Formal mathematical theories are, according to Husserl, theories of objectual domains determinate as to form but indeterminate as to content (which he named *formal domains*) belonging to a corner of formal logic that he called formal ontology.[8] The emergence of formal

---

[3] By "formalization", I do not mean axiomatization within the context of formal-logical systems, but the tendency to privilege purely formal mathematical theories, which characterize their (purely formal) domains independently of any intuition.

[4] The objectual and conceptual varieties of mathematical intuition give us objects and concepts respectively prior to their theories, whose basic truths only express the intuitively given.

[5] Which allows us to invent theories independently of any prior intuition, but maybe answering to other demands, such as the needs of mathematics and science, the desire to generalize already existing theories–Riemann and Cantor being classical cases–, or the pursuit of "esthetic" goals such as intrinsic elegance or beauty, etc.

[6] See, for instance, Riemann's lecture "On the Hypotheses that Lie at the Bases of Geometry". (Riemann 1854)

[7] *Ausdehnungslehre*, basically a forerunner of vector analysis.

[8] For Husserl, a *formal domain* (or manifold) is essentially a structured system of materially indeterminate objects (*formal objects*) functioning in the system as, basically, contentless placeholders. The structuring relations of formal manifolds are characterized only formally, i.e. independently of the nature of particular materially determined (proper) objects that may take the place of formal objects upon interpretation (which can be construed as the filling of formal objects with material content). We can also define a formal domain simply as the abstract aspect common to all models of a formal theory. A formal domain, Husserl says, is the objective correlate of a formal theory (see Chapters 6 and 7 for still another concept of formal domain that can be found in Husserl's writings). Husserl's ideas concerning the fundamental distinction between material mathematical sciences, such as geometry and mechanics, and formal mathematical sciences, such as the theory of quaternions, were strongly influenced by Grassmann's views as put forward in *Die lineale Ausdehnungslehre* (Grassmann 1844). In this work Grassmann makes this

mathematics (which dates back to the dawn of modern mathematics in the 16th century, with the introduction of the so-called imaginary numbers by the Italian algebraists–if not before, with the creation and development of algebraic thought in the Islamic empire), was definitely not, for Husserl, *in itself,* the cause of any crisis in mathematics.[9]

Physics, in turn, had by then seen two groundbreaking developments: the theories of relativity (special, 1905; general, 1916), which Husserl did not mention, and quantum mechanics, which he did. Quantum theory is particularly interesting in its use of mathematics, compared with the traditional mathematical sciences of Nature (and this is something to which we should pay attention, since, as already mentioned, the mathematization of science can, for Husserl, contribute to the cultural crisis phenomenology is called to overcome). Although Born, Jordan, Heisenberg and Dirac had succeeded in formulating quantum mechanics in terms of matrix calculus, this did not follow naturally from an idealized *picture* of the phenomena, a mathematical model of reality, that is, one which required the matrix approach as the *correct* approach. The creators of quantum mechanics relied simply on the *formal* analogies they discerned between empirical rules of calculation and properties of matrix operations.[10]

Other mathematical formulations of quantum mechanics, such as wave mechanics, offered even more striking examples of the new use of mathematics. For example, de Broglie's model of the free electron as a particle accompanied by a purely mathematical wave whose frequency was associated with the electron's energy (and whose velocity was *higher* than the velocity of light, and so was a priori forbidden by relativity theory to represent a real wave) could suggest to the concerned observer (as Husserl surely was) that science had completely lost touch with reality, let alone reality effectively experienced by the senses, in favor of purely mathematical reconstructions of it (Heisenberg's uncertainty principle or the loss of causality and determinacy in the atomic scale but reinforced this perception).

---

distinction clearly, claiming that formal mathematical theories, unlike material ones, are not theories of domains existing independently of them, but *forms defined by these theories themselves* (which, he continues, implies that the axioms of formal theories are not axioms in the proper sense, i.e. fundamental unproved truths, but *definitions*). The concept of truth, according to Grassmann, also changes: in material theories, truth is correspondence with the facts; in formal theories, it simply means consistency. Michael Crowe's *A History of Vector Analysis* (Crowe 1994) contains a detailed exposition of Grassmann's *Ausdehnungslehre*). The similarity to some of Husserl's (and also David Hilbert's) fundamental conceptions concerning the nature of formal mathematics is striking. But, unlike Husserl, who locates formal theories not in mathematics proper, but in formal logic (formal ontology, to be precise), Grassmann does the opposite, reserving the term "mathematics" to formal mathematics and denying physical geometry, mechanics and like material theories mathematical status in the strictest sense.

[9] As we shall see, it is the *use* of purely formal mathematical theories that Husserl saw as a possible source of philosophical problems, if not kept under surveillance so as to avoid degenerating into a form of "technique" void of intuitive content and alienated from life.

[10] Basic principles, such as the principle of correspondence and the technique of quantization, are based mostly on formal analogies.

Husserl was certainly informed of these new trends and one might believe that he thought the "Galilean" mathematical substruction of Nature had gone too far in quantum theory.[11] But this belief cannot be sustained; Husserl did not seem to believe that the crisis was brought about by quantum theory. Although he evidently noticed the obvious differences between the old and the new physics, he did not think quantum theory *in particular* was to be blamed; the problem, he thought, could not be imputed exclusively to contemporary science.[12]

What Husserl meant, in fact, was not a crisis *in science*, but *of science*; it was the *project* of a mathematical science of Nature that he thought was facing a crisis, not in its methods, results, or practical relevance, but in its lack of self-consciousness, by being oblivious to the sources and scope of validity of its concepts and methods. According to Husserl, the winds of crisis had been blowing for some time already—since the first days of modern science, to be more precise—and were stirred by, basically, an alienated conception of Nature and the uncritical use of formal mathematical methods (with the emphasis on "uncritical"). This undesirable situation, he thought, could only be overcome by philosophy, not any philosophy, however, but the kind of philosophy he advocated, in which the very idea of an objective, transcendent Nature, completely determined in itself, which speaks the language of mathematics is traced back to its origins in transcendentally reduced consciousness.[13] Our intuitions and "living experiences", due to their privileged epistemological status, should provide the guarantee for the rightful use and scope of validity of formalist thinking in science. In short, for Husserl, both the diagnosis of a crisis of science and the prophylaxes suggested served the goal of

---

[11] In fact, the mathematical treatment of quantum phenomena does not fit the model that Husserl presented for classical physics: experience → mathematization of experience via abstraction and idealization→mathematical investigation of the properties of mathematized reality → explanation of past experiences and prevision of new ones. The second step is missing. However, since the classical approach was, for him, already crying out for philosophical clarification; and since his theme was not the applicability of mathematics to science in general, but a critique of the scientific conception of Nature and the mathematical methods of science, he did not become involved with the problem that quantum mechanics poses for the philosophical investigation of the scientific applicability of mathematics in general, even though it *does* pose problems for *his* account of the applicability of mathematics in science.

[12] It is important to remark that Husserl was not an enemy of science, modern or traditional. Rather the opposite, he thought that the mathematical sciences of Nature were admirable endeavors, needing only some doses of philosophical criticism in order to be properly understood and circumscribed to their rightful domain of validity (but, unfortunately, Husserl's critique of "technization" would lead to Heidegger's critique of "technique" and eventually to a general post-modernist aversion to science).

[13] "a meticulous intentional analysis, strictly free of prejudices and in absolute evidence... does not deprive in the least the natural conception of the world, that of daily life and also of the exact science of Nature, of its sense, but proceeds to the interpretation of what is effectively and properly contained in this sense". (This quote comes from appendix I of Husserl 1936a; these appendices belong to a selection made by Walter Biemel from the Husserl's manuscripts known as *Forschungsmanuscripte*. The English version is mine).

establishing phenomenology as a first philosophy that could work for a renewal of humanity based on *personal* responsibility.

For Husserl, this crisis was innate to the modern mathematical science of physical Nature whose creation he attributed to Galileo and other scientists of that period. The problem, he thought, did not lie in treating Nature mathematically, but in the following related points: 1) the obliteration of the intentional acts of idealization that allowed this, and, consequently, the naïve attitude of taking as an independent reality what is intentionally constituted, namely, idealized Nature, an entity that *by constitution* completely and satisfactorily escaped apprehension by our senses (a form of Platonism that dislodges the direct intuitive access to Nature from the fundamental position and foundational role Husserl thought it had by right); 2) the enthronization of mathematical methods as the only truly scientific ones, which had to be extended to any intellectual endeavor we call scientific, including psychology and phenomenology, if phenomenology, as Husserl wanted, insisted on maintaining its pretensions to scientific *exactitude*; and 3) the uncritical use of *formal* mathematics, a scientific methodology demanding critical assessment.

For Husserl, only transcendental phenomenology could deal with these problems, by clarifying the intentional constitution of idealized (or mathematized) Nature and determining the domain of legitimacy of the mathematical methods of the science of Nature and their scope[14] (what might entail, for instance, what is "left out" in the idealization process, what is given *directly* in experience may be precisely what is of interest to us in other areas of science such as psychology, the establishment of whose foundations, as we know, Husserl thought of as one of the main goals of phenomenology).[15] By securing, but also restricting, the scope of mathematics in science in general, Husserl hoped to open the way for *non-mathematical*, purely descriptive sciences aiming to remain faithful to the intuitively given *as the fundamentally given*, such as phenomenology itself. Remember that phenomenology, conceived of as a pure science of essences given in eidetic intuition, is out of the reach of mathematical methods (see, for instance, *Ideas* §71*ff*.). In short, Husserl naturally saw a

---

[14] "it is not the case of dominating technically and practically the method as a 'qualified worker' in the field of activity of the physicist, but to understand, by a regression to the ways of thinking of the creators of the method, including their mutations, its proper sense and rightful limits". (Appendix I of Husserl 1936 a–see n. 13. The English version is mine.)

[15] "The present situation in Europe of general collapse of spiritual humanity does not change anything in the results of the science of Nature; and these results, in their autonomous truth, do not contain any reason for the reform of the sciences of Nature. If such reasons exist, they have to do with the relationship of these truths to scientific and extra-scientific humanity and spiritual life. It is what is psychic and spiritual in general, in their collapse that forces us to create an effective and authentic psychology that makes human existence, personal existence, personal life, personal actions and their spiritual consequences, and the personal community that is built upon these actions and these consequences finally comprehensible and then make us see the edifice of a new humanity". (Appendix I of Husserl 1936a–see n. 13. The English version is mine).

critique of the uses of mathematical methods in science, their justification and delimitation, as a much needed prelude to phenomenology understood as a non-mathematical, albeit *scientific*, first philosophy.[16]

My aim here is a very restricted one: to analyze Husserl's criticism of the mathematical science of Nature and check whether his recipe for overcoming the "crisis" of science can be followed in face of the specific nature of scientific methodology. Although not in a prominent position, the philosophical problem of the applicability of mathematics to the physical sciences appears in *Crisis*,[17] and I want to assess the correctness of Husserl's model of how this happens, to assess whether abstracting and idealizing from perceptual experience can *always* provide the mathematical models science needs (the question is whether mathematical models of physical reality are *always* abstract, idealized versions of the reality *perceived*– which, of course, is for Husserl already an intentional construct on the basis of raw sensorial data), whether the genesis of these models, whenever available, can *always* be traced back to the life-world, and whether there is any reason, *as far as science is concerned*, for somehow restricting the use of *purely formal* mathematical methods.[18] If we answer these questions negatively, the project of avoiding formalistic "degeneracy" in the mathematical sciences of Nature by rooting them in the *Lebenswelt* clashes with scientific methodology.[19] Husserl would, then, be interfering with, rather than simply clarifying, the methods of science.

---

[16] For Husserl, even some physical sciences, such as biophysics, cannot be mathematized in the same way physics is: "Mathematical physics is an extraordinary instrument of knowledge of the world in which we effectively live; that of Nature, which always maintains in all its relative changes a concrete and empirical identical unity. It makes a physical technique practically possible. But it has its limits; not in the fact that we do not, empirically, leave the level of approximations, but in the fact that it is only a thin layer of the concrete world that is in this way effectively grasped. Physiology, biophysics, as doctrines of organic bodies in the totality of the concrete organic world, can borrow from physics as often as they want (organisms being in fact able to be idealized as mathematical bodies); it remains true that, as a principle, biophysics cannot ever dissolve into physics. Biophysical reality and causality cannot ever reduce to physical reality and causality". (Appendix IV of Husserl 1936a–see n. 13. The English version is mine.)

[17] There are two senses in which we can say that Husserl identifies an "application" of mathematics to the sciences of Nature: the first, in the construction of mathematical models of Nature via idealization and abstraction; the second, in the use of mathematics for the investigation of these models (which, as Husserl claimed, we erroneously tend to see as inquiring into Nature *itself*, not only ideal *models* of Nature).

[18] My arguments will have no bearing on Husserl's project of restricting the use of formal methods in science for the sake of "spiritual renewal", but they will, however, make it clear that this restriction cannot be justified *methodologically*. I plan to show that, from a strictly scientific perspective, Husserl is wrong in believing that formal methods must always be ready to be instilled with their "original sense" in order to be adequate and safe. More, I think Husserl's criticism is based on a misconception regarding the pattern of use of mathematics in modern physics. If scientific methods must be restricted, for whatever reasons, the development of science will pay the price.

[19] But I shall not touch the question whether Husserl is right or not in seeing the "formalistic alienation" of modern science as part of the "crisis" of meaning of modernity (I do not particularly think it is).

So, let us start by seeing how Husserl thought Nature–the Nature we experience with our senses–becomes a mathematical manifold:

1) First step, idealization, i.e. moving from the intuitively given to ideal reconstructions of it. For instance, magnitudes such as time, distance, speed or temperature, which are, and can only be experienced as *discrete* sets of values mathematically represented by *rational* numbers, are thought as *de facto* continuous magnitudes whose values can only be accurately represented by *real* numbers.[20] Idealization here amounts to taking observed magnitudes as *actually* continuous, and their values as *actually* represented by real numbers, not simply the rational values we observe in measurements (in fact, to say that a rational value of a magnitude is observed is already an idealization; we can at best locate the observed value within a range of values). The idealization proceeds with the identification of all the *possible* values of physical magnitudes with sets of mathematical entities (often, but not always, real numbers), and correlations among the values of these magnitudes as *mathematical functions*. These are usually the first (but by far not the only) moves of mathematical idealization in physics. Many others concur and superpose: bodies are reduced to massive *points*, trajectories to perfect geometrical *lines*, physical space to *geometrical* space, interactions among bodies (gravitational, electric or magnetic) to manifestations of *mathematical fields*, dynamic processes to *differential equations*, etc., etc.[21]

The next step is, for Husserl, the philosophically suspicious misstep. The reality we *actually experience* is degraded to the condition of only being an imperfect approximation of the world of idealized entities we substituted for it. Mathematical manifolds of ideal entities obtained by "exactification"[22] take the place of aspects of the world *we experience* as the *truly real* aspects of the world.[23] Mathematical idealities, in Husserl's

---

[20] At this point mathematization is already well established. Temperature, for example, is no longer the index of what we *feel*, but something we *measure*. It is not *immediate* experience that is idealized, but a first, rough, but already mathematical draft of it.

[21] "Upon the vague 'greater', 'smaller', 'more', 'less', and the vague 'equal' one could determinately superimpose the exact 'so much' greater or less, or 'how many times' greater or less, and the exact 'equal'" (Husserl 1936, 311). Idealization is for Husserl an act of *reason*: "But logical concepts are not concepts taken from what is simply intuitive; they arise through a rational activity proper to them, the development of ideas, *exact developments of concepts*, e.g., through that sort of idealization which produces, out of the empirically straight and curved, the geometrical straight line and circle". (Husserl 1936, 312)

[22] "How can mathematical, true nature be determined through normal appearances [*normal appearances = data of normal sensibility in relation to normal understanding, JJS*]? This occurs through the methods of rendering the continua exact, through the transformation of sensible causalities into mathematical causalities, etc." (Husserl 1936, 306)

[23] "Mathematics and mathematical science, as a garb of ideas, or the garb of symbols of the symbolic mathematical theories, encompasses everything which, for scientists and the educated generally, *represents* the life-world, *dresses it up* as 'objectively actual and true' nature. It is through the garb of ideas that we take for *true being* what is actually a

jargon, provide a "hypothetical substruction" of the world we experience. The substruction is hypothetical in the sense that it is established *ex hypothesi* and must be subjected to experimental confirmation. This, however, can never be definitive, for *ex hypothesi* experience can only approximate true reality. Experimentation is then "contaminated" by the very hypothesis it is called to confirm.[24]

In short, for Husserl, as far as the applicability of mathematics is concerned, the first task of mathematics in science is to provide models of Nature, obtainable by a combined process of abstraction ("focusing" on particular aspects of the given; in this case, formal aspects, which makes the process one of *formal* abstraction) and idealization (exactification). Husserl's criticism at this point is directed mainly at the reversal of priority that science, he thinks, operates: mathematized Nature taking the place of intuited Nature as the truly basic reality.

Husserl's account of the first steps in the process of mathematization of the physical sciences fails to realize, however, that it is not always the case that mathematical models of Nature are idealized abstractions from what is intuitively given. Think of phenomena on the atomic scale that are so notoriously difficult to access intuitively. Physicists are often obliged to rely on mathematical constructs (for instance, Planck's wave mentioned above) that have nothing to do with abstracting and idealizing from experience; and more often than not, quantum physicists must do without any intermediate models at all (like, for example, the mathematization of quantum mechanics in terms of matrices mentioned above). But, to be fair, classical physics largely (but not completely) follows the pattern Husserl traced (which, however, would make contemporary science impossible if strictly enforced, something I shall come back to later).

---

method—a method which is designed for the purpose of progressively improving, *in infinitum*, through 'scientific' predictions, those rough predictions which are the only ones originally possible within the sphere of what is actually experienced and experienceable in the life-world. It is because of the disguise of ideas that the true meaning of the method, the formulae, the 'theories,' remained unintelligible and, in the naive formation of the method, was *never* understood". (Husserl 1936 §9h, 51-2)

[24] "Taking into account all the 'natural laws already proved or in operation as working hypotheses', and on the basis of the whole available formal system of laws belonging to this *mathesis*, they [*the mathematical physicists JJS*] draw the logical consequences whose results are to be taken over by the experimenters. But they also accomplish the formation of the available logical possibilities for new hypotheses, which of course must be compatible with the totality of those accepted as valid at the time. In this way, they see to the preparation of those forms of hypotheses which now are the only ones admissible, as hypothetical possibilities for the interpretation of causal regularities to be empirically discovered through observation and experiment in terms of the ideal poles pertaining to them, i.e., in terms of exact laws. But experimental physicists, too, are constantly oriented in their work toward ideal poles, toward numerical magnitudes and general formulae. Thus in *all* natural scientific inquiry these are at the center of interest. All the discoveries of the old as well as the new physics are discoveries in the formula-world which is coordinated, so to speak, with nature". (Husserl 1936, §9g, 47-48)

2) Then comes, according to Husserl, a decisive move: the mathematical coordination of mathematical idealities by means of mathematical formulae is taken to be the very essence of reality. So, by commanding mathematical models of reality, formulas are believed to command reality itself. All possible intuitions (which, remember, are always conceived only as approximations) are taken as a priori *determined* by formulas (within a certain range of acuity, of course, since "true"–i.e. mathematical–reality always eludes us): we can predetermine *mathematically* the outcome of experiments. Nature is now completely subject to mathematics, for only mathematics can give *accurate* access to Nature (as opposed to the approximations granted to our senses). Husserl sees here another reversal of priority: even though mathematics is "nothing but [only] a particular practice" in the world, mathematics is thought to command the world. Although mathematics has its genesis in some practices of our lives, he thinks, mathematics directs our lives.[25] He says:

> One has the formulae, one already possesses, in advance, the practically desired prediction of what is to be expected with empirical certainty in the intuitively given world of concretely actual life, in which mathematics is merely a special [form of] praxis. Mathematization, then, with its realized formulae, is the achievement which is decisive for life. (Husserl 1936, 9f, 43)

Putting (1) and (2) together, we find that, for Husserl, the mathematical science of Nature begins with (formal) abstraction and idealization (exactification) when the Nature we directly experience is, with respect to its formal aspects, modeled by mathematical manifolds, and then proceeds by subjecting mathematized Nature to the scrutiny of mathematical theories (even purely formal mathematical theories–in which case, he thinks, strict surveillance must be exercised. I shall come back to this later), which dominate the world of experience by dominating the mathematical manifolds we take for Nature itself. One of the main goals of mathematical theories of Nature–besides revealing the architecture and functioning of the machinery of the world, since science "naïvely" presupposes that the mathematical models of the world simply *are* the world–is to establish formulas, equations and the like from which anticipations of experience can be deduced. In short, for Husserl, "Galilean" methodology consists basically in moving from immediate experience–the realm of intuitions–into the formal, onto which is laid the burden of regulating by means of its formulas the whole field of possible intuitions.

---

[25] This observation contains, in a slightly reformulated version, the central question concerning the philosophical problem of the applicability of mathematics to the natural sciences: how is it possible that mathematics, *a product of free human creativity*, can have anything to say about the natural world out there, existing *independently* of us? (The phenomenological answer, is one sentence, is: because Nature, the idealized Nature of science is to some extent already shaped by us).

A question now presses itself upon us: which "formulations", i.e. which mathematical theories can *rightfully* take upon themselves the honorable task of *determining* the entire field of possible experiences with their formulae? Consider the following quote:

> Actually the process whereby material mathematics is put into formal-logical form, where expanded formal logic is made self-sufficient as pure analysis or theory of manifolds, is perfectly *legitimate*, indeed necessary; the same is true of the technization which from time to time completely loses itself in merely technical thinking. But all this can and must be a method which is understood and practiced in a fully conscious way. It can be this, however, only if care is taken to avoid dangerous shifts of meaning by keeping always immediately in mind the original bestowal of meaning *[Sinngebung]* upon the method, through which it has the sense of achieving knowledge about the *world*. Even more, it must be freed of the character of an *unquestioned tradition* which, from the first invention of the new idea and method, allowed elements of obscurity to flow into its meaning. (Husserl 1936, 9g, 47)

Although Husserl admits that *purely formal* mathematical theories are bona fide *logical* theories, he is not willing to admit them into mathematical physics *if certain precautions are not taken*. For him, the reliance on purely formal methods–which necessarily implies some degree of "technization"[26]–is acceptable, provided that: first, we do not lose sight of the fact that it is *this world we live in and intuitively experience* that we want to know, which translates into the requirement that the *meaning* of the symbols and operations with them can be at any moment, at least in principle, recaptured by reactivating the original sense-bestowing acts; and second, we do not allow "moments of obscurity" or "shifts of meaning", as I conjecture, based on Husserl's treatment of the problem of "imaginaries" in mathematics, to happen when "imaginary" entities are introduced and used as technical instruments.

This is important and should be emphasized. For Husserl, the use of mathematics in science risks to degenerate into a "technique" devoid of meaning. This is why he believed we should always be in the position to "reactivate" meaning bestowing acts. Blind symbolic manipulation in which rules (and sometimes no rule at all)[27] take the place of meaning is how technical alienation is manifested in mathematics. If the use of mathematics in natural science is to be a *responsible* one, we should, Husserl thinks, know what we are doing, that is, have the meaning of our calculi either clearly in mind or within reach of intentional clarification. A particularly problematic instance of technization is the enlargement of a calculus that has a meaning by the adjunction of "imaginary" elements

---

[26] Husserl defines "technization" thus: the transformation of its experiencing, discovering way of thinking, which forms, perhaps with great genius, constructive theories, into a way of thinking with transformed concepts, 'symbolic' concepts" (Husserl 1936, §9g, 48). In short, technization means symbolization and "mechanical" manipulation of symbols.

[27] Of course, a problem Husserl never considered probably because he never thought it possible is the *heuristic* use of blind symbolic manipulations in science.

that have none. In this case meaning reactivation is *impossible*. Husserl's recipe for overcoming technization in science, then, is closely tied to his treatment of imaginary entities in mathematics.[28]

Husserl's demands are very strict: the theories we utilize for the investigation of the mathematical models of Nature, and then, derivatively, of Nature itself, even incurring in "technization" for practical reasons, must be capable, if necessary, of being instilled with their original meaning upon reactivation; and, most importantly, in case they are, again for practical reasons, enlarged by the introduction of imaginary entities (i.e. entities that have no counterpart in what is intuitively given), these entities cannot bring in "changes of meaning" (we shall soon see how Husserl thought that could be avoided).

Apparently, and this is an important remark, for Husserl, the *theories* of the abstract, idealized models of Nature (the theories of mathematical physics) must be extracted *intuitively* from these models themselves (like Euclidean geometry, from the space of physical geometry). The problem is that this is not always possible. There are situations in which scientists must *enrich* the structure they manage to extract from the given before theorizing can even begin. At this point, inevitably, using Husserl's own terminology, some "obscurity" is introduced that cannot be eliminated. Science is no longer telling us how Nature, at least in its formal aspects, is idealized *to be*, only at best how it *behaves*. Obscurity is due to the embarrassing fact that to know how Nature behaves we had to resort to theoretical entities that do not correspond to anything in *experienced* Nature. The road back from the methods of science to the living experience of Nature is blocked.

In any case, be that as it may, scientists are almost never content to remain within the boundaries of the *natural* theories extracted from idealized models of Nature, even when they can be obtained. This is, after all, typical of mathematics; mathematical theories are often extended into other mathematical theories and mathematical domains into other mathematical domains. Once an aspect of Nature has been mathematized, it shares the fate of any mathematical manifold: that of being up for grabs, for there is no a priori restriction on which mathematical theory can be summoned to subsume it theoretically, the theory intrinsic to the manifold (its natural theory, whose language, concepts and basic truths are offered directly in intuition), any theory whose domain is obtained from the original manifold by the adjunction of new (theoretical, "imaginary") entities, or even any theory only formally identical to any of the above.

But Husserl would not accept such liberality. The imperative that scientists should not lose sight of the world of experience requires in the first place, as we have seen, that the mathematical models of experience be obtained *from experience*, and then, that we be careful not to subject those models to shifts of meaning (supposedly by immersing them into other

---

[28] For Husserl's treatment of the problem of imaginaries in mathematics see Chapter 6.

manifolds or extending them by the adjunction of new "imaginary" elements, when, Husserl thinks, they lose their sense as models of *this* world).

For Husserl, there are privileged theories in mathematical physics, namely, those extracted from the idealized models of Nature (the mathematical manifolds we substitute for it) by mathematical intuition. These are what I called *natural theories*. Their meaning cannot be jeopardized on the pain of losing science's precious link with "living experience". But, as even Husserl had to admit, mathematical theories often dialogue with one another. So, which theories did he think should be allowed to converse with the privileged natural theories? Can natural domains be smuggled into the domains of mathematical theories other than their own, even purely formal theories? From what Husserl explicitly said, we can conclude that he accepted some mingling of theories, but under strict conditions.[29]

A problem reemerges here that had already occupied Husserl in his first book, *Philosophy of Arithmetic* and other texts of that and later periods (the 1901 Göttingen talks on the transition through the impossible ("imaginary") and the completeness of an axiom system, in particular): How can symbolic manipulations tell us what the facts are in the domains of materially determinate theories? In our case, how can playing with symbols according to rules be given the *right* to tell what the facts are in the mathematical models of Nature, and then, indirectly, in Nature itself, anticipating the (ideal) outcome of experience?

Husserl answers: provided symbolic manipulations are carried out within symbolic theories *formally equivalent* to natural theories (i.e. the formal domain of the formal theory must be isomorphic to the corresponding natural domain) or else *natural theories are logically complete* (i.e. natural domains are *definite*) and symbolic manipulations are carried out in consistent symbolic extensions of them (written obviously in extended languages). (Husserl 1901, 411-13; Husserl 1929, §31; Husserl 1994, 13, 15-16)

The first, although not *explicitly* mentioned in *Crisis*, is the obvious generalization of the justification he presented for symbolic arithmetic in *Philosophy of Arithmetic*; the second is his general "solution" to the problem of imaginaries in mathematics and is explicitly mentioned in *Crisis*. There is, then, at least one thread linking Husserl's first and last published works, *Philosophy of Arithmetic* and *Crisis*, respectively: some reservations concerning "purely" symbolic methods of knowing. It surfaced in *Philosophy of Arithmetic* under the guise of the need for a logical-epistemological justification of symbolic arithmetic; in *Crisis*, in the need

---

[29] With respect to algebraic symbolic reasoning Husserl says: "the powerful elaboration of signs and ways of algebraic thinking, a decisive moment which was, in a sense, pregnant with future fruits, and in another disturbing for our fate" (Husserl 1936, §9f). He is here alluding to the methodological relevance of symbolization and, at the same time, the risk of alienation it poses (the "crisis" for science it potentially carries).

for a justification of the technical methods of the mathematical sciences of Nature, and the task of overcoming their "formalistic alienation" by a philosophical analysis of the genesis and scope of these methods.[30]

But are these restrictions reasonable? Did Husserl really understand the *unrestricted* formal nature of mathematics (even materially determined mathematics like physical geometry) and the *full* extent of the role of mathematics in science? Let us examine these questions. Mathematics is a formal science, meaning that mathematical truths are indifferent to material content–they are, precisely, *formal*. Being formal, there is no reason why a mathematical domain cannot be investigated by a theory that is not *its* own, but whose domain is isomorphic to it (as we investigate geometric domains by algebraic methods). But, as obvious as this fact may be, Husserl never affirmed it in so many words (*even though it is the formal nature of contentual arithmetic that is behind his strategy of justification for symbolic arithmetic in* Philosophy of Arithmetic). He never explicitly said that *the only things* mathematics–even contentual mathematics– cares about are the formal (or structural) properties of its domains, not the objects these domains contain or the concepts governing them. He believed that, although formal, mathematical theories are, or should eventually be, when *scientifically* relevant, theories of well-determined objectual domains and well-determined concepts, science, real science is *always* of objects or concepts (even though the *only* aspects of its domain that any particular science can ever know are the exclusively formal ones).[31]

Husserl managed to justify symbolic arithmetic epistemologically by showing that there is an isomorphism between its formal domain and the domain of conceptual arithmetic, i.e. in the fact that the domain of numerical concepts and conceptual operations is *structurally identical* with the domain of symbolic representations of numerical concepts and operations with these symbols. But, in *Philosophy of Arithmetic*, Husserl missed the opportunity of asking the fundamental question: *Why* can we obtain mathematical knowledge of a domain by investigating another isomorphic to it? The *fact* that this is possible satisfied him. Husserl seemed to think that by manipulating numerical signs we are still, indirectly, manipulating numbers, as puppeteers move their puppets. But the truth, of course, is that moving from numbers and operations with them to signs and symbolic operations is useful and safe *only because the structures of the numerical and the symbolic domains are identical*; what *really* interests *contentual* arithmetic is indifferently instantiated in any domain isomorphic to the numerical domain.

---

[30] See Chapter 3 for Husserl's treatment of the problem of imaginaries in its many guises.
[31] Remember that Husserl distinguishes between mathematical *science*, such as physical geometry, and applied mathematics in general, from purely formal mathematics, which is for him a chapter of formal ontology, and thus pure formal logic. He fails to realize that this distinction, besides *mathematically* irrelevant, risks being philosophically misleading.

Had Husserl drawn all the consequences of this he would have realized that, as far as mathematical interests (or the interests of mathematical physics) are concerned, moving to the level of the purely symbolic is not necessarily tantamount to leaving behind what really matters, the intuitively given for the simplest fact that the formal structure that manifests itself in intuition can also manifest itself symbolically, and the formal structure of the intuitively given is all that mathematics or *objective* science can ever hope to know. If formal properties of reality can be expressed with meaningful symbols (denoting something in Nature), they can do so with meaningless symbols. The meaning of symbols has no role in expressing the formal structural aspects of experience.[32]

Husserl's justification of the use of symbolic methods in general–when there is perhaps no isomorphism between the domain of interest and the symbolic domain–required that the theories to which we bring symbolic

---

[32] I use the terms "structural" and "formal" interchangeably because structural properties are formal. We could define mathematical structures simply as the objective correlates of mathematical theories. We say, for instance, that group theory characterizes the structure of group. Two structures are, in this sense, *equal* if they are correlated to logically equivalent theories. In this sense, a structure has no property that its theory or any theory equivalent to it cannot show. This poses a problem if the theory is not logically (syntactically) complete; structures would then be in general only partially determined. We can remedy this by saying that theories in general, in fact, only characterize *families* of structures. The problem would, then, remain of singling out structures. We could do this by means of *complete* theories, but, even though all the properties of a structure would in this case be determined in principle, this definition makes the notion of structure depend on the logical powers of the underlying logic or the expressive powers of the language we choose. In order to make the notion of structure a purely semantic notion, it seems more convenient to adopt the following definition, more mathematical than logical: mathematical structures are the common aspects of *isomorphic* mathematical domains. This is the *abstract* (or semantic) notion of structure, as opposed to the *theoretical* (or syntactic) notion introduced before. In this sense, structures can be characterized by *categorical* theories. Given a categorical theory **T**, the structure it characterizes is the structure of its models (since they are all isomorphic). Any theorem of **T** establishes a (formal) property of the structure **T** characterizes, but there may be properties of this structure that are not theorems of **T**, if **T** is not complete (not all categorical theories are complete). In particular, the numerical structure is the formal ω-sequence, characterized by second-order arithmetic (which, however, cannot prove everything that is true of it). The interesting aspect of this notion of structure is that, in order to prove no matter what fact about a structure, we are not confined to a particular domain that instantiates it or a particular theory that characterizes it. Husserl had his own definition of structure (which he calls *formal domain*), which he defined as intentional (objective) correlates of formal (i.e. non-interpreted) theories, i.e. structures in the syntactic sense. For him, a formal domain could always be completely characterized, if not by its theory, then by its *complete* extension, which he thought to exist (he wrote these things decades before Kurt Gödel's appearance on the scene). According to Husserl, a formal theory always determines a *unique* formal domain (which shows a certain confusion on Husserl's part between the notions of logical completeness and categoricity), any theorem of a formal theory is true in the formal domain it defines, and all (formal) properties of a formal domain can (ideally) be derived in its theory, since it is (ideally) complete. So, for formal domains, there is no clear distinction between the semantic and the syntactic notions of truth, therefore, no clear distinction either between the syntactic and the semantic notions of completeness (see Chapter 6, where a different notion of formal domain is discussed).

help be logically complete, a condition that he thought to be in principle universally satisfiable. Husserl's injunction that a theory should be "master of its domain", meaning by it that it should be logically complete, i.e. able in principle to decide any question that could be raised concerning its domain, translates Husserl's epistemological (not necessarily logical) optimism. But, of course, since mathematical theories, even interpreted theories, are invariably formal (their truths are invariant under isomorphisms), there is no problem, no possibility of error, in investigating the formal properties of a given domain **D** (say, conveniently idealized empirical Nature) within the context of any theory **T** whose domain extended **D**, i.e. whose domain contained a sub-domain **D'** isomorphic to **D**, *even if **T** was a purely formal theory*. If we were able to show in **T** that φ is true for the elements of **D'**, then φ is also true in **D**. In short, the purely "technical" nature of **T**, despite the shift of meaning it promotes, did not induce any error, a fact established by formal considerations, not by "instilling sense" into **T** (and this is only a very elementary instance of similar formal methods).

This all too common mathematical procedure (for instance, one can show that an equation over the *real* field has *real* solutions by operating with *imaginary* numbers–a fact that reveals the true, formal nature of mathematics) is strongly limited by Husserl. He thinks that, by being "masters of their domains", theories can only get a helping hand from other theories, with different domains (previously stripped of material content and reduced to their formal core) if they do not really need it.[33] Other theories may provide more convenient, but not be more powerful methods (in the sense of methods able to prove more results). Husserl requires theories to be ideally complete because all they need in order to decide any

---

[33] Here are some examples of theories that have given up being "master of their domains", and gained with it: (1) The introduction of points and the line at infinity in the Euclidean plane (merely formal objects without any material meaning, given that they do not correspond, as idealizations, to anything we can experience) by Johannes Kepler, Girard Desargues, Blaise Pascal, Philippe de La Hire and Jean-Victor Poncelet and other creators of Projective Geometry brought about a formal mathematical theory that proved its utility by proving, by new methods, important theorems of Euclidean Geometry that involve only projective properties. It also made possible the introduction of important methodological principles such as the Principle of Continuity (that allowed for a uniform treatment of conics, in which all conics are seen as projective images of circles) and the immensely useful and elegant Principle of Duality. (In a typical mathematical way, we can also treat standard problems of Projective Geometry using other mathematical means, for instance, vectors in the complex plane. (See, Nikolic', 107-116.) (2) The introduction of negative and imaginary (complex) numbers into arithmetic by the Italian algebraists of the 16th century made possible a uniform treatment of algebraic equations. (3) The creation of infinitesimal methods in geometry by, for instance, Bonaventura Cavalieri, allowed for an improvement over the Archimedean method of exhaustion. Whereas Archimedes' method was one of justification rather than discovery–indirect justification, to make matters worse–infinitesimal methods were both of discovery and justification–*direct* justification, to make things better. The apparent "problem" that infinitesimals are not, and could not be geometrical entities properly speaking, but only "mere" formal objects did not seem to bother the creators of the method very much.

question that can be asked about their domains, he thinks, should be extracted directly (intuitively) or indirectly (deductively from intuitive axioms) from these domains themselves. The primacy of the intuitively given–and, then, of the I–over purely symbolic methods that conditions the solution to the problem of imaginaries in mathematics that Husserl presented in the beginning of his philosophical career is still clearly noticeable in the strategy he proposed for bringing science back from alienation and naïveté into full clarity concerning sense and methods.[34]

But, again, why should the formal structure of a given mathematical manifold be displayed only as the structure of *this* domain and not of any other that happens to have a sub-domain isomorphic with it (and so, with the *same* formal properties of the original domain)? If we abandon Husserl's strong conviction that theories should be "masters of their domains", those domains could also, as far as science, particularly empirical science is concerned, which is formal structure, be investigated by other, maybe more resourceful theories.

As I said above, in *Crisis*, Husserl did not disclose the full spectrum of problems concerning the applicability of mathematics in natural science. In particular, he takes it for granted that it is applicable (as a conceptual and symbolic system suitable for the modeling, upon idealization, of our experience, and as a provider of natural theories and suitable extensions of them), but does not question why.[35] Also, as I have just argued, given Husserl's strategy of justification of standard mathematical methods, it is clear that he thought that natural manifolds should be "definite", meaning that they should be defined by a complete system of axioms, corresponding to the "logical-formal idea of a world in general"[36] (Husserl 1936, 9g). Of

---

[34] "The ontic meaning *[Seinssinn]* of the pregiven life-world is a *subjective structure [Gebilde]*, it is the achievement of experiencing, prescientific life. In this life the meaning and the ontic validity *[Seinsgeltung]* of the world are built up—of that particular world, that is, which is actually valid for the individual experiencer. As for the "objectively true" world, the world of science, it is a structure at a higher level, built on prescientific experiencing and thinking, or rather on its accomplishments of validity *[Geltungsleistungen]*. Only a radical inquiry back into subjectivity—and specifically the subjectivity which *ultimately* brings about all world-validity, with its content and in all its prescientific and scientific modes, and into the "what" and the "how" of the rational accomplishments—can make objective truth comprehensible and arrive at the ultimate ontic meaning of the world. Thus it is not the being of the world as unquestioned, taken for granted, which is primary in itself; and one has not merely to ask what belongs to it objectively; rather, what is primary in itself is subjectivity, understood as that which naively pregives the being of the world and then rationalizes or (what is the same thing) objectifies it". (Husserl 1936, §14, 69)

[35] The only philosophical problems related to the applications of mathematics in science that Husserl recognizes can be solved, he thought, by *restricting* the applicability of mathematics. His strategy is *restrictive*, rather than *explicative*.

[36] "The presupposition of classical physics: the subjectively changing natural phenomena, the empirical phenomena with their empirical progress in terms of worse or better approximation (their perfecting), points towards the mathematical idea of a Nature in itself as a unified world of bodies in themselves. This implies a universally valid mathematics of Nature [... and in it] lies the foundation of a causal legality, according to

course, if a definite manifold corresponds to the idea of the logical form of a world in general, the world itself, considered in its formal (abstracted, idealized) aspects, must ideally constitute a coherent system of definite mathematical manifolds. Logical completeness, as an ideal, follows from this.

This may stand as an ideal for science, but since it is far from realized, and there is no guarantee that it could be realized (in fact, if we consider that logical-formal axiomatic systems must be designed according to certain reasonable effectiveness restrictions, Gödel showed that it is in general unrealizable), the mathematical science of Nature must do what it can, just like mathematics itself, that is, freely extend natural domains and theories into the purely symbolic. Formal methods can be secured independently of the restrictions Husserl imposed on them. Can they still be seen as a form of alienation if logically justified? Would this justification meet Husserl's standards? I believe it would, and that Husserl did not take them into consideration only because at the time he did not have the necessary logical instruments.

I want now, to conclude, to consider Husserl's model for the intentional genesis of natural manifolds from the given of perceptual intuition (considerations of this kind fall within the scope of genetic phenomenology).[37] The question is whether Husserl is right in believing that the mathematical models of Nature are "extracted" from immediate perceptual experience by the combined action of abstraction and idealization.

In the genetic analyses of *Crisis*, the relation between perceptual and mathematized Nature is cast in terms of dichotomies such as concrete versus abstract, real versus ideal, intuitively given versus categorically constituted. For Husserl, the first terms of these dichotomies designate starting points, the second, points of arrival; constitution is what happens

---

which all bodies, whose ideal essence consists in their causal, space-temporal being, are calculable. *The general mathematical legality is definite in the sense that it has the form of a finite number of fundamental mathematical laws (the axioms) in which all laws are included, in a purely deductive manner, as consequences* [my emphasis]". (Appendix IV, Husserl 1936a–see n. 13. The English version is mine).

[37] With respect to geometry, Husserl carries out such an investigation in the famous essay "The Origin of Geometry". But the general task of a phenomenological critique of science (in the sense of a radical foundation of science in transcendental subjectivity) is much broader; it consists essentially in questioning the "evidences" belonging to a world that is already "out there", presupposed, with the sense it has, by science and our unquestioned being in the world. Some of these evidences (concerning the world as simply "given") are: 1) it is a coherent unity subjected to strict causality; 2) its being is completely determined; and it corresponds to the being of the world of a system of truths in themselves, the possession of which constitutes the goal of objective science; 3) "the world is–always is in advance–and ...every correction of an opinion, whether an experiential or other opinion, presupposes the already existing world, namely, as a horizon of what in the given case is indubitably valid as existing, and presupposes within this horizon something familiar and doubtlessly certain with which that which is perhaps canceled out as invalid came into conflict" (Husserl 1936, §28, 110), and 4) the experience of the world can be infinitely perfected. To trace the "genesis" of *this* sense of the world in consciousness constitutes the core of a phenomenological critique of the presuppositions of science.

in between. The idea of constitution of the ideal from the real reverses the order of priority in Plato's idea of participation. By adhering to it, Husserl implicitly criticizes the extreme "Platonism" of modern science.[38] For him, the real world of our direct experience is not an *imperfect* copy of the ideal world of science, but *conversely*, it is the ideal world that *originates* in the real world, being *constituted* by intentional acts of consciousness (abstraction and idealization).

For Husserl, as already stressed, mathematics comes into science primarily by substituting the given of experience with mathematical manifolds, i.e. mathematical modeling. The question is whether this covering of the intuitively given with mathematical concepts is univocally determined by experience or, contrarily, mathematization is sub-determined by the given. Does the road from sensibility to understanding admit of bifurcations? Is the covering of perceptual data with categorial formations without alternatives? How did Husserl answer these questions?

Let us consider the geometrization of perceptual space as a paradigmatic case. As I show in another chapter of this book (Chapter 2), this is a complex business, but one thing is clear. For Husserl, sensorial space is constituted from impressions coming from different sources, visual, tactile, kinesthetic, which coalesce in a coherent representation of space *as sensed*. But sensorial space is not yet perceptual space, which contains aspects that are out of reach of sensorial perception. For example, perceptual space is represented as infinite and continuous, and sensorial perception cannot reach into the global and infinitesimally local properties of space. Perceptual space is the sensorial space of an idealized perceiver considered in general, not only me, but others, not real us, but ideal us, not only now, but at any time, not only here, but anywhere. Certain characters of perceptual space that are not directly perceived (such as infiniteness and continuity) are there to make room for the whole field of possible sensorial perceptions of an ideal perceiver in abstracto *in general*. Perceptual space is, then, constituted out of sensorial space, but is not *reducible* to it. Nonetheless, perceptual space is *not* yet a mathematical realm properly speaking. It does not admit the exact determinations of mathematics. Bodies in perceptual space are not geometrical bodies. They are close or far from one another, but not at *precisely* this or that distance measured by a *real* number, etc.

In order to become an object of the mathematical science of Nature, perceptual space must be *idealized* ("exactified"), thus becoming *physical* space, and *abstracted*, no longer a *continuous medium* filled with perceptual content, but an empty *manifold* of null *points*. Formalist alienation is

---

[38] "FOR PLATONISM, the real had a more or less perfect methexis in the ideal. This afforded ancient geometry possibilities of a primitive application to reality. [But] through Galileo's *mathematization of nature, nature itself* is idealized under the guidance of the new mathematics; nature itself becomes—to express it in a modern way—a mathematical manifold *[Mannigfaltigkeit]*". (Husserl 1936, §9, 23)

essentially the belief that perceptual space is *not* the original space, but only an approximation to physical space, the *real* reality to which only mathematics has access.

Mathematical space has a mathematical structure and a mathematical theory. For Husserl, it is the Euclidean structure of Euclidean geometry. Of course, Euclidean geometry cannot properly be said to be justified perceptually. Before being Euclidean, space must be mathematical, which perceptual space is not (it has at best a proto-mathematical structure). But, in any case, for Husserl, Euclidean geometry is the only geometric structure consistent with mathematical space as the abstract, ideal limit of perceptual space.

Now, can we investigate the geometric structure of physical space, 3-dimensional and Euclidean, which has a material content, by means of other mathematical theories, say, abstract n-dimensional Riemannian geometry, which has none? From what we can extract from Husserl's considerations concerning formal mathematics in application, we can, but only if Euclidean geometry is logically complete.[39]

This, however, is not in general the case. If science had to obey Husserl's stricture, it would be severely impaired in its capability to access the structure of perceptual reality (conveniently idealized and reduced to its abstract formal aspects, of course). But there is more. Husserl accepted that science might find it useful to give up Euclidean geometry and to represent physical space differently–for example, scientists may find more appropriate to represent space in the sub-quantum level as discontinuous, or 11-dimensional–provided, however, these counter-intuitive representations were kept at a safe distance from the range of effective perception. Physical space can be non-Euclidean overall, but has to be Euclidean locally. It can be discontinuous or microscopically many-dimensional, but it has to be 3-dimensional and continuous macroscopically. Science has not so far overstepped the limits of perception. It still takes the perceiving subject as ruler in his domain of jurisdiction. But *must* it? Husserl believes it must. Science may not.[40]

---

[39] Euclidean geometry is a rare case in which this can be shown, at least to the extent it is captured by Tarski elementary (first-order) axiomatics, provably complete.

[40] It would be interesting to have known Husserl's assessment of Einstein's relativity theory, since it is the theory that most dramatically strains our perceptual experience of space and time. The absoluteness of simultaneity, in particular, regardless of the space separation of events, seems to sit so naturally with our most basic intuitions that it was a basic presupposition of classical Newtonian science and went unquestioned for centuries. However, some would claim (Hermann Weyl certainly would) that the constitution of the objective world in intersubjectivity, not in a consciousness closed in itself, naturally imposes a principle of relativity in physics. But this principle is not enough. Einstein's theory of relativity also requires an operational conception of the meaning of physical notions (besides the *fact* that the velocity of light does not depend of the state of motion of the source). Husserl's objectivity as intersubjectivity does not lead directly to the theory of relativity.

In any case, immediate perception is still taken as a reliable source of information concerning the structure of physical space, but less so when other aspects of Nature are considered. Science must often rely on whatever patterns it can extract from the mass of raw, usually poorly structured, empirical data and *impose* upon them convenient mathematical structuring with virtually no intuitive grasp on *objective* subjacent structure, if there indeed is such a thing (by "convenient" I mean having some explicative and predictive powers). Any physical or mathematical model, or even no model at all, just any theory able to organize the data of raw experience and make correct predictions will be a strong candidate for the right theory, independently of any, often missing, intuitive guidance. Is the mere awareness of this situation part of Husserl's clarification of the sense of the mathematical science of Nature? I believe it is, but it is not enough in the way of the intuitive grounding of science.

An important fact that must be kept in mind is that intuition is amenable to mathematical treatment–i.e., mathematized–*only* in its formal structural aspects,[41] since mathematics does not care for material content, only for form: this is what is meant by saying that mathematics is a *formal* science. Mathematical modeling consists in *choosing* mathematical manifolds that adequately express the *formal* aspects of empirical experience, and this often leaves considerable room for alternatives.[42]

The hard fact about the mathematical modeling of Nature is that immediate experience only provides initial and boundary conditions. Abstraction and idealization are not, even if not taken as univocally determining the outcome of the process, the only relevant actors in our efforts to represent Nature mathematically. The combined facts that mathematics is a formal science, that physics only cares for the formal aspects of Nature (in *this* consists the essence of Galilean methodology,

---

[41] James Jeans says that, in physics, we "must limit ourselves to describing the *pattern of events* [my emphasis] in mathematical terms", that "one may dig, one may sow, one may reap. But the final harvest will always be a sheaf of mathematical formulae". (Jeans, 15)

[42] An interesting example of the use of a physical and, correlatively, mathematical model exclusively for the sake of deriving formal properties of a *different* empirical context was that of James Clerk Maxwell's "On Physical Lines of Force" (Maxwell 1861-62). Although he did not believe that his treatment of electro-magnetic phenomena in terms of vortices and strains in an elastic mechanical medium corresponded to electromagnetic reality, he adopted this approach in order "to make use of the mathematical analogies of the two problems to assist the imagination in the study of both" (*Ibid.*, Part I, p. 163). This is as good an example as any of the fact that for the physicist only formal properties and formal identity (which Maxwell expressed as "mathematical analogies") matter (and this is why mathematics is so useful to him). Maxwell knew very well that his model was not *materially* faithful to reality; he was not extracting a model of reality from the "given", but using a model of reality that had no more than *some* degree of formal identity with the given in order to derive the equations that would regulate the given. Maxwell's method- logy was so efficient that by purely formal means he discovered the existence of electromagnetic waves (later experimentally confirmed by Heinrich Hertz). This *heuristic* use of mathematics seems to me to be completely alien to Husserl's account of the purely *representational* role of mathematics in science.

in emphasizing the *how* rather than the *why*, the *mathematical form* of the mechanism of the world rather than *what* this mechanism is), and, most importantly, that there is *no a priori* restriction on which mathematical theory can be mobilized so as better to investigate the formal aspects of Nature explain why *downgrading* the role of direct perception (obviously not to the point of ignoring it, which should remain a boundary condition) can actually *foster* the progress of science. "Formalist alienation", ironically, can account for the immense success of the mathematical science of Nature, which thrives *because* large portions of it do *not* have and *cannot* have intuitive foundations.

Husserl's project of "rooting" the mathematical science of Nature in the life-world could arguably jeopardize its development, even if it could bring science back into the sphere of meaningfulness of which he thought it had taken leave. From a strictly scientific perspective, however, that project seems inappropriate, given in particular the strangeness of some of the realms that science is called to investigate compared to the cozy familiarity of the life-world. There are no *necessary* links connecting the life-world to the mathematical manifolds we use to model Nature. But the Husserlian model of the applicability of mathematics to the sciences of Nature, as discernible in *Crisis*, points to the opposite conclusion: abstraction and idealization determining, perhaps not completely, but in any case more determinately than I think it reasonable to suppose, the acceptable mathematical models of intuitively given Nature, which should be left in the care of their ideally complete theories, whose foundations rest on an intuitive grasp of the governing concepts and basic truths of their domains. The history of science, most notably in the last century, has shown that the use of mathematics in science did not and, most importantly, could not follow the pattern Husserl indicated. Nor do I see any reason why it should, since, as I believe I have made clear here, Husserl failed to see, in full clarity, despite his insightful remarks on the nature of mathematical knowledge, the formal nature of the *whole* of mathematics (which he acknowledged only in part) and the *general* pattern of its applicability in science (which he detected only in its more conservative aspects).

The case Husserl explicitly wanted to advance in *Crisis* for the establishment of the scientific rights of a descriptive, non-mathematical rigorous science (phenomenology), with the task of clarifying scientific endeavor in general as to the sense and methods, in either the physical or human realm, or for that matter, mathematical, is not, I think, in the least weakened by the problems I pointed out. It only requires a more thorough investigation of intentional constitution, one that does not overburden abstraction and idealization as the sole ways of ascending from raw sensorial perception to mathematics.

References

Crowe, Michael 1994. *A History of Vector Analysis*. New York: Dover.
Granel, Gérard 1976. "Préface". *La crise des sciences européennes et la phénoménologie transcendantale*. Paris: Gallimard, I-IX.
Grassmann, Hermann 1844. *Die lineale Ausdehnungslehre*. Leipzig: Wiegand, translated by Lloyd Kannenberg as *A New Branch of Mathematics*, Chicago: Open Court, 1995.
Husserl, Edmund 1887-1901. *Philosophy of Arithmetic, Psychological and Logical Investigations with Supplementary Texts from 1887-1901*. Dordrecht: Kluwer, 2003, translation by Dallas Willard of *Philosophie der Arithmetik. Mit ergänzenden Texten (1890-1901)*, Husserliana XII, The Hague: M. Nijhoff, originally published as *Philosophie der Arithmetik*, Halle: Pfeffer, 1891.
Husserl, Edmund 1900. *Prolegomena to Pure Logic,* volume I of his *Logical Investigations*, New York: Humanities Press, 1970, translation of *Prolegomena zur reinen Logik*, Halle: Niemeyer.
Husserl, Edmund 1900-01. *Logical Investigations*. New York: Humanities Press, 1970, translation of *Logische Untersuchungen*.
Husserl, Edmund 1901. "Double Lecture on the Transition through the Impossible ("Imaginary") and the Completeness of an Axiom System". Included as Essay III in Husserl 1887-1901, 409-52.
Husserl, Edmund 1931. *Cartesian Meditations: An Introduction to Phenomenology*. The Hague: M. Nijhoff, 1960, translation of *Cartesianische Meditationen*, Husserliana I, The Hague: M. Nijhoff, 1950 (1931).
Husserl, Edmund 1936. *The Crisis of European Sciences and Transcendental Phenomenology: An Introduction to Phenomenological Philosophy*, Evanston: Northwestern University Press, 1970, translation of *Die Krisis der europäischen Wissenschaften und die transzendentale Phänomenologie*, Husserliana VI, The Hague: M. Nijhoff, 1954.
Husserl, Edmund 1936a. *La crise des sciences européennes et la phénoménologie transcendentale*, Paris: Gallimard, 1976, translation of *Die Krisis der europäischen Wissenschaften und die transzendentale Phänomenologie*.
Husserl, Edmund, 1994. *Early Writings in the Philosophy of Logic and Mathematics*. Dordrecht: Kluwer, translation by Dallas Willard of *Aufsätze und Rezensionen 1890-1910*, Husserliana XXII, The Hague: M. Nijhoff, 1979.
Jeans, James 1981. *Physics and Philosophy*. New York: Dover.
Maxwell, James Clerk 1861-1862. "On Physical Lines of Force". *Philosophical Magazine and Journal of Science* 4th Series, 21 & 23.
Nikolic, Aleksander M. 2004. "Karamata's Products of Two Complex Numbers". *The Teaching of Mathematics*, vol. VII, 2, 107-116.
Riemann, Bernhard 1854. "On the Hypotheses which Lie at the Bases of Geometry". *Nature* vol. VIII nos. 183, 184, 14-17.

APPENDIX I

*Translated by Dr. Ruth Ellen Burke,
California State University, San Bernardino*

NINE LETTERS FROM CANTOR'S LETTER BOOK III (COD. MS. 18 NIEDERSÄCHSISCHE STAATS- UND UNIVERSITÄTSBIBLIOTHEK, ABTEILUNG HANDSCHRIFTEN UND SELTENE DRÜCKE, UNIVERSITY OF GÖTTINGEN)

Cantor Letter 238

To Professor Dr. Jacob Lüroth, University of Freiburg

Halle a/S. 29 November 1895

Dear Friend,

Prof. Edmund Husserl, about whom you are inquiring, is a person highly-valued and generally loved by us because of his peaceable and sterling character. I have known him since 1886, when, in order to prepare the *Habilitation* in Halle, he studied for two additional semesters, attending lectures in psychology with Prof. Stumpf and on the calculus of probabilities with me. He was born in 1859 and studied mathematics and natural science from '76-'78 in Leipzig for three semesters, then in Berlin for six semesters. Then he went to Vienna, where he received his degree under Königsberger for work on the calculus of variations. In '83 he went back once more to Berlin for a semester in order to work on a course on Abelian functions for Weierstrass.

While he was still in Leipzig and Berlin; he had a predilection for attending philosophy lectures given by Wundt and Paulsen and enthusiastically continued those studies under Brentano and Zimmermann in Vienna from '83-'86. His main fields are logic and psychology, although he is also thoroughly proficient in the remaining areas of philosophy, and has also given most of his lectures here in both. As concerns the latter, to be more certain, yesterday I went to my colleague Benno Erdmann, who told me that Husserl had worked here with the greatest success and that students gladly attend his lectures. This is also confirmed by the fact that we repeatedly recommended him for the position of professor extraordinarius in Münster and that last summer, he was also presented along with Riehl as a candidate for the ordinarius by the faculty at Kiel.

He is Protestant; however, I am convinced that with his sense of moderation and tolerance he would prove successful as a teacher of Catholic students. It would give me great pleasure if, due to his in-depth, multifaceted studies and accomplishments, he could obtain the well-deserved position of professor ordinarius in philosophy.

You also write me to say that I have been too critical of Thomae and Veronese. However, I beg you to give reasons for this judgment! If I have not done them justice in some respect I am gladly ready to make amends.

Best wishes, your old friend,

G. Cantor

P. S. What I am writing you in this letter is not confidential, but can be utilized it at your own discretion.

Cantor Letter 240

To Herrn Canon Franz Woker, D. theol. in Paderborn

Halle an der Saale, 30 November 1895

Honorable Herr Canon,

The friendship you have so often shown me induces me to pick up my pen and request an enormous favor of you.

It is a matter of appointing someone to the position of professor ordinarius on the philosophy faculty of Freiburg University in Baden. No doubt you will wonder what business of mine this is. Just a few days ago I received a letter from an old university friend, Professor *J. Lüroth*, mathematics professor in Freiburg, in which he asked me to provide him information about Dr. Edmund Husserl, a Privatdozent at Halle University, because he would be among those included on the list of nominees. I replied yesterday 29 November and in good conscience most warmly recommended Dr. Husserl. I know that still other members of the philosophy faculty in Halle have received similar inquiries and have replied along the same lines.

For that reason, as well as the fact that last summer Dr. H. was nominated in Kiel along with Freiburg professor Dr. A. Riehl (who leaves for Kiel at Easter), it very likely follows that the Freiburg philosophy faculty there will indeed place Dr. H. on the list of nominees.

My request is related to this likelihood.

Inquiries I have made indicate that it is nonetheless *highly unlikely* that Dr. H will obtain the position in question.

That is to say, it must be assumed that the current holder of the position, Prof. Riehl, will see to it that other candidates who are closer to his own viewpoint, first of all, Külpe, ordinarius at Würzburg (student of the Leipzig positivist Wundt), Rickert, extraordinarius in Freiburg (student of the *atheist* Riehl himself) (presumably some kind of exchange deal is virtually to be made between Külpe and Rickert), Marty in Prague (disciple and intimate of the unfortunate Brentano) *will be given precedence over Dr. Husserl.*

Now, Husserl, as I know for certain, and also as undeniably emerges from the courses he has given here on the proofs for the existence of God and contra Darwinism; is a *theist*. And both for this reason and also owing to his peaceable; thoroughly honest character and his tactful nature, he is much, much better suited to be a teacher of Catholic students than the candidates favored by Prof. Riehl. I remember with pleasure meeting your friend Prof. *Heiner* years ago at your beloved rectory in Halle. This led me the idea of awakening interest in Dr. Husserl through your kind intervention with the theology faculty in Freiburg, so that when his and other names came before the philosophy faculty, and after Dr. H. was presented by the phil. fac. with others, they would support him in a suitable manner before the Baden government. Surely the theol. facult. in Freiburg takes a great interest in having as irenically disposed professors as possible active in the rest of the faculties!

May I therefore request your influential endorsement of this matter? It should, however, not become known in Freiburg (the theol. faculty excepted) that the information and request come from me! For you know well that the confederacy of Protestant crows from all over stick closely together; this dark society's local members, well known to you, would peck my eyes out because of it![1]

I have friendly relations and active professional correspondence with your fellow Westphalian, W. Killing. Recently, we met in Berlin at the 80th birthday celebration of another of your fellow Westphalians, C. Weierstrass. Should you go to Münster, please give Killing my regards.

With best wishes to you and your dear sister,

Yours truly,

Georg Cantor

P.S. Prof. Lüroth told me in his letter that the philosophy faculty's deliberations regarding the appointment would not begin until the third of December.

---

[1] Like his father, Cantor was Protestant (not to mention a Rosicrucian). Copenhagen census records show that his paternal grandparents were Jewish. His mother was Catholic. (COH)

Cantor Letter 249

To Herrn Canon Franz Woker in Paderborn

Halle, 10 December 1895

Most Honorable Herr Canon,

By now, you have received my letter of 30 November.

It appears that Freiburg professor Dr. Riehl is working things out so that Husserl will not even appear on the list of candidates at all, for my friend Lüroth has just written me the following:

"The battle is raging around the extraordinarius (in title only) here (N.B. a *Privatdozent* with the title Prof.) *Rickert* (N.B. student of Riehl), who has not done much, but who has been declared by some colleagues to be the only person who could be considered".

If you were to be able to take up the Husserl matter in the manner I wished, then I ask you to see that this information also reaches your friend in Freiburg.

In the last few days, I have repeatedly been with your successor, Schwermer, who is a splendid fellow, and I am happy to be able to assist him in a matter concerning one of the children in his parish (a pupil at the municipal high school here who is to be expelled for really trivial reasons).

With best wishes, your

Respectfully devoted,

G. Cantor

Cantor Letter 250

To Herrn Prof. Dr. Jacob Lüroth, Freiburg i.B.

Halle, 10 December 1895

Dear Lüroth,

Given the conditions that you described, you did well to turn to Benno Erdmann for help, and he will soon reply to you, according to what I heard from him today (he received your letter yesterday evening). What is more, I have no doubt that in that letter he will bestow upon Husserl the same praise that I heard from his lips. I only want (*this time completely confidentially*) to make you aware that Erdmann is a friend of *Riehl*, so that the value of what he says can only be measured in consideration of this aspect. He knows that *Riehl* protects Dr. *Rickert* (son of the parliamentarian) and would not want to offend his friend. This merits consideration when assessing Benno Erdmann's remarks.

Presumably, Erdmann will also recommend old *Avenarius* in Zurich and young *Spitzer* in Graz. The latter is, as Erdmann said to me, a Darwinian, which Riehl is indeed also. If you need further information of any kind in this important affair, I am at your disposal as you wish.

With many greetings from my house to yours.

Your old friend,

G. Cantor

Cantor Letter 253

To Herrn Canon Dr. theol. Franz Woker

Halle a/Salle, 15 December 1895

Esteemed friend,

Many thanks for your friendly intervention in the Husserl matter.

Above all, I am glad to hear that your friend Dr. Heiner has created such an important and significant institute, as the Coll. Sapientiae will be. I wholeheartedly wish him the greatest success. I am likwise happy that Dr. Braig has found a position as professor of philosophy on the theology faculty there. I surely still remember well the beautiful little writings he authored that you gave me to read a long time ago.

In all of this, I however consider that from the Cathol. standpoint, which I myself have favored for years, it is *debatable* whether the theol. faculty in Freiburg had not to take great interest *in having peacefully disposed and competent forces also at work in other faculties*. In particular, it seems to me with the actual unbreakable interconnections among all the faculties, *not a matter of indifference* whether the ordinarius for philosophy of the phil. Faculty is a *theist* or an *atheist*, whether he works *for Darwinism* or *against Darwinism*. This is exactly where the Freiburg theologians had the *worst experiences* with Professor Riehl who is departing for Kiel at Easter, which can easily be repeated with his student, Rickert (son of the well-known Berlin parliamentarian), or with any other candidate recommended by Riehl.

I did not in any way mean that the theol. faculty would want to speak up for Husserl directly, but I was of the opinion, supported by Pastor Schwermer's information about the theol. faculty's influence on the government, that it would not be impossible for them to show some private, *unofficial* encouragement of the candidate I so warmly recommended....

Your respectfully devoted,

G Cantor

Cantor Letter 262

To Herrn Canon Dr. Woker, Spiritual Advisor in Paderborn

Halle a.d.Saale, 30 December 1895

Very Esteemed Friend,

You will have received my postcard of two days ago. I immediately gave the desired information about Prof. Husserl to your friend in Freiburg and I am pleased that, as a result of your kind efforts, interest in H. has been aroused there; I know he is worthy of this and that there will be no reason to regret his appointment, should the authorities in the Grand Duchy of Baden decide the same.

I shall comply with your kind request that I visit you in Paderborn in the course of the coming year, God willing. To be honest, the means that I currently have at my personal disposal are very limited. The six children, of whom the oldest is now 21, the youngest 9 years old, require more and more for their education. With assets on both sides of our family, thank God, we are indeed well-off. It is just that my income has not increased satisfactorily compared with earlier times, so that we have to be very careful to make do with what we have. Unfortunately, I shall receive little from the government, with which I otherwise have a very good relationship; my colleagues in mathematics at Berlin and Göttingen with positions of ordinarius (which I have been for 17 years now) earn more than twice as much as I do. This is primarily because in my academic work I am totally independent of Weierstrass and Kronecker, who only promote and recommend those whom have been affiliated and subordinate to them.

The publications in philosophy and mathematics journals, on the other hand, do not bring in a single penny; rather, they incur only expenses for the offprints that I absolutely need in order to show myself appreciative of all the academic mailings I receive from all over the world. After many years of study, I am now, however, ready to begin to work on another type of publication that I hope will bring me some compensation for my efforts and expenses; these publications will be of an apologetic and irenic nature. My plan to do so has been laid out for a long time, to start with, a new edition of certain works by Jacques André

Emery (Superior General of the Congregation of Saint Sulpice) providing essential additions and explanations is to be prepared. I wish to begin with his extremely important, yet all but unknown, work "Christianisme de Bacon" and then have his "Leibnitz sur la religion et la morale" follow.

Whether I have these publications appear under my own name is a minor matter and will depend upon time and circumstances.

(I have not yet contacted a publisher, however. I would prefer to entrust the matter to a Catholic publisher, but I do not know which. Would you also advise and help me in this matter?)

I believe I have made sufficient progress with the theological understanding of this work with the help of the writings of Cardinal Franzelin and the Italian Perrone; the splendidly thoroughly prepared compendium of your most reverend bishop Dr. Simar, "*Lehrbuch der Dogmatik*", has become most particularly valuable in orienting me with this.

On the other hand, it now looks as if what you know as my long cherished profound desire is going to be fulfilled after so many years, namely that I enter into a friendly, confidential, private correspondence with prominent representatives of Thomist theology and philosophy in Rome in relation to the question as to whether and to what extent my theory of the transfinite is suited to fill certain lacunae St. Thomas that left open in his system by the relatively undeveloped concept of realist actual infinities or transfinites.

For now, I cannot tell you anything more specific about this plan that is still pending. In any case its execution would coincide with the lofty intentions of His Holiness, Leo XIII, in whose encyclical "Aeterni Patris" we read:

> We have no intention of discountenancing the learned and able men who bring their industry and erudition, and, what is more, the wealth of new discoveries, to the service of philosophy; for, of course We understand that this leads to the development of learning.

And it further says:

> if there be anything that ill agrees with the discoveries of a later age, or, in a word, improbable in whatever way–it does not enter Our mind to propose that for imitation to Our age.

Father Tilman Pesch has published a new apology with Herder: "Christliche Lebensphilosophie". Unfortunately, I have not seen it yet. Do you know, perhaps, whether the esteemed Father is in residence in Exaeten?

Once again, my most wholehearted good wishes for the New Year!

In enduring friendship, your devoted,

Georg Cantor

Cantor Letter 264

To Herrn Prof. F. X. Heiner, Reverend in Freiburg-Baden

Halle, 31 December 1895

Very esteemed Colleague:

Wholehearted thanks for your kind letter of the 29th.

To begin with, as concerns Prof. Riehl, it remains absolutely certain that, with the appointment to Kiel, privy councilor Althoff (who is still at the time "omnipotent" in university matters here) has held out as a prospect to him that he will obtain an ordinarius professorship at one of the big Prussian universities the moment such an opportunity arises.

This is hardly a surprise to anyone who, as I am, is familiar with the circumstances in Prussia and knows that the decisions in question de facto do not depend on the Minister one bit. From the Catholic point of view, one must be glad that you are getting rid of Prof. Riehl, who, if I am not mistaken, gained entry to you as a "Catholic", but converted to Protestantism in Freiburg, and one can only hope that none of his like-minded ilk replaces him. This sort can cause a great deal of mischief, as you found out with Riehl himself over the course of many long years.

On the other hand, I would not consider it advisable to take this case to the newspapers. The theologians in Kiel may assure themselves about what they have in him and may see how to contend with him. Besides, we cannot know whether it is not perhaps divine providence that assigns such radical people to Protestant universities in order to hasten the subversion and disintegration of Protestantism. Would we be interested in preventing that? Not at all! May a most brilliant career lie ahead for Herr Prof. Riehl in Prussia! That will not hurt *Catholicism* and for that reason it must be of absolutely no consequence to us what becomes of that gentleman.

If I may be permitted to express my opinion concerning the other side of things, in the matter of the *Separatvotum* that, as you inform me, a representative of your esteemed faculty will submit to the Senate on their behalf, I would just advise that the following be stressed:

If a Catholic will not be appointed, the theol. faculty should have no reason to object to the appointment of Mr. Husserl, because according to the conscientious and thorough inquiries they have conducted not only are his exceptional academic qualifications beyond all question, but also because his personality and character justify the belief that he would discharge his duties with desirable tact.

You see that I do not consider it advisable that, for your part, you emphasize the fact that Dr. H is *"a devout Christian"*. This fact is certainly indeed rightfully decisive in this matter for you yourself, but it does not appear to me to be necessary for it to be said by you. For his opponents could indeed later use precisely *this point against H.* and say: "He's being protected by the contingent on the other side of the Alps because of his religious faith; H. consequently must be a very dangerous fellow!"

You have to know the Protestants as well as I do to know that you can count on such a turnaround from them.

Please accept my warmest wishes for happiness and blessings for the New Year.

With most respectful esteem, your devoted

G. Cantor

P.S. I shall be sincerely grateful for any information on the further progression of the Husserl matter. Here the great respect that Husserl has generated with his writings, namely his *Philosophy of Arithmetic* can be highlighted, in addition, the fact that he has already been active as a academic teacher since 1887 and that last summer he was proposed for the ordinarius in Kiel along with Riehl, also in Halle he was over and over recommended to the Ministry for promotion. Through all this, Husserl's candidacy will gain precedence over that of Dr. H. Rickert, who has less seniority and has written only two small booklets, "Ueber die Definition" and "Über den Gegenstand der Erkenntnis", bearing a Kantian-Windelbandian-Lotzean positivistic imprint....

Cantor Letter 277

To Herrn Prof. Dr. F. X Heiner in Freiburg, i. B.

Halle, 11 January 1896

Esteemed Colleague:

Receipt of your postcard dated 10 January indicates that you would welcome information about those candidates under consideration by the relevant philos. faculty commission. As far as I know from the reports that have reached me, they are as follows:
  2. Avenarius in Zurich (born 1843)
  3. Eucken in Jena (1846)
  1. Siebeck in Giessen (1842)
  4. Natorp in Marburg (1854)
  5. Spitzer in Graz (1854)
  6. Gross in Giessen (1861)
  7. Busse in Marburg (1862)

As much as I could advise you to take an attitude of "*tolerari posse*" with regard to my young friend Husserl (b. 1859), I must express my grave reservations concerning those seven names.

Re: 1. He is one-sided, of historico-philological orientation, and without training in the natural sciences. From his "Lehrbuch der Religionsphilosophie" you can get to know him fully.

Re: 2. A respected man who has thought and developed a *new* modern philosophical system from the innumerably many other ones. He only has a small number of students. *As a teacher he lacks real effectiveness. In Zurich, Kym has a far greater number of students.*

Re: 3. He is strongly anti-Catholic, as his essay, "Die Philosophie des Thomas von Aquino u. die Kultur der Neuzeit" (*Zeitschrift f. Philos. U. philos. Kritik,* Halle a/S, 1885, Vol. 87, p. 161) proves.

Re: 4. A competent man but lacking in teaching skills. He reads his lectures mechanically from his notes. He is an extreme neo-Kantian and student of the Jewish Kantian philologist, Hermann Cohen. He is, while of great integrity, a difficult, unbearable colleague because of his bad temper.

Re: 5. Published a book on Darwinism in the early 80's, since then has written nothing. Is presumably Jewish and a radical liberal in all respects.

Re: 6. Esthetician, without training in the natural sciences; rather unimportant.

Re: 7. Used to be in Tokyo; is not a significant scientific force, since in Marburg, Cohen and Natorp did not even want to permit his *Habilitation*, which only came about through the philosopher Bergmann under pressure from on high (Althoff).

I hope that I have been of some help to you with these facts. I am always at your disposal, should you need further information.

Respectfully yours,

G. C.

Cantor Letter 291

To Herrn Prof. Dr. F. X. Heiner in Freiburg

Halle a.d.Saale, 4 February 1896

Esteemed Colleague:

I am much obliged to you for your letter of 2 February and for the cordial interest that you have developed in this matter. Even if you should not adopt the point of departure I have sought, I shall nonetheless always remain thankful for the opportunity to get to know you better and for having become acquainted with your magnanimous sense of Christian philosophy. Our common friend, Franz Woker, has told me *in detail everything* you have done and still continue to do further along these lines at the University of Freiburg. For that may Almighty God grant you his rich heavenly blessings and may the Church preserve your strength to a ripe old age!

Since it was as clear as can be from your kind letter that for your part you share my viewpoint, I do not hesitate to express my great surprise regarding the attitude of the colleague representing your faculty in the Senate and with whom you are clearly not in agreement. If he has no other more plausible reason for reservations towards Mr. H, other than (as I hear to my surprise—I have just know Mr. H since his arrival in Halle, therefore, since 1886, only as a Christian) than because he brought out the fact that Husserl came from Judaism and became a Christian. So I once again see the sad case, unfortunately occurring frequently in Germany, of a Catholic Christian who in the most flagrant manner acts contrary to the holiest traditions of the Church!

Do I have to object that, leaving completely out of account the most holy person of our Savior, his twelve apostles were all Jews?

Is this very revered colleague then not acquainted with the fact that from time immemorial the Church has ordered that the Jewish question be viewed only as a religious question?

Does he by chance believe that the Church has ever sanctioned the poison of anti-Semitism conspicuous in so many German Catholics?

Or is he by chance of the opinion that at the moment, under the most glorious pontificate of Leo XIII, racial anti-Semitic leanings will be tolerated in Rome?

Has he by chance no knowledge of Prince Archbishop Kohn, residing in Austria, who is the son of an Israelite petty bourgeois?

Has the name of the venerable Alsatian priest Ratisbonne never come to his ears?

Has he still not gotten hold of the great Leo's henotic encyclicals of 20 June 1894 and 14 April 1895?

Does he by chance believe himself able to support the great work for unity that H. Holiness has proposed to himself for the final years of his life by promoting hate-filled, unchristian racial anti-Semitism?

It is not impossible that during these Easter holidays in Rome, I shall still find the opportunity to put into their proper place all the material about this question of Judaism that I have been collecting for years, which I have, though, most painfully encountered in the nonetheless rare cases involving Catholic clergy.

By the way, according to your letter, it is still not totally out of the question that Herr H. will obtain the position. His name is surely on the list, and you yourself give me your word, upon which I am still relying, to make at least another attempt to gain the support from enlightened parliamentarians that was denied you by less enlightened colleagues.

Our highly esteemed friend Woker will apparently, so he writes, be coming to Halle shortly. When he does, I shall also discuss the matter with him. We have not seen each other for two years.

Please keep me in your kind thoughts. I remain

Most respectfully yours,

Your good friend, G. Cantor

APPENDIX II

*Translated by Dr. Ruth Ellen Burke,
California State University, San Bernardino*

DOCUMENTS PERTAINING TO HUSSERL FROM THE NIEDERSÄCHSISCHE STAATS- UND UNIVERSITÄTSBIBLIOTHEK, ABTEILUNG HANDSCHRIFTEN UND SELTENE DRÜCKE AT THE UNIVERSITY OF GÖTTINGEN AND FROM THE GEHEIMES STAATSARCHIV PREUSSISCHER KULTURBESITZ IN BERLIN

Letter from H. E. Müller

Göttingen, 29 May 1900

Esteemed Colleague,

The thoughts awakened in me by your most confidentially made communication concerning the possible appointment of Prof. Husserl to this university are the following:

You are aware that I am at the point of drafting a thorough report (to be sent to the imperial board of trustees the day after tomorrow) for a top ministry concerning the position in which this institute of psychology is placed by the sudden departure in April of Dr. Pilzecker, who was so often a voluntary assistant to me, as well as by the impending loss of his equipment (half of all the equipment used up to this point in the Institute). Since I am now lacking any assistance, and without assistance I can neither prepare more extensive demonstrations, nor arrange for tutorials, nor can I proceed with my own previously initiated experimental research, the machinery has in fact now come to a standstill. If a top ministry wishes to do something for philosophical activity in Göttingen, then it seems obvious to me that, above all, what is already existing is to be preserved and cultivated and a gross step backwards, right in the most modern and most rapidly advancing and most promising discipline of experimental psychology, not be permitted. I rather doubt however that, after appointing a third extraordinarius in philosophy at this university, a top ministry will still be in the position to be able to provide the means that are required for a suitable reequipping of the psychological institute (at least 5000 marks) and for the long-term maintenance of the operation and an assistant's position (700 and 1200 annually).

Should a top ministry not be able to decide upon the allocations mentioned here, then, for the reasons indicated above, I would no longer be in a position to hold the special psychology classes and tutorials I have held up until now and accordingly to switch to especially cultivating in my courses and tutorials, along with you, Reheisch [?] and Peiper [?], and possibly even Husserl, the subjects of logic and theory of knowledge (including natural philosophy and its history) that are basically no less dear to my heart. That, in that case, Prof. Husserl's activity on our faculty will entirely correspond to the expectations he has perhaps entertained cannot be guaranteed.

Let us suppose further that either you or I should quit the faculty here. Then in the case that Mr. Husserl were appointed now, no fewer than three extraordinary professors would be on hand here, but not one of them would be qualified to serve as the successor to the ones departing. I fear, though, that under such circumstances a certain resistance to passing over Mr. Husserl in silence would ensue, although, as I said, he is not in the least qualified to replace either of us. As for me, I would always need a colleague beside me who possesses broad philosophical and pedagogical knowledge—which is characteristic of you—a colleague who would have the depth of disciplinary expertise required to evaluate a *Habilitationsschrift* on a subject in ancient philosophy, for example. About such matters, however, Husserl scarcely understands the least thing anymore than I do. On the other hand, that he also would not replace even my humble self as a psychologist, I may well affirm confidently without being immodest.

I began reading Husserl's "The Philosophy of Arithmetic" (Vol. 1, Halle 1891) in its day—as far as I know nothing further by him exists–, but did not continue because only the first volume exists and it is not worthwhile to spend time reading the mere beginning of such a work. I have already waited nine years for the announced continuation. I doubt very much that promoting Mr. Husserl at this time would serve to hasten the publication of this sequel.

Collegial best wishes

Sincerely yours,

G. E. Müller

Letter from Prof. Baumann

Göttingen, 29 May 1900

Honorable Privy Councilor:

In reference to the matter of appointing Professor Husserl as irregular professor extraordinarius in Göttingen, I have the honor to note: Husserl, now 41 years old, has not continued writing the first volume of his "Philosophical Arithmetic: Psychological and Logical Investigations, 1891". If he were to have in the meantime demonstrated teaching gifts through the courses he has given, then bestowing him an extraordinary professorship is probably to be granted him. Whether Göttingen would be the place for him, however, I would most seriously doubt! What we have wanted for years are Privatdozenten in philosophy. Mr. G. E. Müller and I have, when the opportunity arose, not failed to give encouragement, but what frightened people off was the presence of two extraordinarius professors on the faculty already. –Since Mr. G. E. Müller and I planned to petition the top Ministry at some point during the summer, I have–strictly confidentially–requested that, for the time being, Müller lay out the matter briefly in a letter to me, since this should be of great importance to the question you posed. The letter is enclosed and can be added to the official records.

I remain, Your Excellency, most respectfully yours truly,

Prof. Baumann

Letter from Dr. Höpfner to the Minister

Göttingen, 30 August 1900

The Royal Trustee of Georg-August University
No. 3267
Re: The proposed appointment of Privatdozent Dr. Husserl in Halle to the position of extraordinary Professor at this University
Ref. Decree No. U I 21276 dated 27 August 1900

Your Excellency:

I shall not be remiss in reporting perfectly respectfully on the decree referred to as follows.

At this university, we have in the field of philosophy, besides the professors ordinarius Baumann and Elias Müller, who are both important in their way, two more professors extraordinarius who are not outstandingly effective as teachers, who otherwise are well regarded and who know how to combine their scholarly calling with lively interest in religion and theology, not to mention active participation in church life. In addition, yet another Privatdozent, Dr. Goedeckemcys, has recently come, about whom it is still uncertain how he will work out.

The existing professors extraordinarius hardly have any prospect of attaining a regular professorship. So they tend towards the disgruntled mood that inevitably befalls professors extraordinarius without prospects. Such elements are a burden to the university. The disgruntlement will break out more swiftly and distinctly in those named if yet another extraordinarius is brought on board, no matter whether or not he surpasses them, or whether he, too, were shortly to come to belong to the more or less disgruntled.

Regarding Professor Husserl, in light of the frank comments that Prof. Baumann makes about him, and in light of the extracts from his "Prolegomena to Logic" that he submitted to me, I must nurture the fear that he would worsen the mood among the philosophy teachers here, since this scholar will probably not have a real future, at least hardly at rigorously academic institutions of higher learning. He is said to possess perspicacity and knowledge, as I have learned from others. However, what I, among others, notice about him is a pronounced scholastic form of thought; as a philosopher he gives dictums as if a philosophy pope stood behind him who gave him a mission. In passing, let me note that Husserl's religious stance is totally unknown to us here at Göttingen; after reading the Prolegomena one might guess that he is Catholic, which would hardly make it easier for him to gain influence in his field here.

Professor Elias Müller has not spoken to me about Husserl. Nonetheless, his letter, which Your Excellency shared with me, shows me that he does not desire to have Husserl as a colleague because he would appear out of place here. Baumann's opinion of Husserl seems to have deteriorated since the end of this May, probably from reading the Prolegomena. Nor does Baumann want Husserl as a colleague. Yet Baumann is manifestly more adept than E. Müller making the best of an unpleasant situation; he would perhaps not make things difficult for Husserl. It is only to be feared that if Husserl did not remain completely deadwood on the faculty and should prove to have some effectiveness as a philosopher of "direct insights and truths given with them", Baumann would champion all the more bluntly and ruthlessly his completely different standpoint, while he now enjoins caution and consideration.

My impartial judgment in the matter would therefore be that Husserl would *not* represent a desirable addition to the faculty of Georg-August University.

The enclosures of the decree cited are again included.

Dr. Höpfner

Letter from W. Fleischmann

Göttingen, 12 May 1905

The Philosophy Faculty of Georg-August University
J. No. 242

Pursuant to the request received through Decree No. 1398 dated 18 April 1905 to state whether reservations exist with regard to His Excellency's intended recommendation concerning the appointment of professor extraordinarius Dr. Husserl to this university, the undersigned members of the faculty venture most respectfully to report that, after a hearing with the specialists in the field, the faculty have objections to complying approvingly with the Minister's wishes and upon request are prepared to provide detailed reasons for this.

The Philosophy Faculty

The Dean:

W. Fleischmann

To
The Royal Trustee of Georg-August University
The Senior Privy Councilor, Right Honorable Dr. Höpfner of Göttingen

Letter from the Göttingen Philosophy Faculty

No. 1737

Göttingen, 22 May 1905

Re: The position of the Philosophy Faculty regarding the appointment of Professor Dr. Husserl as Professor Ordinarius
Regarding the decree of 15 April 1905
n I N⁰ 16156

Immediately after the receipt of the decree n I N⁰ 16156 of 15 April of this year, my representative asked the Philosophy Faculty whether they seemed to have objections to the appointment of Professor Husserl as professor ordinarius.

Nevertheless, the faculty's report only arrived here on May 12th. According to the contents of this document enclosed here:

> the undersigned members of the faculty venture most respectfully to report that, after a hearing with the specialists in the field, the faculty have objections to

complying approvingly with the Minister's wishes and upon request are prepared to provide detailed reasons for this

I would have found it seemly had the faculty not merely established the existence of reservations about the wish expressed to them, but had at least also indicated the direction in which their reservations might lie. Since this did not occur, I sought out the Dean of the faculty in order to be apprised of what was behind the way the faculty had proceeded. According to what he shared with me, the faculty would be of the opinion that the reservations that they entertained regarding Professor Husserl's appointment would require a detailed statement, and the faculty did not feel certain as to whether Your Excellency wanted to receive such a statement from them. The decree of 15 April of this year regarding Husserl's appointment had indeed only asked "if", not "why".

Furthermore, the dean confirmed what other trustworthy parties had already told me: that the faculty was afraid that consenting to Your Excellency's wishes would create a difficult situation in the event that one of the ordinarius philosophy professors had to cease his teaching activities. Husserl would then, of course, move into the vacant regular position. In that case, however, probably (since thinking [in terms of human reckoning] in the first place of Baumann's departure), a great lacuna would appear in the teaching of history of philosophy, in that a historian of basic Greek philosophy would be lacking—for which Husserl would not appear qualified.

But since, according to what the Dean had to say, the discussion conducted among the faculty about the matter was as exhaustive as it was impassioned, it seems to me beyond doubt that also underlying the faculty's hesitation to speak out, as would be natural, are unwarranted and unseemly motives, such as, for example, the belief that they found themselves facing a *fait accompli*.

Given that it is an unassailable fact that Husserl's competence as a painstaking scholar and very pleasant teacher has now been fully proven through many years of activity here, the Dean (Privy Councilor Fleischmann) has moreover also declared himself in favor of the faculty's unanimous opinion about Husserl. [The transcriber noted inconsistencies in the text written in old German script.]

Rough Draft of Hilbert's *Separatvotum* for Husserl

My Separatvotum for Husserl   30 July 1908

Your Excellency,

I most respectfully request you to grant my wish to bring a reservation to bear in relation to the report of the faculty here concerning the matter of filling the position of regular ordinarius in philosophy.

Even though, as the faculty's report explains, the three areas of training of scholars within philosophy, namely, systematic, historical and experimental psychology stand today, each in its own right, as independent sciences on account of their scope and the differences in their practices, nonetheless, systematic, i.e., purely theoretical, philosophy, through whose enduring nature and success both those sister disciplines first became possible, has a claim to the central position; in my opinion, failing to recognize this claim, or relegating its requirements to the background would be equally disastrous for science and for teaching (indeed for the entire development of our culture).

Systematic philosophy is represented at our university by our colleague Mr. Husserl, who is viewed in professional circles as one of the most prominent (and creatively most active) scholars in this field, whose work "Logical Investigations" counts as epoch-making. From my students, and indeed the most intelligent of them, I know that Mr. Husserl above all exercises great attraction upon the more mature attendees (and spurs them to do independent research); he has successfully founded a philosophical school here that is highly regarded in the philosophical world.

In light of these facts it is, as I believe, in the overall interest of our faculty and university, and an underlying imperative requirement, to keep our colleague Mr. Husserl permanently at our university. Göttingen's setting, which is especially suited to it owing to the highly abstract interests of the young people studying here, will not drive the Husserlian philosophical school away, and in consideration of the difficult situation in which the faculty would find itself were our colleague Mr. Husserl to be called away to another post, with the total lack of a replacement for the then sole still available position of extraordinarius, I most respectfully propose that Your Excellency wish to assign the regular ordinarius position in philosophy to Mr. Husserl.

The need to strengthen the field of philosophical history given first priority by the faculty in their report would, in my opinion, be satisfied if a younger philologically-historically trained scholar were appointed personal professor ordinarius here, particularly since our colleague Mr. Baumann has not up until now in any way confined his teaching activity relating to the historical-philological field, and besides our colleague Mr. Husserl also has success in the history of philosophy. As a person suitable for such an appointment, may I suggest Dr. E. Cassirer, whose name was also put forward as the faculty's second choice for regular ordinarius. For the rejuvenation of the faculty, Professor P. Natorp (54 years old), I believe, deserves preference and Dr. G. Misch, Privatdozent in Berlin, would likewise be on hand to represent historical philosophy. His history of autobiography, highly prized in professional circles, is based on erudite knowledge of the classical source documents.

Hilbert

Extracts from Hilbert's *Denkschrift* for Leonard Nelson

...In filling chairs of philosophy in Germany, two main trends are evident. One emanates from empirical presuppositions. To give a rough outline of it, it pursues philosophy with the greatest possible dependency on the experimental methods of the natural sciences. Under the influence of people such as Wundt, Helmholtz, Avenarius, Mach and Lipps, one is led therein, above all, to make the psychological conditions and development of mental activities on the basis of which knowledge of truth becomes possible into an object of scientific investigations of an experimental-psychological nature. As indispensable as are the benefits of this way of doing research, which has reached its apogee in proponents like Stumpf, Georg Elias Müller, Külpe, and also in Ach, Messer and Ziehen, the genuine philosophical problems have nevertheless not been settled with it. Even among the ranks of experimental psychologists themselves, reservations are increasing as to whether the mental foundations of thinking and knowing can truly be understood and exhausted through work in laboratories of experimental psychology and whether, in particular, the knowledge's claims to validity can be proven in this way.

On the basis of these reservations, the other main present-day philosophical trends in Germany have once again gained ground in opposition to the "psychologism" of the movement depicted. This second basic trend is linked to the historical traditions embodied in the names Fichte, Schelling and Hegel. It is speculative and pursues philosophy as science of the mind, meaning that it actually abandons any viable relationship to the natural sciences, isolates itself from them, and, in that it rejects every attempt of a psychological analysis of knowing as unessential for the *validity* of truth, tries to grasp its grounds for validity from dialectical theories of a priori knowledge, whereby it then runs the danger of being in contradiction with the findings of the progressive natural sciences. In so far as it does not generally feel drawn to irrationalist philosophizing and comes close to the thought processes of medieval philosophical mysticism—which has repeatedly become apparent in recent times—this movement follows the models it meets with in the series of neo-Platonists from Plotinus to the representatives of post-Kantian philosophy of identity. And so the history of philosophy becomes one of its most essential concerns. Its current representatives have been particularly successful in opposing the application of the methodology of the natural sciences to the field of what is known as philosophy of values, philosophy of history and philosophy of civilization. The names of Simmel, Windelband, Rickert, Joël, Eucken come to mind. The greater the latter trend's sphere of activity, the more outwardly abundantly productive its accomplishments have been up until now, the more subjective and changeable, the less exact and the poorer in results for the exact sciences as compared to the first trend. May I further stress that among younger thinkers of Rickert's circle, the philosophical trend in the Rickert school linked to Fichte has given rise a counter-movement whose goal is a revival the Hegelian system, despite the serious disappointments that that philosophy brought to thought in the natural sciences and economics in the middle of the previous century....

...Above all, Brentano's school has directed its investigations toward the creation of an exact theory of acts of judgment and of logic, with the goal of constructing a theory of science such as Bolzano already had in mind as the foundation of the validity of the exact sciences. From this school that flourished in Austria, only *one* representative came to Germany, namely, Husserl. In contrast to the other representatives of this school, Husserl has not incurred the reproach of "psychologism". He adopts the standpoint of an a priori method and expressly rejects psychologism. From this theoretical stance he not only founded a school, but also befriended the speculative orientation in philosophy while he strengthened it considerably. For, since he actually carried out a far-reaching grounding of logic and the related sciences to it, as a result the odium of sterility with respect to its application to the exact sciences was lifted from speculative dogmatics after Husserl became acquainted with their methods. However, in so doing the problem is only seemingly solved. For in fact Husserl's method is also psychological in nature. And only as a result of a misunderstanding of its true nature did it become possible to post Husserl's successes on the "a priori dogmatism" side of the ledger.

Now, Nelson's manner of investigation is not entirely dissimilar to Husserl's. Like Brentano's school, Nelson also proceeds from the analysis of mental functions. In this sense, his philosophy is not speculative but exact science like the natural sciences. The rigor of the logic, the unambiguousness of the methodology, the austere clarity of the problems that Nelson takes on, and of the systematic solutions that he provides for them, make him appear to be one of the few scholars who practice philosophy as an exact science and who thereby make it possible to deepen and to clarify the foundations of the other exact sciences. In this respect, in so doing, Nelson stands in conscious contrast to Husserl, since he does not share his concessions to the speculative trend in philosophy, expressly places himself on the ground of psychological, empirical investigation. On the other hand, however, he does not share the empiricism into which the rest of Brentano's school fell either, because he claims a priori validity for philosophical knowledge established through psychological criticism. The psychological method exclusively serves him for the purpose of presenting knowledge that is, by its very nature, independent of experience, rational knowledge that, as a psychological fact, is like any other psychological fact, is an *object* of psychology, thus of an empirical science that is analogous to the other natural sciences, whose *content*, however, for that reason, does not need to lie in the field of psychology any more than our knowledge of natural laws or our knowledge of external facts is itself psychological in nature, although as knowledge it can be the object of psychological consideration.

Openly adhering to the view that the critical method of investigating the foundations of our knowing must be clearly and consciously psychological plainly represented a threat to Nelson from the reigning cultural-philosophical-historical movement that Husserl's investigations escaped by adhering to intellectual intuition. Owing to its theoretical position, Nelson did not receive the recognition from that influential movement that his work deserved.

Precisely the clear understanding that the method of the critique of reason is psychological in nature nevertheless guarantees Nelson's work its rigorous unambiguousness based on the exact natural sciences, which then explains the fact that understanding reigns precisely among the foremost representatives of the exact sciences concerning the significance attributable to Nelson's effectiveness when it comes to the analysis and grounding of the validity claims of their axioms and principles....

Extracts from Leonard Nelson's Letter to David Hilbert

Westend-Berlin, 29 December 1916

To Privy Councilor Hilbert, Göttingen,

...You pointed out that Göttingen had the good fortune to possess at its university in Husserl a man who, exceptionally, not tainted by relativism, believed in the possibility of philosophical science and worked on that to make it a reality. He hoped to find here in Göttingen a favorable setting for his work. And, he himself in fact created the groundwork, founded a school and became characteristic of Göttingen. However, people did not see, or did not want to see, how valuable it is for the development of science to secure the continuity of such efforts. They discounted the significance of a milieu created by personage like Husserl and made no attempt to keep him, because they did not want to keep him. Instead of gathering all forces to improve upon the foundations lain, they did not even do the minimum by using the rudiments on hand for the creation of a scientific tradition and by leaving those that are ready to further science, the conditions already on hand for that. People are not afraid to put the interests of science behind any other interests of a more personal nature, and as a result commit blunders that spoil, if not totally destroy, the fruit of decades-long work.

You express the opinion that my present situation is not unlike that into which Husserl was put in that out of a lack of understanding and good will, the groundwork that he himself had begun to prepare for this in Göttingen was not secured for his scholarly activity. I also believe the comparison is probably in order. I am even of the opinion that right now the setting in Göttingen is of still greater value to my activity than it could have been for Husserl, and this is why, and for other reasons, to be mentioned forthwith, I find myself in a far more unfortunate situation than Husserl's was, as long as I lack the prospect that my activity can set down roots for the duration in ground that is in and of itself so beautiful and so fertile.

What makes it so extremely valuable for me to work at Göttingen at the present time is the fact that here the genuinely scientific spirit is indigenous to the place in a way that is only produced by mathematical training. I definitely have the impression that no other German university offers anything comparable to what Göttingen possesses in this academic milieu, which, after its ancient tradition, it above all owes to your influence. I am not a mathematician by profession. But I have acquired enough of an understanding of mathematics to believe myself capable of a definite judgment about where scientific soundness is and is not found. What, in my opinion, is characteristic of your determinant influence, and thereby of that of Göttingen's academic milieu, is the conjunction of scientific rigor, on the one hand, and freedom of outlook, a freedom that, more than the exactitude with which individual investigations are conducted, nevertheless does not lose sight of the major problems that ultimately gave rise to the development of rigorous methods. In your hands, method remains an instrument and does not become an end in itself, as it has for so many who have adopted your rigor but no longer see the purpose above the means and have lost all standards by which to discern the significance of the issues. Still others have surely acquired a broad outlook, but along with that have neglected the need for scientific rigor. Thoroughness and freedom are both necessary requirements. Neither suffices alone. Only in combination do they constitute the essence of the scientific, of the genuinely philosophical spirit.

I find this combination only in you, and the great model that you have always been for me is really what originally drew me to Göttingen. Perhaps I may mention here the circumstance to which I owe my first acquaintance with your name, and which I regard as one of the most fortunate coincidences of my life. While I was in my last year of secondary-school, Hessenberg, who at the time was himself still a young assistant at the technical institute in Charlottenburg, was giving me private lessons, which through discussions about mathematics and philosophy very soon shifted to the sphere of our shared interests. On one occasion—we had just spoken about the outline of a geometric system of axioms that I had submitted to him—expressing enthusiastic reverence for its author, Hessenberg alluded to your just published book on the foundations of geometry and acquainted for the first time with the fact that Göttingen, as it had through Gauss 100 years earlier, had now through you become the center of mathematics research. My longing to go personally on pilgrimage to this holy land of science was reinforced by the piety existing in my family for the place where my great-grandfather Dirichlet had been active, and although I was very far from possessing a specialist's knowledgeable judgment, I nevertheless understood that in no living being other than you would I find embodied to such an extent what was for me worthy of veneration in the scientific personality of my great-grandfather. It was the spirit of Kant, Fries, Dirichlet that drew me to Göttingen, to the place where you are active.

My belief that I would find here, if anywhere at all, academic earnestness and a rigorous sense of truth, and hence have the prerequisites for my work met, did not betray me. It is not just that all the capable students that I have had have been almost exclusively mathematicians and mathematical physicists from your school, but I myself have found no other living teacher besides you. For, although I was not lacking in good will in seeking a philosophy teacher, since, disdaining any obsession with originality, I suffered a great deal from the lack of a living scientific school, so I was on my own to entrust myself to the school's great *dead* members. What I learned from you was not exactly philosophy, but from you I learned what science is, and with mathematics as my model, I have acquired the standard that all intellectual work must satisfy if it claims to be scientific.

As a result, my own work can be appreciated and understood only by those who have passed through the school of mathematics. Not in order imitate slavishly mathematical methods do I require knowledge of them, but in order to know what solid intellectual work is. It is not here a matter of the wealth of mathematical knowledge, if only insight into the workings of methodical thinking succeeds in protecting the one having it from falling victim undiscriminatingly to the first fine dilettante she or he stumbles upon. One can, however, only gain this insight where genuine science is at work and where one is just as far from being bogged down in the mechanical exploitation of methods that have become routine as from losing oneself in the labyrinth of speculations unamenable to scientific methods. For me, it would be a grim prospect to have to forgo student material such as can only be provided me through your influence. Like Plato, I must write on my lecture-hall: "May whoever understands nothing about mathematics remain outside". For whoever understands nothing about mathematics will lack any understanding of my entire manner, of the technique of my academic work, and of the significance of such work in general.

Since that is the case, may I suggest that a milieu such as Göttingen offers is even more valuable to me than it was able to have been for Husserl, who admittedly also originally came from the mathematical school, but who bit by bit turned more and more away from it and turned towards a school of mystical vision, whereby he also deadened the feel in his school for the demands and value of a specifically scientific method. He even goes so far, after his own lack of success with it, as to see a danger in methodological thinking and thinks that it would ruin philosophers for whom the truth only reveals itself in mystical vision. Even though Husserl himself remains protected by certain inhibitions from mystical degeneracy by virtue of strong ties to mathematics that he has not been able to cast off, one must unfortunately nonetheless note with horror that after the school as such had torn down the bridges to mathematics behind them, how they unrestrainedly his students lapsed into every excess of Neo-platonic mysticism, the prevalence of which is all the more dangerous since Husserl's scientific past will unjustifiably carry over to the school the assurance that it is willing to do scientific philosophy in earnest and is capable of this. Here I fundamentally part company with Husserl's circle. For me philosophy is not a matter of mystical vision, but one of the most sober, driest thinking, and if it was a mistake not to keep Husserl here, because he would have been able to create a climate here for his activity, then if something similar befalls me, it will there-fore affect me doubly, because I would not thereby only be losing a favorable setting for my academic activity, but quite simply *the* setting for my activities. For here alone—at the university with the greatest scientific tradition and present—can I find it....

Hilbert's Private Letter to Becker

Göttingen, 30 July 1918.

Dr. Becker, Undersecretary, Ministry of Culture, Berlin

Dear Dr. Becker:

In my opinion, mathematics, physics and philosophy form an interconnected scientific system and I have always seen it a part of my life's work especially to cultivate the relationship between mathematics and philosophy.

Among the philosophers who are not primarily historians or experimental psychologists, Husserl and Nelson appear to me to be the most prominent personalities, and for me, it is no accident that both of them appeared in Göttingen's mathematical setting. For that reason, exactly ten years ago today, in a *Separatvotum* dated 30 July 1908, I ventured to write to the Minister to propose Husserl for the new position of regular ordinarius in philosophy to be filled, while Cassirer should receive the position of extraordinarius that then became available—in order, as I explained at the time, "to keep our colleague Mr. Husserl permanently at our university" and hence the "Husserl school which, owing to the highly abstract interests of the young people studying here, is especially suited to it, will not be driven away from the Gottingen setting".

My fear that we would lose Husserl has unfortunately become a reality and my sole concern is presently directed toward *preventing the "Husserl Case" from repeating itself once again*, by also losing the second just as outstanding systematic philosopher, namely Nelson.

In my opinion, this would be disastrous, since I am convinced that in the future Nelson will attain a dominant position within philosophy and Göttingen, with its excellent *Dozent* and student material, could through Nelson become *a center for systematic philosophy*. Flowing out of this conviction was the 3 March 1917 *Separatvotum* addressed to the Minister by some colleagues and myself that spoke out on Nelson's behalf, with which however the Minister did not comply in filling the position of extraordinary professor of philosophy. In the current matter of the new Ordinarius position presently to be filled, a larger number of colleagues have now once again agreed upon a *Separatvotum* that will shortly be placed in the hands of the Minister.

Given the difficulty of the matter—the legal question regarding the manner of preparing the proposals for the philosophy professor is also controversial—and in face of the great importance of the decision, I am taking the liberty of informing you that some of the signatories of the latest *Separatvotum*: Professors Runge, Debye, Caratheodory and I, are quite prepared to have a personal discussion with you, Herr Privy Councilor, were such a discussion possible in mid-September or later.

With sincere greetings,

Most respectfully,

Hilbert

Letter from H. Weyl,

Degersheim, 13 August 1918

Dear Colleague,

Only in the very broadest outlines can I declare myself in agreement with the portrayal of contemporary state of German philosophy based on the standpoint of the Friesian school and the assessment of Nelson's intellectual work that the *Denkschrift* contains. What is decisive for the significance of a philosopher is not what he *deduces,* but what he *saw,* the depth of direct insights, in which the essence, interrelationship and necessity of the world structure at hand opens up to him or her.

Capacity for concise expression and the forming of this viewpoint is only secondary, in last place, however, the acuity and the consistency of thought that develops a coherent system from such seeds. Now, in the first respect, Nelson's philosophy seems to me poor in basic intuitions; its exactness, however, through which it wins mathematicians so easily over to itself, I consider for the most part to be formal appearance and its deductions idle. Nonetheless, I have to confess that I did not study Nelson's works much more thoroughly once I arrived at the subjective certainty that no new clarity for me was to be obtained on this path. I believe that Husserl and his school, as slowly and tentatively as it approaches the genuinely philosophical problems, has struck out in a more promising path in order to arrive at a philosophy that in depth of insight can compete with Fichte (to name the grandest example known to me), in rigor of thought with the exact sciences. However, the goal is still a long way away.

Despite this standpoint deviating from the *Denkschrift*, I wholeheartedly support the wishes expressed in the petition. For despite any objections I too believe that, after Husserl, Nelson is the most important systematic philosopher among today's thinkers. Borne by an earnest, austere sense of truth, his work attests to a far-reaching, independent, forceful philosophical life well-suited to kindling such a life in those who seek the path to knowledge through them. His thinking is of ironclad consistency, and (what says more), he is one in theory and practice; his person radiates clarity, purity and strength. An inspired and inspiring instructor of the kind that one would like to have very many more of for the scholarly German youth returning home from the war.

Above all, I further support the petition out of an interest in systematic philosophy in general. Psychology, in my opinion, is a science of matters of fact that is certainly of no more importance to philosophy than is, say, mathematics, physics, biology, or history. History of philosophy–the account of what one or another person said and in end also held to be true–has just as little to do with true philosophy (however important heritage may be, so that no strides already made are again lost). For what ultimately matters however is that we ourselves think and understand. Filling chairs of philosophy with psychologists and historians can therefore only be interpreted as an *abandonment* of philosophy. This weary spirit of abandoning insight is however out-of-date today. A fierce yearning to understand not to be satisfied by the jumble of "facts" piled up in all the individual sciences, a tumultuous has been kindled anew. It would be grave form of backwardness if these aspirations did not find expression in the representatives of philosophy in the German universities.

signed H. Weyl,

Professor of Mathematics, Zürich.

David Hilbert, Summer 1918
"For the Minister"

For 15 years I have been fighting for philosophy.
Althoff produced a position of Ordinarius out of thin air.
Here only a little Extraordinarius in accordance with my wish.
So I promise something great: premier center for philosophy.
That works nowhere else but in Göttingen, or else nowhere at all, so that no one from here goes to Freiburg and Heidelberg.
Cultural issues of the highest order is at stake.

Going elsewhere!
With my appointment to Berlin, conviction that it would only be possible to retain N if I obtain more signatures.
I cannot carry out an important part of my teaching program without N.[2]
N. is the leaven; here he will represent a school oriented toward firm principles that will tip the balance.
His appointment is a first-rate cultural feat: reformation of the spirit of the professoriate. Without N, I am a cipher on the faculty.
3 philosophers and not one in systematic philosophy with expertise in mathematics is an unacceptable state of affairs.
Habilitation of young teachers.
Philosopher with universal education.

---

[2] Presumably Leonard Nelson.

SELECTED BIBLIOGRAPHY

Bachelard, S. *A Study of Husserl's Formal and Transcendental Logic*. Evanston: Northwestern University Press, 1968.

Barcan (Marcus), R. "A Functional Calculus of First Order Based on Strict Implication". *The Journal of Symbolic Logic*, 11(1):1-16, 1946.

――――. "The Deduction Theorem in a Functional Calculus of First Order Based on Strict Implication". *The Journal of Symbolic Logic*, 11(1): 115-18, 1946.

――――. "The Identity of Individuals in a Strict Functional Calculus of Second Order". *The Journal of Symbolic Logic*, 12(1): 12-16, 1947.

Bar-Hillel, Y. "Husserl's Conception of a Purely Logical Grammar". In Mohanty (ed.) 1977: 128-37.

Becker, O. "The Philosophy of Edmund Husserl". In R. O. Elveton (ed.), *The Phenomenology of Husserl, Selected Critical Readings*, Chicago: Quadrangle Books, 1970 (1930): 40-72.

Benacerraf, P. "Frege: The Last Logicist". In W. Demopoulos (ed.), *Frege's Philosophy of Mathematics*, Cambridge: Harvard University Press, 1995 (1981): 41-67.

―――― and Putnam, H. (eds.) *Philosophy of Mathematics, Selected Readings*. Cambridge: Cambridge University Press, 2nd ed. rev. 1983 (1964).

Bergson, H. *Ecrits et Paroles*, vol. 3. Paris: PUF, 1959.

Bernays, P. "Hilbert's Significance for the Philosophy of Mathematics". Tr. by Paolo Mancosu, in Mancosu 1998: 189-97 (1922)

――――. "Hilbert, David". In P. Edwards (ed.), *The Encyclopedia of Philosophy*, vol. 3, New York: MacMillan: 496-504, 1967.

Beyer, C. *Von Bolzano zu Husserl. Eine Untersuchung über den Ursprung der phänomenologischen Bedeutungslehre*. Dordrecht: Kluwer, 1996.

Bochenski, J. *A History of Formal Logic*. New York: Chelsea Publishing Co., 1970.

Bolzano, B. *Theory of Science*. Partial translation by R. George, Berkeley: University of California Press, 1972 (1837).

Boolos, G. "Saving Frege from Contradiction". *Proceedings of the Aristotelian Society* N.S. 87: 137-51, 1986/87. Also in Demopoulos (ed.) 1995: 438-52.

――――. "The Consistency of Frege's Foundations of Arithmetic". In *On Being and Saying, Essays in Honor of Richard Cartwright*. J. Jarvis Thomson (ed.), Cambridge: M.I.T. Press, 1987: 3-20. Also in Demopoulos (ed.). 1995: 211-33.

_____. "The Standard of Equality of Numbers". In *Meaning and Method: Essays in Honor of Hilary Putnam*, Cambridge: Cambridge University Press, 1990: 261-77. Also in Demopoulos (ed.), 1995: 234-54.

_____. "Whence the Contradiction?" *Proceedings of the Aristotelian Society*, Supplementary Volume LXVII, (1993): 213-33. Also in Schirn (ed.), 1996: 234-52.

Bourbaki, N. "The Architecture of Mathematics". In F. Lelionnais (ed.), 1971: 23-43.

Brentano, F. "On Attempts at the Mathematicization of Logic". In *Psychology from the Empirical Standpoint*, New York: Humanities Press, 1973: 301-06 (section X of the 1911 appendix), 1991.

Brouwer, L. E. J. "Intuitionism and formalism". In Benacerraf and Putnam (eds.): 77-89 (1912).

_____. "Intuitionist Set Theory". In Mancosu (ed.): 23-27 (1921).

_____. "On the Significance of the Principle of Excluded Middle in Mathematics, Especially in Function Theory". In van Heijenoort (ed.): 335-45 (1928).

_____. "Intuitionist Reflections on Formalism". In Mancosu (ed.): 40-44; in van Heijenoort (ed.): 490-92 (1928).

_____. "Mathematics, Science, and Language". In Mancosu (ed.): 45-53 (1929).

_____. "The Structure of the Continuum". In Mancosu (ed.): 54-63 (1930).

_____. "Consciousness, philosophy, and mathematics". In Benacerraf and Putnam (eds.): 90-96 (1948).

_____. "Historical background, principles, and methods of intuition". In Ewald (ed.): 1197-1207 (1952).

Brück, M. *Über das Verhältnis Edmund Husserls zu Franz Brentano. Vornehmlich mit Rücksicht auf Brentanos Psychologie*. Würzburg: K. Tritsch, 1933.

Burge, T. "Frege on Extensions of Concepts from 1884 to 1903". *The Philosophical Review* XCIII(I): 3-34, 1984.

Cantor, G. *Grundlagen einer allgemeinen Mannigfaltigkeitslehre. Ein mathematisch-philosophischer Versuch in der Lehre des Unendlichen*. Leipzig: Teubner, 1883. In Cantor 1932: 165-246.

_____. "Rezension von Freges *Grundlagen der Arithmetik*'. *Deutsche Literaturzeitung* VI (20): 728-729, 1885.

_____. *Gesammelte Abhandlungen zur Lehre vom Transfiniten*. Halle: C.E.M. Pfeffer, 1890. Also in Cantor 1887/88.

_____. "Mitteilungen zur Lehre vom Transfiniten". *Zeitschrift für Philosophie und philosophische Kritik*, 91 (1887): 81-125; 92 (1888): 240-65. In Cantor 1932: 378-439. Also published as *Gesammelte Abhandlungen zur Lehre vom Transfiniten* 1890.

_____. *Briefbücher I (1884-1888), II (1890-1895), III (1895-1896)* at the Niedersächsische Staats-und Universitätsbibliothek Göttingen, Abteilung Handschriften und Seltene Drucke (Cod. Ms. 18).

_____. *Gesammelte Abhandlungen*. E. Zermelo (ed.), Berlin: Springer, 1932.

_____. "Principien einer Theorie der Ordnungstypen". Dated November 6, 1884, first published by I. Grattan-Guinness in *Acta Mathematica* 124, 1970: 65-107.

_____. *Briefe*. H. Meschkowski and W. Nilson (eds.), Berlin: Springer, 1991.

Casanave, A. L. (ed.) *Symbolic Knowledge from Leibniz to Husserl*. London: College Publications, 2012.

Cassou-Noguès, P. *Hilbert*. Paris: Les Belles Lettres, 2001.

Cavaillès, Jean. *Méthode axiomatique et formalisme*. Paris: Hermann, 1981 (1937).

_____. *Sur la logique et la théorie de science*. Paris: P.U.F., 1947.

_____. *Philosophie Mathématique*. Paris: Hermann, 1962.

Cavallin, J. *Content and Object, Husserl, Twardowski and Psychologism*. Dordrecht: Kluwer, 1997.

Church, A. "On the Law of Excluded-Middle". *Bulletin of the American Mathematical Society* 34(1): 75-78, 1928.

_____. "Review of M. Farber *The Foundations of Phenomenology*". *Journal of Symbolic Logic* 9: 63-65, 1944.

Coffa, A. "Kant, Bolzano and the Emergence of Logicism". *The Journal of Philosophy* 74: 679-89, 1982. Also in Demopoulos (ed.) 1995: 29-40.

Crowe, M. *A History of Vector Analysis*. New York: Dover Books, 1994.

da Silva. J. J. "Husserl's Philosophy of Mathematics". *Manuscrito* 16(2): 121-148, 1993.

_____. "Husserl Phenomenology and Weyl's Predicativism". *Synthese* 110: 277-96, 1997.

_____. "Husserl's Conception of Logic". *Manuscrito* 22(2): 367-97, 1999.

_____. "Husserl's Two Notions of Completeness, Husserl and Hilbert on Completeness and Imaginary Elements in Mathematics". *Synthese* 125: 417-38, 2000.

_____. "The Many Senses of Completeness". *Manuscrito* 23(2): 41-60, 2000.

_____. "Gödel and Transcendental Phenomenology". *Revue Internationale de Philosophie* 234(4): 553-74, 2005.

_____. "Husserl on the Principle of the Excluded Middle". In *Husserl and the Logic of Experience*, G. Banham (ed.), New York: Palgrave Macmillan, 2005: 51-81.

_____. "Mathematics and the Crisis of Science". *Diálogos* 91: 37-58, 2008.

_____. "Beyond Leibniz, Husserl's Vindication of Symbolic Mathematics". In Hartimo 2010: 123-45.

_____. "Husserl on Geometry and Spatial Representation". *Axiomathes* 22: 5-30, 2012.

_____. "Away from the Facts, Symbolic Knowledge in Husserl's Philosophy of Mathematics". In A. L. Casanave (ed.) 2012: 115-35.

Dantan, A. R. G. A. *Bertrand Russell and the Origins of the Set-theoretic 'Paradoxes'*. Basel: Birkhäuser, 1992.

Dauben, J. *Georg Cantor. His Mathematics and Philosophy of the Infinite.* Princeton: Princeton University Press, 1979.

Demopoulos, W. "Frege, Hilbert, and the Conceptual Structure of Model Theory". *History and Philosophy of Logic* 15: 211-25, 1994.

―――――. "Frege and the Rigorization of Analysis". *Journal of Philosophical Logic* 23: 225-46, 1994. Reprinted in Demopoulos 1995: 68-88.

―――――. (ed.) *Frege's Philosophy of Mathematics*. Cambridge: Harvard University Press, 1995.

de Muralt, A. *The Idea of Phenomenology. Husserlian Exemplarism.* Evanston: Northwestern University Press, 1974.

Derrida, J. *Edmund Husserl's 'Origin of Geometry': An Introduction.* Lincoln: University of Nebraska Press, 1989. Translation of *Introduction à "L'Origine da la géométrie" de Husserl*, Paris: PUF, 1962.

Dreyfus, H. L. and H. Hall (eds.). *Husserl. Intentionality and Cognitive Science.* Cambridge: MIT Press, 1982.

Dieudonné, J. *"David Hilbert (1862-1943)"*. In F. Lelionnais (ed.) 1971: 304-11.

Dummett, M. *Truth and Other Enigmas.* Cambridge: Harvard University Press, 1978.

―――――. *Frege: Philosophy of Language.* London: Duckworth, 1981.

―――――. *The Interpretation of Frege's Philosophy.* Cambridge: Harvard University Press, 1981.

―――――. *Frege: Philosophy of Mathematics.* Cambridge: Harvard University Press, 1991.

―――――. *Origins of Analytical Philosophy.* Cambridge: Harvard University Press, 1993.

―――――. "Reply to Boolos". In Schirn (ed.) 1996: 253-60.

Ewald, W. (ed.). *From Kant to Hilbert. Readings on the Foundations of Mathematics.* Volume II, Oxford: Oxford University Press, 1996.

Feferman, S. "Weyl Vindicated: 'Das Kontinuum' 70 Years Later". *Temi e prospettive della logica e della filosofia della scienza contemporanee I*, Bologna, 1988: 59-93. Reprinted, with a Postscript, in Feferman 1998: 249-283.

―――――. *In the Light of Logic.* New York: Oxford University Press, 1998.

Fitch, F. "The Problem of the Morning Star and the Evening Star". *Philosophy of Science* 16: 131-41, 1949.

Føllesdal, D. "Husserl and Frege: A Contribution to Elucidating the Origins of Phenomenological Philosophy". In Haaparanta (ed.): 3-47. Translation by C. O. Hill of his Master's Thesis: *Husserl und Frege. Ein Beitrag zur Beleuchtung der Enstehung des phänomenologische Philosophie.* Oslo: Ascheloug, 1958.

―――――. "Husserl's Notion of Noema". *Journal of Philosophy* 66: 680-87, 1969.

_____."Quine on Modality". In *Words and Objections. Essays on the Work of W. V. Quine*. In D. Davidson and J. Hintikka (eds.), Boston: Reidel, 1969.

_____. "Husserl: fifty years later... the Noema twenty years later". *Proceedings of the Eighteenth World Congress of Philosophy*, 1988.

_____."Bolzano's Legacy". *Grazer philosophische Studien* 53: 1-10, 1997.

Frege, G. "Begriffsschrift, a formula language, modeled upon that of arithmetic for pure thought". In van Heijenoort (ed.) 1967: 1-82 (1879).

_____. *Foundations of Arithmetic*. Oxford: Blackwell, 1986 (1884).

_____. "On Formal Theories of Arithmetic". In his *Collected Papers* 1984: 112-21 (1885).

_____. "Reply to Cantor's Review of *Grundlagen der Arithmetik*". In his *Collected Papers*. 122. Originally published in *Deutsche Literaturzeitung* 6, 28 (1885): 1030.

_____. "Function and Concept". In P. Geach and M. Black (eds.) 1980: 21-41 (1891).

_____."Review of Georg Cantor, Zur Lehre vom Transfiniten: Gesammelte Abhandlungen aus der *Zeitschrift für Philosophie und philosophische Kritik*". In his *Collected Papers*. 178-81. Originally published in *Zeitschrift für Philosophie und philosophische Kritik* 100 (1892): 269-72.

_____. "On Concept and Object". In P. Geach and M. Black (eds.) 1980: 42-55 (1892).

_____. "On Sense and Reference". In P. Geach and M. Black (eds.) 1980: 56-78 (1892).

_____. *Basic Laws of Arithmetic*. Berkeley: University of California Press, 1964 (1893).

_____. "Review of Dr. E. Husserls *Philosophy of Arithmetic*". *Mind* 81, 323 (July 1894): 321-37. Also in his *Collected Papers*: 195-209.

_____. "Frege on Russell's Paradox". In P. Geach and M. Black (eds.) 1980: 214-24 (1903).

_____. *Nachgelassene Schriften*. Hamburg: Meiner, 1969.

_____. *Écrits logiques et philosophiques*. Paris: Seuil, 1971.

_____. *Wissenschaftlicher Briefwechsel*. G. Gabriel et al. (eds.), Hamburg: Meiner, 1976.

_____. *Schriften zur Logik und Sprachphilosophie aus dem Nachlass*. 2nd rev. ed. Hamburg: Meiner, 1978.

_____. "Draft Towards a Review of Cantor's *Gesammelte Abhandlungen zur Lehre vom Transfiniten*". In his *Posthumous Writings*, 1979: 68-71.

_____. *Posthumous Writings*. H. Hermes et al. (eds.), Oxford: Blackwell, 1979.

_____. *Philosophical and Mathematical Correspondence*. Abridged by B. McGuinness. G. Gabriel, et al. (eds.), Oxford: Blackwell, 1980.

_____. *Translations from the Philosophical Writings of Gottlob Frege*. 3rd ed. P. Geach and M. Black (eds.), Oxford: Blackwell, 1980 (1952).

_____. *Collected Papers on Mathematics, Logic and Philosophy*. B. McGuinness (ed.), Oxford: Blackwell, 1984.
_____. "Briefe an Ludwig Wittgenstein". *Grazer philosophische Studien*, 33/34: 8, 1989
_____ and E. Husserl. *Frege-Husserl Correspondance*. Mauvezin: T.E.R., 1987.
Gerlach, H. "Es ist keine Seligkeit 13 Jahre lang Privadocent und Tit. 'prof.' zu sein. Husserls hallesche Jahre 1887 bis 1901". In H. Gerlach and H. Sepp (eds.) 1994: 15-39.
_____ and Sepp, H (eds.). *Husserl in Halle*. Bern: Peter Lang, 1994.
Gödel, K. *Collected Works* I-III. New York: Oxford University Press, S. Fefermann et al. (eds.) 1990-95.
_____. "Russell's Mathematical Logic". In his *Collected Works* II: 119-41 (1944).
_____. "Is Mathematics Syntax of Language?" In his *Collected Works* III: 334-62 (1953).
_____. "Some basic theorems on the foundations of mathematics and their implications". In his *Collected Works* III: 304-23 (1951).
_____. "The Modern Development of the Foundations of Mathematics in the Light of Philosophy". In his *Collected Works* III: 374-87 (1964). The "Introductory Note" by D. Føllesdal, 364-73.
_____. "What is Cantor's continuum problem". In his *Collected Works* II: 254-70 (1964).
Granel, G. "Préface, *La crise des sciences européennes et la phénoménologie transcendantale*". Paris: Gallimard, 1976: I-IX.
Grassmann, H. *Die lineale Ausdehnungslehre*. Leipzig: Wiegand, 1844. Translated into English by Lloyd Kannenberg, *A New Branch of Mathematics*, Chicago: Open Court, 1995.
Grattan-Guinness, I. "The Correspondence Between Georg Cantor and Philip Jourdain". *Jahresbericht der Deutschen Mathematiker-Vereinigung* 73:111-130, 1971.
_____. "Bertrand Russell on His Paradox and the Multiplicative Axiom: An Unpublished Letter to Philip Jourdain". *Journal of Philosophical Logic* 1: 103-10, 1972.
_____. "Preliminary Notes on the Historical Significance of Quantification and the Axiom of Choice in Mathematical Analysis". *Historia Mathematica* 2: 475-88, 1975.
_____. *Dear Russell-Dear Jourdain, A Commentary on Russell's Logic Based on his Correspondence with Philip Jourdain*. London: Duckworth, 1977.
_____. "How Russell Discovered His Paradox". *Historia Mathematica* 5: 127-37, 1978.
_____. "Georg Cantor's Influence on Bertrand Russell". *History and Philosophy of Logic* 1: 61-93, 1980.

_____. *The Search for Mathematical Roots 1870-1940, Logics, Set Theories and the Foundations of Mathematics from Cantor Through Russell to Gödel*. Princeton: Princeton University Press, 2000.
Gray, J. *The Hilbert Challenge*. Oxford: Oxford University Press, 2000.
Haack, S. *Philosophy of Logics*. Cambridge: Cambridge University Press, 1978.
Haaparanta L. (ed.) *Mind, Meaning and Mathematics, Essays on the Philosophical Views of Husserl and Frege*. Dordrecht: Kluwer, 1994.
Hallett, M. *Cantorian Set Theory and Limitation of Size*. Oxford: Clarendon, 1984.
Hartimo, M. (ed.). *Phenomenology and Mathematics*. Dordrecht: Springer, 2010
Heck, R. "The Development of Arithmetic in Grundgesetze". *The Journal of Symbolic Logic* 58: 579-601, 1993. Also in W. Demopoulos (ed.). 1995: 257-94.
Helmholtz, H. von. "On the Factual Foundations of Geometry". In P. Pesic (ed.) 2006: 47-52 (1866).
Helmholtz, H. von. "The Origin and Meaning of Geometrical Axioms". In P. Pesic (ed.) 2006: 53-70 (1870).
Hilbert, D. "Über den Zahlbegriff". *Jahresbericht der Deutschen Mathematiker Vereinigung* 8:180-84, 1900. Translated as "On the Concept of Number" in W. Ewald (ed.): 1089-95.
_____. "On the Foundations of Logic and Arithmetic". In van Heijenoort (ed.): 129-38 (1904).
_____. "Axiomatisches Denken". *Mathematische Annalen* 78: 405-15, 1918. Translated as "Axiomatic Thought" in W. Ewald (ed.): 1105-14.
_____. "The New Grounding of Mathematics, First Report". In P. Mancosu (ed.) 1998: 198-214. Also in W. Ewald (ed.): 1115-33 (1922).
_____. "The logical foundations of mathematics". In W. Ewald (ed.): 1134-47 (1923).
_____. "On the Infinite". In van Heijenoort (ed.): 369-92 (1925).
_____. "The Foundations of Mathematics". In van Heijenoort (ed.): 464-79 (1927).
_____."Problems of the Grounding of Mathematics". In P. Mancosu (ed.): 227-33 (1929).
_____."Logic and the Knowledge of Nature". In W. Ewald (ed.): 1157-65 (1930).
_____."The Grounding of Elementary Number Theory". In P. Mancosu (ed.): 266-67. Also in W. Ewald (ed.) 1996: 1148-56 (1931).
Hill, C O. *La logique des expressions intentionnelles*. Mémoire de Maîtrise, Université de Paris-Sorbonne, 1979. Published online.
_____. *Word and Object in Husserl, Frege and Russell, the Roots of Twentieth Century Philosophy*. Athens: Ohio University Press. 1991.
_____. "Frege Attacks Husserl and Cantor". *The Monist* 77(3): 347-57, 1994. Also in Hill and Rosado Haddock 2000: 95-107.

_____. "Husserl and Frege on Substitutivity". In L. Haaparanta (ed.), 1994: 113-40. Also in Hill and Rosado Haddock 2000: 1-21.

_____. "Husserl and Hilbert on Completeness". In *From Dedekind to Gödel, Essays on the Development of the Foundations of Mathematics*, J. Hintikka (ed.) Dordrecht: Kluwer, 1995. Also in Hill and Rosado Haddock 2000: 143-63.

_____. "Did Georg Cantor Influence Edmund Husserl?" *Synthese* 113: 145-70, 1997. Also in Hill and Rosado Haddock 2000: 137-159.

_____. *Rethinking Identity and Metaphysics, Foundations of Analytic Philosophy*. New Haven: Yale University Press, 1997.

_____. "The Varied Sorrows of Logical Abstraction". *Axiomathes* 1-3: 53-82, 1997. Also in Hill and Rosado Haddock, 2000: 67-93.

_____. "From Empirical Psychology to Phenomenology: Husserl on the Brentano Puzzle". In *The Brentano Puzzle*, R. Poli (ed.), Aldershot: Ashgate, 1998: 151-68.

_____. "Review of Edmund Husserl's *Logik und allgemeine Wissenschaftslehre* (Husserliana XXX)". *History and Philosophy of Logic* 19, 1998: 115-17.

_____. "Review of Edmund Husserl, *Early Writings in the Philosophy of Logic and Mathematics*", *Modern Logic* 8 (1-2): 142-53, 1998-2000.

_____. "Abstraction and Idealization in Georg Cantor and Edmund Husserl". *Abstraction and Idealization. Historical and Systematic Studies, Poznan studies in the philosophy of the sciences and the humanities*, F. Coniglione et al. (eds.) Amsterdam: Rodopi, 1999. Also in Hill and Rosado Haddock 2000: 109-135.

_____. "Circling Gottlob Frege, Review of *Frege Importance and Legacy*, M. Schirn (ed.)". *Diálogos*: 203-13, 1999.

_____. "Review of The New Theory of Reference, P. Humphreys and J. Fetzer (eds.)" *History and Philosophy of Logic* 20, 1999: 125-27.

_____. "Husserl, Frege and 'the Paradox'", *Manuscrito*, 23, 2, October 2000: 101-32.

_____. "Husserl's *Mannigfaltigkeitslehre*". In Hill and Rosado Haddock 2000: 161-177, 2000.

_____. "Review of W. Demopoulos' *Frege's Philosophy of Mathematics* and W. W. Tait's *Early Analytic Philosophy*". *Synthese* 133: 441-52, 2002.

_____. "Phenomenology from the Metaphysical Standpoint" *Diálogos*, XLIII, 91, January, 2008: 19-35.

_____. "Husserl and Phenomenology, Experience and Essence". In *Phenomenology and Existentialism*, Dordrecht: Springer, 2009: 9-22.

_____ and G. E. Rosado Haddock. *Husserl or Frege? Meaning, Objectivity, and Mathematics*, La Salle: Open Court, 2000.

Hintikka, J. *Knowledge and Belief.* Ithaca: Cornell University Press, 1962.

_____. *Models for Modalities*. Dordrecht: Reidel, 1969.

_____. *The Intentions of Intentionality and Other New Models for Modalities*. Dordrecht: Reidel, 1975.

_____. "On the Development of the Model-Theoretic Viewpoint in Logical Theory". *Synthese* 77: 1-36, 1988.

_____."Hilbert Vindicated". *Synthese* 110(1): 15-36, 1997.

_____."Hyperclassical Logic (A.K.A. IF Logic) and its Implications for Logical Theory". *The Bulletin of Symbolic Logic* 8(3): 404-23, 2002.

_____."Independence-friendly logic and axiomatic set theory". *Annals of Pure and Applied Logic* 126: 313-33, 2004.

_____."On Tarski's Assumptions". *Synthese* 142: 353-69, 2004.

_____. "Hilbert was an Axiomatist, Not a Formalist". In *Approaching Truth: Essays in Honor of Ilkka Niiniluoto*, Sami Pihlström (ed.), et al.. Newcastle upon Tyne: College Publications, 33-48, 2008.

_____ and D. Davidson (eds.) *Words and Objections, Essays on the Work of W. V. Quine*. Boston: Reidel, 1969.

Hodges, W. "Truth in a Structure". *Proceedings of the Aristotelian Society* 86: 135-51, 1985/86.

Humphreys, P. and J. Fetzer (eds.) *The New Theory of Reference, Kripke, Marcus, and Its Origins*. Dordrecht: Kluwer, 1998.

Husserl, E. *Über den Begriff der Zahl: Psychologische Analysen*. Heynemansche Buchdrückerei, Halle, 1887. In Husserliana XII: 289-339. Published in English as *On the Concept of Number* in Husserl 2003: 305-57.

_____. "Die formal und die wirkliche Arithmetik". In Husserliana XXI: 21-23 (1889/1890).

_____. "The Concept of General Arithmetic". In Husserl 1994: 1-6 (1890).

_____."Zur Logik der Zeichen". In Husserliana XII: 340-73, published in English as "On the Logic of Signs (Semiotic)" in Husserl 1994: 20-51 (1890).

_____. "Arithmetic as an A Priori Science". In Husserl 1994: 7-11 (1891).

_____. "Letter from Edmund Husserl to Carl Stumpf". In Husserl 1994: 12-19 (1891).

_____. *Philosophie der Arithmetik*. Halle: Pfeffer, 1891. Published in Husserliana XII: 1-283.

_____. *Philosophie de l'arithmétique*. Translation of his *Philosophie der Arithmetik*. Paris: PUF, 1972 (1891).

_____. *Philosophy of Arithmetic, Psychological and Logical Investigations with Supplementary Texts from 1887-1901*. Dordrecht: Kluwer, 2003. Tr. by D. Willard of his *Philosophie der Arithmetik, mit ergänzenden Texten* (1890-1901), Husserliana XII, The Hague: M. Nijhoff, 1970.

_____. "Zur Lehre der Inbegriff". In Husserliana XII: 385-407 (1891).

_____. "Memorandum of a Verbal Communication from Zermelo to Husserl". In Husserl 1994: 442 (1902).

_____. Husserl, Edmund 1903. "A Report on German Writings in Logic from the Years 1895-1899, Third Article". In Husserl 1994 (1903): 246-59

_____. "Personal Notes". In Husserl 1994: 490-500, (1906).

_____. *Logik, Vorlesung 1896.* Dordrecht: Kluwer, 2001 (1896).

_____. "Double Lecture on the Transition through the Impossible ("Imaginary") and the Completeness of an Axiom System". In Husserl 2003: 409-52 (1901).

_____. "The Domain of an Axiom System/Axiom System–Operation System". In Husserl 2003: 475-92 (1901).

_____. "Husserl's Excerpts from an Exchange of Letters Between Hilbert and Frege". In Husserl 2003: 468-73 (1899).

_____."Besprechung von E; Schröder, Vorlesungen über die Algebra der Logik I". In Husserliana XII: 3-43, published in English as "Review of Ernst Schröder's Vorlesungen über die Algebra der Logik" in Husserl 1994: 52-91.

_____."Drei Studien zur Definitheit und Erweiterrung eines Axiomensystems". Husserliana XII, 452-69, n.d.

_____. *Logische Untersuchungen.* 2nd rev. ed. Halle: Niemeyer, 1913 (1900-1901).

_____. *Logical Investigations.* New York: Humanities Press, 1970. Translation of his *Logische Untersuchungen,* also published as Husserliana XVIII, XIX/I-II, The Hague: M. Nijhoff, 1975, 1984 (1900/1901).

_____. *Prolegomena to Pure Logic. Logical Investigations* vol. I, New York: Humanities Press, 1970. Translation of *Prolegomena zur reinen Logik, Logische Untersuchungen, Erster Teil,* Halle: Niemeyer, 1901.

_____. *Allgemeine Erkennthistheorie, Vorlesung 1902/03.* E. Schuhmann (ed.) Dordrecht: Kluwer, 2001 (1902/1903).

_____. *Logik, Vorlesung 1902/03.* E. Schuhmann (ed.), Dordrecht: Kluwer, 2001 (1902/1903).

_____. "Husserl and Brentano, 27. III. 1905". *Briefwechsel, Die Brentanoschule I.* Dordrecht: Kluwer, 1994 (1905).

_____. *Introduction to Logic and Theory of Knowledge.* Dordrecht: Springer, 2008, tr. by C. O. Hill of his *Einleitung in die Logik und Erkenntnistheorie, Vorlesungen 1906/07,* U. Melle (ed.) Husserliana XXIV, Dordrecht: M. Nijhoff, 1984.

_____. *Vorlesungen über Bedeutungslehre.* Husserliana XXVI, The Hague: M. Nijhoff, 1987 (1908).

_____. *Alte und neue Logik, Vorlesung 1908/09.* E. Schuhmann (ed.) Dordrecht: Kluwer, 2003 (1908/1909).

_____. *Introduction to the Logical Investigations, A Draft of a Preface to the Logical Investigations.* E. Fink (ed.). Translated by P. Bossert and C. Peters, The Hague: M. Nijhoff, 1975, published in German in Husserliana XX/1, 272-329 (1913).

_____. *Introduction à la logique et à théorie de la connaissance* (1906-1907). Translation by L. Joumier of Husserliana XXIV. Paris: Vrin, 1998 (1984).

———— . *Thing and Space: Lectures of 1907*. Husserliana XVI. Translated and edited by R. Rodcewicz, Dordrecht: Kluwer, 1997 (1907).

———— . *Ideas: A General Introduction to Pure Phenomenology*. London: Allen & Unwin, 1931 Translation by W. R. Boyce Gibson of *Ideen zu einer reinen Phänomenologie und phänomenologischen Philosophie I*, The Hague: M. Nijhoff, 1950 (1913).

———— . *Idées directrices pour une phénoménologie*. Translation by P. Ricoeur. Paris: Gallimard, Collection Tel, 1993.

———— . *Logik und allgemeine Wissenschaftstheorie, Vorlesungen 1917/18, mit ergänzenden Texten aus der ersten Fassung 1910/11*. U. Panzer (ed.) Husserliana XXX, Dordrecht: Kluwer, 1996 (1917/1918).

———— . "Recollections of Franz Brentano". In Husserl 1981: 342-49 (1919). Also translated by L. McAlister in her *The Philosophy of Brentano*, London: Duckworth, 1976: 47-55.

———— . *Erste Philosophie (1923-24), Erster Teil: Kritische Ideengeschichte*. R. Boehm (ed.) Husserliana VII, The Hague: M. Nijhoff, 1956 (1923/24).

———— . *Kant e l'idea della filosofia trascendentale*. Milan: Mondadori, 1990. Translation by C. La Rocca of Husserliana VII: 208-287, 350-408.

———— . *Philosophie première 1: Histoire critique des idées*. Translation by A. L. Kelkel of Husserliana VII, Paris: PUF, 1990.

———— . *Experience and Judgment*. London: Routledge and Kegan Paul, 1973. Translation of his *Erfahrung und Urteil*, Hamburg: Meiner, 1972 (1939).

———— . *Expérience et jugement*. Paris: PUF, 1970. Translation of *Erfahrung und Urteil*, Hamburg: Glassen & Goverts, 1954.

———— . *Formal and Transcendental Logic*. The Hague: M. Nijhoff, 1969 (1929).

———— . *Cartesian Meditations: An Introduction to Phenomenology*. The Hague: M. Nijhoff, 1960. Translation of *Cartesianische Meditationen* Husserliana I, The Hague: M. Nijhoff, 1950 (1931).

———— . *Méditations Cartésiennes*. Translation by G. Pfeiffer and E. Lévinas. Paris: Vrin, 1992.

———— . *The Crisis of European Sciences and Transcendental Phenomenology: An Introduction to Phenomenological Philosophy*. Evanston: Northwestern University Press, 1970. Translation of *Die Krisis der europäischen Wissenschaften und die transzendentale Phanomenologie*, Husserliana VI, The Hague: M. Nijhoff, 1954 (1936).

———— . *La crise des sciences européennes et la phénoménologie transcendentale*. Paris: Gallimard, 1976, tr. of *Die Krisis der europäischen Wissenschaften und die transzendentale Phänomenologie*,

———— . *Phenomenology and the Crisis of Philosophy*. Contains: "Philosophy as Rigorous Science" (1911), *Logos* I: 289-341 (1910-11), and "Philosophy and the Crisis of European Man", The Vienna Lecture

(1935). Translation by Q. Lauer. New York: Harper & Row, 1965. The Vienna lecture is also in Husserl 1970: 269-99.

———. "The Origin of Geometry". In Husserl 1970: 353-78. English translation of "Der Unsprung der Geometrie als intentionalhistorisches Problem", Beilage III in Husserliana VI: 365-86, 1936.

———. *Articles sur la logique (1890-1913)*. Translated by Jacques English, Paris: PUF, 1975.

———. *Husserl: Shorter Works*. P. McCormick and F. Elliston (eds.), Notre Dame: University of Notre Dame Press, 1981.

———. *Studien zur Arithmetik und Geometrie. Texte aus den Nachlass, 1886-1901*. I. Strohmeyer (ed.) Husserliana XXI, The Hague: M. Nijhoff, 1983.

———. *Early Writings in the Philosophy of Logic and Mathematics*. Dordrecht: Kluwer, 1994. Translated by D. Willard of *Aufsätze und Rezensionen* 1890-1910, Husserliana XXII, The Hague: M. Nijhoff, 1979.

———. *Ms A 1 35*, manuscript on set theory available in the Husserl Archives in Leuven, Cologne and Paris.

Husserl, M. "Skizze eines Lebensbildes von E. Husserl". *Husserl Studies* 5: 105-25, 1988.

Illemann, W. *Husserls vorphänomenologische Philosophie*. Leipzig: Hirzel, 1932.

Jeans, J. *Physics and Philosophy*. New York: Dover Books, 1981.

Jourdain, P. "The Development of the Theories of Mathematical Logic and the Principles of Mathematics". *The Quarterly of Pure and Applied Mathematics* 48: 219-315, 1912.

———. "The Development of the Theory of Transfinite Numbers". In *Selected Essays on the History of Set Theory and Logic*, I. Grattan-Guinness (ed.) Bologna: CLUEB, 1991.

Jones, H. *A Critical Account of the Philosophy of Lotze, the Doctrine of Thought*. Glasgow: James Maclehouse and Sons, 1895.

Kant, I. *Critique of Pure Reason*. London: Macmillan & Co, 1963. Translation of *Kritik der reinen Vernunft*. Riga: Hartknoch, 1781.

———. *Logic*. New York: Dover Books, 1974. Translation of his *Logik*, G. Jäsche (ed.), Berlin: Heimann, 1800.

Kaplan, D. and R. Montague. "A Paradox Regained". *Notre Dame Journal of Formal Logic* 1(3): 79-90, 1960.

Kilmister, C. W. *Russell*. London: The Harvester Press, 1984.

Kline, M. "The Instillation of Rigor in Analysis". *Mathematical Thought from Ancient to Modern Times*, Oxford: Oxford University Press, 1971: 947-78.

Kreiser, L. "Review of *Nachgelassene Schriften*". *Deutsche Zeitschrift für Philosophie* 21: 523, 1973.

———. "Zur Geschichte des wissenschaftlichen Nachlasses Gottlob Freges". *Ruch Filozoficznej* 33(1): 42-47, 1974.

_____. *Gottlob Frege, Leben–Werk–Zeit.* Hamburg: Meiner, 2001.
Kripke, S. *Naming and Necessity.* Oxford: Blackwell, 1980 (1972).
_____. "Outline of a Theory of Truth". *Journal of Philosophy* 72: 690-716, 1975. Also in Martin (ed.) 198: 53-81.
Leibniz, G. W. *Die Grundlagen des logischen Kalküls.* Hamburg: Felix Meiner Verlag, 2000.
Lelionnais, F. (ed.), *Great Currents of Mathematical Thought. Vol. 1 Mathematics: Concepts and Development,* New York: Dover, 1971.
Levinas, E. *Théorie de l'intuition dans la phénomenologie de Husserl.* Paris: J. Vrin, 1930.
Linke, P. *Grundfragen der Wahrnehmungslehre,* Munich: Reinhardt, 1929.
_____. *Niedergangserscheinungen in der Philosophie der Gegenwart und die Wege zu ihrer Ueberwindung,* Munich: Reinhardt, 1961.
_____. "Gottlob Frege as Philosopher". Translated by C. O. Hill, *The Brentano Puzzle.* R. Poli (ed.), Aldershot Ashgate, 1998 (1947): 45-72.
Linsky, L. (ed.). *Reference and Modality.* Oxford: Oxford University Press, 1971.
Lotze, H. *Logic.* New York: Garland, 1980. Reprint of B. Bosanquet's translation of *Logik,* 1888.
Mancosu, P. *From Brouwer to Hilbert.* New York: Oxford University Press, 1998.
Majer, Ulrich 1997. "Husserl and Hilbert on Completeness: A Neglected Chapter in Early Twenty Century Foundation of Mathematics". *Synthese* 110: 37-56.
Marcus, R. B. "Extensionality". *Mind* 69: 55-62, 1960. Also in Linsky (ed.): 44-51.
_____. *Modalities.* New York: Oxford University Press, 1993.
Martin, R. (ed.) *Recent Essays on Truth and the Liar Paradox.* Oxford Clarendon Press, 1984.
Marty, A. "Über subjektlose Sätze und das Verhältnis der Grammatik zur Logik und Psychologie". *Vierteljahrsschrift für wissenschaftliche Philosophie,* 19: 19-87, 263-334, 1895.
Maxwell, J. C. "On Physical Lines of Force". *Philosophical Magazine and Journal of Science,* Fourth Series, 21 & 23, 1861-1862.
Mohanty, J. N. *Edmund Husserl's Theory of Meaning.* 3rd ed. The Hague: M. Nijhoff, 1976 (1964).
_____ (ed.). *Readings on Husserl's Logical Investigations.* The Hague: M. Nijhoff, 1977.
_____. "Husserl and Frege: a New Look at their Relationship". *Research in Phenomenology* 4: 51-62, 1974.
_____. *Husserl and Frege.* Bloomington: Indiana University Press, 1982.
_____. *The Philosophy of Edmund Husserl, A Historical Development.* New Haven: Yale University Press, 2008.

Montague, R. "Syntactical Treatments of Modality, with Corollaries on Reflexion Principles and Finite Axiomatizability". *Acta Philosophica Fennica* 16: 153-67, 1963.

Moran, D. *Edmund Husserl-Founder of Phenomenology*. Cambridge: Polity Press, 2005.

Mulligan, K. (ed.). *Mind, Meaning and Metaphysics, the Philosophy and Theory of Language of Anton Marty*. Dordrecht: Kluwer, 1990.

Nikolic, A. M.. "Karamata's Products of Two Complex Numbers". *The Teaching of Mathematics* VII(2): 107-116, 2004.

Osborn, A. *Edmund Husserl and his Logical Investigations*. 2nd ed. Cambridge: Cambridge University Press, 1949 (1934).

Parsons, C. "Platonism and mathematical intuition in Kurt Gödel's thought". *The Bulletin of Symbolic Logic* 1: 44-74, 1995.

Peckhaus, V. *Hilbertprogramm und Kritische Philosophie. Das Göttinger Modell interdisziplinärer Zusammenarbeit zwischen Mathematik und Philosophie*. Göttingen: Vandenhoeck & Ruprecht, 1990.

_____ and R. Kahle. "Hilbert's Paradox". Report No. 38, 2000/2001. Institut Mittag-Leffler, The Royal Swedish Academy of Sciences. Published on the Internet, 2000/2001.

Pesic, P. *Beyond Geometry: Classic Papers from Riemann to Einstein*. New York: Dover Books, 2006.

Picker, Bernold. "Die Bedeutung der Mathematik für die Philosophie Edmund Husserls". *Philosophia Naturalis* 7, 1962: 266-355. His 1955 Münster dissertation.

Plessner, Helmuth. *Husserl in Göttingen*. New York: Garland, 1980.

Poincaré, H. *La science et l'hypothèse*. Paris: Flammarion, 1968 (1902). Translated as *Science and Hypothesis*. New York: Dover Books, 1952.

Prior, A. N. "On a Family of Paradoxes". *Notre Dame Journal of Formal Logic* II(1): 16-32, 1961.

Quine, W. v. O. "Review of Ruth C. Barcan, 'The Deduction Theorem in a Functional Calculus of First Order Based on Strict Implication'". *Journal of Symbolic Logic* 12: 95, 1947.

_____. "Review of Ruth C. Barcan, 'The Identity of Individuals in a Strict Functional Calculus of Second Order'". *Journal of Symbolic Logic* 12: 95-96, 1947.

_____. "The Problem of Interpreting Modal Logic". *Journal of Symbolic Logic* 12(2): 43-48, 1947.

_____. "Reference and Modality". *Journal of Symbolic Logic* 19(1): 137-38, 1954.

_____. "Quantifiers and Propositional Attitudes". *Journal of Philosophy* 53: 177-87, 1956.

_____. *Word and Object*. Cambridge: M.I.T. Press, 1960.

_____. *From a Logical Point of View*. New York: Harper & Row, 1961 (1953).

_____. "Two Dogmas of Empiricism". In Quine 1961: 20-46 (1953).

_____. *Ontological Relativity and Other Essays*. New York: Columbia University Press, 1969.

_____. *Ways of Paradox*. Cambridge: Harvard University Press, 1976.

_____. "Three Grades of Modal Development". In Quine, 1976: 158-76 (1953).

_____. "Reply to Professor Marcus". In Quine 1976: 177-84 (1962).

_____."What Price Bivalence". *The Journal of Philosophy* 78(2): 90-95, 1981.

_____. "Promoting Extensionality". *Synthese* 98: 143-51, 1994.

Rang, B. and W. Thomas. "Zermelo's Discovery of Russell's Paradox". *Historia Mathematica* 8: 16-22, 1981.

Reid, C. *Hilbert*. New York: Springer, 1970.

_____. *Courant in Göttingen and New York*. New York: Springer, 1976

Riemann, B. 1854. "On the Hypotheses which Lie at the Bases of Geometry". Translated by W. K. Clifford, *Nature* VIII (183): 184, 14-17. Also in *A Source Book in Mathematics*, D. E. Smith (ed.), New York: Dover Books: 411-25, 1959 and Pesic 2006, 23-40.

Rollinger, R. *Husserl's Position in the School of Brentano*. Dordrecht: Kluwer, 1999.

Rosado Haddock, G. E. *Edmund Husserl's Philosophie der Logik und Mathematik im Lichte der gegenwärtigen Logik und Grundlagenforschung*. Doctoral Thesis, Rheinischen Friedrich-Wilhelms-Universität zu Bonn, 1973.

_____. "Remarks on Sense and Reference in Frege and Husserl". *Kant-Studien* 73(4): 425-39, 1982. Also in Hill and Rosado Haddock 2000: 23-40.

_____. "On Husserl's Two Notions of Sense". *History and Philosophy of Logic* 7(1): 31-41, 1986. Also in Hill and Rosado Haddock 2000: 53-66.

_____. "On Husserl's Distinction Between States of Affairs (*Sacherverhalt*) and Situation of Affairs (*Sachlage*)". In *Phenomenology and the Formal Sciences*, T. Seebohm et al. (eds.), Dordrecht: Kluwer, 1991: 35-48. Also in Hill and Rosado Haddock 2000: 253-62.

_____. "Interderivability of Seemingly Unrelated Mathematical Statements". *Diálogos* 59: 121-34, 1992. In Hill and Rosado Haddock 2000: 241-52.

_____. "On Antiplatonism and its Dogmas". *Diálogos* 67(7-38), 1996. In Hill and Rosado Haddock 2000: 263-90.

_____. "On the Semantics of Mathematical Statements". *Manuscrito* 19(1): 149-75, 1996.

_____. "Husserl's Relevance for the Philosophy and Foundations of Mathematics". *Axiomathes* 8 (1-3), 1997: 125-32.

_____. "To be a Fregean or to be a Husserlian: That is the Question for Platonists". *Advances in Contemporary Logic and Computer Science*. W. Carnielli and I. D'Ottaviano (eds.) American Mathematical Society, 1999. In Hill and Rosado Haddock 2000: 199-220.

Russell, B. *Principles of Mathematics*. London: Norton, 1903.

———. "On Denoting". In Russell 1956: 41-56, 1905.

———. "On the Substitutional Theory of Classes and Relations". In Russell 1973: 165-89 (1906).

———. "On 'Insolubilia and their Solution by Symbolic Logic". In Russell 1973: 190-214 (1906).

———. "On Some Difficulties in the Theory of Transfinite Numbers and Order Types". In Russell 1973: 135-64 (1906).

———. "The Theory of Logical Types". In Russell 1973: 215-52 (1910).

———. "L'importance philosophique de la logistique." *Revue de Métaphysique et de Morale* 19: 281-291, 1911. In English, "The Philosophical Implications of Mathematical Logic" in *The Monist* 22: 481-93, 1913 and in *Essays in Analysis*: 284-94. Partially translated in Husserl Ms A I 35.

———. "The Philosophy of Logical Atomism". In Russell 1956: 177-281, 1918.

———. *Introduction to Mathematical Philosophy*. London: Allen & Unwin, 1919.

———. *Principia Mathematica to *56*. 2nd ed., Cambridge: Cambridge University Press, 1964 (1927).

———. "My Mental Development." In Schilpp ed. 1944: 3-20.

———. *A History of Western Philosophy*. London: Allen & Unwin, 1946.

———. *Logic and Knowledge*. London: Allen & Unwin, 1956.

———. *My Philosophical Development*. London: Allen & Unwin, 1985 (1959).

———. *Essays on Analysis*. D. Lackey (ed.). London: Allen & Unwin, 1973.

Scanlon, J. "Tertium non datur: Husserl's Conception of a Definitive Multiplicity". In *Phenomenology and the Formal Sciences*, T. M. Seebohm et al. (eds.), Dordrecht: Kluwer, 1991: 139-47.

Schilpp, P. (ed.). *The Philosophy of Bertrand Russell*. Evanston: Northwestern University Press, 1944

Schirn, M. (ed.) *Frege: Importance and Legacy*. Berlin: de Gruyter, 1996.

———. "On Frege's Introduction of Cardinal Numbers as Logical Objects". M. Schirn (ed.) 1996: 114-73.

Schmit, R. *Husserls Philosophie der Mathematik: platonische und konstructivische Moment in Husserls Mathematik Begriff*. Bonn: Bouvier Verlag, 1981.

Schoen, H. *La Métaphysique de Hermann Lotze, ou la philosophie des actions et des réactions réciproques*. Paris: Fischbacher, 1902.

Scholz, H. "Briefe an Husserl 8. III 1936". In E. Husserl's *Briefwechsel, vol. VI, Philosophenbriefe*. Dordrecht: Kluwer, 1994 (1936): 379-80.

Scholz, H. and F. Bachmann. "Der wissenschaftliche Nachlass von Frege". In *Actes du congrès international de philosophie scientifique, Vol. VIII: Histoire de la logique et de la philosophie scientifique*. Paris: Hermann, 1936: 24-30.

Schuhmann E. and K. Schuhmann. "Husserl Manuskripte zu seinem Göttinger Doppelvortrag von 1901". *Husserl Studies* 17: 87-123, 2001.
Schuhmann, K. *Husserl-Chronik*. The Hague: M. Nijhoff, 1977.
Sebestik, J. *Logique et mathématique chez Bernard Bolzano*. Paris: Vrin, 1992.
_____. "Postface." In Cavaillès, 1997.
Simons, P. "Unsaturatedness". *Grazer Philosophische Studien* 14: 73-96, 1981.
_____. "Wittgenstein, Schlick and the A Priori". In *Philosophy and Logic in Central Europe from Bolzano to Tarski. Selected Essays*, Dordrecht: Kluwer, 1992: 361-76.
Sluga, H. *Gottlob Frege*. London: Routledge & Kegan Paul, 1980.
Smith, B. *Austrian Philosophy: the Legacy of Franz Brentano*. Chicago: Open Court, 1994.
_____. "Putting Semantics Back into the World". *Grazer Philosophische Studien* 43 (1993): 91-109.
_____ (ed.). *Parts and Moments, Studies in Logic and Formal Ontology*. Munich: Philosophia Verlag, 1982.
_____ and D. W. Smith (eds.). *The Cambridge Companion to Husserl*. Cambridge: Cambridge University Press, 1995.
Smullyan, A. "Modality and Description". *Journal of Symbolic Logic* 13: 31-37, 1948.
_____. "Review of Quine's 'The problem of interpreting modal logic'". *Journal of Symbolic Logic* 12: 139-41, 1947.
Sokolowski, R. "Le concept husserlien d'intuition catégoriale". *Études Phénoménologiques* 19: 39-61, 1994.
_____. "The Logic of Parts and Wholes in Husserl's *Logical Investigations*". J. N. Mohanty (ed.) 1977: 94-111.
Tieszen, R. *Mathematical Intuition: Phenomenology and Mathematical Knowledge*. Dordrecht: Kluwer, 1989.
_____. "Kurt Gödel and Phenomenology". *Philosophy of Science* 59: 176-94, 1992.
_____. "Mathematical Realism and Gödel's Incompleteness Theorems". *Philosophia Mathematica* 3(2): 177-201, 1994.
_____. "Gödel's Path from the Incompleteness Theorems (1931) to Phenomenology (1961)". *The Bulletin of Symbolic Logic* 4(2): 181-203, 1998.
_____. "Gödel's Philosophical Remarks on Logic and Mathematics: Critical Notice of Kurt Gödel: Collected Works, Vols. I, II, III". *Mind* 107: 219-32, 1998.
van Atten, M. *Brouwer Meets Husserl. On the Phenomenology of Choice Sequences*. Dordrecht: Springer, 2007.
van Heijenoort, J. (ed.). *From Frege to Gödel. A Source Book in Mathematical Logic, 1879-1931*. Cambridge: Harvard University Press, 1967.

Veraart, A. "Geschichte des wissenschaftlichen Nachlasses Gottlob Freges und seiner Edition. Mit einen Katalog des ursprünglichen Bestands der nachgelassenen Schriften Freges". In *Studies on Frege*, vol. 1, M. Schirn (ed.), Stuttgart, Bad Cannstatt: Frommann-Holzboog: 49-106, 1976.

Wang, H. *From Mathematics to Philosophy*. London: Routledge & Kegan Paul, 1974.

―――. *Beyond Analytic Philosophy*. Cambridge: M.I.T. Press, 1986.

―――. *Reflections on Kurt Gödel*. Cambridge: M.I.T. Press, 1987.

―――. *A Logical Journey, From Gödel to Philosophy*. Cambridge: M.I.T. Press, 1996.

Wartofsky, M. (ed.) *Proceedings of the Boston Colloquium for the Philosophy of Science 1961/1962*. Dordrecht: Reidel, 1963.

Weyl, H. "Der circulus vitiosus in der heutigen Begründung der Analysis". *Jahresbericht der Deutschen Mathematikervereinigung* 28. In English, "The Circulus Vitiosus in the Current Foundation of Analysis". In *The Continuum: A Critical Examination of the Foundation of Analysis* (1918).

―――. *Das Kontinuum, Kritische Untersuchungen über die Grundlagen der Analysis*. In English, *The Continuum: A Critical Examination of the Foundation of Analysis*, tr. by S. Pollard and T. Bole, New York: Dover Books, 1994 (1918).

―――. *Space, Time, Matter*. New York: Dover Books, 4th ed., 1988. Translation of his *Raum, Zeit, Materie*, Berlin: Springer Verlag, 7th ed., 1918.

―――. "David Hilbert and His Mathematical Work". *Bulletin of the American Mathematical Society* 50: 612-54, 1944.

―――. *Philosophy of Mathematics and Natural Science*. New York: Atheneum, 1963 (1949).

―――. "Erkenntnis und Besinnung (Ein Lebensrückblick)". *Studia Philosophica* 15. In English, "Insight and Reflection" in *Mind and Nature, Selected Writings on Philosophy, Mathematics and Physics*, Princeton: Princeton University Press, 2009 (1955).

―――. *Philosophy of Mathematics and Natural Science*. New York: Attheneum, 1963.

Willard, D. "Husserl on a Logic That Failed." *The Philosophical Review* 89(1): 46-64, 1980.

―――. *Logic and the Objectivity of Knowledge*. Athens: Ohio University Press, 1984.

―――. "Translator's Introduction" to Husserl 1994, VII-XLVIII, 1994.

INDEX

A posteriori, 36, 38, 44, 290, 314
A priori, 11, 12, 32-40, 44, 52-56, 62, 74-78, 84-89, 96, 99, 101-02, 105-10, 137, 147, 199, 205, 215, 245-48, 261, 266, 268, 273-75, 286-97, 301, 307-09, 312-21, 326, 347, 353, 355, 365, 386-87
A priori, synthetic, 2, 11, 37, 44n., 82
Abelian functions, 367
Abstract entities, 24, 222-23, 230, 326, 335
Abstract impressions, 335-39
Abstract objects, 88, 175, 335-39
Abstracta, 71, 138, 338
Abstraction process, 6, 7, 15, 46, 70, 77, 85, 95, 106-07, 132, 142, 186, 197-98, 271-72, 313, 348n., 350, 352-53, 361-65
Abstractions, 15, 183, 234, 259, 301
Abstrusities, 10, 311, 319
Absurdity, 19, 61-62, 69, 74, 96, 103, 118-19, 199, 207, 215, 219, 225, 229, 231, 237, 243, 248, 254, 314, 330-32
Acts, 23, 39, 53, 56, 65, 83-84, 86, 89, 94-95, 102, 107-08, 200, 203, 255, 268, 281-82, 287-88, 311, 316, 326, 328, 340, 343, 349, 351, 354, 362, 387
Addition, 78, 98, 116, 118, 199
Adjectives, 228, 246-48
Aggregates, 86, 181, 196, 212-13, 215n., 317
Agnosticism, 301

Alchemists, 302, 307
Algebra, 13, 17, 52, 62-64, 68, 71-72, 74, 78, 101, 115n., 131-32, 246, 313, 346-47, 356n., 357, 359n.
Algebraists, 66, 74, 77, 87, 347, 359n.
Algorithms, 63-67, 69-70, 78
Alienation, 38n., 46, 54, 65, 73-75, 278n., 287, 345n., 347n. 348, 350n. 354, 356n., 357, 360-62, 365
Althoff, Friedrich, 374, 376, 394
Ambiguity, 86, 104, 122, 133, 139, 179, 187, 190, 202, 204, 208, 213, 230, 234-35, 237, 245, 254, 260, 317, 319, 387-88
Analogy, 41, 43-44, 47, 51, 53, 66, 77-78, 105, 183, 190, 197, 199, 209, 221-22, 247-48, 253, 260, 278, 337, 347, 364, 388
Analysis, mathematical, 2-3, 7-9, 202, 252, 256, 265-66, 268, 271, 277-78, 280-83, 354
Analytic logic, 11-12, 24, 84, 96, 105, 108-09, 313
Analytic propositions, 11, 14, 38, 81-85, 99
Analytic philosophy, XI, 2, 12, 81, 93, 110, 153, 188-90, 195, 212n., 231, 233, 238-39, 262, 299-300, 320, 328, 331-32
Analytic-synthetic distinction, 11, 81, 83, 137
Analyticity, 3, 11-13, 24, 40, 43, 57, 81-90, 102-05, 108-11, 133,

141, 203-04, 208, 215, 248, 249n., 313
Angles, 10, 33, 41, 71n., 84, 235, 294, 311
Anti-metaphysical animus, 300-02, 317-18
Anti-semitism, 377
Antinomies, 16-17, 172, 177, 189, 219, 237, 275, 305. See also Contradictions; Paradoxes.
Apodicticity, 12, 75, 84, 100, 216, 307, 338
Apollonius, 33, 77
Apophantics, 73, 121, 125, 130, 286, 291n.
Apostles, 377
Appearances, 35, 174, 176, 230, 231, 331, 351n., 393
Apples, 98, 314
Applicability, 3, 22, 24, 65n., 68n., 70, 73-77, 97, 195, 206, 321, 348n., 350, 352, 353n., 360, 362-63, 365, 386-87
Applied geometry, 37-39, 50-52
Applied mathematics, XII, 65n., 358, 363
Arbitrariness, 7, 19, 24, 41, 45, 50, 57, 65, 68, 78, 85, 86, 98, 101, 104, 107, 118, 120, 122, 126, 202, 209, 220, 256, 273, 317
Archimedean axiom, 48n.
Archimedes, 33, 359
Arguments, of a function, 55, 220, 227, 230, 232; 237, 253-57
Aristotle, 6, 31, 102-03, 303-04, 318, 320, 331, 333
Arithmetica universalis, 19, 68,
Arithmetics, 104, 209
Arithmetization, 2, 3, 7, 268
Arithmoi noetoi, 6, 304
Astrology, 302, 307
Astronomy, 35, 180
Atemporality, 97, 312
Atheism, 369, 371
Atoms, 38n., 41n., 347, 352

Attributes, 224-25, 228, 237, 248, 271, 273-74, 318, 320, 329
Ausdehnungslehre, 346n., 347n.
Austerity, 4, 89, 306, 316, 387, 393
Austin, J. L., 223n.
Autograph collection, 154-55, 157-58, 161-62, 166
Avenarius, Richard, 371, 375, 386
Awkwardness, 221, 255, 308
Axiom of completeness, 48n., 109, 116-17, 130-31, 134, 138, 141-42, 145-46, 149
Axiom of reducibility, 184-185, 190, 236, 280
Axiomatic systems, 18, 21, 51, 52, 58, 67, 81, 88, 103-04, 109, 115-20, 123-25, 128-31, 133, 137-38, 141-42, 146, 162, 208-09, 216, 317, 356, 361
Axiomatization, 18, 23, 31, 33, 48, 50-52, 57, 85, 89, 93-94, 106-110, 140-42, 162, 167, 216, 316, 335, 346n.

Bachelard, Suzanne, 131n.
Bachmann, Friedrich, 160
Basic Law V, 158-64, 168, 172, 179, 185, 190, 210, 226-27, 231, 235-36, 256-57, 261
Barcan, Ruth. See Marcus, Ruth Barcan.
Bauch, Bruno, 166-67
Baudin, Abbé, 331n.
Baumann, Julius, 381-85
Beauty, 64, 269, 346n., 371, 389
Being, 9-10, 83-85, 97, 106, 265, 274-79, 290n., 293, 296, 306-11, 319-21, 327, 330-31, 340-43, 351n., 360n., 362
Benacerraf, Paul, 83
Bergson, Henri, 54n.
Berkeley, California, 45
Berkeley, George, 45, 72, 331,

Berlin, 2-5, 151, 166, 221, 255, 367, 369, 371-72, 379, 385, 388, 391, 394
Bernays, Paul, 165
Betweenness, 48, 51, 56, 58
Biemel, Walter, 348n.
Binocularity, 39n, 40
Biophysics, 350n.
Bivalence, 285-97. See also Law of Excluded Middle. *Tertium non datur*;
Blindness, 17, 62, 64, 191, 200, 211, 238, 312, 317, 354
Bodies, 32-46, 50, 53, 56-57, 291, 294, 326, 334, 350n., 351, 360-62
Bodies, rigid, 32, 34, 38n., 41-42, 45, 56-57
Bolyai, Janos, 34
Bolzano, Bernhard, 3, 9, 10, 81, 82, 84, 96, 244n., 300, 311, 387
Bombs, 155-56, 158, 160, 164, 166, 168, 243, 320
Boole, George, XIII, 63n.-64n., 68
Born, Max, 347
Boston Colloquium 1962, 187, 188n., 189n.
Bourbaki, Nicolas, 87-88, 117
Boyce Gibson, W. R., 326
Bracketing. See *Epoché*.
Bradley, F. H., 299
Braig, Carl, 371
Breeding, 222-23
Brentano, Franz, 3-9, 12, 13, 19, 63n., 81, 83, 84, 89, 93, 152, 197, 198, 200, 300, 306, 307, 310, 367, 369, 387
Brouwer, L. E. J., XII, 93-95, 97, 100, 106-08, 110, 281-83, 295, 272
Brunelleschi, Filippo, 77
Burge Tyler, 164,
Burgess, John, 189n.

Burke, Dr. Ruth Ellen, XIII, 368, 379
Busse, Ludwig, 375

Cabbalists, 302
Calculi, XIII, 63-64, 67-68, 70, 137, 185, 354
Calculi, extensional, 68, 187-88, 200, 318-19
Calculus, logical, XIII, 24, 63, 68-69, 100
Calculus of classes, XIII, 17, 63n., 68, 69, 115, 200, 212, 306
Calculus of variations, 4, 367
Cantor, Georg, XI, XII, 1, 3, 5-7, 9, 13, 15, 16, 81, 82, 86, 110, 151-52, 160-61, 196, 200-02, 209n., 210n., 213n., 299-300, 302-05, 310, 316-17, 321, 334, 336, 346, 367-78
Cantorian set theory, 18, 85, 160, 196, 281, 313
Caratheodory, Constantin, 392
Cardinality, 305
Carnap, Rudolf, XII, 165, 299
Carrots, 289
Carton, Sydney, 176-77
Cassirer, Ernst, 385, 391
Categorematic expressions, 219, 221, 222
Categorematic meanings, 222, 225
Categoricity, 70n., 358n.
Categories, 229, 245-46, 276, 296, 312, 320, 329
Catholicism, 368, 369, 373, 374, 377, 382
Cats, 15-16
Cauchy sequences, 73
Causality, 12, 84, 85, 307, 309, 312, 347, 350n., 361
Cephalomotor system, 39n.
Certainty, 3, 23, 82, 99, 293n., 314, 332, 353, 393
Chaos, 88, 245, 300

Chess, 20, 102, 165, 203, 205
Christ, Jesus, 300
Christianity, 302, 374
Church, Alonzo, 168, 295
Circles, 33, 49, 351, 359n.
Circles, square, 69n., 71n., 329
City, 88, 221, 255
Clarity, X, 13, 32, 83, 89, 95, 187, 191, 197, 206, 211, 306, 308, 313, 319, 360, 365, 387, 393
Class of all classes, 220, 235, 256, 261. See also Set of all sets.
Cohen, Hermann, 376
Cogitata, 201, 334, 341-42
Cohen, Paul J., 282
Collective combination, 95, 197, 198, 200
Color, 15, 43, 53, 85, 175, 224, 225, 231, 248, 272, 294
Columbus, Christopher, 174
Completeness, 18-22, 48n., 71-78, 103-04, 108-09, 115-34, 137-49, 208, 216, 275, 290n., 356-65. See also Definiteness.
Completeness, axiom of. See Axiom of completeness.
Completeness, ontological, 293-96
Completeness, semantic, 128, 131, 138, 141, 145n., 147
Completeness, syntactic, 115, 128, 131, 138-49, 358n.
Concatenation, 22, 86, 202, 317
Concept horse, 221, 248, 255
Concept words, 153, 164, 172, 205, 220, 223, 229-33, 235, 252-53, 255, 258
Conciergerie, 176-77, 180
Concreteness, 12, 22, 24, 85, 86, 95, 102, 175, 186, 197-98, 222, 225, 291, 294, 307, 316, 325, 331, 350n., 353, 361
Conflicts, 12, 69-70, 96, 127, 268, 285, 287-88, 295, 314, 361
Confusion, 12, 25, 96, 131, 138, 149, 153, 187, 189, 202, 211, 219, 229-30, 232-33, 235, 237, 243, 258, 266, 296, 300-01, 317, 339, 341, 358n.
Congruence, 41-44, 48-58, 69n., 274, 294
Conics, 77, 359n.
Consciousness, XII, 11, 12, 23, 38, 47, 52, 56, 97, 105-08, 139, 268-72, 279-82, 289, 297, 313, 327-31, 334-43, 348, 361n., 362, 363n.
Conservativeness, 71, 73
Consistency, 5, 19, 21, 31, 70-78, 87-88, 102, 108, 118-34, 138, 145, 186, 189, 201n., 204, 213-16, 237, 248, 312, 347n., 356, 363, 392-93
Constants, 122-25, 145n., 147, 148, 184
Constitution, XI, 25, 32, 38-45, 52-58, 63-64, 76, 105, 144, 148, 182, 266-72, 275-76, 279-81, 286-96, 329-30, 332, 335, 338-39, 341-42, 349, 361-65
Constructivism, 266, 272, 279, 281-83, 332
Contingency, 56, 83, 188, 375
Continuity, 8, 41, 48n., 83, 89, 304, 306, 319, 359n., 362
Continuum, 38, 41n., 47, 107, 198, 265-83, 334, 336
Continuum hypothesis, 336
Contradiction Burali-Forti's, 160
Contradictions, 14, 16-18, 20, 21, 69, 85, 87, 96, 99, 101-02, 104, 106, 108, 119, 126, 129, 131, 133, 140, 157-63, 166, 171-72, 179-84, 188-89, 196, 200-11, 213-16, 220, 227-28, 231-38, 243, 249-50, 258-61, 266, 273-74, 293, 301, 303, 305-06, 314-16, 321
Contradictions, thicket of, 15, 180, 210, 233, 258, 321

Contradictions-in-terms, 85, 96, 198, 208, 214-16, 314-16
Convenience, 34, 36, 53-54, 66, 88, 146, 153, 175, 183-84, 236, 260, 287, 318, 334, 358n., 359, 363-64
Conventionalism, 35, 45, 53, 270-71, 281, 333
Copernican Revolution, 329, 343
Copula, 224, 338
Corporeality, 282, 327
Correlates, 37, 39, 46, 48, 143-44, 148, 271, 276, 286, 327, 329-30, 334, 338, 340, 342, 346n., 358n.
Cosmos, 302
Counterfactuals, 319
Courses of values, 225-26,
Courtine, Jean-François, 201n., 242n.
Couturat, Louis, 152
Creation, 34, 70, 73-74, 230, 267, 287, 310, 329, 337-38, 346-47, 349, 359n., 387-88
Creativity, 5, 21, 69, 70, 87, 102, 109, 204, 353n., 385
Crisis, 74, 200, 301, 305, 318, 345-46
Crisis of meaning, 345-46
Critique of knowledge, 308
Critique of reason, 12, 105, 388
Crowe, Michael, 374n.
Crows, 369
Cults, 303
Curvature, 34-35, 38, 41-42, 44, 56

Damnation, 238
Danger, 81, 183, 281, 278n., 318, 354, 375, 380, 391
Darkness, creatures of, 89, 239, 318
Darmstaedter, Ludwig, 154

Darnay, Charles, 176-77, 181, 183
Darwinism, 369, 371
Dauben, Joseph, 304, 305
de Broglie, Louis, 347
de dicto, 337
de la Hire, Philippe, 359n.
de Muralt, André, 65n.
de re, 337
de St. Evrémonde, Marquis, 176-78, 181, 183
Death blow, 233, 258
Debye, Peter, 392
Decidability, 71n., 125-32, 141, 143, 216, 275, 287n., 289, 290n. 295, 359-60
Dedekind, Richard, 17, 22, 64n.
Dedekind cuts, 73
Deduction, 7, 21; 36, 67-69, 72n., 78, 86-87, 99-105, 115, 118, 148, 202-04, 207, 209, 232, 261, 314, 316, 360, 361n., 393
Deductive systems, 72, 86, 100, 102, 105, 202, 207, 316
Deep nature of things, 219-22, 227, 235-36, 245-46, 251, 255-57, 262
Defects, 11, 14, 83, 159, 167, 210, 236, 238
Definability, 95, 132-33, 139, 184, 267, 279, 281-82
Definite article, 173, 180, 182, 224, 226, 233
Definiteness, 71, 115-34, 137-49. See also Completeness.
Definiteness, absolute, 115, 129-34, 144
Definiteness, relative, 115-16, 125-34, 144
Degeneracy, 65, 347n., 350, 354, 391
Dehumanization, 55
Delgado, Monica, XIII
Demopoulos, William, 190, 238

Denotation, 20, 48, 61, 64-69, 122-26, 143, 145n., 148, 173, 180-82, 273-74, 291, 358
Denumerability, 282
Dependency, 40, 54, 102, 139, 219-39, 241-62, 268, 271-72, 281-82, 296, 329, 337-38, 342, 386
Dependent meanings, 219-39, 241-62
Derivability, 14, 115n., 124, 133, 140, 148, 207n., 210, 213, 216, 228, 231, 234, 250, 252, 259, 261-62
Derrida, Jacques, 290n.
Desargues, Girard, 77, 359n.
Descartes, René, 4, 331
Description, phenomenological, 267-69, 340
Description theory of names, 172, 181, 189
Descriptions, 122-23, 129, 132, 134, 175-76, 180-90, 226, 237, 248, 260-61, 278, 339
Descriptions, theory of definite, 181-83, 188, 190-91, 260
Descriptions, eliminative theory of, 172, 181, 189
Diagnosis, 345, 348
Diagonalization, 305
Dialogue, 328, 356
Dickens, Charles, 176
Dilthey, Wilhelm, XI
Dimensionality, 37-44, 50, 53, 57, 72, 75-76, 363
Dingler, Hugo, 156n.
Dirac, Paul, 347
Direct insight, 382, 392
Directions, 49-50, 55, 98, 175, 212
Dirichlet, Peter, 282, 390
Disaster, 82, 155, 319, 385, 392
Discontinuity, 41-42
Disgust, 6, 89, 301
Distances, 48-50, 57, 351, 362
Distortions, 45, 266

Divine Intellect, 6, 303
Dogmatism, 301, 387
Domain, apophantic, 25, 130, 286
Domain, natural, 119n., 356, 361
Domain, ontological, 122, 125-27, 130, 132, 291
Domains, 3, 11, 21, 24, 34, 38-39, 64-78, 84, 86, 101-04, 115-34, 137-49, 203-09, 249, 267, 269, 272, 274-81, 286, 288, 290-96, 327, 341, 346-49, 355-65
Domains, formal, 77n., 115-25, 128, 137-39, 142-49, 346, 356-58n.
Domains, objective, 70n., 131, 139, 147-48
Doubt, 6, 8, 13, 15, 31, 42n., 57, 83, 95, 106, 159, 163, 174, 182, 197, 199-200, 211-12, 225, 256, 261, 267, 286, 301, 306, 309, 319, 331-32, 361n., 368, 370, 379-81, 384
Drugs, 45, 176
Dualism, 301
Dummett, Michael, 18, 151-52, 157, 163, 168, 238, 285, 295, 331
Duty, 210n., 304, 319
Dynamics, 46, 62, 63n., 78, 117, 277, 289

Eden, 171, 318. See also Paradise.
Ego, XII, 40, 50, 57, 266, 268, 277-78, 286, 290, 295, 327, 329-32, 334-35, 360
Egyptians, 33
Eidetic, 11, 76, 324, 326, 349
Eidetikoi, 6, 82, 304
Eidos, 6, 267, 304
Einstein, Albert, 42n., 45, 54n., 363n.
Electromagnetism, 46, 364n.,
Electrons, 347
Elegance, 77, 296, 318-19, 346n., 359n.

Emery, Fr. Jacques-André, 373
Emotions, 89, 302
Empirical facts, 34, 35, 38, 88
Empirical persons, 84, 86, 89, 102, 108, 203, 316
Empirical psychology. See Psychology, empirical.
Empirical sciences, X, 12, 32, 84, 307, 309-10, 318, 360
Empiricism, 7, 12, 53, 81-84, 96, 296, 302, 306-07, 311, 314, 387
Emptiness, 12, 39, 48, 58, 69, 78, 85, 88, 102, 105, 107, 123, 142, 204, 220, 222, 230, 252-53, 260, 277-78, 301, 362
Encyclicals, 373, 377
Enemies, XII, 82, 176, 348n.
Energy, 206, 304, 347
Enigmas, 1, 296
Entailment, 214-15, 275, 312, 315, 319, 349
Epiphany, 294
Epistemology, XI, XII, 9, 12, 22, 23, 35-37, 47, 52, 62, 70, 74, 76, 78, 85, 89, 106, 115-17, 126-27, 133, 137, 147, 149, 172, 190, 265-66, 268, 270-72, 277-79, 282, 285, 287n., 289, 296, 309, 336, 340, 343, 348, 356-57, 359
Epoché, XII, 267, 260, 271, 326-31, 340-43
Equality, 33, 41, 46-50, 53, 63n., 69, 71n., 98, 130, 158-59, 164-65, 172, 174-80, 182, 185, 187, 191, 205, 212, 225-26, 232, 246, 256, 281, 294, 301, 303, 320, 330, 351n., 358, 385
Equations, 19, 20, 52, 71n., 72, 76, 103, 116, 119, 131, 205, 207, 220, 252, 285, 293, 295, 351, 353, 359, 364
Equivalence, 11, 34, 41, 68, 72, 104, 115n., 124, 129, 131, 138-45, 148, 175, 182-84, 187, 191, 208, 212, 279, 291, 335, 356, 358n.
Erdmann, Benno, 368, 370, 371
Error, 14-17, 82, 159-60, 185, 191, 200, 210, 229, 259, 262, 300, 359
Essence, laws of, 12, 56, 84, 96, 214, 245-46, 307, 309, 315
Essencelessness, 72
Essences, analyses of, 11, 83, 84, 96, 108, 215, 309, 313-17
Essences, 83, 84, 96, 105-06, 197, 213-16, 251, 267, 309, 314-18, 320, 326, 329, 343, 349, 361n., 392
Essentialism, 89, 187-89, 318, 320, 343
Esthetics, 287, 318, 346n., 376. See also Beauty.
Ethics, 105, 273, 318-19
Eucken, Rudolf, 376, 386
Euclid, 6, 31, 33, 48, 124, 304, 333
See also Geometry Euclidean; Geometry non-Euclidean.
Euclid's postulates, 31, 34, 41, 49
Evasion, 180, 234
Evening star, 162, 186
Evidence, 7, 11, 14, 24, 45, 75n., 82-83, 88, 96-99, 108, 116, 131, 146, 189, 210, 225, 256, 261, 282, 288-93, 314, 327-330, 336, 343, 348n., 361
Exact sciences, 25, 216, 348n., 387-88, 393
Exactness, 16-19, 25, 48, 63, 67, 83, 100-01, 108, 138, 154, 160, 171, 190, 197, 216, 227, 243, 257, 269, 301, 306-10, 348n., 349, 351-53, 362, 371, 386-93
Excluded middle, law of. See Law of excluded middle; *Tertium non datur*. Bivalence.
Exoticism, 37
Experience, 10, 22, 24-25, 35-58, 65, 81-84, 89, 94-99, 107, 110,

197, 206, 265-80, 285-96, 311, 314, 325-29, 332-40, 347-64, 387
Extensibility, 44, 130-31, 142, 145-46, 149
Extensional calculi. See Calculi, extensional.
Extensionality, 106, 110, 158, 162, 164, 180, 186-90, 228, 231, 235-38, 279-81. See also Principles of extensionality.
Extensions of concepts, 13-15, 68-69, 106, 159-64, 172, 178-85, 190, 195, 198, 205-06, 210-13, 216, 225, 227, 232, 236-38, 256, 258, 273, 346, 356, 360
Extraterrestrials, 44
Eyes, 16, 32, 40, 77, 167, 224, 305, 328, 340, 369

Fallacies, 185, 219, 228, 235, 237
Families, 156, 176, 222, 243, 281, 358n., 372, 390
Fantasy, 74
Fatal tendency, 153, 233, 255
Featherless biped, 182, 223
Fermat's conjecture, 335
Fichte, Johann Gottlieb, 55, 266, 268, 386-87, 393
Fiction, 36, 74, 182-83, 229, 234
Fig leaves, 318
Findlay, J.N., 243, 244n.
Finiteness, 5, 33, 39n., 41, 44, 50, 77, 104, 107, 124, 133-34, 139-43, 161, 199, 208, 210n., 275, 282, 361n.
First philosophy, 320, 349, 350
Fitch, Frederic, 186, 190
Flatness, 38, 39n., 41-44, 46, 53, 228, 230, 236,
Fluidity, 44
Føllesdal, Dagfinn, 93, 188, 189n., 319, 332, 337n.

Form, logical, 140, 147, 171, 188, 222, 312, 319, 354, 361
Formal systems, 19, 22, 24, 63, 70n., 73, 104, 115-19, 121-25, 128, 133, 137-38, 145-46, 190, 209, 216, 352n.
Formalism, XII, 18-20, 38n., 65, 73-75, 109, 115n., 117, 118n., 128, 133, 154, 165, 278n., 348, 350, 357, 362, 365
Formalization, 12, 20, 23, 24, 62, 68n., 100, 102-03, 105, 141-42, 147, 204, 206, 246, 280, 313, 346
Forms, theory of, 21, 72n., 73, 76, 86, 102-03, 142, 204, 208, 316
Forms, Platonic theory of. See Ideas, Platonic theory of.
Fractions, 8, 19, 104, 206, 209
Franzelin, Cardinal J-B., 373
Free mobility, principle of, 34
Freedom, X, 7, 10, 12, 17, 21, 23-24, 34, 35, 37, 39, 42, 45, 50, 53, 57-58, 70, 74, 82, 84, 87, 96, 101-05, 109, 122, 127-28, 131-32, 149, 167, 186, 203-09, 235, 246, 275, 279, 307, 329, 331, 337, 339, 343, 346-48, 353n., 354, 361, 389
Frege, Alfred, 154-56, 163
Frege, Gottlob, *Basic Laws*, 14, 15, 157-65, 171-72, 179, 189-90, 195, 210, 213, 225, 227, 250, 258, 261
––*Begriffsschrift*, 153, 157, 162, 168, 171
––*Foundations of Arithmetic*, 13-15, 18, 20, 159, 161-65, 172-73, 195, 210, 213n., 221, 223-25, 238
––"Function and Concept", 164, 219, 222, 225-26, 229, 236, 241-42, 251-56, 261

Index   421

—"On Concept and Object", 161, 164, 219, 221-22, 225-26, 229, 236, 241, 247, 251-57, 261
—"On Sense and Meaning", 164
—"Review of Husserl", 14-16, 117, 195n., 211
Freiburg, 368-77, 394
Fries, Jakob Friedrich, 390
Friesian school, 392
FRQL, 89, 219, 222-23, 231, 235-39, 262
Fulfillment, 72, 244, 254, 272-78
Functions, 2, 9, 55, 122, 167, 179, 182-87, 219-22, 226-32, 236-37, 243, 250-61, 268, 279, 281-82, 319, 321, 351, 368
Furniture, ultimate, 317

Galileo Galilei, 349, 362n.
Games, XII, 74, 76, 102, 203. See also Chess.
Gaul, 173, 220, 252, 253
Gauss, Carl Friedrich, 33, 34, 305, 390
Genesis, XII, 1, 32, 57, 95, 117, 267, 295, 350, 353, 357, 361
Genetic analyses, 36, 117, 137, 361
Genetic phenomenology, 43, 361
Genetic progression, 38-39, 58
Genetic psychology, 36, 332
Geometrization, 2, 362
Geometry, 2, 3, 6-7, 12, 13, 19, 31-58, 63n., 64, 69n., 71n., 74, 76-77, 87, 98, 105, 138, 142, 153, 196, 201 02, 249, 267n., 268, 287, 294, 304, 312, 317, 331, 346, 347n., 351, 355, 357, 359n., 361-63, 390
Geometry, applied, 37,
Geometry, elliptic, 34
Geometry, Euclidean, 19, 31, 34-36, 39-44, 48, 50, 52, 55, 57-58, 64, 134, 287, 304, 355, 359n., 363

Geometry, foundations of, 153, 389
Geometry, hyperbolic, 34
Geometry, metric, 35
Geometry, non-Euclidean, 34-35, 43-44, 268, 287, 363
Geometry, physical, 32, 46-47, 51-53, 76, 347, 355, 357
Geometry of physical space, 32, 34, 39, 53
Ghosts, 72, 128
Gloves, 78
God, 6, 255, 304, 330, 369, 372, 377
Gödel, Kurt, XI, XII, XIII, 1, 22, 23, 110, 132, 172, 228, 325-43, 358n., 361
Gödel's Gibbs Lecture, 333
Gödel's incompleteness theorem, 22, 132, 228, 290n.
Goodness, 319
Goodrick-Clarke, Nicholas, 302
Goosens, Bernard, 201n., 242n.
Göttingen, 17, 18, 20, 48n., 61, 62n., 68, 70-73, 75, 108-09, 115, 117, 118n., 133, 138, 142-47, 151, 155, 162, 191, 201, 206, 209, 216, 265, 315, 356, 367, 372, 379, 381-83, 385, 388-94
Göttingen Mathematical Society, 20, 68, 191, 206
Göttingen talks, 70-73, 115n., 117, 133, 138, 143-47, 209, 356
Grammar, 21, 104, 208, 220-21, 230, 237, 242-47, 255, 258, 294-95, 321
Grammar, logical, 67n., 78, 148
Granel, Gérard, 345n.
Grass, blades of, 295
Grassmann, Hermann, 148, 346-47
Grattan-Guinness, Ivor, 157
Gravity, 96, 312, 314, 350

Greeks, 3, 31, 33, 71n., 74, 77, 384
Group theory, 358n.
Guillotine, 176-77, 180

Habilitation, 5, 13, 61, 201n., 305, 321, 367, 376, 380, 394
Hall, Harrison, 332
Halle, 1, 5, 6, 10, 16, 109, 151, 200, 305, 316, 367-78, 381
Hallucination, 45
Hamilton, William Rowan, 63, 346
Hardy, G. H., 160,
Harmony, 287n., 288, 295, 325
Harvard Yard, 295
Hasse, Helmut, 155n.
Heart and kidneys, 223
Heck, Richard, 173, 206, 223
Hegel, Georg F. W., 299-300, 386-87
Heidegger, Martin, XI, 299, 348n.
Heiner, Fr. F. X., 369, 371, 374-76
Helmholtz, Hermann von, 34-37, 42n., 43, 44n., 53-57, 386
Helmholtz-Lie theorem, 42
Henkin, Leon, 122n., 123
Hermite, Charles, 5, 6, 303
Heisenberg, Werner, 347
Hessenberg, Gerhard, 389
Heuristic, 52, 354, 364n.
Hierarchy, 222-23, 227, 229, 232, 235-36, 280, 338
Hilbert, David, XI, XII, XIII, 1, 3, 16-22, 48-52, 76, 87-88, 94, 108-10, 115-134, 138-49, 151-57, 162-68, 201n., 216, 231, 237, 277, 289, 304, 305, 315, 347n., 384-92, 394
Hilbert's circle, 18, 109, 118n., 162, 201n., 315
Hintikka, Jaakko, 188, 189n., 319

Hobgoblins, 307
Hodgepodge, 312
Hönigswald, Richard, 167
Hoffmann, Arthur, 166
Holism, 303
Homeomorphism, 43n.
Homo sapiens, 222-23
Homogeneity, 39n., 42n., 43, 55, 56
Horizons, 41, 61, 287, 290n., 361
How many?, 68, 75, 98-99, 106, 111, 116, 119
Human beings, 35, 100, 106, 132, 222-23, 226, 308, 320
Human spirit, 224, 225
Hume, David, 11, 83, 174, 176
Hume's principle, 174
Huntington, Edward, 154n., 161
Husserl, Edmund, *Cartesian Meditations*, X, 266n.
--*Experience and Judgment*, X, 25, 276, 285-86, 296
--*Formal and Transcendental Logic*, X, 11, 21, 23, 25, 96, 106, 108, 134, 197, 203, 211, 237, 274, 278, 286, 293-94, 305
--*Ideas I*, X, 56, 103, 108, 115n., 116n., 130, 133, 138, 139-42, 144, 149, 203, 208, 265-66, 269, 276, 283, 326, 328, 330, 342n.
--*Logical Investigations*, X, XII, 5, 10-11, 21, 32, 50, 61-62, 72-73, 78, 93, 103, 110, 118, 137, 143-49, 199, 201, 206, 208, 212, 244n., 259n., 265-66, 276, 286, 306, 385
--"On the Concept of Number", 5, 7, 13, 62, 84, 95-96, 100, 107, 109, 116, 196, 197-99, 207n., 305
--"The Origin of Geometry", 32, 296, 361n.

—*Philosophy of Arithmetic*, 5-7, 8, 13-16, 20, 33, 61-66, 69-70, 78, 84, 86, 93-96, 100, 106-10, 115-17, 132, 149, 162, 195-99, 202, 206-07, 211-12, 305-06, 356-57, 375, 380-81

—*Prolegomena*, XI, XII, 61, 71-73, 109, 118n., 202, 286, 382

—"Review of Schröder", 17, 62-64, 68, 70-71, 115n., 200-01, 306

—*Thing and Space 1907*, 32, 38n.

Husserl, Malvine, 305

Husserl Archives, X, XII, 94, 201n., 242n., 260n., 315

Hyletic data, 38n., 40, 331, 332, 338-39

Hypnotism, 107, 302

I. See Ego.

Ideal constructions, 33, 51

Ideal elements, 277, 281

Ideal entities, 84, 96, 100, 268, 286, 311, 335, 351

Ideal meanings, 97, 312

Ideal objects, 25, 268, 277, 280-82, 314

Idealism, 6, 9-11, 219, 300-03, 307, 310, 317, 320-21, 326-27, 330-32, 339, 341

Ideality, 25, 33, 37, 58, 83, 97, 311-12

Idealization, 36-39, 46-54, 58, 198, 274, 280, 286-89, 293, 347-53, 355-56, 359-65

Ideas, 6, 9-10, 12, 84, 214, 251, 303, 307, 310-11, 315, 318, 336-37, 351n., 352

Ideas, Platonic theory of, 10, 310

Ideas-in-themselves, 9, 10, 82

Identity, 14, 58, 67, 69, 70, 89, 101, 106, 110, 131, 153, 158, 162-65, 172-91, 205, 210-11, 223-26, 232, 261-62, 364n., 386

Identity of indiscernibles, 185, 189

Illusion, 9, 18, 45, 82, 151, 173, 221, 308, 313, 331

Imaginaries, 19-21, 51, 61-78, 103, 105, 109, 115-33, 137-38, 146-49, 195, 199, 206-09, 347, 354-60

Imagination, 3, 19, 39, 44n. 47-48, 65, 103, 141, 173, 207, 233, 252, 259, 330, 346, 364n.

Immanence, 11, 268, 279, 283, 313

Incompatibility, 273,

Incomplete meanings, 219

Incomplete symbols, 183-84, 230, 234, 241-62

Independence, 54, 76, 87, 102, 128-29, 139, 219-39, 241-62, 268, 271-72, 274-83, 286, 288, 292, 296, 303, 310, 319, 327-28, 332-39, 342, 346, 347n., 349, 353n., 361, 372, 385, 387, 393

Independence-friendly logic, 319

Indeterminate objects, 148,

Individuals, 21, 85, 185-86, 198, 222, 229, 234, 302

Induction, 36, 74, 75n., 97, 268, 272, 280-81, 314, 333

Infections, 181

Inference, modes of, 17, 231-33, 237

Inference, 17, 22, 67, 84, 100-01, 161, 171, 173, 176, 179, 187, 191, 201, 216, 225, 227, 231-33, 235, 237

Infinite, actual, 6, 34, 303, 373

Infinite regress, 271, 276

Infinitely distant points, 161

Infinitesimals, 74, 302, 359n.

Infinity, 9, 12, 19, 33-34, 41, 43-44, 50-51, 53, 74-77, 87, 98, 105, 107, 124, 140, 142, 198-99,

205, 209, 268, 274-75, 280, 282, 290n., 293, 301, 303, 306, 359n., 362, 373
Inkfish, 107
Innateness, 40, 349
Inner architecture, 108
Inner voice, 304, 305
Integers, 76-78, 119, 131n., 267
Intellect, 303, 333
Intensional logics, 172, 185, 187, 189, 191, 262, 318
Intensionality, 106, 172, 185-89, 191, 236, 238, 262, 318
Intensions, 89, 106, 187, 191, 212, 237, 238, 262, 315, 318, 320
Intentional genesis, XII, 32, 361
Intentionality, XI-XII, 32-33, 38n., 53, 65, 78, 117, 139, 267-69, 272-79, 286-96, 327-40, 348-50, 354, 358n. 361-62, 365
Interconnections, 21, 86, 204, 316, 371
Intersubjectivity, 22, 266n., 334, 340-41, 363n.
Intuition, XII, 2, 4, 7, 13, 31, 36-40, 47-58, 67, 75-78, 82, 95, 97, 106-07, 132, 189, 198, 266-82, 291n., 326, 333-43, 346, 349, 355-58, 361, 364, 388
Intuition, conceptual, XII, 47, 75-77, 326, 342
Intuition, eidetic, 326, 349
Intuition, geometrical, 31, 38-39, 47, 50-53, 58
Intuition, mathematical, 333-40, 342, 346n., 356
Intuition, perceptual, 31, 39, 47, 51, 53, 55, 58, 361
Intuition, primordial, 94, 106, 107, 280
Intuition of essences, 267, 326, 343
Intuitionism, XI, XII, 59, 93-94, 97, 107-08, 270, 272, 276, 278, 283, 287, 289-90, 294

Intuitions of space, 2, 7, 13, 39-40, 47, 52, 55-56, 58, 82
Intuitions of time, 2, 3, 7, 107
Intuitive experience, 270, 272, 274, 276, 279-80, 340
Invention, 34, 50, 75, 77, 82, 88, 139, 156n., 157, 165, 167, 235, 237, 311, 312, 346n., 354
Inversions, 78
Inviolability, 219, 221-22, 225, 235-36, 261-62
Irrationality, 270, 321, 386
Irreferentiality, 233
Islamic empire, 347
Isotropism, 43, 55, 56

Jeans, James, 364n.
Jewishness, 369n., 376-77
Jena, 151, 375
Joël, Karl, 386
Jones, E. E. Constance, 154n.
Jones, Henry, 300-01
Jordain, Pascual, 347
Jourdain, Philip, 154n., 156n., 159, 160, 166
Judaism, 377
Judgments, 11, 13, 37, 48, 67, 69, 82-83, 96, 121, 203, 268, 271-78, 285-95
Julius Caesar problem, 173, 190
Jungles, 89, 318-20
Jupiter, 174, 212, 224

Kahle, R., 201n.
Kanger, Stig, 188, 189n.
Kant, Immanuel, XII, 7, 11, 36-38, 44, 53-54, 78, 81-83, 107, 110, 268, 281, 299-301, 308, 336, 341, 390
Kantianism, 2, 7, 34, 54, 64, 81-83, 302, 308, 320, 330, 340-41, 375-76, 386
Kennedy, John F., 210-11
Kepler, Johannes, 74, 77, 359n.

Keys, 10, 21, 72, 103, 115, 118n., 126, 208, 234, 236, 238, 262, 299, 311
Killing, Wilhelm, 369
Kinesthetics, 38, 40, 287, 362
King of France, 183
Kinship, XII, 5, 18, 21, 88, 94, 108-09, 216
Klein, Felix, 20
Königsberger, Leo, 367
Kohn, Archbishop Theodor, 377
Korselt, Alwin, 161-66
Kreiser, Lothar, 152
Kripke, Saul, 181, 188, 189n., 191
Kronecker, Leopold, 3, 4, 74, 372
Külpe, Oswald, 369, 386
Kym, Andreas Ludwig, 376

Labyrinths, 88, 390
Lauer, Quentin, 342
Laundry lists, 152, 154
Law, Leibniz', 14, 175, 185, 189, 225
Law, unprovable, 256, 261. See also Basic Law V.
Law of Excluded Middle, 97, 101, 125, 134, 215, 251. See also *Tertium non datur*, Bivalence.
Laws, 11, 12, 14-15, 20-21, 36, 44n., 51, 56, 58, 63n., 70-71, 75, 83-87, 96-101, 106-07, 133-34, 149, 159, 161, 163, 175, 177, 181, 185-86, 189-90, 202, 204, 206, 214-16, 226-27; 245-51, 256, 261-62, 277, 290, 293-94, 304, 307-16, 352, 361n., 388
Laws, essential. See Essence, laws of.
Laws, logical, 11, 83, 97, 133-34, 161, 215, 226-27, 248-49, 251, 256, 261, 277

Laws, mathematical, 12, 84, 96, 214, 307, 315, 361n.
Laws, natural, 352, 388
Laws of meaning, 245-51
Lebenswelt, 54, 65, 286-87, 297, 345n., 350, 351n., 352n., 360n., 365
Leibniz, Gottfried, XII, 11, 14, 61, 62n., 64, 83, 175, 185, 189, 225, 325, 373
Lie, Sophus, 63, 346
Liebmann, Heinrich, 156n., 157n., 162, 232
Life-world, see *Lebenswelt*.
Light, 45, 57, 347, 363n.
Lindström, Sten, 189n.
Lines, straight, 31, 33-34, 41, 48-49, 56, 212, 351n.
Linke, Paul, 154n., 165
Lobachevski, Nicholas, 34
Logic, transcendental, X, 1, 11, 21, 23-25, 96, 106, 108, 134, 197, 203, 211, 237, 274, 278, 286, 293n., 294, 305
Lohmar, Dieter, 201n., 242n.
Lotze, Hermann, 10, 38, 95, 98, 300-01, 310-11, 375
Löwenheim, Leopold, 153-56, 165-66, 168
Löwenheim-Skolem theorem, 142
Luft, Sebastian, 201n., 242n.
Lüroth, Jacob, 367-70

Mach, Ernst, 386
Magic, 16, 302, 304
Manifold, spatial, 40, 44
Manifolds, 4, 6, 18, 21, 40-46, 50-57, 63, 70-74, 81, 85-89, 95, 98, 101-05, 109, 117, 118n., 120-22, 124, 127-29, 133, 138-49, 195-96, 201-04, 205n., 208, 212-16, 304, 313, 316-17, 346, 351, 353-56, 360-65. See also *Mannifaltigkeiten*.

Manifolds, theories of, 5, 6, 63, 72, 81, 86-87, 95, 102-05, 196, 201-05, 208-09, 215-16, 313-17, 354

Mannigfaltigkeiten, 5, 6, 117n., 196, 199, 201-06, 209, 215, 313-17, 362n., See also Manifolds.

Mannigfaltigkeitslehre, Cantor's, 5, 6, 196, 201-02, 209, 316-17

Mannigfaltigkeitslehre, Husserl's, 81, 86-87, 102-05, 196, 201-05, 208-09, 215-16, 313-17, 354

Marcus, Ruth Barcan, 185-91, 319-20

Marty, Anton, 156n., 157n., 244n., 369

Marx, Karl, 299

Material content, 42n., 70n., 72, 107, 346n., 357, 359, 363-64

Materialism, 247, 300-02

Mathematical attention, 106-07

Mathematics, applied, XII, 357n.

Mathematics, modern, 203, 313, 347, 349

Mathesis, 215, 352n.

Matters of fact, 11, 12, 43, 63, 81-85, 96, 99, 107, 265, 273, 288, 290, 293, 296, 307, 314, 393

Maximality, 117, 138, 142, 147, 149

Mc Guinness, Brian, 152-54

Meaning, shifts of, 215, 250, 316, 354-55, 359

Meaning, theory of, 1, 242, 248, 268, 273-276n.

Meaning, unity, 329, 334

Meaninglessness, XII, 20-21, 24, 64, 66, 70, 103-05, 119, 199, 208-09, 235, 242, 244-46, 251-54, 258-59, 273, 275, 327, 330, 358

Meanings, incomplete, 183-84, 219-221, 224, 230, 232, 234, 242-62

Meanings, independent, 219-39, 241-62

Meanings, realm of, 224, 229, 245-46, 248, 249n., 257-58

Meanings, science of, 246,

Measurement, 35, 41, 46, 51, 351

Mechanical procedures, 20, 23, 66, 102-03, 203, 206, 354n., 376, 390

Mechanics, 3, 98, 312, 346n., 347, 348n., 352

Mechanics, wave, 347

Medicine, 231, 318

Mereology, 47, 51

Metaphor, 72, 289, 291, 296, 329, 333, 334

Methodology, 35, 45, 54, 58, 71-75, 161, 191, 313, 340, 345, 349-50, 353, 356n., 359n., 364, 386-87, 391

Mice, 16

Mikton, 6, 304

Mind, 15, 43, 95, 100, 107, 132, 198, 310, 312, 326-27, 333, 338-39, 341, 386

Misch, Georg, 385

Misinterpretation, 9, 12, 96, 137

Mittag-Leffler, Gösta, 7, 304

Mittag-Leffler Institute, 157, 160

Modal logic, 171n., 172, 185-91, 262, 318-19

Modal logic, quantified, 185-89

Modalities, 187, 318

Model theory, 81, 87, 88, 129, 189

Models, 37, 39, 46, 79, 73-77, 81, 87-88, 122n.-23, 129, 131, 138, 141-42, 189, 346-47, 350-65

Modernity, 302, 345, 350

Mohanty, J. N., 93, 241

Molds, 21, 38, 40, 43, 45, 86, 102, 188, 204, 316, 319, 332

Moments, 10, 40, 94-95, 106, 272, 311
Monism, 300
Moons, 174, 212, 224, 255
Morass, 15, 210, 233, 259
Morning star, 186, 220
Morphology, 42, 46, 48, 269, 286, 291
Motion, 41n., 42, 56, 57, 363n.
Müller, Georg Elias, 379-82, 386
Multiplicities, 40, 43n., 72n., 95, 97-98, 107, 117, 119-20, 122, 124, 130, 196-97, 201n., 204, 215n., 313. See also Manifolds; *Mannigfaltigkeiten*.
Mystery, 77, 152, 234, 236, 260, 271, 301, 320
Mysticism, 302, 304, 307, 343, 386, 391
Myths, 89
Mythical entities, 10, 84, 311, 319

Naïvety, XIII, 10, 14, 32, 106, 211, 281, 301, 311, 330, 349, 352n., 353, 360
Names, 20, 95, 121-23, 147, 153, 164, 172, 174, 180-83, 186-90, 198, 205, 220-23, 226-35, 248-59, 273, 282, 321
National Endowment for the Humanities, XIII
Natorp, Paul, 375-76, 385
Naturalism, 9, 82, 296, 300, 302, 306, 319, 325, 334
Nature, 12, 25, 37, 46-47, 84, 296, 300, 307-09, 314, 327-28, 345-65
Nazism, 166, 302
Necessity, 7, 8, 14, 21, 24, 35, 37-38, 41, 43, 44, 50-53, 55-58, 66, 75n., 78, 85, 97, 102, 104, 127, 130, 133, 137, 139-40, 144, 149, 173, 176, 181-85, 188-90, 196-99, 204, 208, 210-11, 222, 230, 236, 245-46, 254-55, 260-61, 267-69, 279, 291-96, 300-03, 309-10, 312, 334, 354, 365, 389, 392
Negation, 120, 125-29, 133, 140, 327
Nelson, Leonard, 386-88, 391-94
Neo-Brouwerians, XII
Neo-Kantianism, 341, 376
Neo-Platonism, 386, 391
Neurath, Otto, 239
Nietzsche, Friedrich, 299
Noemata, 93, 272, 279, 282, 332, 337
Noetics, 272, 280, 282, 286, 331, 332
Nominalism, XII, 281
Nomology, 134, 290n.
Non-existence, 31, 42n., 69, 96, 103, 105, 116, 195, 205-08
Nonsense, 14, 20-21, 75, 103-04, 178, 206, 209-13, 219, 225, 235, 237, 242, 245-46, 248, 339
Nothingness, 37
Noumena, 54, 340
Numbers, cardinal, 3, 8, 13, 20, 75, 85, 97-99, 102, 104, 161, 197, 199, 200n., 201n., 203, 207n., 209, 305-06, 313-14
Numbers, complex, 8, 66, 71-77, 116, 207n., 210n., 359n.
Numbers, imaginary. See Imaginaries.
Numbers, irrational, 19, 104, 166, 206, 209, 279, 281
Numbers, natural, 94, 97-98, 116-20, 132, 199, 267-69, 271-72, 274-76, 279-82, 335
Numbers, negative, 8, 19, 66, 104, 116, 118-20, 206, 207n., 359n.
Numbers, ordinal, 85, 97, 107, 207n., 313
Numbers, rational, 73, 131, 351

Numbers, real, 5, 6, 71-73, 109, 138, 141-42, 279-83, 304, 351, 362
Numbers, transfinite, 5-7, 304-05, 346
Numbers, whole, 2, 5-8, 20, 82, 104, 119, 207, 209-10, 303

Objectivities, higher-order, 271, 272, 277
Objects, concrete, 22, 175, 186, 291
Objects, formal, 73n., 116-17, 121-25, 130, 134, 139, 143-46, 346n., 359n.
Objects, higher-order, 86, 102, 203, 282
Objects, impossible, 19, 62n., 72, 74, 103, 105, 116, 195, 205-08, 254, 356
Objects, spatial, 31, 52, 105, 329
Objectuality, 68-69n., 72-76, 138-39, 271, 332, 346, 357
Obligation, 319
Occultism, 302, 304, 305, 307
Oculomotor system, 39n.
Odium, 9, 89, 306, 319, 320, 387
Ogden, C. K. 167
Olympus, Mount, 308
ω-sequences, 76
One and one and one, 95, 98, 198
Ontology, 11, 50, 54, 71-75, 77n., 83, 85, 89, 105, 115-16, 121-22, 125-27, 130, 132, 147-49, 186-87, 245, 270-72, 275-81, 286, 291-96, 308-09, 313, 318-20, 331-32, 334, 339, 346, 347n., 357n.
Ontology, bizarre, 187
Ontology, formal, 50, 71-75, 77n., 85, 121-22, 125, 130, 132, 147-49, 278, 286, 346, 347n., 357n.
Ontology, idealistic, 89, 186, 318-19
Optics, 63n.

Optimism, 289, 336, 359
Origins, 8, 33-36, 39, 83, 107-08, 153, 158, 171-72, 179, 189, 195, 207, 286, 332, 341, 348
Osborn, Andrew, 2-4
Oswald, Lee Harvey, 210-11

Palmistry, 302
Paradise, 239, 259, 304, 305. See also Eden.
Paradox, Cantor's, 201n.
Paradox, Grelling's, 274
Paradox, "Russell's", 14-18, 106, 153, 157-65, 172, 179-80, 182, 201n., 206, 213, 227, 237, 241, 256-58
Paradox, Zermelo's, 17, 162-63, 201n., 213, 231
Paradoxes of set theory, 17-18, 153, 158, 162-63, 167-68, 172, 180, 190, 201, 214-15, 233, 241, 249-51, 257-62, 315-16, 321, See also Set of all sets; Russell's Paradox
Paradoxes of the infinite, 9
Parsons, Charles, 326, 336n., 337
Particulars, 231, 234-35
Pasch, Moritz, 48n., 154n., 101
Passivity, 38, 43, 54, 55, 58, 270, 272, 285, 297
Patterns, formal, 233, 312, 350n., 352, 364-65
Paulsen, Friedrich, 367
Peacefulness, 367, 369, 371
Peano, Giuseppe, 6, 152, 154, 156n., 160, 303
Peckaus, Volker, 201n.
Perception, 31, 34-48, 51, 54-58, 107, 198, 269-73, 283, 287, 314, 325, 328-29, 332-39, 343, 362-63, 365
Perception, doors of, 45
Pernet, Alain, 201n., 242n.
Perrone, Giovanni, 373

Pessimists, 300
Phenomenological method, 340-41
Phenomenological reduction, 326-34, 340, 343
Phenomenology, X, XI, XII, 1-3, 11, 12, 23, 25, 43, 56, 81-84, 93, 94, 96, 105-06, 110, 196, 265-79, 283, 300, 325-43, 345, 347, 349-50, 361, 365
Phenomenology, genetic, 43, 361
Phenomenology, transcendental, XI, XII, 23, 25, 105-06, 325-43, 349
Philipse, Herman, 332
Phrenology, 302
Physical organ, 325, 335, 338-39
Physics, 12, 29, 35, 38, 57, 74, 105, 268-69, 345-64, 391, 392, 393
Physics, mathematical, 12, 25, 105, 350, 352n., 354-58, 364n.
Physiology, 55, 350n.
Pilzecker, Alfons, 379
Planets, 186-87, 220
Plato, 6, 10, 31, 304, 325, 362, 390
Platonism, 1, 6, 9, 10, 11, 82, 281, 303-04, 307, 310, 326, 332-33, 336n., 349, 362
Playfair axiom, 34
Plotinus, 386
Poincaré, Henri, 35, 37, 45, 54, 132, 265, 268, 270
Poincaré's ball, 35
Politics, 318
Poncelet, Jean-Victor, 359n.
Pope, 373, 377, 382
Pope Leo XIII, 373, 377
Positivism, 37, 82, 301-02, 369, 375
Possible worlds logic, 188
Post-modernism, 348n.
Pragmatism, 74, 75, 236, 295
Pre-predicativity, 38, 276, 286, 297

Pre-science, 32, 37, 38n., 41, 43, 46, 47, 54
Predicaments, 221, 254-56
Predicates, 82, 86, 102, 173, 184, 203, 220-21, 225, 227-29, 232, 237, 243, 250, 253, 255, 257-58, 270, 273, 276, 291
Predication, 86, 102, 162, 165, 175, 203, 220-30, 235, 237, 248-49, 252, 254-55, 257-58, 276, 286-87, 291-92, 295, 297
Prejudices, XIII, 9, 128, 267, 318, 348n.
Presence and absence, 277-78
Present emperor of France, 249
Present king of England, 183
Present king of France, 183
Presentation. See Representation.
Presuppositions, 37, 53, 75, 140, 171, 189, 274, 279, 285-86, 288-89, 292-93, 296, 307, 310, 360-63n., 386
Prey, 107
Primitiveness, 3, 31, 49n., 99, 183, 186, 229, 244-46, 271-72, 281, 362n.
Principia Mathematica, 166, 181-86, 228, 236, 238, 299-300
Principle of complete induction, 272
Principle of comprehension, 281
Principle of continuity, 359n.
Principle of definition by recursion, 270, 272, 280
Principle of duality, 76n. 359n.
Principle of free-mobility, 34
Principle of non-contradiction, 87, 97, 101-02, 133
Principle of reasoning, 274
Principle of substitutivity of identicals, 14, 175, 211, 231-32
Principles of extensionality, 106, 110, 158, 186-87, 190, 228, 231, 235-38. See also Basic Law V.

Probabilities, calculus of, 367
Probability, 6, 18, 33, 44, 74, 96, 117, 142, 152, 154, 163, 166, 190, 238, 266, 268, 272, 277, 278, 304, 313-14, 319, 335, 339, 342, 354n., 367, 373, 381-82, 384, 389
Proclus, 31
Projective techniques, 51
Proper names, 20, 153, 164, 172, 174, 180, 182, 186-88, 205, 220-23, 226, 229-35, 252, 254-55, 258-59
Prophylaxes, 345, 348
Propositions-in-themselves, 10, 82, 84, 311
Protestantism, 368-69, 374-75
Pseudo-objects, 180, 233-34, 259
Psychologism, XI, XII, 1, 3, 7-8, 10-12, 23, 82-84, 95, 96, 109, 116-17, 137, 279, 302, 329, 333, 341, 380, 386-87, 391-93
Psychology, descriptive, 36, 327-29, 341
Psychology, empirical, 9, 19, 83, 89, 310
Psychology, experimental, 379, 385-86
Psychophysical functions, 36, 40, 41n., 43-44, 53-54, 287
Puppeteers, 357
Pure arithmetic, 12, 95, 96, 99, 105, 107-08, 310, 313
Pure geometry, 12, 32, 36, 38-39, 46-47, 50-54, 105, 249n.
Pure logic, XII, 2, 10-12, 24, 25, 61, 82-87, 96-98, 102, 104, 118n., 127, 133, 139-40, 159, 161, 196, 202-03, 208, 248, 307, 311-14
Pure mathematics, XII, 12, 47, 50, 84, 94, 96-97, 107-08, 200, 261, 307, 313-14, 347
Pure reason, 301
Pure thought, 39

Puzzles, 1, 9, 120, 132, 142, 181, 187-89

Quantification, 132, 185-87, 189n., 191, 231-33, 275, 277, 282
Quantification, existential, 187, 231-32
Quantitative analyses, 103, 206
Quantity, 65, 71, 76, 100, 104, 116, 132, 196, 199, 209
Quantization, 347n.
Quantum theory, 347-48, 352, 363
Quaternions, 63, 74, 207n., 346
Quine, Willard, 81-82, 89, 106, 110, 185-91, 219, 238-39, 262, 285, 295-96, 299, 318-20,

Rabbit parts, 89
Racism, 377
Radicality, 1, 2, 3, 4, 7, 10, 17, 24, 45, 82, 108, 162, 223, 232, 234, 286, 306, 308, 329, 360n., 361n., 374, 376
Rationalism, 270, 336, 386
Ratisbonne, Fr. A-M., 377
Rawness, 38-40, 44, 53, 94, 350, 364-65
Real spaces, 33, 38, 43
Realism, XI, 1, 6, 37, 38, 261, 271, 295-96, 301, 303, 326-35, 341-42, 373
Reality, XIII, 6, 24, 32, 35, 37, 38n., 41, 43, 45-46, 54, 58, 84, 86, 88-89, 97, 102, 108, 188, 203, 268-70, 274, 285, 287-88, 294, 296, 301, 303-04, 308-10, 312-13, 316-18, 321, 328-33, 336, 338-39, 342, 347-53, 358, 362-64
Reality, absolute, 329-30
Reality, ultimate structure of, 89, 108, 188, 316-17

Reason, 2, 12, 23, 46, 48, 54, 55, 87, 103, 105, 205, 280, 301, 312, 320, 350n., 388
Rebellion, 229-30
Receptivity, 9, 285
Recipes, 182, 225, 282, 350, 355
Recursion, 116, 270, 272, 280
Reference, 20, 24, 71, 89, 103-06, 110, 127, 142, 145-48, 172-91, 195, 205-10, 223, 227, 233, 246, 248, 251, 253, 262, 270, 327, 331-32, 339
Reference, New Theory of, 172, 176, 188, 191
Reference, "Old" Theory of, 176, 181
Reference, "Very Old" Theory of, 176
Referential opacity, 187, 191
Reign of Terror, 176, 181
Relations of ideas, 11, 83, 99
Relativity, theory of, 187, 347, 363n.
Religion, 299, 302, 304, 319, 373, 375-77, 382
Representability, 291n., 292
Representation, 22, 61-62, 78, 248, 266, 289, 293, 312, 339, 357, 362
Representations, intuitive, 32, 35, 37, 41, 51, 54-57, 363
Representations Symbolic, 58, 62, 198, 266, 268, 357
Responsibility, 319, 349
Retinas, 40
Reversibility, 76n., 220, 232
Rickert, Heinrich, 369-71, 375, 386-87
Ricoeur, Paul, 329
Riddles, 77, 187, 189
Riehl, Alois, 368-71, 374-75
Riemann, Bernard, 34, 39, 50, 53-54, 57, 63, 76, 86, 148, 202, 317, 346, 363

Rigid bodies, 32, 34-36, 38n., 41-45, 48, 56-57
Rigid reference, 181, 185, 189-90
Rigor, XI, 2-9, 41, 67, 82, 99, 101, 154, 197, 204, 280, 302, 306-10, 314, 326, 343, 365, 382, 387-90, 393
Risk, XII, 38., 75, 106, 211, 354, 356n., 357n.
River stages and kinship, 89
Rohlfing, Dr. Helmut, XIII
Romanticism, 307, 308
Roots, 1, 2, 3, 7, 19, 39, 56, 78, 82, 89, 98, 184, 185, 238, 247, 254, 285, 296, 297, 300, 302, 345n., 350, 365, 389
Rosado Haddock, Guillermo, 93, 201n., 242
Roses, 255
Rosicrucians, 302, 369n.
Round squares, 249, 329, 331
Rules, 19, 20, 24, 62, 63, 66, 69, 70, 87, 101-04, 108, 118, 149, 183-84, 204-09, 214, 229-30, 232, 248, 261, 273, 280-81, 291, 293, 318, 347, 354, 356
Runge, Carl, 392
Russell, Bertrand, 1, 3, 15-18, 89, 93, 106, 110, 151-68, 172, 179-91, 195, 200-02, 205, 213, 219, 227-31, 234-37, 241, 249-50, 252, 256-57, 259-62, 265, 273, 280, 299, 305, 317, 321, 334-35
Russell's paradox. See Paradox, Russell's.

Safety, 41, 52, 73, 106, 145, 155, 211, 237, 340, 350, 357, 363
Sailors, 239, 318
Salva veritate, 82, 164
Sartre, Jean-Paul, 299
Schelling, Friedrich von, 386
Schirn, Matthias, 171
Schoen, Henri, 301-02

Scholasticism, 247, 262, 382
Scholz, Heinrich, 152, 155-67
Schönflies, Arthur, 160, 163-64
Schröder, Ernst, XIII, 17, 62n., 63, 64, 68-71, 115n., 152, 164, 200-01, 306
Science, exact, 25, 216, 348n., 387, 388, 393
Science, natural, 9, 11, 12, 37, 38, 83-85, 96, 105, 301, 302, 306-10, 314, 353n., 354, 360, 367, 376, 386-88
Science, physical, 54, 58, 333, 345, 350, 352
Science, theory of, 288, 313, 387
Sebestik, Jan, 131n.
Segments, 41n., 48-50, 175, 225
Self-subsistence, 172-174, 219, 223, 229
Semantics, 71n., 128, 131, 138, 141, 145n., 147, 149, 238, 241, 257, 262, 291-92, 294-95, 358n.
Sensations, 43, 56, 85, 100, 270, 331-32, 336
Sense, 10, 14, 43, 61, 84, 89, 126, 191, 286, 318. See also Meaning.
Sense data, 38, 40, 44, 45, 56, 271, 330, 337, 350, 351n.
Sense giving (*Sinngebung*) 43, 277, 329, 331, 354
Sense impressions, 40, 44n., 335, 337
Senselessness, 125, 209, 221, 242, 254, 256, 294, 320, 328, 330. See also Meaninglessness.
Senses, 31, 35n., 39-43, 44n., 52-56, 82, 109, 276, 287, 303, 347, 349, 351, 353
Sensibility, 53, 74, 77, 351n., 362
Sensualism, 82, 302
Serpent. See Snakes.
Set essence, 214, 251, 315

Sets of all sets, 14, 200, 201n., 213-15, 227, 250, 305. See also Class of all classes.
Set theory, 5, 16-18, 82, 102, 110, 153, 158, 160, 162-63, 167-68, 180, 195-216, 228, 233, 242n., 249, 258, 261, 281-82, 299-300, 303-96, 313-17, 321, 336, 343
Sets, infinite, 19, 198-99, 280-82, 306
Sets, transfinite, 5, 304
Shells, 222, 230, 260
Shock, 15, 159, 163, 238, 256, 310
Siebeck, Hermann, 375
Signs, 19-20, 63, 67, 69, 78, 101-03, 108, 179, 191, 198, 203-07, 227, 229, 232, 244-45, 257, 272, 356n., 357, See also Symbols.
Simar, Bishop Hubert T., 373
Simmel, Georg, 386
Simons, Peter, 81, 244n., 259n.
Simplicity, 7, 34, 35, 45, 62, 72, 98-99, 175, 177-78, 183, 231, 233, 237, 242, 259, 276, 296, 303, 311, 315, 319, 338, 358
Sin, 187, 232, 325
Sixth sense, 338
Size and shape, 42, 46, 48
Skepticism, 82, 301, 302
Sluga, Hans, 18
Smith, Quentin, 188n., 189n.
Smullyan, Arthur, 186
Snakes, 107, 171
Soames, Scott, 188n., 189n., 190
Socrates, 228
Solipsism, 268, 271
Something in general, 85, 98
Somatomotor system, 39n.
Souls, 5, 108, 303, 304, 314
Soviet army, 156
Space Euclidean, 34, 39, 41n., 44n., 363

Space, intuitive, 37, 39, 45-47, 53, 55-58
Space perceptual, 46, 58, 362-63
Space, physical, 31-47, 50-58, 287, 325, 351, 362-63
Space real, 33, 38, 43
Space, sensorial, 287, 362
Space, transcendent, 32, 38, 39, 41, 43, 54
Space-constituting functions, 38n., 41, 44-45, 53-55
Space-representing functions, 53, 56
Spaces, mathematical, 50, 57, 58, 363
Species, 6, 10, 21, 198, 222-23, 303, 311-12
Speculative philosophy, 386-87
Speiser, Hans, 154n.
Spies, 176
Spirit, 5, 62, 63, 73, 225, 300-03, 349n., 350n.
Spiritism, 302, 307
Spitzer, Hugo, 371, 375
Square circle, 69n., 229
Stars, 162, 180, 186, 220, 221, 233
States of affairs, 147, 271, 273, 274n., 288, 294, 338
Sterility, 14, 34, 64, 73, 74, 178, 188, 210-11, 225, 313, 318, 387
Strangeness, 5, 7, 45, 94, 95, 229, 300, 310, 313, 321, 365
Structure, geometrical, 31, 36, 46-47, 51, 58
Structure spatial, 31, 35, 39, 50, 52
Structure of reality, 89, 108, 188, 316-18
Structures, 19, 22-25, 31, 35-36, 39, 42-52, 57-58, 64, 70-72, 76-81, 87-89, 102, 105, 108, 109n., 121, 131-33, 188, 209, 222, 225-38, 244-48, 269, 276n., 282, 297, 316-26, 334, 337-38, 346n., 355-58, 360, 363-64, 392
Structures perceptual, 47, 58
Stumpf, Carl, 5, 8, 13, 19, 62n., 103, 162, 207, 367, 386
Subjectivity, X, XII-XIII, 10, 22-25, 83, 245, 267-72, 279-82, 285, 288-97, 300, 326-36, 339-42, 360-61, 386, 393
Substantives, 165, 174, 224, 237
Substitution, 14, 69, 77, 99, 175-80, 184-90, 211, 212n., 226, 231-33, 262
Substrata, 107, 276, 294n.
Superstition, 100, 302, 307
Surfaces, 39n., 40, 41, 46-48, 175, 225
Survival of the fittest, 74
Syllogisms, 7, 13, 85-86, 94, 313
Symbolic reasoning, 32, 62, 63n., 67-68, 71, 132, 137-38, 146-47, 149, 278, 356n.
Symbols, 20-21, 23, 32, 47, 51-52, 58, 62-70, 73, 75, 78, 87, 103, 120-22, 126, 128, 130, 132, 145n., 146-48, 174, 179, 183-84, 204, 206, 228-35, 241, 243, 257, 259-62, 351n., 354, 356-58
Symptoms, 238, 274
Syncategorematic expressions, 241, 243-45, 247-48, 251
Syncategorematic meanings, 244, 247
Syntactic objectivities, 271, 276
Syntax, 71, 230, 247-48, 271, 276, 282, 291-95, 358n. See also Completeness, syntactic.
Synthetic a priori, 2, 11, 37, 44n., 82
Synthetic judgments, 2, 11, 81, 83
Systematic philosophy, X, 385, 392-94

Tannery, Paul, 303
Tarski, Alfred, 123, 363n.
Tartarus, 308
Technization, 65n., 74, 348, 354-55
Teleology, 73, 274
Temperature, 35, 351
Temptation, 43n., 93, 122n., 211, 212, 229, 304, 318, 337
Terms, singular, 165, 232-33
Terminology, 64-65, 73, 121, 196, 242, 244n., 279, 304, 355
Tertium non datur, 133, 273, 274-75, 277, 285-86, 287n., 295. See also Bivalence; Law of the excluded middle.
Thales' theorem, 41
Theologians, 371, 374
Theology, 5, 6, 303-04, 369, 371, 373, 374, 382
Theory of theories, 117n.
Theosophy, 302
Third realm, 333-34
Thomae, Carl Johannes, 165, 368
Thomas Aquinas, 373
Thomism, 373
Three-ity, 107
Triangles, 10, 41, 84, 212, 294, 311
Truth value, 232
Truths in themselves, 10, 289, 293-95, 310, 361n.
Tieszen, Richard, 93, 326
Time, 2, 7, 13, 54n., 57, 106-07, 265-69, 279-93, 309, 314, 325, 328, 343, 351, 363
Titans, 308
Titz, Marga, 160
Topology, 46, 48, 50
Torment, 95, 200, 301, 306, 313
Torsion, 41-44,
Totality, 6, 40, 44, 98, 104, 125, 130, 133, 139, 143, 196, 208, 212, 214-16, 235, 251, 272, 275, 295, 303-04, 315, 317, 331, 350n., 352
Touch, 35n., 38, 40, 362
Tractatus, 166-67
Transcendent reality, 35, 37, 39, 41n., 43, 293n., 339, 348
Transcendent space, 32, 38-39, 41, 43, 52, 54
Transcendental attitude, 327-30
Transcendental consciousness, 334-35, 337, 339
Transcendental constitution, 329
Transcendental ego, XII, 44, 290, 295, 327, 329, 334-35
Transcendental idealism, 37, 299, 271, 320, 330-32, 339, 341
Transcendental logic, 1, 24
Transcendental phenomenology. See Phenomenology, transcendental.
Transcendental reductions, 326-27, 330, 335, 348
Transcendental subjectivity, 44, 295, 326-27, 329-32, 340-42, 361n.
Transcendental turn, X
Transfinite realm, 5-7, 275, 303-05, 346, 373
Transparency, 7, 202, 326
Triangles, 10, 41, 84, 212, 294, 311
Truth, 4, 9, 10, 12, 19, 20, 24, 31, 37, 39, 47, 53-54, 58, 63, 65, 67n., 68, 70n., 73, 75, 81-84, 87, 95-97, 99-100, 102, 105-08, 119n., 133, 140, 148, 172, 175, 181, 189, 191, 204-05, 223, 225, 231-32, 237, 246, 248, 251, 253-54, 261, 270-80, 285-90, 293-95, 301, 310-14, 321, 337-39, 346n., 347, 349n., 355, 357-61, 365, 382, 386, 390-93
Truth, intrinsic, 274-75, 285, 289-90

Truth and falsehood, 20, 21, 96, 104, 125, 128, 131, 133, 140, 181, 185, 191, 208, 226, 248-49, 251, 253-54, 274, 285-89, 293-95, 312, 315, 335
Truth in itself, 10, 289, 292-95, 310, 361n.
Truth values, 20, 82, 232, 253-54, 274-75, 286-90
Two-ity, 94, 106-07
Two-oneness, 97, 107
Types, logical, 179, 221, 228-37, 247-48, 273, 282, 291-92
Types, theory of, 229, 235, 265

Unboundedness, 34, 41, 44, 53, 56
Uncertainty principle, 347
Unfilled, 220, 221, 243. See also Unsaturatedness.
Unity, 8, 65, 67, 77, 83, 89, 95, 97, 197, 200, 202, 214, 245, 271, 279, 281, 294-95, 303-06, 315, 319, 329, 334, 350n. 361, 377
Universals, 10, 84, 89, 231, 237, 261-62, 311, 313, 318, 321, 326
Unsaturatedness, 121-22, 220n., 243, 252-53, 258-59
Use and mention, 187

Vaccine, 76
Vacuousness, 74, 75
Vagueness, 13, 100, 142, 145n., 186, 269, 280-82, 301, 351n.
Validity, 10, 11, 13, 19, 21, 22, 34, 36, 42n., 51, 58, 63n., 69, 84 87, 97, 99, 102-04, 119, 124-25, 130-34, 148, 171, 179, 186, 200-09, 215-16, 227, 232, 236, 249-51, 266-67, 271, 274-75, 277, 281, 285-90, 291n., 294-96, 306, 311-12, 314, 327, 329, 343, 348, 352n., 360, 361n., 386-88

Values, 34-35, 41, 56, 122, 182, 184-87, 226, 228, 233, 236, 253-54, 256, 351
Values, courses of, 158-59, 179, 226, 351
van Atten, Mark, 93
Variables, 85, 121-22, 126-27, 129-30, 145n.-46, 148, 181-82, 184, 186-87, 233, 247, 275
Vector analysis, 346n., 347n.
Vectors, 359
Velocity, 347, 363n.
Vengeance, 191, 301
Venn, John, 64n.
Venus, 162, 186, 221, 253
Veraart, Albert, 156n., 164-67
Verifiability, 42, 58, 103, 132, 285, 289, 206, 132, 285, 290-96
Vérités de fait, 11
Vérités de raison, 11
Veronese, Giuseppe, 303, 368
Vesuvius, 221, 255
Vicious-circle fallacy, 185, 266, 271
Vienna Circle, 300
Vision, 35n., 38, 40, 45, 55-56, 77, 273, 287, 318, 362, 391

Wandering spirits, 73
Wang, Hao, 22-23, 335, 341-42
War, 12, 18, 96, 155-56, 161-63, 166, 301, 393
Warning sign, 233, 259
Weber, Dr. Marielène, XIII
Weber, Rev. Fritz, XIII
Weierstrass, Karl, XI, 1-9, 19, 81-82, 110, 151, 200, 207n., 367, 369, 392
Wesenschau, 342-43
Weyl, Hermann, XII, 42n. 46, 55-58, 265-83, 363n., 392-93
Whitehead, Alfred North, 110, 168, 237

Wholes, 38n., 43, 52, 57, 77, 85, 86, 180, 197, 202, 213-14, 220, 224, 227, 233, 243-48, 250-53, 258, 304, 312, 314-17, 321, 328
Widersinnigkeit, 196, 208, 214-15, 242-43, 249-51, 316
Will to live, 106
Willard, Dallas, 67n., 185, 285, 295, 299, 318
Windelband, Wilhelm, 375, 386
Wineglasses, 230
Witchcraft, 302
Wittgenstein, Ludwig, XI, 153-57, 166-67, 237
Woker, Canon Franz, 368, 370-72, 377-78
World, 24-25, 31, 34-35, 42n., 44-45, 57, 82, 89, 107, 148, 188, 234-35, 259, 261, 266, 268, 285-97, 300-03, 308-10, 314, 318, 321, 325-35, 338, 340-41, 348n., 350-56, 360-65, 392. See also Life-world.
World, concrete, 24, 350n.
World, external, 57, 334
World-logic, 292
Worms, can of, 238, 262
Worthlessness, 14, 66, 212
Wundt, Wilhelm, 376, 369, 386

Zermelo, Ernst, 17, 151, 160, 162-63, 167-68, 201n., 213, 231, 270
Zermelo-Fraenkel, 110
Zero, 34, 35, 38, 41, 42
Zimmermann, Robert von, 367
Zsigmondy, Karl, 154n., 165

www.ingramcontent.com/pod-product-compliance
Lightning Source LLC
Chambersburg PA
CBHW071309150426
43191CB00007B/563